Bud Scott

MATHEMATICAL STRUCTURES FOR COMPUTER SCIENCE

SECOND EDITION

MATHEMATICAL STRUCTURES FOR COMPUTER SCIENCE

SECOND EDITION

Judith L. Gersting

Indiana University—
Purdue University at Indianapolis

W. H. FREEMAN AND COMPANY
New York

Cover design by Renee Kilbride Edelman.

Design adapted from a graphic by Benoit B. Mandelbrot and reproduced by his kind permission. From *The Fractal Geometry of Nature* by Benoit B. Mandelbrot. W. H. Freeman and Company. Copyright © 1982.

Library of Congress Cataloging-in-Publication Data

Gersting, Judith L.
  Mathematical structures for computer science.

  (A Series of books in the mathematical sciences)
  Includes index.
  1. Mathematics—1961-   . 2. Mathematical models.
3. Electronic data processing—Mathematics.   I. Title.
II. Series.
QA39.2.G47 1986   510   86-9974
ISBN 0-7167-1802-2

Printed in the United States of America

3 4 5 6 7 8 9 HA 6 5 4 3 2 1 0 8 9

To Adam and Jason,
two beloved discrete structures

Contents

PREFACE

Since the publication of the first edition of this book in 1982, courses in discrete mathematics or discrete structures have been rapidly gaining in popularity. More and more colleges and universities are offering the course as part of programs in computer science and mathematics. As might be expected, educators continue to debate the nature of the course and its place in the curriculum. Many variations in content and level of instruction are found from campus to campus.

The first edition of this book enjoyed wide use, and we felt we could extend its appeal even further. We asked many users of the first edition for recommendations for changes and examined many new course outlines. Our findings showed remarkable agreement in the desired changes. With the improvements we have made in the second edition, we feel the book is now complete enough and flexible enough to be used in courses with a wide variety of emphases.

CHANGES IN THE SECOND EDITION

Reviewers and users suggested that we add several key topics and expand others. The resulting major changes to the second edition are these:

Expanded coverage of logic to include a rather complete introduction to first-order predicate logic (Sections 1.1 and 1.2)

Expanded coverage of mathematical induction (Section 1.4)

Inclusion of a section on recursion and recurrence relations (Section 1.5)

Inclusion of some material on countable and uncountable sets (in Section 2.1 and several other places)

Expanded coverage of combinatorics (Sections 2.2 and 2.3)

Inclusion of a short section on the binomial theorem as an application of combinatorics (Section 2.4)

Inclusion of a short section on matrices and their operations (Section

3.3) in preparation for using matrices as examples of algebraic structures

Expanded coverage of graphs to include linked list representations and path and traversal algorithms (Chapter 4)

The addition of many new examples and exercises illustrating computer science applications or extensions of the concepts presented. These include

PROLOG (predicate logic)
program verification (predicate logic and induction)
analysis of algorithms (recurrence relations)
Programmable Logic Arrays (Boolean algebra)
2's complement circuit (Boolean algebra)
encoding circuit for group codes (sequential networks)
cyclic codes (generalizations of Hamming codes)
computer network protocol models (finite-state machines)

Altogether, in addition to new Practice Problems as appropriate, there are almost 500 new exercises, many with multiple parts.

FEATURES OF THE BOOK

In preparing this edition, I was careful to preserve those features that made the first edition a useful learning tool. As before, I have attempted to write in a style that is clear, conversational, and informal, avoiding the theorem-proof presentation that intimidates so many students, both in computer science and mathematics. The text provides motivation and develops results before giving formal statements, an approach that seems more natural to the student and, indeed, parallels the historical process.

Numerous Practice Problems are again provided throughout the text. They provide immediate reinforcement of some idea or notation just introduced, a feature my students have found very helpful. These problems are simple, and the student should work them as they are encountered. Answers to all Practice Problems are presented at the back of the book.

Each section ends with a list of the definitions, techniques, and main ideas of the section. Students find that this checklist helps them to review the material and sort out what is important. The best way to clarify and reinforce understanding of the concepts is to have students work as many exercises as time permits; therefore, I have provided a wealth of exercises. Most of the exercises are straightforward, but others introduce extensions of the text material. Answers to selected exercises, denoted in the text by $\star$, are given at the back of the book.

A complete solutions manual is available to the instructor from the publisher.

OUTLINE OF THE SECOND EDITION

The expanded topic coverage and increase in the number of chapters make the new edition suitable for both one-term and longer courses. For those instructors not planning on covering the complete text, the outline offers suggestions on how they can adapt the chapters and sections to suit their courses.

Chapter 1 of the second edition covers logic, proof techniques, and recursion. Sections 1 and 2 present propositional and predicate logic; students enjoy this material, and it is a relatively painless introduction to formalism. However, Section 2, covering formal proofs, can be omitted without any subsequent difficulties. Section 3 discusses various proof techniques (direct, contrapositive, etc.), while Section 4 concentrates on mathematical induction. The last section looks at recursion and recurrence relations.

Chapter 2 discusses sets and combinatorics, and includes a section on the binomial theorem as an application of combinatorics. **Chapter 3** covers relations, functions, and matrices. Much of the material in Chapters 2 and 3 will be review and can be treated as quickly as the instructor feels the students can proceed. In particular, Sections 2.4 ("The Binomial Theorem") and 3.3 ("Matrices") may be unnecessary. It has been my experience, however, that students have quite fuzzy notions about sets and functions and that a thorough presentation of these topics is warranted.

Chapter 4 covers graphs and trees, with a discussion of various alternatives for computer representation of graphs and a presentation of algorithms for Euler path, shortest path, depth-first search, tree traversal, and others.

Chapter 5 discusses notions of models and homomorphisms, using Boolean algebras as a starting point. **Chapter 6** presents material on gating networks.

Chapter 7 considers algebraic structures and homomorphisms. Here again, the depth of coverage can be varied to suit the class. Most of the spirit of this chapter carries over into Chapter 9, but no specific information from Chapter 7 is required in Chapter 9. However, Chapter 7 is a prerequisite for **Chapter 8,** which introduces coding theory.

Chapter 9 covers finite-state machines, their homomorphisms, and their recognition capabilities (Kleene's Theorem). **Chapter 10** is a look at finite-state machine "hardware" and, of course, depends upon Chapter 9. In addition, Chapter 6 is a prerequisite for Section 10.2, and Chapter 7 is a prerequisite for Section 10.4.

Chapter 11 covers Turing machines, unsolvability results, and a bit of computational complexity. **Chapter 12** introduces formal languages and relates finite-state machines and Turing machines to languages. Section 12.1 is independent, but Section 12.2 requires Sections 9.3, 11.1, and 11.2.

Except as noted above, the chapters are relatively independent, so maximum flexibility in preparing a course outline is possible. A computer science discrete structures course might cover most of Chapters 1 through 5, 7, 9, and 11, or if the students already have the background of Chapters 1 through 4, then Chapters 5, 6, 7, 9, 11, and 12 would provide a good core of material. Courses with a stronger engineering emphasis would probably include Chapters 6, 8, and 10. A discrete mathematics course could cover Chapters 1 through 5, 7, and 8. A one-year discrete structures course could proceed sequentially through the entire text. Again, because of the relative independence of the topics and the variety of student backgrounds, there are many other options for organizing the material into a one-semester or one-year course.

ACKNOWLEDGMENTS

A few of the Practice Problems and exercises in this book previously appeared in *Abstract Algebra: A First Look* (Joseph E. Kuczkowski and Judith L. Gersting, New York: Marcel Dekker, 1977) and are used here with permission of Marcel Dekker, Inc.

My thanks go to the reviewers of the second edition for their thoughtful comments: Stefan Burr, City University of New York; Chris Caldwell, University of Tennessee; William H. Caldwell, University of North Florida; Midge Cozzens, Northeastern University; Robert Dillon, Aurora College; Robert Gayvert, Monroe Community College; Akhiro Kanamori, Boston University; and Larry Kost, University of Vermont.

Reviewers of the first edition were Allen Acree, William Dorn, Victor Klee, Yale Patt, and Charles Swart, and their helpful suggestions are still appreciated.

Finally, my family deserves a special thank-you. My husband John is ever patient with my ''projects.'' Our two children have graduated from toy cars to soccer balls but remain considerate when ''Mom has to work now.''

June 1986   **Judith L. Gersting**

NOTE TO THE STUDENT

This book contains several features that will assist you in learning the material; among the most important are the many Practice Problems. These problems are generally not difficult and are meant to be worked as soon as you get to them. Answers are given at the back of the book. You'll find learning much easier if you give these problems your best effort as you go along.

Judith L. Gersting

Chapter 1

Logic, Induction, and Recursion

As time goes on, computer science is becoming more of a science. Its theoretical framework is becoming firmer, and computer scientists are increasingly aware of the need for a precise vocabulary and the rigors of mathematical thinking. The purpose of this book is to provide a better understanding of the mathematical tools, language, and thought processes used by computer scientists.

The first two sections of Chapter 1 cover some important concepts in mathematical logic. In addition to providing a foundation for the organized, precise method of thinking that characterizes any scientific investigation, logic has direct applications to computer science. Circuit logic (that is, the logic governing computer circuitry) is a direct analog of statement logic (see Chapter 6). Logic is also used in program verification, where it is proved that the output of a given computer program will always comply with certain predetermined conditions (see Section 1.2).

In Section 1.3, various methods of mathematical proof are presented. Section 1.4 concentrates on mathematical induction, a proof technique with particularly wide application in computer science. The last section discusses recursion, which is closely related to mathematical induction and is important in algorithms and their analysis.

SECTION 1.1   STATEMENTS AND QUANTIFIERS

In general communication we often express ourselves in English by using questions, exclamations, and so forth, but to communicate facts or information, we use statements. Technically, a **statement** (or **proposition**) is a sentence that is either true or false.

1.1 Example   Consider the following:

    (a) Ten is less than seven.
    (b) How are you?
    (c) She is very talented.
    (d) There are life forms on other planets in the universe.

Sentence (a) is a statement because it is false. Because item (b) is a question and cannot be considered either true or false, it is not a statement. In sentence (c), the word "she" is a variable, and the sentence is neither true nor false because "she" is not specified; therefore (c) is not a statement. Sentence (d) is a statement because it is either true or false; we do not have to be able to decide which.   ☐

Connectives and Truth Values

To vary our conversations, we do not confine ourselves to simple statements. We combine simple statements with connecting words to make compound statements. The truth value of a compound statement depends on the truth values of its components and the connecting words used. A common connective is the word "and." (Words like "but" and "also" express different shades of meaning but have the same effect on truth values.) If we combine the two true statements "Elephants are big" and "Baseballs are round," we would consider the resulting statement, "Elephants are big and baseballs are round," to be true. In logic, we use the symbol ∧ to denote "and" and capital letters to denote statements. We agree, then, that if A is true and B is true, $A \wedge B$ (read "A and B") is also true.

1.2 Practice   If A is true and B is false, what truth value would you assign to $A \wedge B$?
If A is false and B is true, what truth value would you assign to $A \wedge B$?
If A and B are both false, what truth value would you assign to $A \wedge B$?
       (Answers to practice problems are in the back of the text.)   ☐

A	B	$A \wedge B$
T	T	T
T	F	F
F	T	F
F	F	F

Figure 1.1

A	B	$A \vee B$
T	T	T
T	F	
F	T	
F	F	

Figure 1.2

The statement "$A \wedge B$" is called the **conjunction** of A and B. We can summarize the effects of conjunction by the **truth table** presented in

Figure 1.1. In each row of the truth table, truth values are assigned to the statement letters, and the resulting truth value for the compound statement is shown.

Another connective is the word "or." The statement "A or B," symbolized $A \vee B$, is called the **disjunction** of statements A and B. If A and B are both true statements, then $A \vee B$ would be considered true, giving us the first line of the truth table for disjunction (see Figure 1.2).

1.3 Practice   Use your understanding of the word "or" to complete the truth table for disjunction. □

Statements A and B may be combined in the form "if A, then B," symbolized by $A \rightarrow B$. This may also be read as "A implies B." The connective here is **implication,** and it conveys the meaning that the truth of A causes the truth of B. There are several other ways to express $A \rightarrow B$, such as "A is a sufficient condition for B," "A only if B," "B follows from A," and "B is a necessary condition for A." In the compound statement $A \rightarrow B$, A is called the **antecedent** and B the **consequent.**

1.4 Example   The statement "Fire is a necessary condition for smoke" can be restated as "If there is smoke, then there is fire." The antecedent is "there is smoke," and the consequent is "there is fire." □

1.5 Practice   Name the antecedent and consequent in each of the following statements. (Hint: rewrite each statement in if–then form.)

(a) If the eggs are fresh, then they will not spoil.
(b) A sufficient condition for using a word processor is that the Great American Novel is to be written.
(c) Susan will pass her physics course only if she is bright and studies hard.
(d) Good combustion is a necessary condition for high gasoline mileage. □

The truth table for implication is less obvious than that for conjunction or disjunction. To understand its definition, let's suppose you hear your roommate remark, "If I graduate this spring, then I'll take a vacation in Florida." If your roommate graduates in the spring and then takes a vacation in Florida, the remark was true. If both A and B are true, we consider the implication $A \rightarrow B$ to be true. If your roommate graduates and then does not take a vacation in Florida, the remark was a false statement. When A is true and B is false, we consider $A \rightarrow B$ to be false. Now suppose your roommate doesn't graduate. Whether he or she takes a vacation in Florida or not, you could not accuse your roommate of making a false statement. By default, we accept $A \rightarrow B$ as true if A is false.

1.6 Practice   Summarize this discussion by writing the truth table for $A \rightarrow B$.   □

The **equivalence** connective, $A \leftrightarrow B$, is shorthand for the statement $(A \rightarrow B) \wedge (B \rightarrow A)$. We can write the truth table for equivalence by constructing, one piece at a time, a table for $(A \rightarrow B) \wedge (B \rightarrow A)$, as in Figure 1.3. Notice from this truth table that $A \leftrightarrow B$ is true exactly when A and B have the same truth values. The statement $A \leftrightarrow B$ is often read "A if and only if B."

A	B	$A \rightarrow B$	$B \rightarrow A$	$(A \rightarrow B) \wedge (B \rightarrow A)$
T	T	T	T	T
T	F	F	T	F
F	T	T	F	F
F	F	T	T	T

Figure 1.3

The connectives we've seen so far are called **binary connectives** because they join two statements together to produce a third statement. Now let's consider a **unary** connective, a connective acting on one statement to produce a second statement. The **negation** connective, A', is a unary connective and is read "not A," "it is false that A," or "it is not true that A." This does not mean that A' always has a truth value of false, but that its truth value is opposite to the truth value of A. (Some texts denote A' by $\neg A$.)

1.7 Practice   Write the truth table for A'. (It will require only two rows.)   □

1.8 Example   If A is the statement "It will rain tomorrow," then A' is the statement "It is not true that it will rain tomorrow," which may be read "It will not rain tomorrow." Finding the negation of a compound statement can be trickier. If P is the statement "Peter is tall and thin," then P' is the statement "It is false that Peter is tall and thin," which may be read "Peter is not tall or he is not thin." However, this statement is *not* the same as "Peter is short and fat." If P is the statement "The river is shallow or polluted," then P' is the statement "It is false that the river is shallow or polluted," which may be read "The river is neither shallow nor polluted" or even "The river is deep and unpolluted." However, P' is *not* expressed by "The river is not shallow or not polluted."   □

Using connectives we can string lots of statements together to form a complex statement. In order to reduce the number of parentheses required, we stipulate which connectives should be applied first. The order of precedence is $'$, $\wedge$, $\vee$, $\rightarrow$, and $\leftrightarrow$. This means that, in the absence of parentheses, the connective $\wedge$ will be applied before the connective $\vee$, for example. Thus, $A \vee B \wedge C$ means $A \vee (B \wedge C)$, not $(A \vee B) \wedge C$; similarly, $A \vee B \rightarrow C'$ means $(A \vee B) \rightarrow (C)'$. We write the truth tables for complex statements much as we did for $(A \rightarrow B) \wedge (B \rightarrow A)$.

1.9 Example   The truth table for the statement $A \vee B' \rightarrow (A \vee B)'$ is given in Figure 1.4.   □

A	B	B'	$A \vee B'$	$A \vee B$	$(A \vee B)'$	$A \vee B' \rightarrow (A \vee B)'$
T	T	F	T	T	F	F
T	F	T	T	T	F	F
F	T	·F	F	T	F	T
F	F	T	T	F	T	T

Figure 1.4

If a statement has n statement letters and we are making a truth table for that statement, how many rows will the truth table have? From truth tables done so far we know that a statement with only one statement letter has two rows in its truth table, and a statement with two statement letters has four rows. The number of rows equals the number of true–false combinations possible among the statement letters. The first statement letter has two possibilities, T and F. For each of these possibilities, the second statement letter has two possible values. Figure 1.5a pictures this two-level "tree" with four "branches," showing the four possible combinations of T and F for two statement letters. For n statement letters, we extend the tree to n levels, as in Figure 1.5b. The total number of branches then equals 2^n. The total number of rows in a truth table for n statement letters is also 2^n.

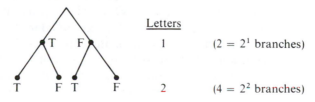

Letters

1          $(2 = 2^1$ branches$)$

2          $(4 = 2^2$ branches$)$

Figure 1.5 (a)

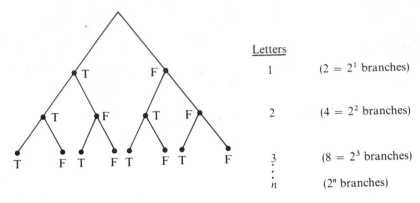

Figure 1.5 (b)

1.10 Practice   Construct truth tables for the following statements:

(a) $(A \to B) \leftrightarrow (B \to A)$ (Remember that $C \leftrightarrow D$ is true precisely when C and D have the same truth values.)

(b) $(A \vee A') \to (B \wedge B')$

(c) $((A \wedge B') \to C')'$

(d) $(A \to B) \leftrightarrow (B' \to A')$ □

The logical operators AND, OR, and NOT are available in many programming languages. These operators act in accordance with the truth tables we have defined upon combinations of true or false expressions to produce an overall truth value. Such truth values provide the decision-making capabilities fundamental to the flow of control in computer programs. Thus at a conditional branch in a program, if the overall truth value of the conditional expression is true, the program will next execute one section of code; if the value is false, the program will next execute a different section of code.

Tautologies

A statement like item (d) of Practice 1.10, whose truth values are always true, is called a **tautology.** A statement like item (b), whose truth values are always false, is called a **contradiction.**

When a compound statement of the form $P \leftrightarrow Q$ is a tautology, as in Practice 1.10(d), the truth values of P and Q agree for every row of the truth table. In this case, P and Q are said to be **equivalent statements.**

Suppose that P and Q are equivalent and that P appears as a component in some larger statement R. What happens when we replace P by Q? As far as truth values go, there should be no change. Let's denote by

R_Q the statement obtained by replacing P with Q everywhere in R. If we construct the truth tables for R and R_Q, then at each row the truth value of P agrees with the truth value of Q, so at each row the truth value of R agrees with the truth value of R_Q. Thus R_Q is equivalent to the original statement R.

1.11 Example   Let R be $(A \to B) \to B$, and P be $A \to B$. From Practice 1.10(d), we know that P is equivalent to $Q = B' \to A'$. Replacing P with Q, we get $R_Q = (B' \to A') \to B$. The truth tables for R and R_Q are shown in Figure 1.6. The truth values of $A \to B$ and $B' \to A'$ agree for every row; hence the truth values of R and R_Q agree for every row. Therefore R and R_Q are equivalent. $\square$

A	B	$A \to B$	$(A \to B) \to B$
T	T	T	T
T	F	F	T
F	T	T	T
F	F	T	F

A	B	A'	B'	$B' \to A'$	$(B' \to A') \to B$
T	T	F	F	T	T
T	F	F	T	F	T
F	T	T	F	T	T
F	F	T	T	T	F

Figure 1.6

We will list a number of particular tautologies, prove one or two of them, and leave the rest as exercises. In this list, the equality sign stands for the equivalence connective. Each of these tautologies, then, tells us that two statements are equivalent. The use of the equality sign is justified because equivalent statements always have equal truth values. We will use 0 to stand for any contradiction and 1 for any tautology.

Some Tautologies

1a. $A \lor B = B \lor A$
2a. $(A \lor B) \lor C = A \lor (B \lor C)$
3a. $A \lor (B \land C) =$
$\quad (A \lor B) \land (A \lor C)$
4a. $A \lor 0 = A$
5a. $A \lor A' = 1$

1b. $A \land B = B \land A$
2b. $(A \land B) \land C = A \land (B \land C)$
3b. $A \land (B \lor C) =$
$\quad (A \land B) \lor (A \land C)$
4b. $A \land 1 = A$
5b. $A \land A' = 0$

(Note that 2a allows us to write $A \lor B \lor C$ with no need for parentheses; similarly, 2b allows us to write $A \land B \land C$.)

1.12 Example   The truth table in Figure 1.7 verifies tautology 1a, and that in Figure 1.8 verifies 4b. Note that only two rows are needed for Figure 1.8 because 1 (a tautology) cannot take on false truth values.   □

A	B	$A \vee B$	$B \vee A$	$A \vee B \leftrightarrow B \vee A$
T	T	T	T	T
T	F	T	T	T
F	T	T	T	T
F	F	F	F	T

Figure 1.7

A	1	$A \wedge 1$	$A \wedge 1 \leftrightarrow A$
T	T	T	T
F	T	F	T

Figure 1.8

1.13 Practice   Verify tautology 5a.   □

The tautologies in our list are grouped into five pairs. In each pair, one statement can be obtained from the other by replacing $\wedge$ with $\vee$, $\vee$ with $\wedge$, 0 with 1, or 1 with 0. Each statement in a pair is called the **dual** of the other. This list of tautologies appears in a more general setting in Chapter 5.

Two additional tautologies that are very useful are DeMorgan's Laws:

$$(A \vee B)' = A' \wedge B' \qquad \text{and} \qquad (A \wedge B)' = A' \vee B'$$

Each is the dual of the other. DeMorgan's Laws help in expressing the negation of a compound statement, as in Example 1.8.

If we have a statement of the form $P \to Q$ where P and Q are compound statements, we can use a quicker procedure than constructing a truth table to determine whether $P \to Q$ is a tautology. We assume that $P \to Q$ is *not* a tautology—that it can take on false values. This can happen only when P is true and Q false. By assigning P true and Q false, we determine possible truth values for the statements making up P and Q. We continue assigning the truth values so determined until all occurrences of statement letters have a truth value. The statement $P \to Q$ is a tautology if and only if some statement letter is assigned both true and false values by this process.

What we have described is a set of instructions, a procedure, for carrying out the task of determining whether $P \to Q$ is a tautology. This procedure can be executed by mechanically following the instructions; in a finite amount of time, we have the answer. In computer science terms, the procedure is an *algorithm*. Algorithms comprise the very heart of computer science, and we will have more to say about them throughout this book. You are probably already aware that the major task in writing a computer program to solve a problem consists of devising an algorithm (a procedure) to produce the solution.

Algorithms are often described in a form that is a middle ground between a purely verbal description in paragraph form (as we gave above for deciding if $P \rightarrow Q$ is a tautology) and a computer program (that, if executed, would actually carry out the steps of the algorithm) written in a programming language. This compromise form to describe algorithms is called **pseudocode**. An algorithm written in pseudocode should not be hard to understand even if you know nothing about computer programming. A pseudocode form of the algorithm to determine whether $P \rightarrow Q$ is a tautology follows:

Algorithm Tautology:$P \rightarrow Q$

1. assign P true
2. assign Q false
3. repeat
       assign determined truth values to components of each compound statement with already assigned truth value
   until all occurrences of statement letters have truth values
4. if some letter has two truth values
   then $P \rightarrow Q$ is a tautology
   else $P \rightarrow Q$ is not a tautology

The steps in the algorithm are numbered to indicate that they are followed sequentially and also to allow easy reference. Steps 1 and 2 assign the truth values true to P and false to Q. Step 3 is an example of a *loop*, where a sequence of steps is repeated until some condition is met. The steps to be repeated are distinguished by being indented and by being framed by the words "repeat" and "until." In algorithm *Tautology:$P \rightarrow Q$*, truth assignments are made to smaller and smaller components of the original P and Q until the *stop loop* condition—all occurrences of statement letters have truth values—is attained. Then the algorithm proceeds to step 4, where the decision of whether $P \rightarrow Q$ is a tautology is made.

1.14 Example   Consider a statement of the form $(A \rightarrow B) \rightarrow (B' \rightarrow A')$. Here P is $A \rightarrow B$ and Q is $B' \rightarrow A'$. Using algorithm *Tautology:$P \rightarrow Q$*, we assign values

$$A \rightarrow B \quad \text{true}$$
$$B' \rightarrow A' \quad \text{false}$$

This completes steps 1 and 2 of the algorithm. Moving on to step 3, the assignment of false to the compound statement $B' \rightarrow A'$ requires the further assignments

     B' true

     A' false

or

     B false

     A true

The combination of A true and $A \rightarrow B$ true requires the assignment

     B true

At this point all occurrences of statement letters have truth values, as follows:

$$\underbrace{\overset{A-T \qquad\quad B-T}{(A \;\longrightarrow\; B)}}_{T} \;\longrightarrow\; \underbrace{\overset{B-F \qquad\quad A-T}{(B' \;\longrightarrow\; A')}}_{F}$$

This terminates the loop at step 3. At step 4, B now has an assignment of both T and F, and therefore $(A \rightarrow B) \rightarrow (B' \rightarrow A')$ is a tautology. We can verify this tautology by constructing its truth table.   □

Algorithm *Tautology:P→Q* is an algorithm to decide whether statements of a certain form are tautologies. However, the process of building a truth table and then examining all the truth values in the final column constitutes an algorithm to determine whether an arbitrary statement is a tautology. This second algorithm is therefore more powerful because it solves a more general problem.

Quantifiers

We have already noted that a variable in a sentence prevents the sentence from being a statement, since the truth value of the sentence is unspecified. For example, the expression "$x > 0$" has no fixed truth value; even if we assume that x represents numeric values, as seems likely, they could be positive or negative. However, expressions containing variables can be made into statements by adding *quantifiers*. Quantifiers are phrases such as "for every" or "for each" or "for some" that tell in some sense *how many* objects have a certain property.

The **universal quantifier** is symbolized by an upside down A, $\forall$, and is read "for all," "for every," "for each," or "for any." Thus the expression

$$(\forall x)(x > 0)$$

is read "for every x, x is greater than zero."

In order to determine the truth value of this expression, we have to know the domain of objects in which we are interpreting this expression, that is, the collection of objects from which x may be chosen. Thus, if the domain of interpretation consists of the positive integers, the expression has the truth value true because every possible value for x has the required property of being greater than zero. If the domain of interpretation consists of all the integers, the expression has the truth value false, because not every possible value for x has the required property. The domain of interpretation is required to contain at least one object so that we are not talking about a trivial case.

1.15 Practice   What is the truth value of $(\forall x)(x > 0)$ if the domain of interpretation consists of

(a) the minimum temperatures for each day last winter in Minnesota
(b) the suggested retail prices for the software at your local computer store   ☐

A generalization of the expression $(\forall x)(x > 0)$ would look like $(\forall x)P(x)$, where $P(x)$ is some unspecified property of x. An interpretation for this expression would consist of not only the collection of objects from which x could take its value but also the particular property that $P(x)$ represents in this domain. Thus an interpretation for $(\forall x)P(x)$ could be the following: the domain consists of all the books in your local library, and $P(x)$ is the property that x has a red cover. $(\forall x)P(x)$, in this interpretation, says that every book in your local library has a red cover. The truth value of this expression, in this interpretation, is undoubtedly false.

The **existential quantifier** is symbolized by a backwards E, $\exists$, and is read "there exists one," "for at least one," or "for some one." Thus the expression

$$(\exists x)(x > 0)$$

is read "there exists an x such that x is greater than zero."

Again, the truth value of this expression depends upon the interpretation. If the domain of interpretation contains a positive number, the expression has the value true; otherwise it has the value false. The truth value of $(\exists x)P(x)$, if the domain consists of all the books in your local library and $P(x)$ is the property that x has a red cover, is true if there is at least one book in the library with a red cover.

1.16 Practice   (a) Construct an interpretation (i.e., give the domain and the meaning of $P(x)$) in which $(\forall x)P(x)$ has the value true.
(b) Construct an interpretation in which $(\exists x)P(x)$ has the value false.

(c) Can you find one interpretation in which both $(\forall x)P(x)$ is true and $(\exists x)P(x)$ is false?

(d) Can you find one interpretation in which both $(\forall x)P(x)$ is false and $(\exists x)P(x)$ is true?   □

Symbols such as P in the expressions $(\forall x)P(x)$ and $(\exists x)P(x)$ are called **predicates**; in particular, they are called **unary predicates** since they involve one variable and are interpreted as properties of single objects. Predicates can be **binary,** with two variables, or **n-ary,** with n variables, which are interpreted as properties of two objects at a time or n objects at a time, respectively. Additional quantifiers can be added to expressions with n-ary predicates.

1.17 Example   The expression $(\forall x)(\exists y)Q(x, y)$ is read "for every x there exists a y such that $Q(x, y)$." In the interpretation where the domain consists of the integers and $Q(x, y)$ is the property that $x < y$, this just says that for any integer, there is a larger integer. The truth value of the expression is true. In the same interpretation, the expression $(\exists y)(\forall x)Q(x, y)$ says that there is a single integer y that is larger than any integer x. The truth value here is false.   □

Example 1.17 illustrates that the order in which the quantifiers appear is important.

In expressions such as $(\forall x)P(x)$ or $(\exists x)P(x)$, x is a *dummy variable*; that is, the truth values of the expressions remain the same in a given interpretation if they are written, say, as $(\forall y)P(y)$ or $(\exists z)P(z)$, respectively. Similarly, the truth value of $(\forall x)(\exists y)Q(x, y)$ is the same as that of $(\forall z)(\exists w)Q(z, w)$ in any interpretation. However, $(\forall x)(\exists x)Q(x, x)$ says something quite different. In the interpretation of Example 1.17, for instance, $(\forall x)(\exists x)Q(x, x)$ says that for every integer x, there is an integer x such that $x < x$. This statement is false, even though $(\forall x)(\exists y)Q(x, y)$ was true in this interpretation. We cannot collapse separate variables together into one without changing the nature of the expression we obtain.

Constants are also allowable in expressions. A constant symbol (a, b, c, etc.) is interpreted as some specific object in the domain. This specification is part of the interpretation. For example, the expression $(\forall x)Q(x, a)$ is false in the interpretation where the domain consists of the integers, $Q(x, y)$ is the property $x < y$, and a is assigned the value 7; it is not the case that every integer is less than 7.

Now we can sum up what is required in an interpretation.

1.18 Definition   An **interpretation** for an expression involving quantifiers consists of the following:

(a) a collection of objects, called the **domain** of the interpretation, which must include at least one object;

(b) an assignment of a property of the objects in the domain to each predicate in the expression; and

(c) an assignment of a particular object in the domain to each constant symbol in the expression. ☐

Expressions can be built using quantifiers together with our previous connectives.

$$(\forall x)[(\exists y)(P(x, y) \wedge Q(x, y)) \to R(x)] \tag{1}$$

and

$$(\exists x)S(x) \vee (\forall y)T(y) \tag{2}$$

are legitimate expressions. Parentheses and brackets help identify the **scope** of a quantifier, the section of the expression to which the quantifier applies. In expression (1) above, the scope of $(\exists y)$ is $P(x, y) \wedge Q(x, y)$, while the scope of $(\forall x)$ is the entire bracketed expression following it. In (2) above, the scope of $(\exists x)$ is $S(x)$ and the scope of $(\forall y)$ is $T(y)$.

Scope is somewhat like order of precedence of connectives in that the truth value of the expression in any particular interpretation, or even whether the expression has a truth value, can be affected if the scope of a quantifier is misunderstood.

1.19 Example   Consider the expression

$$(\forall x)(\exists y)(P(x, y) \wedge Q(x, y))$$

Here the scope of $(\exists y)$ is all of $P(x, y) \wedge Q(x, y)$. The scope of $(\forall x)$ is $(\exists y)(P(x, y) \wedge Q(x, y))$; parentheses or brackets can be eliminated when the scope is clear. In the interpretation where the domain consists of the positive integers, $P(x, y)$ is the property "$x \leq y$," and $Q(x, y)$ is the property "x divides y," the expression is true. For any x that is a positive integer, there does exist a positive integer y, for example $y = 2x$, such that $x \leq y$ and x divides y.

Now suppose the expression is

$$(\forall x)((\exists y)P(x, y) \wedge Q(x, y))$$

Here the scope of $(\exists y)$ is only $P(x, y)$. If we use the same interpretation as before, no truth value is determined for this expression. Given any x, we can choose a y (such as $y = 2x$) so that $(\exists y)P(x, y)$ is true, but in $Q(x, y)$, y is free to take on any value in the domain of the interpretation, regardless of what we chose for y to satisfy $(\exists y)P(x, y)$. For values of y that are multiples of x, $Q(x, y)$ is true, and so is the entire expression; for other values of y, $Q(x, y)$ is false, and so is the entire expression. ☐

In the second expression of Example 1.19 above, $(\forall x)((\exists y)P(x, y) \wedge Q(x, y))$, there is an occurrence of a variable, namely the last occurrence of y, that does not fall within the scope of a quantifier involving that variable. If a variable has such an occurrence, then the variable is said to be a **free variable** in the expression. As in Example 1.19, an expression with free variables will not in general have a truth value in a given interpretation. Rather, the expression will be true for some choices of values of the free variable and false for others. To be consistent with our earlier definition of a statement (a sentence that has a truth value), we will call an expression involving quantifiers a **statement** only if it has no free variables, and we will be interested primarily in statements from now on.

1.20 Practice   What is the truth value of the statement

$$(\exists x)[A(x) \wedge (\forall y)(B(x, y) \rightarrow C(y))]$$

in the interpretation where the domain consists of all integers, $A(x)$ is "$x > 0$," $B(x, y)$ is "$x > y$," and $C(y)$ is "$y \leq 0$"? Construct another interpretation with the same domain in which the statement has the opposite truth value.   □

Many English language sentences can be expressed as statements containing quantifiers. For example, "Every parrot is ugly" is really saying that for any thing that is a parrot, that thing is ugly. Letting $P(x)$ denote "x is a parrot" and $U(x)$ denote "x is ugly," we see that the sentence can be symbolized as

$$(\forall x)(P(x) \rightarrow U(x))$$

Other English language variations that take the same symbolic form are "All parrots are ugly" and "Each parrot is ugly."

Similarly, "There is an ugly parrot" becomes

$$(\exists x)(P(x) \wedge U(x))$$

Variations here are "Some parrots are ugly" and "There are ugly parrots."

In representing these English language sentences in statement form, we used $(\forall x)$ with implication and $(\exists x)$ with conjunction. The other two possible combinations almost never express what we would normally want to say. The statement $(\forall x)(P(x) \wedge U(x))$ says that everything is an ugly parrot; the statement $(\exists x)(P(x) \rightarrow U(x))$ is true as long as there is something, call it x, that is not a parrot, because then $P(x)$ is false and the implication is true.

1.21 Practice   Using the symbols $S(x)$, $I(x)$, and $M(x)$, write statements that express the following:

(a) All students are intelligent.
(b) Some intelligent students like music.
(c) Everyone who likes music is a stupid student.   □

Validity

Statements with no quantifiers are true or false depending upon the truth values assigned to the statement letters. A statement with quantifiers is true or false in a given interpretation. Choosing an interpretation for a statement with quantifiers can thus be considered analogous to choosing truth values in a statement without quantifiers, except that there are an infinite number of possible interpretations for a quantified statement and only 2^n possible rows in the truth table for an unquantified statement with n statement letters.

For unquantified statements, a tautology is true for all rows of the truth table. The analog to tautology for quantified statements is *validity*— a statement is **valid** if it is true in all possible interpretations. The validity of a valid statement must be derived from the form of the statement itself, since validity is independent of any particular interpretation.

We know that constructing the truth table for an unquantified statement and examining all possible truth assignments constitutes an algorithm for deciding "tautologism." However, because we clearly cannot look at all possible interpretations, how can we go about deciding validity? As it turns out, no algorithm to decide validity exists. We must simply use reasoning to determine whether the form of a statement makes the statement true in all interpretations. Of course, we can prove that a statement is invalid by finding a single interpretation in which the statement has the truth value false.

The following table summarizes parallels and differences between unquantified and quantified statements.

Unquantified Statements	*Quantified Statements*
1. True or false, depending on truth value assignments to statement letters	1. True or false, depending on interpretation
2. Tautology—true for all truth value assignments	2. Valid statement—true for all interpretations
3. Algorithm (truth table) to determine whether statement is a tautology	3. No algorithm to determine whether statement is valid

Now let's try our hand at determining validity.

1.22 Example   (a) The statement $(\forall x)P(x) \to (\exists x)P(x)$ is valid. In any interpretation, if every element of the domain has a certain property, then there exists an element of the domain that has that property. (Here we make use of the fact that the domain of any interpretation must have objects in it.) Therefore, whenever the antecedent is true, so is the consequent, and the implication is therefore true.

(b) The statement $(\forall x)P(x) \to P(a)$ is valid because in any interpretation, *a* is a particular member of the domain and therefore has the property that is shared by all members of the domain.

(c) The statement

$$(\forall x)(P(x) \wedge Q(x)) \leftrightarrow (\forall x)P(x) \wedge (\forall x)Q(x)$$

is valid. If both *P* and *Q* are true for all the elements of the domain, then *P* is true for all elements and *Q* is true for all elements, and vice versa.   □

1.23 Example   The statement $(\exists x)P(x) \rightarrow (\forall x)P(x)$ is not valid. For example, in the interpretation where the domain consists of the integers and $P(x)$ means that *x* is even, it is true that there exists an integer that is even, but it is false that every integer is even. The antecedent of the implication is true and the consequent is false, so the value of the implication is false.   □

We do not, of course, have to go to a mathematical context to construct an interpretation in which a statement is false, but it is frequently easier to do so because the relationships among objects are relatively clear.

1.24 Practice   Is the statement

$$(\forall x)(P(x) \vee Q(x)) \rightarrow (\forall x)P(x) \vee (\forall x)Q(x)$$

valid or invalid? Explain.   □

Negations of statements with quantifiers also require some care. The negation of the statement "Everything is beautiful" is "It is false that everything is beautiful" or "Something is nonbeautiful." Symbolically,

$$((\forall x)A(x))' \leftrightarrow (\exists x)(A(x))'$$

is valid. Note that "Everything is nonbeautiful," or $(\forall x)(A(x))'$, says something *stronger* than the negation of the original statement.

The negation of "Something is beautiful" is "Nothing is beautiful" or "Everything fails to be beautiful." Symbolically,

$$((\exists x)A(x))' \leftrightarrow (\forall x)(A(x))'$$

is valid. In English, the statement "Everything is not beautiful" can be misinterpreted as "Not everything is beautiful" or "There is something nonbeautiful." However, this misinterpretation, symbolized by $(\exists x)(A(x))'$, is *not as strong* as the negation of the original statement.

✔ Checklist

Definitions

statement (*p. 1, 14*)	equivalent statements (*p. 6*)
proposition (*p. 1*)	dual of a statement (*p. 8*)
conjunction (*p. 2*)	pseudocode (*p. 9*)
truth table (*p. 2*)	universal quantifier (*p. 10*)
disjunction (*p. 3*)	existential quantifier (*p. 11*)
implication (*p. 3*)	predicate (*p. 12*)
antecedent (*p. 3*)	unary predicate (*p. 12*)
consequent (*p. 3*)	binary predicate (*p. 12*)
equivalence (*p. 4*)	*n*-ary predicate (*p. 12*)
binary connective (*p. 4*)	interpretation (*p. 12*)
unary connective (*p. 4*)	domain (*p. 12*)
negation (*p. 4*)	scope (*p. 13*)
tautology (*p. 6*)	free variable (*p. 14*)
contradiction (*p. 6*)	valid statement (*p. 15*)

Techniques

Construct truth tables for complex statements.

Recognize tautologies and contradictions.

Determine the truth value of a quantified statement in a given interpretation.

Translate English language sentences into statements involving quantifiers, and vice versa.

Recognize a valid statement and explain why it is valid.

Recognize a nonvalid statement and construct an interpretation in which it is false.

Main Ideas

Statements and how they can be combined (conjunction, disjunction, etc.).

Truth values for compound statements that depend upon the truth values of their components.

A list of basic tautologies.

Quantified statements, whose truth value depends upon the interpretation considered.

Valid statements, which are quantified statements that are true in all interpretations and whose validity is therefore inherent in the form of the statement itself.

Exercises Section 1.1

Answers to starred items are given at the back of the book.

★ 1. Which of the following are statements?
   (a) The moon is made of green cheese.
   (b) He is certainly a tall man.
   (c) Two is a prime number.
   (d) Will the game be over soon?
   (e) Next year interest rates will rise.
   (f) Next year interest rates will fall.
   (g) $x^2 - 4 = 0$

2. What is the truth value of each of the following statements?
   (a) 8 is even or 6 is odd.
   (b) 8 is even and 6 is odd.
   (c) 8 is odd or 6 is odd.
   (d) 8 is odd and 6 is odd.
   (e) If 8 is odd, then 6 is odd.
   (f) If 8 is even, then 6 is odd.
   (g) If 8 is odd, then 6 is even.
   (h) If 8 is odd and 6 is even, then $8 < 6$.

★ 3. Find the antecedent and consequent in each of the following statements:
   (a) Healthy plant growth follows from sufficient water.
   (b) Increased availability of microcomputers is a necessary condition for further technological advances.
   (c) Errors will be introduced only if there is a modification of the program.
   (d) Fuel savings implies good insulation or all windows are storm windows.

4. Several forms of negation are given for each of the following statements. Which are correct?
   (a) The answer is either 2 or 3.
       (1) Neither 2 nor 3 is the answer.
       (2) The answer is not 2 or not 3.
       (3) The answer is not 2 and it is not 3.
   (b) Cucumbers are green and seedy.
       (1) Cucumbers are not green and not seedy.
       (2) Cucumbers are not green or not seedy.
       (3) Cucumbers are green and not seedy.
   (c) $2 < 7$ and 3 is odd.
       (1) $2 > 7$ and 3 is even.
       (2) $2 \geq 7$ and 3 is even.
       (3) $2 \geq 7$ or 3 is odd.
       (4) $2 \geq 7$ or 3 is even.

5. Let A, B, and C be the following statements:

   A: Roses are red.
   B: Violets are blue.
   C: Sugar is sweet.

   Translate the following compound statements into symbolic notation.
   (a) Roses are red and violets are blue.
   (b) Roses are red, and either violets are blue or sugar is sweet.
   (c) Whenever violets are blue, roses are red and sugar is sweet.
   (d) Roses are red only if violets aren't blue or sugar is sour.
   (e) Roses are red and, if sugar is sour, then either violets aren't blue or sugar is sweet.

6. With A, B, and C defined as in Exercise 5, translate the following statements into English:
   (a) $B \vee C'$
   (b) $B' \vee (A \rightarrow C)$
   (c) $(C \wedge A') \leftrightarrow B$
   (d) $C \wedge (A' \leftrightarrow B)$
   (e) $(B \wedge C')' \rightarrow A$
   (f) $A \vee (B \wedge C')$
   (g) $(A \vee B) \wedge C'$

★ 7. Using letters for the component statements, translate the following compound statements into symbolic notation:
   (a) If prices go up, then housing will be plentiful and expensive; but if housing is not expensive, then it will still be plentiful.
   (b) Either going to bed or going swimming is a sufficient condition for changing clothes; however, changing clothes does not mean going swimming.
   (c) Either it will rain or it will snow but not both.
   (d) If Janet wins or if she loses, she will be tired.
   (e) Either Janet will win or, if she loses, she will be tired.

8. Construct truth tables for the following statements, where A, B, and C are statements. Note any tautologies or contradictions.
   ★ (a) $(A \rightarrow B) \leftrightarrow A' \vee B$
   ★ (b) $(A \wedge B) \vee C \rightarrow A \wedge (B \vee C)$
   (c) $A \wedge (A' \vee B')'$
   (d) $A \wedge B \rightarrow A'$
   (e) $(A \rightarrow B) \rightarrow [(A \vee C) \rightarrow (B \vee C)]$
   (f) $A \rightarrow (B \rightarrow A)$
   (g) $A \wedge B \leftrightarrow B' \vee A'$
   (h) $(A \vee B') \wedge (A \wedge B)'$
   (i) $((A \vee B) \wedge C') \rightarrow A' \vee C$

★ 9. A memory chip from a microcomputer has 2^4 bistable (ON–OFF) memory elements. What is the total number of ON–OFF configurations?

10. Consider the following fragment of a Pascal program:

```
for count: = 1 to 5 do
    begin
        read(a);
        if ((a < 5.0) and (2*a < 10.7)) or (sqrt(5.0*a) > 5.1)
        then
            writeln(a)
    end
```

The input values for a are 1.0, 5.1, 2.4, 7.2, and 5.3. What are the output values?

11. Suppose that A, B, and C represent conditions that will be true or false when a certain computer program is executed. Suppose further that you want the program to carry out a certain task only when A or B is true (but not both) and C is false. Using A, B, and C and the connectives AND, OR, and NOT, write a statement that will be true only under these conditions.

12. Verify by constructing truth tables that the statements in the list on page 7 are tautologies. (We have already verified 1a, 4b, and 5a.)

13. Verify by constructing truth tables that the following statements are tautologies, where the equality sign stands for the equivalence connective:

★ (a) $A \lor A'$ (b) $(A')' = A$

★ (c) $A \land B \to B$ (d) $A \to A \lor B$

(e) $(A \lor B)' = A' \land B'$ } DeMorgan's Laws
(f) $(A \land B)' = A' \lor B'$ }

14. Use algorithm *Tautology*:$P \to Q$ to prove that the following are tautologies:

★ (a) $(B' \land (A \to B)) \to A'$ (b) $((A \to B) \land A) \to B$

(c) $(A \lor B) \land A' \to B$ (d) $(A \land B) \land B' \to A$

15. In each case, construct compound statements P and Q so that the given statement is a tautology.

(a) $P \land Q$ (b) $P \to P'$ (c) $P \land (Q \to P')$

16. The truth table for $A \lor B$ shows that the value of $A \lor B$ is true if A is true, if B is true, or if both are true. This use of the word "or" where the result is true if both components are true is called the **inclusive or**. It is the inclusive or that is understood in the sentence "We may have rain or drizzle tomorrow." Another use of the word "or" in the English language is the **exclusive or,** in which the result is false when both components are true. The exclusive or is understood in the sentence "At the intersection, you should turn north or south." Exclusive or is symbolized by A ⊕ B.

(a) Write the truth table for the exclusive or.

(b) Show that A ⊕ B ↔ $(A \leftrightarrow B)'$ is a tautology.

17. Every compound statement is equivalent to a statement using only

the connectives of conjunction and negation. To see this, we need to find equivalent statements for $A \vee B$ and $A \rightarrow B$ that use only $\wedge$ and $'$. These new statements can replace, respectively, any occurrences of $A \vee B$ and $A \rightarrow B$. (The connective $\leftrightarrow$ was defined in terms of other connectives, so we already know that it can be replaced by a statement using these other connectives.)

(a) Show that $A \vee B$ is equivalent to $(A' \wedge B')'$.

(b) Show that $A \rightarrow B$ is equivalent to $(A \wedge B')'$.

18. Show that every compound statement is equivalent to a statement using only the connectives of

(a) disjunction and negation

(b) implication and negation

(Hint: see Exercise 17.)

19. Prove that there are compound statements that are not equivalent to any statement using only the connectives $\rightarrow$ and $\vee$.

20. The binary connective $|$ is defined by the following truth table:

A	B	$A|B$
T	T	F
T	F	T
F	T	T
F	F	T

Show that every compound statement is equivalent to a statement using only the connective $|$. (Hint: use Exercise 17 and find equivalent statements for $A \wedge B$ and A' in terms of $|$.)

21. The binary connective $\downarrow$ is defined by the following truth table:

A	B	$A \downarrow B$
T	T	F
T	F	F
F	T	F
F	F	T

Show that every compound statement is equivalent to a statement using only the connective $\downarrow$. (Hint: see Exercise 20.)

22. In a certain country, every inhabitant is either a truth teller who always tells the truth or a liar who always lies. Traveling in this country, you meet two of the inhabitants, Percival and Llewellyn. Percival says, "If I am a truth teller, then Llewellyn is a truth teller." Is Percival a liar or a truth teller? What about Llewellyn?

23. What is the truth value of each of the following statements in the interpretation where the domain consists of the integers?

★ (a) $(\forall x)(\exists y)(x + y = x)$
★ (b) $(\exists y)(\forall x)(x + y = x)$
★ (c) $(\forall x)(\exists y)(x + y = 0)$
★ (d) $(\exists y)(\forall x)(x + y = 0)$
  (e) $(\forall x)(\forall y)(x < y \lor y < x)$
  (f) $(\forall x)(x < 0 \rightarrow (\exists y)(y > 0 \land x + y = 0))$
  (g) $(\exists x)(\exists y)(x^2 = y)$
  (h) $(\forall x)(x^2 > 0)$

24. Give the truth value of each of the following statements in the interpretation where the domain consists of the states of the United States, $Q(x, y)$ is "x is north of y," $P(x)$ is "x starts with the letter M," and a is "Mississippi."
    (a) $(\forall x)P(x)$
    (b) $(\forall x)(\forall y)(\forall z)(Q(x, y) \land Q(y, z) \rightarrow Q(x, z))$
    (c) $(\exists y)(\forall x)Q(y, x)$
    (d) $(\forall x)(\exists y)(P(y) \land Q(x, y))$
    (e) $(\exists y)Q(a, y)$

25. For each statement, find an interpretation in which the statement is true, and one in which it is false.
    ★ (a) $(\forall x)[(A(x) \lor B(x)) \land (A(x) \land B(x))']$
      (b) $(\forall x)(\forall y)(P(x, y) \rightarrow P(y, x))$
      (c) $(\forall x)(P(x) \rightarrow (\exists y)Q(x, y))$
      (d) $(\exists x)(A(x) \land (\forall y)B(x, y))$
      (e) $[(\forall x)A(x) \rightarrow (\forall x)B(x)] \rightarrow (\forall x)(A(x) \rightarrow B(x))$

26. Identify the scope of each of the quantifiers and indicate any free variables in the following expressions:
    (a) $(\forall x)(P(x) \rightarrow Q(y))$
    (b) $(\exists x)(A(x) \land (\forall y)B(y))$
    (c) $(\exists x)((\forall y)P(x, y) \land Q(x, y))$
    (d) $(\exists x)(\exists y)(A(x, y) \land B(y, z) \rightarrow A(a, z))$

27. Using the predicate symbols shown and appropriate quantifiers, write each English language sentence as a symbolic statement.

> $D(x)$ is "x is a day." M is "Monday."
> $S(x)$ is "x is sunny." T is "Tuesday."
> $R(x)$ is "x is rainy."

    ★ (a) All days are sunny.
    ★ (b) Some days are not rainy.
    ★ (c) Every day that is sunny is not rainy.
      (d) Some days are sunny and rainy.
      (e) No day is both sunny and rainy.
      (f) It is always a sunny day only if it is a rainy day.
      (g) No day is sunny.

(h) Monday was sunny, therefore every day will be sunny.
(i) It rained both Monday and Tuesday.
(j) If some day is rainy, then every day will be sunny.

28. Using the predicate symbols shown and appropriate quantifiers, write each English language sentence as a symbolic statement.

$C(x)$ is "x is a Corvette." $P(x)$ is "x is a Porsche."
$F(x)$ is "x is a Ferrari." $S(x, y)$ is "x is slower than y."

★ (a) Nothing is both a Corvette and a Ferrari.
  (b) Some Porsches are slower than only Ferraris.
★ (c) Only Corvettes are slower than Porsches.
  (d) All Ferraris are slower than some Corvette.
  (e) Some Porsches are slower than no Corvette.
  (f) If there is a Corvette that is slower than a Ferrari, then all Corvettes are slower than all Ferraris.

29. If

$L(x, y)$ is "x loves y." j is "John."
$H(x)$ is "x is handsome." k is "Kathy."
$M(x)$ is "x is a man." $W(x)$ is "x is a woman."
$P(x)$ is "x is pretty."

give English language translations of the following symbolic statements:

★ (a) $H(j) \land L(k, j)$
★ (b) $(\forall x)(M(x) \rightarrow H(x))$
  (c) $(\forall x)[W(x) \rightarrow (\forall y)(L(x, y) \rightarrow M(y) \land H(y))]$
  (d) $(\exists x)(M(x) \land H(x) \land L(x, k))$
  (e) $(\exists x)[W(x) \land P(x) \land (\forall y)(L(x, y) \rightarrow H(y) \land M(y))]$
  (f) $(\forall x)(W(x) \land P(x) \rightarrow L(j, x))$

30. Explain why each statement is valid.
  (a) $(\forall x)(\forall y)A(x, y) \leftrightarrow (\forall y)(\forall x)A(x, y)$
  (b) $(\exists x)(\exists y)A(x, y) \leftrightarrow (\exists y)(\exists x)A(x, y)$
  (c) $(\exists x)(\forall y)P(x, y) \rightarrow (\forall y)(\exists x)P(x, y)$
  (d) $A(a) \rightarrow (\exists x)A(x)$
  (e) $(\forall x)(A(x) \rightarrow B(x)) \rightarrow ((\forall x)A(x) \rightarrow (\forall x)B(x))$

31. Give interpretations to prove that each of the following statements is not valid:
★ (a) $(\exists x)A(x) \land (\exists x)B(x) \rightarrow (\exists x)(A(x) \land B(x))$
  (b) $(\forall x)(\exists y)P(x, y) \rightarrow (\exists x)(\forall y)P(x, y)$
  (c) $(\forall x)(P(x) \rightarrow Q(x)) \rightarrow ((\exists x)P(x) \rightarrow (\forall x)Q(x))$
  (d) $(\forall x)(A(x))' \leftrightarrow ((\forall x)A(x))'$

32. Decide whether each of the following statements is valid or invalid. Justify your answer.

(a) $(\exists x)A(x) \leftrightarrow ((\forall x)A(x)')'$
(b) $(\forall x)P(x) \vee (\exists x)Q(x) \rightarrow (\forall x)(P(x) \vee Q(x))$
(c) $(\forall x)A(x) \leftrightarrow [(\exists x)(A(x))']'$
(d) $(\forall x)(P(x) \vee Q(x)) \rightarrow (\forall x)P(x) \vee (\exists y)Q(y)$

33. Several forms of negation are given for each of the following statements. Which are correct?
    (a) Some people like mathematics.
        (1) Some people dislike mathematics.
        (2) Everybody dislikes mathematics.
        (3) Everybody likes mathematics.
    (b) Everyone loves ice cream.
        (1) No one loves ice cream.
        (2) Everyone dislikes ice cream.
        (3) Someone doesn't love ice cream.
    (c) All people are tall and thin.
        (1) Someone is short and fat.
        (2) No one is tall and thin.
        (3) Someone is short or fat.
    (d) Some pictures are old or faded.
        (1) Every picture is neither old nor faded.
        (2) Some pictures are not old or faded.
        (3) All pictures are not old or not faded.

34. PROLOG (for PROgramming in LOGic) is a programming language based on statements with quantifiers, predicates, and logical connectives. It is particularly useful in artificial intelligence applications, since it more closely parallels human thought processes than more conventional programming languages. Suppose that a PROLOG data base has been created with the following entries:

Wendy's sells Frosty.	McDonald's sells Coke.
Wendy's sells root beer.	McDonald's sells lemonade.
Wendy's sells Pepsi.	McDonald's sells tea.
Wendy's sells milk.	McDonald's sells milk.
Wendy's sells hamburger.	McDonald's sells chicken.
Wendy's sells cheeseburger.	McDonald's sells hamburger.
Wendy's sells fries.	McDonald's sells fries.
Wendy's sells fish.	McDonald's sells cookies.

Frosty drink	hamburger food
root beer drink	cheeseburger food
Pepsi drink	fries food
milk drink	fish food
Coke drink	chicken food
lemonade drink	cookies food
tea drink	
Sprite drink	

What do you think will be the results of each of the following PROLOG queries? (Hint: each answer is a list of items.)

(a) which (x: Wendy's sells x and x food)

(b) all (x: x drink and McDonald's sells x and Wendy's sells x)

(c) all (x: (either Wendy's sells x or McDonald's sells x) and x drink)

(d) all (x: x food and Wendy's sells x and not McDonald's sells x)

SECTION 1.2   PROPOSITIONAL LOGIC AND PREDICATE LOGIC

Formal Systems

Mathematical results are often called theorems. In formal logic systems, which use statements of the type we studied in the previous section of this chapter, the word "theorem" has a very precise meaning. In such systems, certain statements are accepted as **axioms**—statements that do not need to be proved. An axiom should therefore be a statement whose "truth" is self-evident. At the least, then, an axiom should be a tautology or, if it involves quantifiers, a valid statement. In addition to axioms, formal systems may contain rules of inference. A **rule of inference** is a convention that allows a new statement of a certain form to be inferred, or deduced, from one to two other statements of a certain form. A sequence of statements in which each statement is either an axiom or the result of applying one of the rules of inference to earlier statements in the sequence is called a **proof sequence.** A **theorem** is the last entry in such a sequence; the sequence is the **proof** of the theorem.

The following outline is typical of the proof of a theorem.

s1	(an axiom)
s2	(an axiom)
s3	(inferred from s1 and s2 by a rule of inference)
s4	(an axiom)
s5	(inferred from s4 by a rule of inference)
s6	(inferred from s3 and s5 by a rule of inference)

Statement s6, the last statement in the sequence, is the theorem, and the entire sequence constitutes its proof. (Of course, the other statements also could be theorems—we would just stop the proof sequence at that point.)

Another requirement that we place on our system, in addition to the one that axioms be tautologies or valid statements, is that there be as few axioms and as few rules of inference as possible. The advantage to this is somewhat like the advantage of an alphabet over pictographs. To form words, we use the alphabet and follow certain rules for combining alphabetic characters; thus we build words out of a very small collection of symbols, rather than devising a new symbol for each word. In the same way, we would like to build our theorems from a minimal collection of axioms and rules of inference.

Choosing a set of axioms and rules of inference must be done with care, however. If we choose too few axioms and too few or too weak rules of inference, then we will not be able to prove some statements that are "true" and therefore should be theorems. On the other hand, if we choose too many axioms and too many or too strong rules of inference, then we will be able to take almost any statement and prove that it is a theorem, including statements that are not "true," and therefore should not be theorems. How do we get just the right things to be theorems?

Propositional Logic

To make sense of the intuitive word "true," we consider two separate formal systems—one for statements without quantifiers and one for statements with quantifiers. The former case, where we deal only with un-quantified statements, is called **propositional logic, statement logic,** or **propositional calculus.** In this system, a "true" statement means a tautology. We therefore want our axioms and rules of inference to allow us to prove all tautologies, and only tautologies, as theorems.

We will take the following statements to be axioms, where P, Q, and R can be compound statements:

1. $P \rightarrow (Q \rightarrow P)$
2. $(P \rightarrow (Q \rightarrow R)) \rightarrow ((P \rightarrow Q) \rightarrow (P \rightarrow R))$
3. $(Q' \rightarrow P') \rightarrow (P \rightarrow Q)$

Each of these can be shown to be a tautology. There is only one rule of inference: from statements P and $P \rightarrow Q$, we can infer statement Q. (This rule of inference is known by its Latin name of **modus ponens,** meaning "method of assertion.")

Because P, Q, and R can be compound statements, each axiom given above is really a statement form, or **schema,** for an infinite number of statements. Thus

$$(A \rightarrow B) \rightarrow ((C \wedge D) \rightarrow (A \rightarrow B))$$

is an axiom because it fits axiom schema 1, where P is the statement $A \rightarrow B$ and Q is the statement $C \wedge D$. But doesn't this mean that we don't have a small number of axioms after all, that we really have an infinite number of axioms? Yes—but there are only three *forms* for the axioms.

1.25 Example   The statement $A \rightarrow A$ is a theorem. A proof sequence follows:

1. $[A \rightarrow ((A \rightarrow A) \rightarrow A)] \rightarrow$ (Axiom 2 with $P = A$, $Q =$
   $[(A \rightarrow (A \rightarrow A)) \rightarrow (A \rightarrow A)]$ $A \rightarrow A$, $R = A$)

2. $A \rightarrow ((A \rightarrow A) \rightarrow A)$ (Axiom 1 with $P = A$, $Q = A \rightarrow A$)

3. $(A \rightarrow (A \rightarrow A)) \rightarrow (A \rightarrow A)$ (from 1 and 2 by modus ponens)

4. $A \rightarrow (A \rightarrow A)$ (Axiom 1 with $P = A$, $Q = A$)

5. $A \rightarrow A$ (from 3 and 4 by modus ponens)

The justification for each step in the proof sequence is given, although these are not really part of the proof sequence. A similar proof sequence could show that $P \rightarrow P$ is a theorem for any statement P, so we really have a theorem schema. In general, we will prove theorem schemata rather than individual theorems.   □

The axioms that we have chosen involve only implication and negation. For statements that involve the connectives of disjunction or conjunction, we use the equivalences

$$A \vee B \leftrightarrow A' \rightarrow B \quad \text{and} \quad A \wedge B \leftrightarrow (A \rightarrow B')'$$

and content ourselves with proving the resulting equivalent statements. In fact, we could have defined disjunction and conjunction in terms of implication and negation. Then all of our statements would have involved only the connectives of implication and negation.

Although we will not prove it here, this system of axioms and one rule of inference does exactly what we want—every tautology is a theorem (i.e., has a proof), and vice versa. This property is described by saying that our formal system is **complete** (everything that should be a theorem is) and **correct** (nothing is a theorem that should not be).

We allow shortcuts in proof sequences by using already proved theorems. Once T has been proved to be a theorem, then T can serve as a statement in another proof sequence. This is because T has its own proof sequence, which could be substituted into the proof sequence that we are constructing.

Deductions

We often want to prove that statements of the form $P \rightarrow Q$, where P and Q are compound statements, are theorems. P is called the **hypothesis** of the theorem, and Q the **conclusion.** If $P \rightarrow Q$ is a theorem, it must be a tautology, and whenever P is true, Q must be true also. Intuitively, we think of being able to deduce Q from P. Formally, we define a **deduction** of Q from P as a sequence of statements ending with Q where each statement is an axiom, or is the statement P, or is derivable from earlier statements by the rules of inference. In effect, this is a proof of a theorem, where we allow P as an axiom. It can be shown that $P \rightarrow Q$ *is indeed a* *theorem if and only if Q is deducible from P.* Our technique for proving theorems of the form $P \rightarrow Q$ is therefore to include the hypothesis as one of the statements in the sequence and to conclude the sequence with Q.

1.26 Example   Using propositional logic, prove the theorem

$$(P \rightarrow (P \rightarrow Q)) \rightarrow (P \rightarrow Q)$$

A proof sequence follows, where we denote the use of the theorem schema proved in Example 1.25 by writing "1.25."

1. $P \rightarrow (P \rightarrow Q)$	(hypothesis)
2. $(P \rightarrow (P \rightarrow Q)) \rightarrow ((P \rightarrow P) \rightarrow (P \rightarrow Q))$	(Axiom 2)
3. $(P \rightarrow P) \rightarrow (P \rightarrow Q)$	(1, 2, modus ponens)
4. $P \rightarrow P$	(1.25)
5. $P \rightarrow Q$	(3,4, modus ponens) $\square$

1.27 Practice   Using propositional logic, prove the theorem

$$P' \rightarrow (P \rightarrow Q)$$ $\square$

More generally, if the hypothesis of a theorem is a series of conjunctions, $P_1 \wedge P_2 \cdots \wedge P_n$, we simply include each conjunct of the hypothesis as a statement in the proof sequence and deduce the conclusion from them. Finally, if the conclusion itself is an implication, $R \rightarrow S$, we can put R into the proof sequence, making it part of the hypothesis, and just deduce S.

To summarize, we have the following possibilities:

1. To prove the theorem $P \rightarrow Q$, deduce Q from P.
2. To prove the theorem $P_1 \wedge P_2 \wedge \cdots \wedge P_n \rightarrow Q$, deduce Q from $P_1, P_2, \ldots, P_n$.
3. To prove the theorem $P_1 \wedge P_2 \wedge \cdots \wedge P_n \rightarrow (R \rightarrow S)$, deduce S from $P_1, P_2, \ldots, P_n, R$.

1.28 Example   A proof of the theorem

$$(P' \rightarrow Q') \wedge (P \rightarrow S) \rightarrow (Q \rightarrow S)$$

using propositional logic is

1. $P' \rightarrow Q'$	(hypothesis)
2. $P \rightarrow S$	(hypothesis)
3. Q	(hypothesis)
4. $(P' \rightarrow Q') \rightarrow (Q \rightarrow P)$	(Axiom 3)
5. $Q \rightarrow P$	(1,4, modus ponens)

6. P (3, 5, modus ponens)

7. S (2, 6, modus ponens)   □

A theorem of the form $P_1 \wedge P_2 \wedge \cdots \wedge P_n \rightarrow Q$ tells us that Q can be deduced from $P_1, P_2, \ldots, P_n$. Thus, for example, in a proof sequence containing the statements $P' \rightarrow Q'$ and $P \rightarrow S$, we can insert the statement $Q \rightarrow S$ and cite the theorem of Example 1.28 as the justification.

1.29 Practice   Use propositional logic to prove the theorem

$$(P \rightarrow Q) \wedge (Q \rightarrow R) \rightarrow (P \rightarrow R)$$ □

Valid Arguments

An argument in English (an attorney's trial summary, an advertisement, or a political speech) is often presented as a series of statements $P_1, P_2, \ldots, P_n$ followed by a conclusion Q. The argument is a **valid argument** if the conclusion is a logical deduction of the conjunction $P_1 \wedge P_2 \wedge \cdots \wedge P_n$—in other words, if $P_1 \wedge P_2 \wedge \cdots \wedge P_n \rightarrow Q$ is a theorem.

1.30 Example   Is the following argument valid? "My client is left-handed, but if the diary was not missing, then my client is not left-handed; therefore, the diary was missing." There are only two simple statements involved here, so we symbolize them as follows:

L: My client is left-handed.

D: The diary was missing.

The argument is then

$$(L \wedge (D' \rightarrow L')) \rightarrow D$$

The validity of the argument is established by the following proof:

1. L (hypothesis)

2. $D' \rightarrow L'$ (hypothesis)

3. $(D' \rightarrow L') \rightarrow (L \rightarrow D)$ (Axiom 3)

4. $L \rightarrow D$ (2, 3, modus ponens)

5. D (1, 4, modus ponens)

Notice that the validity of the argument is a function only of its logical form and has nothing to do with the truth of any of its components. We still have no idea whether the diary was really missing. Furthermore, the argument "Skooses are pink, but if Gingoos does not like perskees, then skooses are not pink; therefore Gingoos does like perskees," which has the same logical form, is also valid, even though it does not make sense.   □

Because any tautology is also a theorem in propositional logic, we may insert a tautology at any step in a proof sequence.

1.31 Example   Consider the argument "The Federal discount rate will drop and interest rates will drop. If interest rates drop, the housing market will improve. Either the Federal discount rate will drop or the housing market will not improve. Therefore the Federal discount rate will drop." Using

> F: The Federal discount rate will drop.
> I: Interest rates will drop.
> H: The housing market will improve.

the argument is $((F \land I) \land (I \to H) \land (F \lor H')) \to F$. A proof sequence to establish validity is

> 1. $F \land I$ (hypothesis)
> 2. $F \land I \to I$ (tautology)
> 3. I (1, 2, modus ponens)
> 4. $I \to H$ (hypothesis)
> 5. H (3, 4, modus ponens)
> 6. $F \lor H'$ (hypothesis)
> 7. F (tautology $(F \lor H') \land H \to F$)

The justification at step 7 is that the tautology given shows that F can be deduced from $F \lor H'$ and H.   □

1.32 Practice   Show that the following argument is valid, using statement letters P, M, and C: "If the product is reliable, then the market share will rise. Either the product is reliable or costs will increase. The market share will not rise. Therefore costs will increase."   □

By now, you may have had the following thought: if theorems are the same as tautologies, and if we can insert tautologies into proof sequences anyway, then to show that an argument is valid, why don't we just show that it is itself a tautology? Establishing that an argument is a tautology can be done by constructing its truth table or using algorithm *Tautology:P→Q* of the previous section. This is a far more mechanical task than constructing a proof sequence. Why have we talked about proofs at all?

If we were to work only within propositional logic, we would indeed not need the idea of a formal proof. However, we know that not all statements can be symbolized within propositional logic, and that there is no mechanical procedure (that corresponds to building truth tables) to establish the validity of statements involving quantifiers. A formal proof technique must be used; what we've done here for propositional logic serves as a basis for what will be done in predicate logic.

Predicate Logic

The formal logic system that allows quantified as well as unquantified statements is called **predicate logic** or **predicate calculus.** In this system, "true" means valid, that is, true in all possible interpretations. We want the axioms and rules of inference to allow us to prove all valid statements, and only valid statements, as theorems.

All tautologies are valid statements, so we want to be able to prove all the theorems of propositional logic within predicate logic. We therefore keep the previous axioms and rule of inference and add some new axioms and a new rule of inference to deal with the quantifiers. The axioms for predicate logic are:

1. $P \rightarrow (Q \rightarrow P)$
2. $(P \rightarrow (Q \rightarrow R)) \rightarrow ((P \rightarrow Q) \rightarrow (P \rightarrow R))$
3. $(Q' \rightarrow P') \rightarrow (P \rightarrow Q)$
4. $(\forall x)(P(x) \rightarrow Q(x)) \rightarrow ((\forall x)P(x) \rightarrow (\forall x)Q(x))$
5. $(\forall x)P(x) \rightarrow P(x)$ or $(\forall x)P(x) \rightarrow P(a)$ where a is a constant
6. $(\exists x)P(x) \rightarrow P(t)$ where t is a constant or variable name not previously used in the proof sequence
7. $P(x) \rightarrow (\exists x)P(x)$ or $P(a) \rightarrow (\exists x)P(x)$ where a is a constant and x does not appear in $P(a)$
8. $((\exists x)P(x))' \leftrightarrow (\forall x)(P(x))'$

The new Axiom 4 is quite straightforward. Axiom 5 says that if a predicate is true for all elements of the domain, it is true for an arbitrary x or for a constant a. Axiom 6 says that if an object exists for which P is true, we may as well name that object; however, the name must be arbitrary, not one we have already used earlier in the proof sequence. (This requirement means that we will want to use Axiom 6 as early as possible in the proof sequence, since other axioms do not have such restrictions.) For example, from Axiom 6 and modus ponens, we can deduce $(\exists y)Q(a, y)$ from $(\exists x)(\exists y)Q(x, y)$ if a is a new constant name. Axiom 7 says that if P holds for a particular value, then there is some member of the domain for which it holds. Axiom 8 agrees with our intuitive understanding of the meaning of the universal and existential quantifiers.

With these axioms, we can remove and insert existential quantifiers and remove universal quantifiers. A new rule of inference allows us to insert universal quantifiers but only under the proper circumstances. The rules of inference for predicate logic are

1. **Modus ponens:** Q can be inferred from P and $P \rightarrow Q$.
2. **Generalization:** If Q has been deduced from $P_1, P_2, \ldots, P_n$, that is, Q is the last step in a deduction sequence, then we can infer $(\forall x)Q$, provided
   a. x is not free in $P_1, P_2, \ldots,$ or P_n, and

b. x is not free in $(\exists y)Q$, where Axiom 6 was then applied to get Q.

In particular, if Q is a theorem, so is $(\forall x)Q$.

Without these restrictions on generalization, we would be able to deduce that predicates are true for all objects in the domain from the truth of these predicates only for specified objects.

Axioms 6 to 8 are not really necessary; we could have simply defined the existential quantifier in terms of the universal quantifier (the essence of Axiom 8) and then proved Axioms 6 and 7. However, adding these axioms makes life simpler.

1.33 Example   Using predicate logic, prove the theorem

$$(\forall x)(P(x) \wedge Q(x)) \rightarrow (\forall x)P(x) \wedge (\forall x)Q(x)$$

A proof sequence is

1.	$(\forall x)(P(x) \wedge Q(x))$	(hypothesis)
2.	$P(x) \wedge Q(x)$	(1, Axiom 5, modus ponens)
3.	$P(x)$	(2, tautology $A \wedge B \rightarrow A$, modus ponens)
4.	$Q(x)$	(2, tautology $A \wedge B \rightarrow B$, modus ponens)
5.	$(\forall x)P(x)$	(3, generalization)
6.	$(\forall x)Q(x)$	(4, generalization)
7.	$(\forall x)P(x) \wedge (\forall x)Q(x)$	(from tautology $A \wedge B \rightarrow A \wedge B$, $A \wedge B$ can be deduced from A, B)

At each of steps 2, 3, and 4, we have combined several steps into one. Such shortcuts are quite acceptable as long as they are perfectly clear. In step 5, generalization was applied to $P(x)$, which was deduced from $(\forall x)(P(x) \wedge Q(x))$. Because x is not free in $(\forall x)(P(x) \wedge Q(x))$, condition (a) of generalization is satisfied. Condition (b) does not apply. Step 6 is also a legal application of generalization. The justification in step 7 is useful, since we often want to conclude $A \wedge B$ where A and B have appeared earlier in the proof sequence.   □

1.34 Practice   Use predicate logic to prove the theorem

$$(\forall x)P(x) \rightarrow (\exists x)P(x).$$ □

It can also be shown that if $P \leftrightarrow Q$, then Q can be substituted for P within an expression in a deduction or proof sequence. Again, use of this rule simplifies proofs.

1.35 Example   A proof of the theorem

$$(\forall x)(P(x) \vee Q(x)) \rightarrow (\exists x)P(x) \vee (\forall x)Q(x)$$

using predicate logic is

1. $(\forall x)(P(x) \vee Q(x))$ (hypothesis)
2. $(\forall x)[(P(x))' \rightarrow Q(x)]$ (substitution from equivalence
                                               $A \vee B \leftrightarrow A' \rightarrow B$ into 1)
3. $(\forall x)(P(x))' \rightarrow (\forall x)Q(x)$ (2, Axiom 4, modus ponens)
4. $(\forall x)(P(x))' \leftrightarrow ((\exists x)P(x))'$ (Axiom 8)
5. $((\exists x)P(x))' \rightarrow (\forall x)Q(x)$ (substitution from 4 into 3)
6. $(\exists x)P(x) \vee (\forall x)Q(x)$ (5, tautology $(A' \rightarrow B) \rightarrow A \vee B$,
                                          modus ponens)   □

Predicate logic, like propositional logic, is also complete and correct—every valid statement is a theorem, and every theorem is a valid statement. We will not prove this fact, but it is the case that many of the theorems seem reasonable. The theorem of Example 1.35, for instance, says that if every element of the domain has either property P or property Q, then at least one element must have property P, or else all elements have property Q. We of course are pleased that our formal system allows us to prove results that seem so much like common sense; in the next section, we will talk about less formal proofs, where we will take these commonsensical derivations for granted, rather then spelling out each step.

Valid Arguments

To prove the validity of an argument containing quantified statements, we proceed much as before. We cast the argument in symbolic form and show that the conclusion can be deduced from the hypotheses. However, this time we use the rules of predicate logic instead of just propositional logic.

1.36 Example   Show that the following argument is valid: "Every microcomputer has a serial interface port. Some microcomputers have a parallel port. Therefore some microcomputers have both a serial and a parallel port." Using

$M(x)$: x is a microcomputer.
$S(x)$: x has a serial port.
$P(x)$: x has a parallel port.

the argument is

$$(\forall x)(M(x) \rightarrow S(x)) \wedge (\exists x)(M(x) \wedge P(x)) \rightarrow (\exists x)(M(x) \wedge S(x) \wedge P(x))$$

A proof sequence is

1. $(\forall x)(M(x) \rightarrow S(x))$ (hypothesis)
2. $(\exists x)(M(x) \wedge P(x))$ (hypothesis)
3. $M(a) \wedge P(a)$ (2, Axiom 6, modus ponens)
4. $M(a) \rightarrow S(a)$ (1, Axiom 5, modus ponens)
5. $M(a)$ (3, tautology $A \wedge B \rightarrow A$, modus
                                            ponens)

6. $S(a)$ (4, 5, modus ponens)

7. $M(a) \wedge P(a) \wedge S(a)$ (3, 6, $A \wedge B$ can be deduced from A, B)

8. $M(a) \wedge S(a) \wedge P(a)$ (substitution from equivalence $A \wedge B \leftrightarrow B \wedge A$ into 7)

9. $(\exists x)(M(x) \wedge S(x) \wedge P(x))$ (8, Axiom 7, modus ponens)

Once again, it is the *form* of the argument that matters, not the content.

$\square$

1.37 Practice   Show that the following argument is valid: "All rock music is loud music. Some rock music exists, therefore some loud music exists." Use predicates $R(x)$ and $L(x)$. $\square$

Program Verification

Although the ideas of predicate logic were developed long before people were writing computer programs, these ideas have found a use in the area of **program verification,** or **program correctness.** Program verification is an attempt to prove that a computer program is correct by using the techniques of a formal logic system. "Correctness" does not necessarily mean that the program solves the problem that it was intended to solve. A program is **correct** if, given any input variables that satisfy certain specified predicates or properties, the output variables, after execution of the program, satisfy other specified properties. Whether these specified properties are appropriate and sufficient to cause the program to solve the problem is another issue. Thus correctness has a narrower definition than everyday usage might imply.

Program verification should not be confused with program testing. **Program testing** seeks to show that particular input values produce acceptable output values; testing is done for a variety of input values. The goal of program testing is to check the program's performance on a wide, representative collection of input values: in general, testing for all input values is not possible. Program verification seeks to show that any acceptable input values must produce acceptable output values. For example, a correct program for computing the length c of the hypotenuse of a right triangle, given positive values a and b for the lengths of the legs, would guarantee that the output values satisfy the predicate $a^2 + b^2 = c^2$. Testing such a program would require taking various particular values for a and b, computing the resulting c, and checking that $a^2 + b^2 = c^2$ for each case.

More formally, let us denote one group of input values by X. The corresponding group of output values, Y, is produced from X by whatever transformations the program works on the data. For a particular program

P, we denote these transformations by $Y = P(X)$. A predicate Q describes conditions that the input values are supposed to satisfy. For example, if the program is supposed to find the square root of a positive number, then we have one input value, x, and $Q(x)$ might be "$x > 0$." A predicate R describes conditions that the output values are supposed to satisfy. These conditions often involve the input values as well; thus in our square root case, if y is the single output value, then we want $R(x, y)$ to be "$y^2 = x$." The program P is correct if

$$(\forall X)[Q(X) \rightarrow R(X, P(X))]$$

Rather than simply having an initial predicate and a final predicate, a program is broken down into individual statements, with predicates inserted between statements as well as at the beginning and end. These predicates are also called **assertions** because they assert what is supposed to be true about the program variables at that point in the program. Thus there is a series of assertions, $R_1, R_2, \ldots, R_n = R$. At each point, the program statement s_i between assertion R_i and R_{i+1} should allow the proof of an implication such as

$$(\forall X_i)[R_i(X_i) \rightarrow R_{i+1}(X_i, s_i(X_i))] \tag{1}$$

where X_i is the collection of program variable values before the execution of s_i. Also

$$(\forall X)[Q(X) \rightarrow R_1(X, s_0(X))] \tag{2}$$

should hold. The intermediate assertions are often obtained by working backwards from the output assertion R. When the intermediate assertions have been determined, implications (1) and (2) can be proved using a formal logic system of axioms and rules of inference.

One axiom justifies an implication such as (1) when the intervening program statement is an assignment statement. This axiom assumes that the program statement is of the form $\mathbf{x} \leftarrow \mathbf{e}$; that is, the variable x takes on the value of e, where e is some expression. The axiom says that (1) will hold if $R_i(X_i)$ is the predicate R_{i+1} with e substituted everywhere for x.

1.38 Example   If the program statement is $\mathbf{x} \leftarrow \mathbf{x} - \mathbf{1}$ and the assertion after this statement is $x > 0$, then the assertion before the statement should be $x - 1 > 0$. Then for every x, if $x - 1 > 0$ before the statement is executed (note that this says that $x > 1$), then after the value of x is reduced by 1, it will be the case that $x > 0$.   □

Program verification involves a lot of detailed work, making it a difficult tool to apply to already existing large programs. It is generally easier to verify programs if the verification is done hand in hand with the program development. Indeed, the list of assertions from beginning to end describes the intended program behavior and can be used early in the program design. In addition, the assertions serve as valuable documentation after the program is complete.

✔ Checklist

Definitions

axiom (*p. 25*)
rule of inference (*p. 25*)
proof sequence (*p. 25*)
theorem (*p. 25*)
proof of a theorem (*p. 25*)
propositional logic (*p. 26*)
statement logic (*p. 26*)
propositional calculus (*p. 26*)
modus ponens (*p. 26, 31*)
axiom schema (*p. 26*)
complete formal system (*p. 27*)
correct formal system (*p. 27*)

hypothesis (*p. 27*)
conclusion (*p. 27*)
deduction (*p. 27*)
valid argument (*p. 29*)
predicate logic (*p. 31*)
predicate calculus (*p. 31*)
generalization (*p. 31*)
program verification (*p. 34*)
program correctness (*p. 34*)
correct program (*p. 34*)
program testing (*p. 34*)
assertion (*p. 35*)

Techniques

Prove theorems in propositional logic.

Use propositional logic to prove that an English language argument is valid.

Prove theorems in predicate logic.

Use predicate logic to prove that an English language argument is valid.

Main Ideas

Proofs in formal systems, consisting of sequences of axioms or statements inferred from axioms by rules of inference.

A proof of a theorem of the form $P \rightarrow Q$ is a deduction of Q from P, where P can be taken as a statement in the deduction sequence.

Valid arguments, where the conclusion can be deduced (by the methods of the formal system) from the conjunction of the other statements.

Any proof in the propositional logic is also a proof in the predicate logic, but not vice versa.

Predicate logic has applications in program verification.

Exercises Section 1.2

★ 1. Justify each of the steps in the following proof sequence of $(P \to Q) \wedge (P \to (Q \to R)) \to (P \to R)$:
1. $P \to Q$
2. $P \to (Q \to R)$
3. $(P \to (Q \to R)) \to ((P \to Q) \to (P \to R))$
4. $(P \to Q) \to (P \to R)$
5. $P \to R$

2. Justify each of the steps in the following proof sequence of $((P \to Q) \to P) \to ((P \to Q) \to Q)$:
1. $(P \to Q) \to (P \to Q)$
2. $((P \to Q) \to (P \to Q)) \to$
   $[((P \to Q) \to P) \to ((P \to Q) \to Q)]$
3. $((P \to Q) \to P) \to ((P \to Q) \to Q)$

In Exercises 3 through 10, prove that each statement is a theorem of propositional logic. You may make use of any previously proved theorems, including preceding exercises.

★ 3. $(P')' \to P$

4. $P \to (P')'$

5. $P \to (P \vee Q)$

6. $(P \to Q) \to (Q' \to P')$

7. $P \vee Q \to Q \vee P$

8. $(P \to (Q \to R)) \to (Q \to (P \to R))$

★ 9. $P' \wedge (P \vee Q) \to Q$

10. $P' \wedge (Q \to P) \to Q'$

Using propositional logic, prove that each argument in Exercises 11 through 14 is valid. Use the statement letters shown.

11. If the program is efficient, it executes quickly: the program is either efficient, or it has a bug. However, the program does not execute quickly. Therefore it has a bug. (*E, Q, B*)

12. The crop is good, but there is not enough water. If there is a lot of rain or not a lot of sun, then there is enough water. Therefore the crop is good and there is a lot of sun. (*C, W, R, S*)

★ 13. Russia was a superior power, and either France was not strong or Napoleon made an error. Napoleon did not make an error, but if the army did not fail, then France was strong. Hence the army failed and Russia was a superior power. (*R, F, N, A*)

14. It is not the case that if electric rates go up, then usage will go down, nor is it true that either new power plants will be built or bills will

not be late. Therefore usage will not go down and bills will be late. (*R, U, P, B*)

In Exercises 15 through 22, prove that each statement is a theorem of predicate logic.

★ 15. $(\forall x)P(x) \rightarrow (\forall x)(P(x) \vee Q(x))$

16. $(\forall x)P(x) \wedge (\exists x)Q(x) \rightarrow (\exists x)(P(x) \wedge Q(x))$

17. $(\exists x)(\exists y)P(x, y) \rightarrow (\exists y)(\exists x)P(x, y)$

18. $(\forall x)(\forall y)Q(x, y) \rightarrow (\forall y)(\forall x)Q(x, y)$

★ 19. $(\exists x)(A(x) \wedge B(x)) \rightarrow (\exists x)A(x) \wedge (\exists x)B(x)$

20. $(\exists x)(R(x) \vee S(x)) \rightarrow (\exists x)R(x) \vee (\exists x)S(x)$

21. $(\exists x)(\forall y)Q(x, y) \rightarrow (\forall y)(\exists x)Q(x, y)$

22. $(\forall x)(A(x) \rightarrow B(x)) \rightarrow ((\exists x)A(x) \rightarrow (\exists x)B(x))$

23. The statement $(\forall y)(\exists x)Q(x, y) \rightarrow (\exists x)(\forall y)Q(x, y)$ is not valid. (Can you find an interpretation to show this?) What is wrong with the following "proof" of $(\forall y)(\exists x)Q(x, y) \rightarrow (\exists x)(\forall y)Q(x, y)$?

    1. $(\forall y)(\exists x)Q(x, y)$ (hypothesis)
    2. $(\exists x)Q(x, y)$ (1, Axiom 5, modus ponens)
    3. $Q(a, y)$ (2, Axiom 6, modus ponens)
    4. $(\forall y)Q(a, y)$ (3, generalization)
    5. $(\exists x)(\forall y)Q(x, y)$ (4, Axiom 7, modus ponens)

Using predicate logic, prove that each argument in Exercises 24 through 27 is valid. Use the predicate symbols shown.

24. There is an astronomer who is not nearsighted. Everyone who wears glasses is nearsighted. Furthermore, everyone either wears glasses or wears contact lenses. Therefore some astronomer wears contact lenses. ($A(x)$, $N(x)$, $G(x)$, $C(x)$)

★ 25. Every member of the board comes from industry or government. Everyone from government who has a law degree is in favor of the motion. John is not from industry, but he does have a law degree. Therefore if John is a member of the board, he is in favor of the motion. ($M(x)$, $I(x)$, $G(x)$, $L(x)$, $F(x)$, j)

26. There is some movie star who is richer than everyone. Anyone who is richer than anyone else pays more taxes than they do. Therefore there is a movie star who pays more taxes than anyone. ($M(x)$, $R(x, y)$, $T(x, y)$)

27. Every computer science student works harder than somebody, and everyone who works harder than someone else gets less sleep than that person. Maria is a computer science student. Therefore Maria gets less sleep than someone else. ($C(x)$, $W(x, y)$, $S(x, y)$, m)

28. You are attempting to verify the correctness of a program. If a statement in the program is the assignment statement **x ← x + 1** and the assertion after this statement is $x = y - 1$, what should the assertion before this statement be? Explain why this works.

29. You are attempting to verify the correctness of a program. If a statement in the program is the assignment statement **x ← 2 * x** (where * indicates multiplication), and the assertion after this statement is $x > y$, what should the assertion before this statement be? Explain why this works.

SECTION 1.3   PROOF TECHNIQUES

Theorems

Mathematical results are often expressed as theorems of the form "if P, then Q," or $P \to Q$, where P and Q may be compound statements. In a theorem of this form, we try to deduce Q from P using axioms and rules of logical inference. If we can do this using only purely logical axioms (valid statements that are true in all interpretations), then the theorem is also true in all interpretations. In this case, the theorem is true because of its form or structure, and not because of its content or the meaning of any of its component statements.

However, we usually want to prove theorems where the meaning is important, because we are discussing a particular subject—number theory or geometry or Boolean algebra or whatever. Here we may take as axioms statements that, though not universally valid, are facts about the particular subject, such as definitions or previously proved theorems. We try to deduce Q from P by a logical progression of statements that begins with P and ends with Q; each step in the sequence is either P, a valid statement (a logical axiom), a subject-specific axiom, or a statement that can be logically inferred from previous statements in the sequence.

It may not be easy to recognize which subject-specific facts will be helpful or to arrange a sequence of steps that will logically lead from P to Q. Unfortunately, there is no formula for constructing proofs, no algorithm or computer program for general theorem proving. Experience is helpful, not only because you get better with practice, but also because a proof that works for one theorem may be modified to work for a new but similar theorem. As you continue to study a subject, your storehouse of facts that can be used in proving theorems increases. One purpose of this textbook is to help you accumulate such facts about theoretical computer science.

Theorems are often expressed and proved in a somewhat less formal way than the logic statements we have been studying in the first two sections of this chapter. For example, a theorem may be a fact about all objects in the domain of interpretation, that is, the subject matter under discussion. In this case, the formal statement of the theorem would begin with at least one universal quantifier; for example, it might look like $(\forall x)(P(x) \rightarrow Q(x))$. This would be informally stated as $P(x) \rightarrow Q(x)$. We can justifiably remove the universal quantifier if we treat x as an arbitrary element of the domain; if we can prove $P(x) \rightarrow Q(x)$ for an arbitrary x, we have then proved $(\forall x)(P(x) \rightarrow Q(x))$. (This is making use of Axiom 5 and the rule of generalization of predicate calculus, discussed in the previous section.) But all of this formal justification is generally omitted.

Similarly, proofs are usually not written a step at a time with formal justifications. Instead, the important steps and their rationale are outlined in narrative form. Such a narrative, however, could be translated into a formal proof if required. In fact, the value of a formal proof is that it serves as a sort of insurance—if a narrative proof *cannot* be translated into a formal proof, it should be viewed with great suspicion.

Before we discuss theorem proving further, let's digress for a few moments. Unfortunately, the reader of a textbook is looking at the static result of a dynamic process and cannot share the adventure of developing new ideas. The textbook may say, "Prove the following theorem," and the reader will know that the theorem is true and, furthermore, that it is probably stated in its most polished form. But the researcher does not suddenly acquire a vision of a perfectly worded theorem together with the absolute certainty that it is true and that all he or she must do is work to find a proof. Before a theorem and its proof reach the final, polished form, a combination of inductive and deductive reasoning comes into play. Suppose you're the researcher trying to formulate and then prove a theorem. Say, for example, that you have examined a lot of cases in which whenever P is true, Q is also true. (Thus, you may have looked at seven or eight numbers divisible by 6 and observed that all the numbers were divisible by 3 as well.) On the basis of these experiences, you may conjecture: if P, then Q (if a number is divisible by 6, then it is also divisible by 3). And the more cases you find where Q follows from P, the more confident you are in your conjecture. This process illustrates **inductive reasoning,** drawing a conclusion based on experience.

No matter how reasonable the conjecture sounds, however, you will not be satisfied until you have applied **deductive reasoning** to it as well. In this process, you try to verify the truth or falsity of your idea. You produce a proof of $P \rightarrow Q$ (thus making it a theorem), or else you find an example that disproves the conjecture. Often it is very difficult to decide which of these two approaches you should try! Suppose you decide to try to disprove, or refute, the conjecture. You then search for an example in which P is true but Q is false—you look for a **counterexample** to your conjecture. A single counterexample to a conjecture is sufficient to disprove it. Thus, you could refute our example conjecture by finding a single number divisible by 6 but not by 3. Since our example is true, such a number does not exist. Of course, hunting for a counterexample and being unsuccessful does *not* constitute a proof that the conjecture is true.

1.39 Example   Consider the statement "Every number less than 10 is bigger than 5," or, expressed as an implication, "If a number is less than 10, then it is also bigger than 5." A counterexample to this implication is the number 4. Four is less than 10, but it is not bigger than 5. Of course, there are other counterexamples, but one is sufficient to disprove the statement.

□

1.40 Practice   Provide counterexamples to the following statements:

(a) All animals living in the ocean are fish.
(b) Input to a computer is always given by means of punched paper tape.   □

Methods of Attack

Suppose you decide to try to prove your conjecture $P \rightarrow Q$. Although a single counterexample is sufficient to refute a conjecture, in general *many* examples do not prove a conjecture—they only strengthen your inclination to look for a proof. The one exception to this situation occurs when you are making an assertion about a finite collection. In this case, the assertion can be proved true by showing that it is true for each member of the collection. For example, the statement "If a whole number between 1 and 20 is divisible by 6, then it is also divisible by 3" can be proved by simply showing it to be true for all the whole numbers between 1 and 20.

Direct Proof

In general, how can you prove that $P \rightarrow Q$ is true? The obvious approach is the **direct proof**—assume the hypothesis P and deduce the conclusion Q.

1.41 Example   We will give a direct proof of our example theorem, "If a number is divisible by 6, then it is also divisible by 3." The theorem asserts something about an arbitrary number; its real form is

$$(\forall x)(x \text{ divisible by } 6 \rightarrow x \text{ divisible by } 3)$$

Thus we let x represent an arbitrary number and prove

x divisible by 6 $\rightarrow$ x divisible by 3

To do this, we assume that the hypothesis, x is divisible by 6, is true, and then we must deduce that the conclusion, x is divisible by 3, is true. We have to make use of the definition of divisibility—a is divisible by b if a equals the product of an integer times b—and also other arithmetic properties.

Hypothesis: x is divisible by 6.

$x = k \cdot 6$ for some integer k	(definition of divisibility)
$6 = 2 \cdot 3$	(number fact)
$x = k(2 \cdot 3)$	(substitution)
$x = (k \cdot 2)3$	(property of multiplication)
$k \cdot 2$ is an integer	(known fact about integers)

Conclusion: x is divisible by 3 (definition of divisibility).   ☐

1.42 Practice   Give a direct proof of the theorem "If a number is divisible by 6, then twice that number is divisible by 4." Show each step in going from hypothesis to conclusion.   ☐

1.43 Example   A direct proof that the product of two even integers is even is: Let $x = 2m$ and $y = 2n$, where m and n are integers. Then $xy = (2m)(2n) = 2(2mn)$, where $2mn$ is an integer. Thus xy has the form $2k$, where k is an integer, and xy is therefore even.

This proof is even less formal than Example 1.41; it does not explicitly state the hypothesis, and it makes implicit use of the definition of an even integer. However, this proof would be perfectly acceptable in most circumstances.   ☐

Contraposition

If you have tried diligently but failed to produce a direct proof of your conjecture $P \rightarrow Q$, and you still feel that the conjecture is true, you might try some variants on the direct proof technique. If you can prove the theorem $Q' \rightarrow P'$, you can conclude $P \rightarrow Q$ by making use of the tautology $(Q' \rightarrow P') \rightarrow (P \rightarrow Q)$. $Q' \rightarrow P'$ is the **contrapositive** of $P \rightarrow Q$. The technique of proving $P \rightarrow Q$ by doing a direct proof of $Q' \rightarrow P'$ is called **proof by contraposition.** We have already discussed direct proofs, so the only new idea here is figuring out what the contrapositive is.

1.44 Example   The contrapositive of the theorem "If a number is divisible by 6, then it is also divisible by 3" is "If a number is not divisible by 3, then it is not divisible by 6." The easiest proof of the theorem is the direct proof given in Example 1.41. A proof by contraposition is subtler.

Hypothesis: x is not divisible by 3.

$x \neq k \cdot 3$ for *any* integer k (This is the negation of
                                             divisibility by 3.)

$x \neq (2k_1)3$ for *any* integer k_1 ($2k_1$ would be an integer k,
                                             ruled out above.)

$x \neq k_1(2 \cdot 3)$ for *any* integer k_1 (properties of multiplication)
$x \neq k_1 \cdot 6$ for *any* integer k_1 (number fact)

Conclusion: x is not divisible by 6 (negation of divisibility by 6). □

1.45 Practice   Write the contrapositive of each statement in Practice 1.5.   □

1.46 Example   Prove that if the square of an integer is odd, then the integer must be odd.

The theorem is n^2 odd $\rightarrow n$ odd. We do a proof by contraposition, and prove n even $\rightarrow n^2$ even. Let n be even. Then $n^2 = n(n)$ is even by Example 1.43. □

Practice 1.10(a) showed that the statements $A \rightarrow B$ and $B \rightarrow A$ are not equivalent. $B \rightarrow A$ is the **converse** of $A \rightarrow B$. If an implication is true, its converse may be true or false. Therefore, you cannot prove $P \rightarrow Q$ by looking at $Q \rightarrow P$.

1.47 Example   The implication "If $a > 5$, then $a > 2$" is true, but its converse "If $a > 2$, then $a > 5$" is false. □

1.48 Practice   Write the converse of each statement in Practice 1.5. □

Theorems are often stated in the form P if and only if Q, meaning P if Q and P only if Q, or $Q \rightarrow P$ and $P \rightarrow Q$. To prove such a theorem, you must prove both an implication and its converse. Remember that any "if and only if" theorem requires you to prove both directions.

1.49 Example   Prove that the product xy is odd if and only if both x and y are odd integers.

We first prove that if x and y are odd, so is xy. A direct proof will work. Suppose that both x and y are odd. Then $x = 2n + 1$ and $y = 2m + 1$, where m and n are integers. Then $xy = (2n + 1)(2m + 1) = 4nm + 2m + 2n + 1 = 2(2nm + m + n) + 1$. This has the form $2k + 1$ where k is an integer, so xy is odd.

Next we prove that if xy is odd, both x and y must be odd, or

xy odd $\rightarrow x$ odd and y odd

A proof by contraposition works well here, so we will prove

$(x$ odd and y odd$)' \rightarrow (xy$ odd$)'$

By the tautology $(A \wedge B)' \leftrightarrow A' \vee B'$, we see that this can be written as

$$x \text{ even or } y \text{ even} \rightarrow xy \text{ even} \tag{1}$$

The hypothesis "x even or y even" breaks down into three cases. We consider each case in turn.

(a) x even, y odd: here $x = 2m$, $y = 2n + 1$, and then $xy = (2m)(2n + 1) = 2(2mn + m)$, which is even.

(b) x odd, y even: this works just like (a).

(c) x even, y even: then xy is even by Example 1.43.

This completes the proof of (1) and thus of the theorem.   □

Contradiction

In addition to contraposition and direct proof, another proof technique you might use is **proof by contradiction.** Again we will let 0 stand for any contradiction, that is, any statement whose truth value is always false. ($A \wedge A'$ would be such a statement.) Once more, suppose you are trying to prove $P \rightarrow Q$. By constructing a truth table, we see that

$$(P \wedge Q' \rightarrow 0) \rightarrow (P \rightarrow Q)$$

is a tautology, so to prove the theorem $P \rightarrow Q$, it is sufficient to prove $P \wedge Q' \rightarrow 0$. Therefore, in a proof by contradiction you assume both the hypothesis and the negation of the conclusion to be true and then try to deduce some contradiction by using these assumptions.

1.50 Example   Let's use proof by contradiction on the statement "If a number added to itself gives itself, then the number is 0." Let x represent any number. The hypothesis is $x + x = x$ and the conclusion is $x = 0$. To do a proof by contradiction, assume $x + x = x$ and $x \neq 0$. Then $2x = x$ and $x \neq 0$. Because $x \neq 0$, we can divide both sides of the equation $2x = x$ by x and arrive at the contradiction $2 = 1$. Hence, $x + x = x \rightarrow x = 0$.   □

1.51 Example   A well-known proof by contradiction shows that $\sqrt{2}$ is not a rational number. Recall that a rational number is one that can be written in the form p/q, where p and q are integers, $q \neq 0$, and p and q have no common factors (other than ± 1).

Let us assume that $\sqrt{2}$ is rational. Then $\sqrt{2} = p/q$, and $2 = p^2/q^2$, or $2q^2 = p^2$. Then 2 divides p^2, so 2 must divide p. This means that 2 is a factor of p, hence 4 is a factor of p^2, and the equation $2q^2 = p^2$ can be written as $2q^2 = 4x$, or $q^2 = 2x$. We see from this equation that 2 divides q^2, hence 2 divides q. At this point, 2 is a factor of q and a factor of p, which contradicts the statement that p and q have no common factors. Therefore $\sqrt{2}$ is not rational.   □

1.52 Practice   Prove by contradiction that the product of odd integers is odd. (We did a direct proof of this in Example 1.49.)   □

Proof by contradiction can be a valuable technique, but it is easy to think that we have done a proof by contradiction when we really haven't. For example, suppose we assume $P \wedge Q'$ and deduce Q without using the assumption Q'. Then we claim $Q \wedge Q'$ as a contradiction. What really happened here is a direct proof of $P \rightarrow Q$, and the proof should be rewritten in this form.

Another fraudulent case of proof by contradiction occurs when we assume $P \wedge Q'$ and deduce P' without using the assumption P. Then we claim $P \wedge P'$ as a contradiction. What really happened here is a direct proof of $Q' \rightarrow P'$, and we have constructed a proof by contraposition.

We have not yet discussed a proof method that is especially useful in computer science—mathematical induction. That will come in the next section.

⌲ Checklist

Definitions

inductive reasoning (p. 40)   contrapositive (p. 42)
deductive reasoning (p. 40)   proof by contraposition (p. 42)
counterexample (p. 40)   converse (p. 43)
direct proof (p. 41)   proof by contradiction (p. 44)

Techniques

Look for a counterexample.

Attempt direct proofs, proofs by contraposition, and proofs by contradiction.

Main Ideas

Inductive reasoning to formulate a conjecture based upon experience.

Deductive reasoning either to refute a conjecture by finding a counterexample or to prove the conjecture.

In proving a conjecture, logical facts and facts about the particular subject can be used.

If we cannot directly prove a conjecture, we can try proof by contraposition or contradiction.

Exercises Section 1.3

★ 1. Write the converse and the contrapositive of each statement in Exercise 3, Section 1.1.

2. Provide counterexamples to the following statements:
   (a) Every geometric figure with four right angles is a square.
   (b) If a real number is not positive, then it must be negative.
   (c) All people with red hair have green eyes or are tall.
   (d) All people with red hair have green eyes and are tall.

3. Give a direct proof that the sum of even integers is even.

4. Prove by contradiction that the sum of even integers is even.

★ 5. Prove that the sum of two odd integers is even.

6. Prove that the product of any two consecutive integers is even.

7. Prove that the sum of an integer and its square is even.

8. Prove by contraposition that if a number x is positive, so is $x + 1$.

★ 9. Let x and y be positive numbers, and prove that $x < y$ if and only if $x^2 < y^2$.

10. Prove that the sum of three consecutive integers is divisible by 3.

★ 11. Prove that the square of an odd integer equals $8k + 1$ for some integer k.

12. Prove that the difference of two consecutive cubes is odd.

13. Prove that the sum of the squares of two odd integers cannot be a perfect square. (Hint: use Exercise 11.)

14. Suppose you were to use the steps of Example 1.51 to attempt to prove that $\sqrt{4}$ is not a rational number. At what point would the proof not be valid?

15. Prove that $\sqrt{3}$ is not a rational number.

16. Prove that $\sqrt[3]{2}$ is not a rational number.

★ 17. Prove or disprove: the product of any three consecutive integers is even.

18. Prove or disprove: the sum of any three consecutive integers is even.

19. Prove or disprove: the product of an integer and its square is even.

20. Prove or disprove: the sum of an integer and its cube is even.

21. Prove or disprove: For a positive integer x, $x + \frac{1}{x} \geq 2$.

SECTION 1.4   INDUCTION

The Method

There is one final proof technique that applies to certain situations. To illustrate how the technique works, imagine that you are climbing an infinitely high ladder. How do you know whether you will be able to reach an arbitrarily high rung? Suppose we make the following two assertions about your climbing abilities:

1. You can reach the first rung.
2. Once you get to a rung, you can always climb to the next one up. (Notice that this assertion is an implication.)

If both statement 1 and the implication of statement 2 are true, then by statement 1 you can get to the first rung and therefore by statement 2 you can get to the second; by statement 2 again, you can get to the third rung; by statement 2 again you can get to the fourth; and so on. You can climb as high as you wish. Both assertions here are necessary. If only statement 1 is true, you have no guarantee of getting beyond the first rung, and if only statement 2 is true, you may never be able to get started. Let's assume the rungs of the ladder are numbered by positive integers—1, 2, 3, ... Now think of a specific property a number might have. Instead of "reaching an arbitrarily high rung," we can talk about an arbitrary, positive integer having that property. We will use the shorthand notation $P(n)$ to mean that the positive integer n has the property P. How can we use the ladder-climbing technique to prove that for all positive integers n, we have $P(n)$? The two assertions we need to prove are

1. $P(1)$ (1 has property P)
2. for any positive integer k,   (if any number has property P,
   $P(k) \rightarrow P(k + 1)$ so does the next number)

If we can prove both assertions 1 and 2, then $P(n)$ holds for any positive integer n, just as you could climb to an arbitrary rung on the ladder.

To prove assertion 1, we need only show that the property holds for the number 1, usually a trivial task. To prove assertion 2, an implication, we assume $P(k)$ is true and show, based on this assumption, that $P(k + 1)$ is true. You should convince yourself that assuming that the property holds for the number k is not the same as assuming what we ultimately want to prove (a frequent source of confusion when one first encounters proofs of this kind). It is merely the way to proceed to prove that implication 2 is true.

The technique we have described here is **proof by mathematical induction.** Below is a summary of this proof technique in pseudocode:

1. if $P(1)$ false

2. then write "theorem is false" and stop
3. else begin
        assume $P(k)$ for arbitrary k
        if $P(k + 1)$ false
        then write "theorem is false" and stop
        else write "theorem is true" and stop
        end of else

Mathematical induction is suggested whenever a theorem begins in the form "For all positive integers n, show that . . . ."

In doing a proof by induction, establishing the truth of assertion 1, $P(1)$, is called the **basis,** or **basis step,** for the inductive proof. Establishing the truth of assertion 2, $P(k) \rightarrow P(k + 1)$, is called the **inductive step.** When we assume $P(k)$ to be true in order to prove the inductive step, $P(k)$ is called the **inductive assumption,** or **inductive hypothesis.**

All of the proof methods we have talked about are techniques for deductive reasoning—ways to prove a conjecture that perhaps was formulated by inductive reasoning. Mathematical induction is also a *deductive* technique, not a method for inductive reasoning (don't get confused by the terminology here). For the other proof techniques, we can begin with a hypothesis and string facts together until we more or less stumble upon a conclusion. In fact, even if our conjecture is slightly incorrect, we might see what the correct conclusion is in the course of doing the proof. In mathematical induction, however, we must know right at the outset the exact form of the property $P(n)$ that we are trying to establish. Mathematical induction, therefore, is not an exploratory proof technique—it can only confirm a correct conjecture.

Inductive Proofs

1.53 Example   Prove that the equation

$$1 + 3 + 5 + \cdots + (2n - 1) = n^2$$

is true for any positive integer n. (The left side of this equation is the sum of all the odd integers from 1 to $2n - 1$.)

Although we can verify the truth of the above equation for any particular value of n by substituting that value for n, we cannot substitute all possible positive integer values. Thus a proof by example does not work. A proof by mathematical induction is called for.

The basis step is to establish $P(1)$, which is the equation $1 = 1^2$. This is certainly true. For the inductive hypothesis, we assume $P(k)$.

$$1 + 3 + 5 + \cdots + (2k - 1) = k^2$$

and try to show $P(k + 1)$,

$$1 + 3 + 5 + \cdots + [2(k + 1) - 1] = (k + 1)^2$$

The left side of $P(k + 1)$ can be rewritten to show the next-to-last term:

$$1 + 3 + 5 + \cdots + (2k - 1) + [2(k + 1) - 1]$$

This expression contains the left side of $P(k)$ as a subexpression, and we can substitute the right side of $P(k)$ for this subexpression. Thus,

$$
\begin{aligned}
1 + 3 + 5 + \cdots + [2(k + 1) - 1] \\
= 1 + 3 + 5 + \cdots + (2k - 1) + [2(k + 1) - 1] \\
= k^2 + [2(k + 1) - 1] \\
= k^2 + [2k + 2 - 1] \\
= k^2 + 2k + 1 \\
= (k + 1)^2
\end{aligned}
$$

This verifies $P(k + 1)$ and completes the proof. □

1.54 Example   Prove that

$$1 + 2 + 2^2 + \cdots + 2^n = 2^{n+1} - 1$$

for any $n \geq 1$.

Again, induction is appropriate. $P(1)$ is the equation

$$1 + 2 = 2^{1+1} - 1 \quad \text{or} \quad 3 = 2^2 - 1$$

which is true. We take $P(k)$

$$1 + 2 + 2^2 + \cdots + 2^k = 2^{k+1} - 1$$

as the inductive hypothesis, and try to establish $P(k + 1)$:

$$1 + 2 + 2^2 + \cdots + 2^{k+1} = 2^{k+1+1} - 1$$

Again, rewriting the sum on the left side of $P(k + 1)$ reveals how the inductive assumption can be used:

$$
\begin{aligned}
1 + 2 + 2^2 + \cdots + 2^{k+1} \\
= 1 + 2 + 2^2 + \cdots + 2^k + 2^{k+1} \\
= 2^{k+1} - 1 + 2^{k+1} \quad \text{(from the inductive assumption } P(k)) \\
= 2(2^{k+1}) - 1 \\
= 2^{k+1+1} - 1
\end{aligned}
$$

□

1.55 Practice   Prove that for any positive integer n,

$$1 + 2 + 3 + \cdots + n = \frac{n(n + 1)}{2}$$

□

Not all proofs by induction involve formulas with sums.

1.56 Example   Prove that for any positive integer n, $2^n > n$. $P(1)$ is the assertion $2^1 > 1$, which is surely true. Now we assume $P(k)$, $2^k > k$, and try to conclude $P(k + 1)$, $2^{k+1} > k + 1$. Beginning with the left side of $P(k + 1)$, we note that $2^{k+1} = 2^k \cdot 2$. Using the inductive assumption $2^k > k$ and multiplying both sides of this inequality by 2, we get $2^k \cdot 2 > k \cdot 2$. We complete the argument

$$2^{k+1} = 2^k \cdot 2 > k \cdot 2 = k + k \geq k + 1$$

or

$$2^{k+1} > k + 1 \qquad \square$$

1.57 Example   Prove that for any positive integer n, the number $2^{2n} - 1$ is divisible by 3.

The basis step is to show $P(1)$, that $2^{2(1)} - 1 = 4 - 1 = 3$ is divisible by 3. Clearly this is true.

We assume that $2^{2k} - 1$ is divisible by 3, which means that $2^{2k} - 1 = 3m$ for some integer m, or $2^{2k} = 3m + 1$. We want to show that $2^{2(k+1)} - 1$ is divisible by 3.

$$
\begin{aligned}
2^{2(k+1)} - 1 &= 2^{2k+2} - 1 \\
&= 2^2 \cdot 2^{2k} - 1 \\
&= 2^2 (3m + 1) - 1 \qquad \text{(by the inductive hypothesis)} \\
&= 12m + 4 - 1 \\
&= 12m + 3 \\
&= 3(4m + 1) \quad \text{where } 4m + 1 \text{ is an integer}
\end{aligned}
$$

Thus $2^{2(k+1)} - 1$ is divisible by 3. $\qquad \square$

It may be appropriate for the first step of the induction process to begin at 0 or at 2 or 3, instead of at 1. The same principle applies, no matter where you first hop on the ladder.

1.58 Example   Prove that $n^2 > 3n$ for $n \geq 4$.

Here we should use induction and begin with a base step of $P(4)$. (Testing values of $n = 1, 2,$ and 3 shows that the inequality does not hold for these values.) $P(4)$ is the inequality $4^2 > 3(4)$, or $16 > 12$, which is true. The inductive hypothesis is that $k^2 > 3k$ and that $k \geq 4$, and we want to show that $(k + 1)^2 > 3(k + 1)$.

$$
\begin{aligned}
(k + 1)^2 &= k^2 + 2k + 1 \\
&> 3k + 2k + 1 \qquad \text{(by the inductive hypothesis)} \\
&\geq 3k + 8 + 1 \qquad \text{(since } k \geq 4) \\
&> 3k + 3 \\
&= 3(k + 1) \qquad\qquad\qquad\qquad\qquad\qquad \square
\end{aligned}
$$

1.59 Practice   Prove that $2^{n+1} < 3^n$ for all $n > 1$. □

An induction proof may be called for when its application is not as obvious as in the above examples. This usually arises when there is some quantity in the statement to be proved that can take on arbitrary non-negative integer values.

1.60 Example   A programming language might be designed with the following convention regarding multiplication: a single factor requires no parentheses, but the product "a times b" must be written as $(a)b$. We want to show that any product of factors can be written with an even number of parentheses. The proof is by induction on the number of factors. For a single factor, there are 0 parentheses, an even number. Assume that for any product of k factors there are an even number of parentheses. Now consider a product P of $k + 1$ factors. P can be written in the form r times s where r has k factors and s is a single factor. By the inductive hypothesis, r has an even number of parentheses. Then we write r times s as $(r)s$, adding 2 more parentheses to the even number of parentheses in r, and giving P an even number of parentheses. □

Fraudulent proofs by induction are also possible. Such a proof occurs when we prove the truth of $P(k + 1)$ without relying on the truth of $P(k)$. What really happened here is a direct proof of $P(k + 1)$ where $k + 1$ is arbitrary. The proof should be rewritten to show that it is a direct proof of $P(n)$ for any n, not a proof by induction.

Verifying Loop Invariants

In Section 1.2, we discussed program verification, an attempt to formally prove that a program is correct by showing that its variables satisfy specified assertions at various points in the program. If the program contains a loop, then we include an assertion that gets checked at each pass through the loop. More specifically, we check that the assertion holds before entering the loop and that the assertion still holds after each iteration of the loop. The assertion is a predicate, or relationship, involving the program variables.

Suppose we can show that a relationship holds among the values of the program variables before the execution of a loop iteration and that it still holds among the values after execution of an iteration. Then the *relationship* among these variables is unaffected by a loop iteration, even though the values themselves may be changed. Such a relationship is called a **loop invariant**. A loop invariant, because it holds before entering the loop and after each loop iteration, will also hold upon exiting the loop.

We denote by $P(n)$ the statement that a proposed loop invariant is true after n iterations of the loop. To prove that we indeed have a loop invariant, we want to show that $P(n)$ is true for all $n \geq 0$. (The value of $n = 0$ corresponds to the assertion upon entering the loop, after zero loop iterations.) A proof by induction should come to the rescue.

1.61 Example   Consider the following pseudocode program, which is supposed to compute $x \cdot y$ for nonnegative integers x and y.

1. given $x, y \geq 0$
2. $i \leftarrow 0$ (this notation means "assign i the value 0")
3. $j \leftarrow 0$
4. while $i \neq x$ do
   begin
   $j \leftarrow j + y$
   $i \leftarrow i + 1$
   end
5. write j

There is a loop at step 4. The quantities x and y are given and remain unchanged throughout the program; values of i and j change. We let i_n and j_n denote the values of i and j, respectively, after n iterations of the loop. If $P(n)$ is the statement $j_n = i_n \cdot y$, then we claim that $P(n)$ is a loop invariant. We prove by induction that $P(n)$ holds for all $n \geq 0$.

$P(0)$ is the statement

$$j_0 = i_0 \cdot y$$

which is true because after zero iterations of the loop (i.e., when we first get to step 4), the value of i is 0 and the value of j is 0, because those values were just assigned in steps 2 and 3.

Assume $P(k)$: $j_k = i_k \cdot y$

Show $P(k + 1)$: $j_{k+1} = i_{k+1} \cdot y$

Between the time when j and i have the values j_k and i_k and when they have the values j_{k+1} and i_{k+1}, one iteration of the loop takes place. In that iteration, j is changed by adding y to the previous value, and i is changed by adding 1. Thus

$$j_{k+1} = j_k + y \tag{1}$$

$$i_{k+1} = i_k + 1 \tag{2}$$

Then

$$j_{k+1} = j_k + y \qquad \text{(by (1))}$$
$$= i_k \cdot y + y \qquad \text{(by the inductive hypothesis)}$$

$$= (i_k + 1)y$$
$$= i_{k+1} \cdot y \qquad \text{(by (2))}$$

We have proved that $P(n)$ is a loop invariant.

So what? What did proving that $P(n)$ is a loop invariant have to do with showing that the program is correct? We know that after every loop iteration, $j = i \cdot y$. In particular, when the loop terminates, i has the value x. Therefore, at this point $j = x \cdot y$, exactly what the program was intended to compute. □

Example 1.61 illustrates that loop invariants say something stronger about the program than we actually want to show; what we want to show is the special case of the loop invariant upon termination of the loop.

We did not, however, prove that the loop in Example 1.61 actually does terminate. What we proved was **partial correctness**—the program gives us the correct answer, given that execution does terminate. Of course, Example 1.61 is also unrealistic; after all, we could undoubtedly compute $x \cdot y$ with a single program statement. However, the same techniques apply to more meaningful loops.

1.62 Practice   Show that the following program computes $x + y$ for nonnegative integers x and y by proving the loop invariant $P(n)$: $j_n = x + i_n$ and evaluating $P(n)$ at loop termination.

1. given $x, y \geq 0$
2. $i \leftarrow 0$
3. $j \leftarrow x$
4. while $i \neq y$ do
     begin
       $j \leftarrow j + 1$
       $i \leftarrow i + 1$
     end
5. write j □

✔ **Checklist**

Definitions

proof by mathematical induction (*p. 47*)
basis step (*p. 48*)
inductive step (*p. 48*)
inductive assumption (*p. 48*)
inductive hypothesis (*p. 48*)
loop invariant (*p. 51*)
partial correctness (*p. 53*)

Techniques

Do proofs by mathematical induction.

Verify loop invariants.

Main Ideas

Mathematical induction, a technique to prove properties of positive integers.

An inductive proof need not begin with 1.

Induction can be used to prove statements about quantities whose values are arbitrary nonnegative integers.

A loop invariant, proved by induction on the number of iterations of the loop, can be used to prove that a program loop behaves correctly.

Exercises Section 1.4

In Exercises 1 through 14, use mathematical induction to prove that the statements are true for every positive integer n.

★ 1. $2 + 6 + 10 + \cdots + (4n - 2) = 2n^2$

2. $2 + 4 + 6 + \cdots + 2n = n(n + 1)$

★ 3. $1 + 5 + 9 + \cdots + (4n - 3) = n(2n - 1)$

4. $1 + 3 + 6 + \cdots + \dfrac{n(n + 1)}{2} = \dfrac{n(n + 1)(n + 2)}{6}$

5. $4 + 10 + 16 + \cdots + (6n - 2) = n(3n + 1)$

6. $5 + 10 + 15 + \cdots + 5n = \dfrac{5n(n + 1)}{2}$

7. $1^2 + 2^2 + \cdots + n^2 = \dfrac{n(n + 1)(2n + 1)}{6}$

8. $1^3 + 2^3 + \cdots + n^3 = \dfrac{n^2(n + 1)^2}{4}$

★ 9. $1^2 + 3^2 + \cdots + (2n - 1)^2 = \dfrac{n(2n - 1)(2n + 1)}{3}$

10. $1^4 + 2^4 + \cdots + n^4 = \dfrac{n(n + 1)(2n + 1)(3n^2 + 3n - 1)}{30}$

11. $(1 + 2 + \cdots + n)^2 = \dfrac{n^2(n + 1)^2}{4}$

(Hint: you will need the result of Practice 1.55.)

12. $1 + a + a^2 + \cdots + a^{n-1} = \dfrac{a^n - 1}{a - 1}$ for $a \neq 0,\ a \neq 1$

★ 13. $\dfrac{1}{1 \cdot 2} + \dfrac{1}{2 \cdot 3} + \dfrac{1}{3 \cdot 4} + \cdots + \dfrac{1}{n(n + 1)} = \dfrac{n}{n + 1}$

14. $\dfrac{1}{1 \cdot 3} + \dfrac{1}{3 \cdot 5} + \dfrac{1}{5 \cdot 7} + \cdots + \dfrac{1}{(2n - 1)(2n + 1)} = \dfrac{n}{2n + 1}$

★ 15. Prove that $n^2 > n + 1$ for $n \geq 2$.

16. Prove that $n^2 > 5n + 10$ for $n > 6$.

17. Prove that $n! > n^2$ for $n \geq 4$, where $n!$ is the product of the positive integers from 1 to n.

★ 18. Prove that $2^n < n!$ for $n \geq 4$.

19. Prove that $(1 + x)^n > 1 + x^n$ for $n > 1,\ x > 0$.

20. Prove that $\left(\dfrac{a}{b}\right)^{n+1} < \left(\dfrac{a}{b}\right)^n$ for $n \geq 1$ and $0 < a < b$.

21. Prove that $1 + 2 + \cdots + n < n^2$ for $n > 1$.

22. (a) Try to use induction to prove that
$$1 + \frac{1}{2} + \frac{1}{4} + \cdots + \frac{1}{2^n} < 2 \qquad \text{for } n \geq 1$$

What goes wrong?

(b) Prove that
$$1 + \frac{1}{2} + \frac{1}{4} + \cdots + \frac{1}{2^n} = 2 - \frac{1}{2^n} \qquad \text{for } n \geq 1$$

thus showing that
$$1 + \frac{1}{2} + \frac{1}{4} + \cdots + \frac{1}{2^n} < 2 \qquad \text{for } n \geq 1$$

For Exercises 23 through 32, prove that the statements are true for every positive integer.

★ 23. $2^{3n} - 1$ is divisible by 7.

24. $3^{2n} + 7$ is divisible by 8.

25. $7^n - 2^n$ is divisible by 5.

26. $13^n - 6^n$ is divisible by 7.

27. $2^n + (-1)^{n+1}$ is divisible by 3.

28. $2^{5n+1} + 5^{n+2}$ is divisible by 27.

29. $3^{4n+2} + 5^{2n+1}$ is divisible by 14.

30. $7^{2n} + 16n - 1$ is divisible by 64.

★ 31. $10^n + 3 \cdot 4^{n+2} + 5$ is divisible by 9.

32. $x^n - 1$ is divisible by $x - 1$ for $x \neq 1$.

33. Use induction to prove that the product of any three consecutive positive integers is divisible by 3.

34. Suppose that exponentiation is defined by the equation

    $$x^j \cdot x = x^{j+1}$$

    for any $j \geq 1$. Use induction to prove that $x^n \cdot x^m = x^{n+m}$ for $n \geq 1$, $m \geq 1$. (Hint: do induction on m for a fixed, arbitrary value of n.)

35. Let A and $B_1, B_2, \ldots, B_n$ be statement letters. Prove that

    $$A \wedge (B_1 \vee B_2 \vee \cdots \vee B_n)$$
    $$\leftrightarrow (A \wedge B_1) \vee (A \wedge B_2) \vee \cdots \vee (A \wedge B_n)$$

    is a tautology for $n \geq 2$.

36. Let $B_1, B_2, \ldots, B_n$ be statement letters. Prove that

    $$(B_1 \wedge B_2 \wedge \cdots \wedge B_n)' \leftrightarrow B_1' \vee B_2' \vee \cdots \vee B_n'$$

    is a tautology for $n \geq 2$.

37. A string of 0s and 1s is to be processed and converted to an even-parity string by adding a parity bit to the end of the string. The parity bit is initially 0. When a 0 character is processed, the parity bit remains unchanged. When a 1 character is processed, the parity bit is switched from 0 to 1 or from 1 to 0. Prove that the number of 1s in the final string, that is, including the parity bit, is always even. (Hint: consider various cases.)

38. What is wrong with the following "proof" by mathematical induction? We will prove that for any positive integer n, n is equal to 1 more than n. Assume that $P(k)$ is true,

    $$k = k + 1$$

    Adding 1 to both sides of this equation, we get

    $$k + 1 = k + 2$$

    thus

    $$P(k + 1) \text{ is true}$$

39. What is wrong with the following "proof" by mathematical induction? We will prove that all computers are built by the same manufacturer. In particular, we will prove that in any collection of n computers where n is a positive integer, all of the computers are built by the same manufacturer. We first prove $P(1)$, a trivial process because in any collection consisting of one computer, there is only one manufacturer. Now we assume $P(k)$; that is, in any collection of k computers, all the computers were built by the same manufacturer. To prove $P(k + 1)$, we consider any collection of $k + 1$ computers. Pull

one of these $k + 1$ computers (call it HAL) out of the collection. By our assumption, the remaining k computers all have the same manufacturer. Let HAL change places with one of these k computers. In the new group of k computers, all have the same manufacturer. Thus, HAL's manufacturer is the same one that produced all the other computers, and all $k + 1$ computers have the same manufacturer.

40. The induction principle used in this section can be summarized as follows:

$$\left. \begin{array}{l} 1.\ P(1)\ \text{true} \\ 2.\ P(k)\ \text{true} \rightarrow P(k + 1)\ \text{true} \end{array} \right\} \rightarrow P(n)\ \text{true for all } n$$

We call this the principle of **weak induction,** as opposed to the principle of **strong induction,** which is

$$\left. \begin{array}{l} 1'.\ P(1)\ \text{true} \\ 2'.\ P(r)\ \text{true for all } r \leq k \rightarrow P(k + 1)\ \text{true} \end{array} \right\} \begin{array}{l} \rightarrow P(n)\ \text{true} \\ \text{for all } n \end{array}$$

Notice that in implication $2'$ there may be many facts available to help us conclude that $P(k + 1)$ is true, but that in implication 2, the only available fact is that $P(k)$ is true. Thus, implications $2'$ and 2 are not directly equivalent. If 2 can be proved, so can $2'$, but not conversely. However, the two induction principles themselves are equivalent; that is, the two major implications are equivalent. In other words, if we accept weak induction as a valid principle, then strong induction is valid, and conversely. The converse is easy to see. If we accept strong induction as valid reasoning, then weak induction is surely valid because we can say that we concluded $P(k + 1)$ from $P(r)$ for all $r \leq k$ even though we used only the single condition $P(k)$. However, the real difficulty arises when we accept weak induction as a valid principle and want to show that strong induction is valid. Why may we conclude that $P(n)$ is true for all n when we used so much more evidence just to conclude $P(k + 1)$? We attack this problem by pointing out that, using the principle of weak induction, we can prove that every nonempty set of positive integers has a smallest member (the Well-Ordering Principle). Now we assume that $1'$ and $2'$ have been shown, and we want to conclude $P(n)$ for all n. Complete the proof by letting T be the subset of the positive integers defined by

$$T = \{t \,|\, P(t) \text{ is not true}\}$$

and showing that $T = \varnothing$ (the empty set).

41. Prove that n is a prime number or a product of two or more prime numbers for $n \geq 2$. (Hint: use strong induction. See Exercise 40.)

In Exercises 42 through 45, prove the loop invariant $P(n)$ and show that the program is correct by evaluating $P(n)$ at loop termination. All variables have integer values.

42. Program to compute x^2 for $x \geq 1$.

   1. given $x \geq 1$
   2. $i \leftarrow 1$
   3. $j \leftarrow 1$
   4. while $i \neq x$ do
      begin
      $j \leftarrow j + 2i + 1$
      $i \leftarrow i + 1$
      end
   5. write j

   $P(n): j_n = (i_n)^2$

★ 43. Program to compute $x!$ for $x \geq 1$.

   1. given $x \geq 1$
   2. $i \leftarrow 2$
   3. $j \leftarrow 1$
   4. while $i \neq x + 1$ do
      begin
      $j \leftarrow j \cdot i$
      $i \leftarrow i + 1$
      end
   5. write j

   $P(n): j_n = (i_n - 1)!$

44. Program to compute x^y for $x, y \geq 1$.

   1. given $x, y \geq 1$
   2. $i \leftarrow 1$
   3. $j \leftarrow x$
   4. while $i \neq y$ do
      begin
      $j \leftarrow j \cdot x$
      $i \leftarrow i + 1$
      end
   5. write j

   $P(n): j_n = x^{i_n}$

45. Program to compute quotient q and remainder r when x is divided by y, $x \geq 0$, $y \geq 1$.

   1. given $x \geq 0$, $y \geq 1$
   2. $q \leftarrow 0$
   3. $r \leftarrow x$
   4. while $r \geq y$ do
      begin
      $q \leftarrow q + 1$

$$r \leftarrow r - y$$
$$\text{end}$$
5. write q, r

$$P(n): x = q_n \cdot y + r_n$$

SECTION 1.5   RECURSION AND RECURRENCE RELATIONS

Recursive Definitions

Suppose we are trying to describe which objects belong to a collection C of objects. Sometimes it is convenient to do this by making three statements.

1. Some specific objects belong to C.
2. Other objects that are obtained from objects in C by doing certain operations are also in C.
3. The only objects that are in C are those that belong to C by statements 1 or 2 above. (This third statement is generally omitted—only a logician thinks to ask at the end of a list of statements if this is the end of the list!)

1.63 Example S consists of a sequence of objects where the kth object in the sequence is denoted by $S(k)$. S is defined by

1. $S(1) = 2$
2. $S(n) = 2 \cdot S(n - 1)$ for $n \geq 2$

By statement 1, $S(1)$, the first object in S, is 2. Then by statement 2, the second object in S is $S(2) = 2 \cdot S(1) = 2(2) = 4$. By statement 2 again, $S(3) = 2 \cdot S(2) = 2(4) = 8$. Continuing in this fashion, we can see that S is the sequence

2, 4, 8, 16, 32, . . .   □

How we generated the members of S in Example 1.63 is similar to how we "climbed the ladder" in discussing proofs by induction. There is a basis step—2 is in S. There is an inductive step—given that we have found the $(n - 1)$st member of S, we can generate the nth member of S. A definition such as that for S in Example 1.63 is called an **inductive definition** or a **recursive definition**. Equation 2 in Example 1.63, where $S(n)$ is defined in terms of an earlier S-value, is an example of a **recurrence relation**.

The structure of a general recursive definition consists of a basis step, where some specific information is given, and an inductive or recursive step, where new information is given in terms of already known information. A recursive definition may define a sequence of objects, where the order of the objects is meaningful, as in Example 1.63, or it may define an unordered collection of objects, as in the next example.

1.64 Example   We know that statement letters can be combined with logical connectives to form compound statements. Although we did not discuss this in Section 1.1, there are combinations that we would consider legal, such as $(A \lor B)'$, and combinations that we would not consider legal, such as $\land \land A''B$. A legitimate combination is called a **well-formed formula,** or **wff.** The collection of well-formed formulas for statements with statement letters and logical connectives can be defined recursively by

1. Any statement letter is a wff.
2. If P and Q are wffs, so are $(P \land Q)$, $(P \lor Q)$, $(P \to Q)$, P', and $P \leftrightarrow Q$.

We often omit parentheses when doing so causes no confusion; thus we write $(P \lor Q)$ as $P \lor Q$.

By beginning with statement letters and repeatedly using rule 2, any wff can be built.   □

1.65 Practice   The sequence T is defined recursively as follows:

1. $T(1) = 1$
2. $T(n) = T(n - 1) + 3$ for $n \geq 2$

Write the first five values in the sequence T.   □

Many important constructs in computer science, including elements of many programming languages, are defined recursively.

Recursive Algorithms

Example 1.63 gives us a recursive definition for a sequence S. Suppose we want to write a computer program to evaluate $S(n)$ for some positive integer n. We can use either of two approaches. If we want to find $S(12)$, for example, we can begin with $S(1) = 2$ and from there compute $S(2)$, $S(3)$, and so on, much as we did in Example 1.63, until we finally get to $S(12)$. A pseudocode algorithm using this approach is shown below. (The notation "$S(n)$" indicates that the algorithm computes a quantity S given the input value n; that is, it computes $S(n)$.)

Algorithm $S(n)$

1. if $n = 1$
2. then $S \leftarrow 2$

```
3. else begin
4.     i ← 2
5.     S ← 2
6.     while i ≤ n do
7.         begin
8.             S ← 2 · S
9.             i ← i + 1
10.        end while
11.    end else
```

Lines 1 and 2 deal with the simple case where n is 1. The loop at lines 6 through 10 computes $S(n)$ for successively larger values of n until the correct upper limit is reached. You can trace the execution of this algorithm for a few values of n to convince yourself that it works, or you can prove the correctness by finding an appropriate loop invariant (see Section 1.4.)

The second approach to computing $S(n)$ uses the recursive definition of S directly. A pseudocode version is given below. (The notation "$SS(n)$" indicates that this algorithm computes a quantity SS given the input value of n; we've called this quantity SS to distinguish this algorithm from the preceding one, but it still computes values for the sequence S.)

Algorithm SS(n)

```
1. if n = 1
2. then SS ← 2
3. else begin
4.     SS ← 2 · SS(n − 1)
5.     end
```

To understand how algorithm $SS(n)$ works, let's consider how we could compute $S(4)$, for example. We can find the value of $S(4)$ if we know the value of $S(3)$, but to compute $S(3)$, we must first compute $S(2)$, and to do this we must first compute $S(1)$. Aha!—this we can do, by the basis step. Knowing the value of $S(1)$, we can then find the value of $S(2)$, then $S(3)$, and finally $S(4)$.

Now suppose we start to execute algorithm $SS(n)$ with an input value of $n > 1$. Lines 1 and 2 are passed over because $n > 1$. When the algorithm gets to line 4, it temporarily suspends activity on computing SS with an input value of n and invokes itself with a smaller input value. The execution of algorithm SS with an input value of $n - 1$, if $n - 1 > 1$, will pass over lines 1 and 2 and then invoke algorithm SS with an input value of $n - 2$. This process will continue, with successive invocations, until the input value is finally 1 and the output value, 2, can be computed by the basis step, lines 1 and 2. This final invocation of the algorithm will then give this output value to the second-to-last invocation, which can now compute its output and give it to the previous invocation, and so on.

Finally, the original invocation of SS can be completed. (Although this sounds more complex than what happens in algorithm $S(n)$, we have described what the algorithm does automatically. We can write algorithm $SS(n)$ easier than we can algorithm $S(n)$ and let the algorithm itself carry out all the work we had to do in writing the loop of algorithm $S(n)$.)

Algorithm $SS(n)$ is an example of a **recursive algorithm,** one that invokes itself. Many programming languages allow such recursion, and it is very natural to use a recursive algorithm to compute a sequence that has been defined recursively.

1.66 Practice   Write a recursive algorithm to compute $T(n)$ for the sequence T defined in Practice 1.65.   ☐

Solving Recurrence Relations

There is an easier procedure for computing the sequence $S(n)$ of Example 1.63 than either algorithm $SS(n)$ or its nonrecursive version, algorithm $S(n)$. Recall that

$$S(1) = 2 \tag{1}$$
$$S(n) = 2 \cdot S(n - 1) \qquad \text{for } n \geq 2 \tag{2}$$

Because

$$S(1) = 2 \ = 2^1$$
$$S(2) = 4 \ = 2^2$$
$$S(3) = 8 \ = 2^3$$
$$S(4) = 16 = 2^4$$

we can see that

$$S(n) = 2^n \tag{3}$$

Using equation (3), we can plug in a value for n and compute $S(n)$ without computing all the lower values of S first. An equation such as (3), where we can substitute a value and get the output value back directly, is called a **closed-form solution** to the recurrence relation (2) subject to the basis step (1). Fiinding a closed-form solution is called **solving** the recurrence relation. Clearly, it is very nice to find closed-form solutions, if possible.

A number of techniques for solving recurrence relations have been developed, some of them very similar to approaches used to solve differential equations. We will not go into these techniques, however; our only solution method will be the "expand, guess, verify" approach.

1.67 Example   Consider again the basis step and recurrence relation for the sequence S of Example 1.63:

$$S(1) = 2 \qquad\qquad\qquad\qquad\qquad\qquad\qquad (4)$$
$$S(n) = 2 \cdot S(n - 1) \qquad \text{for } n \geq 2 \qquad\qquad (5)$$

Let's pretend we don't already know the closed-form solution and use the "expand, guess, verify" approach to find it. Beginning with $S(n)$, we expand by using the recurrence relation repeatedly:

$$
\begin{aligned}
S(n) &= 2 \cdot S(n - 1) \\
&= 2(2 \cdot S(n - 2)) \;\; = 2^2 \cdot S(n - 2) \\
&= 2^2(2 \cdot S(n - 3)) = 2^3 \cdot S(n - 3)
\end{aligned}
$$

$$\vdots \qquad\qquad\qquad \vdots$$

By looking at the developing pattern, we guess that the general term has the form

$$S(n) = 2^k \cdot S(n - k)$$

This expansion of S-values in terms of lower S-values must stop when $n - k = 1$, that is, when $k = n - 1$. At that point,

$$S(n) = 2^{n-1} \cdot S(n - (n - 1)) = 2^{n-1} \cdot S(1) = 2^{n-1} \cdot 2 = 2^n$$

which expresses the closed-form solution.

We are not yet done, however, because we guessed at the general term. We now confirm our closed-form solution by induction on the value of n. The statement we want to prove is therefore $S(n) = 2^n$ for $n \geq 1$.

For the basis step, $S(1) = 2^1$. This is a true equation by (4). We assume that $S(k) = 2^k$. Then

$$
\begin{aligned}
S(k + 1) &= 2 \cdot S(k) \qquad \text{(by (5))} \\
&= 2 \cdot 2^k \qquad \text{(by the inductive hypothesis)} \\
&= 2^{k+1} \qquad\qquad\qquad\qquad\qquad\qquad \square
\end{aligned}
$$

1.68 Practice   Find a closed-form solution for the recurrence relation, subject to the basis step, for sequence T:

1. $T(1) = 1$
2. $T(n) = T(n - 1) + 3 \qquad$ for $n \geq 2$

(Expand, guess, and verify.) $\qquad\qquad\qquad\qquad\qquad\qquad\qquad\qquad \square$

Analysis of Algorithms

Often several algorithms exist that do the same task. In comparing algorithms, several criteria can be used to judge which is the "better" algorithm. We might ask, for example, which is easiest to understand or which runs most efficiently. One way to judge the efficiency of an algorithm is to estimate the number of operations that must be performed by the algorithm. We only count the operations that are basic to the task at hand, excluding "housekeeping" operations that only contribute a small percentage to the total work required.

For example, suppose that the task is to search an ordered list of words or numbers for a particular item x. Since any possible algorithm seems to require comparing elements from the list with x until a match is found, the basic operation to count is comparisons. One algorithm is simply to compare x with each entry in the list until we either find x in the list or exhaust the list. If there are n items in the list, this algorithm could require n comparisons in the worst case, that is, where x is either the last item in the list or is not in the list at all.

A more efficient technique is the recursive **binary search algorithm,** which appears below. In the first step in this algorithm, the middle item in a list of n items is to be found. If n is an odd number, there is indeed a middle item. For example, in a list of 5 items, item number 3 is the middle one. If n is even, there is no middle item, but we define item $n/2$ to be the middle one.

Algorithm BinarySearch

1. given an ordered list of n items, find the middle item
2. compare the middle item with x
3. if the middle item equals x
   then we have found x
4. else if $x <$ middle item
   then if first half of list is not empty, do *BinarySearch* on first half of list
5. else if second half of list is not empty, do *BinarySearch* on second half of list

1.69 Example   Suppose the list of items is

> 3, 7, 8, 10, 14, 18, 22, 34

and x is the number 25. There are 8 items in the list, so $n = 8$. Then $n/2 = 4$, so the fourth item, 10, is taken as the middle item. At line 2, we compare the middle item with x. Because $x > 10$, we invoke the search on the second half of the list, namely the items

> 14, 18, 22, 34

This time the number of items has been reduced to 4, so we pick $\frac{4}{2} = 2$ as the location of the middle item. Because $x > 18$, the middle item, we search the second half of this list, namely the items

22, 34

Now the middle item is number $\frac{2}{2} = 1$, or 22. Because $x > 22$, we search the second half of this list, namely

34

This is a 1-element list, in which the middle item is the only item. We compare 34 with x. Because $x < 34$, we would search the first half of this list, but the first half is empty. Thus we are done—we know that x is not in the list. There were four comparisons required in all.   □

The study of the efficiency of algorithms, in terms of the number of operations they might require, is part of what is called **analysis of algorithms.** We will look at analysis of algorithms more formally in Chapter 11, but we are interested at this point because it often involves solving a recurrence relation.

In the case of the binary search, for example, one comparison is done at the middle value, and then the process is repeated on half the list. If the original list is n elements long, then half the list is at worst $n/2$ elements long. (In Example 1.69, for example, when 10 is the middle value, the right "half" of the list has 4 elements, but the left "half" has only 3.) If we are to keep cutting the list in half, it is convenient to consider only the case where we get an integer value each time we cut in half, so we will assume that $n = 2^m$ for some $m \geq 0$. If $C(n)$ stands for the maximum number of comparisons required for an n-element list, then we see that

$$C(n) = 1 + C\left(\frac{n}{2}\right)$$

(1 comparison at the middle value, plus however many comparisons are needed for half the list)

The basis step is

$$C(1) = 1$$

since it would require only 1 comparison to search a 1-element list.

1.70 Example   Solve the recurrence relation

$$C(n) = 1 + C\left(\frac{n}{2}\right) \qquad \text{for } n \geq 2, \, n = 2^m$$

subject to the basis step

$$C(1) = 1$$

Expanding, we get

$$C(n) = 1 + C\left(\frac{n}{2}\right)$$

$$= 1 + \left(1 + C\left(\frac{n}{4}\right)\right)$$

$$= 1 + 1 + \left(1 + C\left(\frac{n}{8}\right)\right)$$

.
.
.

and the general term seems to be

$$C(n) = k + C\left(\frac{n}{2^k}\right)$$

The process stops when $2^k = n$, or $k = \log_2 n$. (We'll omit the base 2 notation from now on—$\log n$ will mean $\log_2 n$.) Then

$$C(n) = \log n + C(1) = 1 + \log n$$

Now we will use induction to show that $C(n) = 1 + \log n$ for all $n \geq 1$, $n = 2^m$. This is a somewhat different form of induction because the only values of interest are powers of 2. We still take 1 as the basis step for the induction, but then we prove that if our statement is true for a value k, it is true for $2k$. The statement will then be true for $1, 2, 4, 8, \ldots$, which is just what we want.

$$C(1) = 1 + \log 1 = 1 + 0 = 1 \qquad \text{true}$$

Assume that $C(k) = 1 + \log k$. Then

$$
\begin{aligned}
C(2k) &= 1 + C(k) && \text{(by the recurrence relation)} \\
&= 1 + 1 + \log k && \text{(by the inductive hypothesis)} \\
&= 1 + \log 2 + \log k && (\log 2 = 1) \\
&= 1 + \log 2k && \text{(property of logarithms)}
\end{aligned}
$$

This completes the inductive proof. $\qquad\qquad\qquad\qquad\qquad$ $\square$

By Example 1.70, the maximum number of comparisons required to do a binary search on an n-element list, with $n = 2^m$, is $1 + \log n$. Note that in Example 1.69 $n = 8$, and 4 comparisons ($1 + \log 8$) were required in the worst case, where x was not in the list.

✔ Checklist

Definitions

inductive definition (*p. 59*)	closed-form solution (*p. 62*)
recursive definition (*p. 59*)	solving a recurrence relation
recurrence relation (*p. 59*)	(*p. 62*)
well-formed formula (wff) (*p. 60*)	binary search algorithm (*p. 64*)
recursive algorithm (*p. 62*)	analysis of algorithms (*p. 65*)

Techniques

Generate values in a sequence defined recursively, or recognize values in a collection defined recursively.

Write a recursive algorithm to generate a sequence defined recursively.

Solve recurrence relations by the "expand, guess, verify" technique.

Main Ideas

Recursive definitions for sequences and collections of objects, where basis information is given and new information depends on already known information

Recursive algorithms, a natural way to compute sequences defined recursively

Certain recurrence relations can be solved by expanding, guessing at the solution, and verifying the solution by induction

Analysis of algorithms, often leading to recurrence relations

Exercises Section 1.5

For Exercises 1 through 6, write the first five values in the sequence.

★ 1. $S(1) = 10$
$S(n) = S(n - 1) + 10$ for $n \geq 2$

2. $A(1) = 2$
$A(n) = \dfrac{1}{A(n - 1)}$ for $n \geq 2$

3. $B(1) = 1$
$B(n) = B(n - 1) + n^2$ for $n \geq 2$

4. $S(1) = 1$

   $S(n) = S(n - 1) + \dfrac{1}{n}$ for $n \geq 2$

★ 5. $F(1) = 1$
   $F(2) = 1$
   $F(n) = F(n - 1) + F(n - 2)$ for $n > 2$ *

6. $M(1) = 2$
   $M(2) = 2$
   $M(n) = 2M(n - 1) + M(n - 2)$ for $n > 2$

7. A collection T of numbers is defined recursively by

   1. 2 belongs to T.
   2. If X belongs to T, so does $X + 3$ and $2 \cdot X$.

   Which of the following belong to T?
   (a) 6   (b) 7   (c) 19   (d) 12

8. A collection M of numbers is defined recursively by

   1. 2 and 3 belong to M.
   2. If X and Y belong to M, so does $X \cdot Y$.

   Which of the following belong to M?

   (a) 6   (b) 9   (c) 16   (d) 21
   (e) 26   (f) 54   (g) 72   (h) 218

★ 9. A collection S of strings of characters is defined recursively by

   1. a and b belong to S.
   2. If X belongs to S, so does Xb.

   Which of the following belong to S?
   (a) a (b) ab (c) aba (d)$aaab$ (e) $bbbbb$

10. A collection W of strings of symbols is defined recursively by

   1. a, b, and c belong to W.
   2. If X belongs to W, so does $a(X)c$

   Which of the following belong to W?
   (a) $a(b)c$ (b) $a(a(b)c)c$ (c) $a(abc)c$
   (d) $a(a(a(a)c)c)c$ (e) $a(aacc)c$

★ 11. In an experiment, a certain colony of bacteria initially has a population of 50,000. A reading is taken every 2 hours, and at the end of each 2-hour interval, there are 3 times as many bacteria as before.

---

*This defines the **Fibonacci sequence** of numbers. A closed-form solution for the Fibonacci sequence is

$$F(n) = \frac{\sqrt{5}}{5} \left(\frac{1 + \sqrt{5}}{2}\right)^n - \frac{\sqrt{5}}{5} \left(\frac{1 - \sqrt{5}}{2}\right)^n$$

(a) Write a recursive definition for $A(n)$, the amount of bacteria present at the beginning of the nth time period.

(b) Which interval is starting when there are 1,350,000 bacteria present?

12. An amount of $500 is invested in an account paying 10% interest compounded annually.

(a) Write a recursive definition for $P(n)$, the amount in the account at the beginning of the nth year.

(b) After how many years will the account balance exceed $700?

In Exercises 13 through 18, given the sequence S, write a recursive algorithm to compute $S(n)$.

13. 1, 3, 9, 27, . . .

14. 2, 1, $\frac{1}{2}$, $\frac{1}{4}$, . . .

★ 15. 1, 2, 4, 7, 11, 16, . . .

16. 2, 4, 16, 256, . . .

17. a, b, $a + b$, $a + 2b$, $2a + 3b$, . . .

18. p, $p - q$, $p + q$, $p - 2q$, $p + 2q$, $p - 3q$, . . .

In Exercises 19 through 24, solve the given recurrence relation subject to the basis step. (Use "expand, guess, and verify.")

★ 19. $S(1) = 5$
$S(n) = S(n - 1) + 5$ for $n \geq 2$

20. $F(1) = 1$
$F(n) = n \cdot F(n - 1)$ for $n \geq 2$

21. $T(1) = 1$
$T(n) = 2 \cdot T(n - 1) + 1$ for $n \geq 2$
(Hint: see Example 1.54.)

22. $S(1) = 1$
$S(n) = S(n - 1) + 2n - 1$ for $n \geq 2$
(Hint: see Example 1.53.)

23. $T(1) = 1$
$T(n) = 2 \cdot T\left(\dfrac{n}{2}\right) + n$ for $n \geq 2$, $n = 2^m$

24. $P(1) = 1$
$P(n) = 2 \cdot P\left(\dfrac{n}{2}\right) + n^2$ for $n \geq 2$, $n = 2^m$

Chapter 2

Sets and Combinatorics

Set theory is one of the cornerstones of mathematics. Many concepts in mathematics and computer science can be conveniently expressed in the language of sets. Operations can be performed on sets to generate new sets. Although most sets of interest to computer scientists are finite or *countable,* there are sets with so many members that they cannot be enumerated. Set theory is discussed in Section 2.1.

If a set is finite, it is often of interest to count the number of elements in the set. This may not be a trivial task. Section 2.2 provides some ground rules for counting, where the set to be counted consists of the number of outcomes of an event. Counting the elements in a set can be made manageable by breaking the event down into a sequence of events, or into disjoint events that have no outcomes in common. Section 2.3 provides formulas for counting the number of ways to arrange objects from a set and to select objects from a set.

Section 2.4 discusses the binomial theorem, an algebraic result that can also be viewed as a consequence of the counting formulas.

SECTION 2.1   SETS

Definitions are important in any science, because they contribute to precise communication. However, if we look up a word in the dictionary, the definition is expressed using other words, which are defined using still other words, and so on. We have to have a starting point for our definitions; ours will be the idea of a **set**. We will not formally define a set. Nonetheless, we will informally describe a set as a collection of objects characterized by some defining property; thus any given object either does or does not have the property and, thus, either does or does not belong to the set.

Notation

We use capital letters to denote sets and the symbol $\in$ to denote membership in a set. Thus $a \in A$ means that object a is a member, or element, of set A, and $b \notin A$ means that object b is not an element of set A. Braces { } are used to indicate a set.

2.1 Example   If A = {violet, chartreuse, burnt umber}, then chartreuse $\in A$ and magenta $\notin A$.   □

Two sets are **equal** if they contain the same elements. (In a definition, "if" really means "if and only if," thus two sets are equal if and only if they contain the same elements.) The order of the elements in a set does not matter; for example, {violet, chartreuse, burnt umber} = {chartreuse, burnt umber, violet}. Also, each element of a set is listed only once; it is redundant to list it again.

In describing a particular set, we have to identify the property characterizing its elements. For a finite set, we do this simply by listing the elements, as in set A of Example 2.1. It is impossible to list all elements of an infinite set. For some infinite sets, however, we can indicate a pattern for listing elements indefinitely. Thus, we might write {2, 4, 6, . . .} to express the set of all positive even integers. Although this is a common practice, the danger exists that the reader will not see the pattern that the writer had in mind. It is better to give the defining property in words and write {2, 4, 6, . . .} as $\{x \mid x$ is a positive even integer}, read as "the set of all x such that x is a positive even integer."

In general, a set whose elements are characterized as having property P is described as $\{x \mid P(x)\}$. Property P here is a unary predicate, used the same way as in statements with quantifiers (see Section 1.1).

2.2 Practice   List the elements of each of the following sets:

(a) $\{x \mid x$ is an integer and $3 < x \leq 7\}$
(b) $\{x \mid x$ is a month with exactly 30 days}
(c) $\{x \mid x = y^3$ for $y \in \{0, 1, 2\}\}$ □

2.3 Practice   Give the defining property of each set below.

(a) {1, 4, 9, 16}
(b) {the butcher, the baker, the candlestick maker}
(c) {2, 3, 5, 7, 11, 13, 17, . . .}   □

It is convenient to name certain standard sets so that we can refer to them easily. We will use

$\mathbb{N}$ = the set of all nonnegative integers   (note that $0 \in \mathbb{N}$)
$\mathbb{Z}$ = the set of all integers

$\mathbb{Q}$ = the set of all rational numbers
$\mathbb{R}$ = the set of all real numbers
$\mathbb{C}$ = the set of all complex numbers

Sometimes we will also want to talk about the set with no elements (the **empty set,** or **null set**), denoted by $\varnothing$ or { }.

Relationships Between Sets

For $A = \{2, 3, 5, 12\}$ and $B = \{2, 3, 4, 5, 9, 12\}$, every member of A is also a member of B. When this happens, A is said to be a subset of B.

2.4 Practice   Complete the definition: A is a **subset** of B if $x \in A \rightarrow$ _____.   □

If A is a subset of B, we denote this by $A \subseteq B$. If $A \subseteq B$ but $A \neq B$ (there is at least one element of B that is not an element of A), then A is a **proper subset** of B, denoted by $A \subset B$.

2.5 Example   Let

$A = \{1, 7, 9, 15\}$
$B = \{7, 9\}$
$C = \{7, 9, 15, 20\}$

Then the following statements (among others) are all true:

$B \subseteq C$
$B \subseteq A$
$B \subset A$
$A \not\subseteq C$
$15 \in C$
$\{7, 9\} \subseteq B$
$\{7\} \subset A$
$\varnothing \subseteq C$ ($x \in \varnothing \rightarrow x \in C$ is true because $x \in \varnothing$ is false.)   □

2.6 Practice   Let

$A = \{x \mid x \in \mathbb{N} \text{ and } x \geq 5\}$
$B = \{10, 12, 16, 20\}$
$C = \{x \mid x = 2y \text{ for } y \in \mathbb{N}\}$

Which of the following statements are true?

(a) $B \subseteq C$ (b) $B \subset A$
(c) $A \subseteq C$ (d) $26 \in C$

(e) $\{11, 12, 13\} \subseteq A$

(f) $\{11, 12, 13\} \subset C$

(g) $\{12\} \in B$

(h) $\{12\} \subseteq B$

(i) $\{x \,|\, x \in \mathbb{N} \text{ and } x < 20\} \nsubseteq B$

(j) $5 \subseteq A$

(k) $\{\varnothing\} \subseteq B$

(l) $\varnothing \notin A$ □

We know that A and B are equal sets if they have the same elements. We can restate this equality in terms of subsets: $A = B$ if and only if $A \subseteq B$ and $B \subseteq A$. Proving set inclusion in both directions is the usual way to establish the equality of two sets.

2.7 Example   We will prove that $\{x \,|\, x \in \mathbb{N} \text{ and } x^2 < 15\} = \{x \,|\, x \in \mathbb{N} \text{ and } 2x < 7\}$.

Proof: Let $A = \{x \,|\, x \in \mathbb{N} \text{ and } x^2 < 15\}$ and $B = \{x \,|\, x \in \mathbb{N} \text{ and } 2x < 7\}$. To show that $A = B$, we show $A \subseteq B$ and $B \subseteq A$. For $A \subseteq B$, we must choose an arbitrary member of A, that is, anything satisfying the defining property of A, and show that it also satisfies the defining property of B. Let $x \in A$. Then x is a nonnegtive integer satisfying the inequality $x^2 < 15$. The nonnegative integers with squares less than 15 are 0, 1, 2, and 3, so these are the members of A. The double of each of these nonnegative integers is a number less than 7. Hence, each member of A is a member of B, and $A \subseteq B$. Any member of B is a nonnegative integer whose double is less than 7. These numbers are 0, 1, 2, and 3, each of which has a square less than 15, so $B \subseteq A$.   □

2.8 Practice   Let $A = \{x \,|\, \cos(x/2) = 0\}$ and $B = \{x \,|\, \sin x = 0\}$. Prove that $A \subseteq B$. □

Sets of Sets

For a set S, we can form a new set whose elements are all of the subsets of S. This new set is called the **power set** of S, $\mathcal{P}(S)$. $\mathcal{P}(S)$ will always have at least $\varnothing$ and S itself as members, since $\varnothing \subseteq S$ and $S \subseteq S$ are always true.

2.9 Practice   For $A = \{1, 2, 3\}$, what is $\mathcal{P}(A)$?   □

In Practice 2.9, A has three elements and $\mathcal{P}(A)$ has eight elements. Try finding $\mathcal{P}(S)$ for other sets S until you can guess the answer to the following Practice problem.

2.10 Practice   If S has n elements, then $\mathcal{P}(S)$ has _____ elements. (Does your answer work for $n = 0$ too?)   □

There are several ways we can show that for a set S with n elements, $\mathcal{P}(S)$ will have 2^n elements. A formal proof would use induction. For the basis step of the induction, we let $n = 0$. The only set with 0 elements is $\varnothing$. The only subset of $\varnothing$ is $\varnothing$, so $\mathcal{P}(\varnothing) = \{\varnothing\}$, a set with $1 = 2^0$ elements. We assume that for any set with k elements, the power set has 2^k elements.

Now let S have $k + 1$ elements and put one of these elements, call it x, aside. The remaining set has k elements, so by our inductive assumption, its power set has 2^k elements. Each of these elements is also a member of $\mathcal{P}(S)$. The only members of $\mathcal{P}(S)$ not counted by this procedure are those including element x. All the subsets including x can be found by taking all those subsets not including x (of which there are 2^k) and throwing in the x; thus, there will be 2^k subsets including x. Altogether, there are 2^k subsets without x and 2^k subsets with x, or $2^k + 2^k = 2 \cdot 2^k = 2^{k+1}$ subsets. Therefore, $\mathcal{P}(S)$ has 2^{k+1} elements.

Analogy with the truth tables of Section 1.1 is another way to show that $\mathcal{P}(S)$ has 2^n elements for a set S with n elements. There we had n statement letters and showed that there were 2^n true–false combinations among these letters. But we can also think of each true–false combination as representing a particular subset, with T indicating membership and F indicating nonmembership in that subset. (For example, the row of the truth table with all statement letters F corresponds to the empty set.) Thus, the number of true–false combinations among n statement letters equals the number of subsets of a set with n elements; both are 2^n.

Binary and Unary Operations

By itself a set is not very interesting until we *do* something with its elements. For example, we can perform several arithmetic operations on elements of the set $\mathbb{Z}$. We might subtract two integers, or we might take the negative of an integer. Subtraction is a binary operation on $\mathbb{Z}$; taking the negative of a number is a unary operation on $\mathbb{Z}$. A binary operation acts on two numbers, and a unary operation acts on one number. To see exactly what is involved in a binary operation, let's look at subtraction more closely. For any two integers x and y, $x - y$ will produce an answer, and only one answer, and that answer will always be an integer. Finally, subtraction is performed on an **ordered pair** of numbers. For example, $7 - 5$ does not produce the same result as $5 - 7$. An ordered pair is denoted by (x, y), where x is the first component of the ordered pair and y is the second component. Order is important in an ordered pair; thus, the sets $\{1, 2\}$ and $\{2, 1\}$ are equal, but the ordered pairs $(1, 2)$ and $(2, 1)$ are not. Two ordered pairs (x, y) and (u, v) are equal only when $x = u$ and $y = v$.

2.11 Practice   Given that $(2x - y, x + y) = (7, -1)$, solve for x and y.   □

2.12 Practice   Let $S = \{3, 4\}$. List all the ordered pairs (x, y) of elements of S.   □

We will generalize these properties of subtraction on the integers to define a binary operation $\circ$ on a set S.

2.13 Definition $\circ$ is a **binary operation** on a set S if for every ordered pair (x, y) of elements of S, $x \circ y$ exists, is unique, and is a member of S.   □

That the value $x \circ y$ always exists and is unique is described by saying that the binary operation $\circ$ is **well-defined**. The property that $x \circ y$ always belongs to S is described by saying that S is **closed** under the operation $\circ$.

2.14 Example   Addition, subtraction, and multiplication are all binary operations on $\mathbb{Z}$. For example, when we perform addition on the ordered pair of integers (x, y), $x + y$ exists and is a unique integer.   □

A candidate $\circ$ for an operation can fail to be a binary operation on a set S in any of three ways: (1) there are elements $x, y \in S$ for which $x \circ y$ does not exist; (2) there are elements $x, y \in S$ for which $x \circ y$ gives more than one result; or (3) there are elements $x, y \in S$ for which $x \circ y$ does not belong to S.

2.15 Example   Division is not a binary operation on $\mathbb{Z}$ because $x \div 0$ does not exist.   □

2.16 Example   Subtraction is not a binary operation on $\mathbb{N}$ because $\mathbb{N}$ is not closed under subtraction. (For example, $1 - 10 \notin \mathbb{N}$.)   □

2.17 Example   Define $x \circ y$ on $\mathbb{N}$ by

$$x \circ y = \begin{cases} 1 & \text{if } x \geq 5 \\ 0 & \text{if } x \leq 5 \end{cases}$$

Then, by the first part of the definition for $\circ$, $5 \circ 1 = 1$, but by its second part, $5 \circ 1 = 0$. Thus, $\circ$ is not well-defined on $\mathbb{N}$.   □

For $\#$ to be a **unary operation** on a set S means that for any $x \in S$, $x^{\#}$ is well-defined, and S is closed under $\#$; in other words, for any $x \in S$, $x^{\#}$ exists, is unique, and is a member of S. We do not have a unary operation if any of these conditions is not met.

2.18 Example   Let $x^{\#}$ be defined by $x^{\#} = -x$, so that $x^{\#}$ is the negative of x. Then $\#$ is a unary operation on $\mathbb{Z}$, but not on $\mathbb{N}$ because $\mathbb{N}$ is not closed under $\#$.   □

From these examples it is clear that whether ∘ (or #) is a binary (or unary) operation depends not only on its definition but also on the set involved.

2.19 Practice Which of the following are neither binary nor unary operations on the given sets? Why not?

(a) $x \circ y = x \div y$; S = set of all positive integers
(b) $x \circ y = x \div y$; S = set of all positive rational numbers
(c) $x \circ y = x^y$; $S = \mathbb{R}$
(d) $x \circ y$ = maximum of x and y; $S = \mathbb{N}$
(e) $x^{\#} = \sqrt{x}$; S = set of all positive real numbers
(f) $x^{\#}$ = solution to equation $(x^{\#})^2 = x$; $S = \mathbb{C}$ □

2.20 Example Let S be the set of all unquantified statements of logic (propositions), as discussed in Section 1.1. Then the connectives $\wedge$, $\vee$, $\rightarrow$, and $\leftrightarrow$ are binary operations on S. The connective $'$ is a unary operation on S. □

So far, all of our binary operations have been defined by means of a description or an equation. Suppose S is a finite set, $S = \{x_1, x_2, \ldots, x_n\}$. Then a binary operation ∘ on S can be defined by an array, or table, where element i, j (ith row and jth column) denotes $x_i \circ x_j$.

2.21 Example Let $S = \{2, 5, 9\}$, and let ∘ be defined by the array

∘	2	5	9
2	2	2	9
5	5	9	2
9	5	5	9

Thus, $2 \circ 5 = 2$ and $9 \circ 2 = 5$. ∘ is a binary operation on S. □

Operations on Sets

We have generally seen operations where the elements operated upon are numbers, but we can also operate on sets. Given an arbitrary set S, we can define some binary and unary operations on the set $\mathcal{P}(S)$. (S is called the **universal set.**) A binary operation on $\mathcal{P}(S)$ must act on any two subsets of S to produce a unique subset of S. There are at least two natural ways in which this could happen.

2.22 Example Let S be the set of all students at Silicon U. Then the members of $\mathcal{P}(S)$ are sets of students. Let A be the set of computer science majors, and let B be the set of business majors. Both A and B belong to $\mathcal{P}(S)$. A new set of students can be defined as consisting of everybody who is majoring in either computer science or business (or both), and is called the union of A and B. Another new set can be defined as consisting of everybody who is majoring in both computer science and business. This set (which might be empty) is called the intersection of A and B.   □

**2.23
Definition**

Let $A, B \in \mathcal{P}(S)$. The **union** of A and B, denoted by $A \cup B$, is $\{x \mid x \in A$ or $x \in B\}$.   □

2.24 Practice

Let $A, B \in \mathcal{P}(S)$. Complete the following definition: The **intersection** of A and B, denoted by $A \cap B$, is $\{x \mid$ _____$\}$.   □

2.25 Example

Let $A = \{1, 3, 5, 7, 9\}$ and $B = \{3, 5, 6, 10, 11\}$. Here we may consider A and B as members of $\mathcal{P}(\mathbb{N})$. Then $A \cup B = \{1, 3, 5, 6, 7, 9, 10, 11\}$ and $A \cap B = \{3, 5\}$. Both $A \cup B$ and $A \cap B$ are members of $\mathcal{P}(\mathbb{N})$.   □

We can use *Venn diagrams* to visualize the binary operations of union and intersection. The shaded areas in Figures 2.1 and 2.2 indicate the set that results from performing the binary operation on the two given sets.

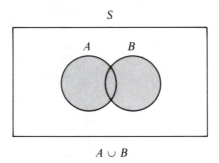

Figure 2.1   Figure 2.2

We will define one unary operation on $\mathcal{P}(S)$.

**2.26
Definition**

For a set $A \in \mathcal{P}(S)$, the **complement** of A, A', is $\{x \mid x \in S$ and $x \notin A\}$.   □

2.27 Practice

Illustrate A' in a Venn diagram.   □

Another binary operation on sets A and B in $\mathcal{P}(S)$ is **set difference**: $A - B = \{x \mid x \in A$ and $x \notin B\}$. This operation can be rewritten as $A - B = \{x \mid x \in A$ and $x \in B'\}$ and, finally, as $A - B = A \cap B'$.

2.28 Practice

Illustrate $A - B$ in a Venn diagram.   □

Two sets A and B such that $A \cap B = \varnothing$ are said to be **disjoint**. Thus, $A - B$ and $B - A$, for example, are disjoint sets.

2.29 Example   Let

$$A = \{x \,|\, x \text{ is an even nonnegative integer}\}$$
$$B = \{x \,|\, x = 2y + 1, y \in \mathbb{N}\}$$
$$C = \{x \,|\, x = 4y, y \in \mathbb{N}\}$$

be subsets of $\mathbb{N}$. Then A and B are disjoint sets, $A \cup B = \mathbb{N}$, $A \cup C = A$, $A' = B$, $A - C = \{x \,|\, x = 4y + 2, y \in \mathbb{N}\}$, and $B \subset C'$.   □

2.30 Practice   Let

$$A = \{1, 2, 3, 5, 10\}$$
$$B = \{2, 4, 7, 8, 9\}$$
$$C = \{5, 8, 10\}$$

be subsets of $S = \{1, 2, 3, 4, 5, 6, 7, 8, 9, 10\}$. Find

(a) $A \cup B$
(b) $A - C$
(c) $B' \cap (A \cup C)$ □

Set Identities

There are many set equalities involving the operations of union, intersection, difference, and complementation that are true for all subsets of a given set S. Because they are independent of the particular subsets used, these equalities are called set identities. Some basic set identities follow.

Basic Set Identities

1a. $A \cup B = B \cup A$	1b. $A \cap B = B \cap A$
2a. $(A \cup B) \cup C =$ $\quad A \cup (B \cup C)$	2b. $(A \cap B) \cap C =$ $\quad A \cap (B \cap C)$
3a. $A \cup (B \cap C) =$ $\quad (A \cup B) \cap (A \cup C)$	3b. $A \cap (B \cup C) =$ $\quad (A \cap B) \cup (A \cap C)$
4a. $A \cup \varnothing = A$	4b. $A \cap S = A$
5a. $A \cup A' = S$	5b. $A \cap A' = \varnothing$

(Note that 2a allows us to write $A \cup B \cup C$ with no need for parentheses; 2b allows us to write $A \cap B \cap C$.)

2.31 Example   Let's prove identity 3a. We might draw a Venn diagram for each side of the equation and see that they look the same. However, identity 3a is supposed to hold for all subsets A, B, and C, and whatever picture we draw cannot be completely general. Thus, if we draw A and B disjoint, that's a special case, but if we draw A and B not disjoint, that doesn't take care of the case where A and B are disjoint. To avoid drawing a picture for each case, let's prove set equality by proving set inclusion in each direction. Thus, we want to prove

$$A \cup (B \cap C) \subseteq (A \cup B) \cap (A \cup C)$$

and also

$$(A \cup B) \cap (A \cup C) \subseteq A \cup (B \cap C)$$

To show that $A \cup (B \cap C) \subseteq (A \cup B) \cap (A \cup C)$, we let x be an arbitrary member of $A \cup (B \cap C)$. Then we can proceed as follows:

$$x \in A \cup (B \cap C) \rightarrow x \in A \text{ or } x \in (B \cap C)$$
$$\rightarrow x \in A \text{ or } (x \in B \text{ and } x \in C)$$
$$\rightarrow x \in A \text{ or } x \in B \text{ and } x \in A \text{ or } x \in C$$
$$\rightarrow x \in A \cup B \text{ and } x \in A \cup C$$
$$\rightarrow x \in (A \cup B) \cap (A \cup C)$$

To show that $(A \cup B) \cap (A \cup C) \subseteq A \cup (B \cap C)$, we reverse the above argument. ☐

2.32 Practice   Prove identity 4a. ☐

These identities may have reminded you of the list of tautologies in Section 1.1, page 7. (If not, check back and compare.) We will discuss this similarity further in Chapter 5.

Once we have proved the set identities in this list, we can use them to prove other set identities, much as we use algebraic identities like $(x - y)^2 = x^2 - 2xy + y^2$ to rewrite algebraic expressions.

2.33 Example   We can use the basic set identities to prove

$$(A \cup (B \cap C)) \cap [(A' \cup (B \cap C)) \cap (B \cap C)'] = \varnothing$$

for A, B, and C any subsets of S. In the following proof, the number to the right is that of the identity used to validate each step.

$(A \cup (B \cap C)) \cap [(A' \cup (B \cap C)) \cap (B \cap C)']$

$$= [(A \cup (B \cap C)) \cap (A' \cup (B \cap C))] \cap (B \cap C)' \qquad \text{(2b)}$$
$$= [((B \cap C) \cup A) \cap ((B \cap C) \cup A')] \cap (B \cap C)' \qquad \text{(1a twice)}$$
$$= [(B \cap C) \cup (A \cap A')] \cap (B \cap C)' \qquad \text{(3a)}$$
$$= [(B \cap C) \cup \varnothing] \cap (B \cap C)' \qquad \text{(5b)}$$
$$= (B \cap C) \cap (B \cap C)' \qquad \text{(4a)}$$
$$= \varnothing \qquad \text{(5b)} \qquad ☐$$

The **dual** for each set identity in our list also appears in the list. The dual is obtained by interchanging $\cup$ and $\cap$ and interchanging S and $\varnothing$. The dual of the identity in Example 2.33 is $(A \cap (B \cup C)) \cup [(A' \cap (B \cup C)) \cup (B \cup C)'] = S$, which we could prove true by replacing each basic set identity used in the proof of Example 2.33 with its dual. Because this method always works, any time we have proved a set identity by using the basic identities, we have also proved its dual.

2.34 Practice (a) Using the basic set identities, establish the set identity

$$(C \cap (A \cup B)) \cup ((A \cup B) \cap C') = A \cup B$$

(A, B, and C are any subsets of S.)

(b) State the dual identity that you now know is true. ☐

There are two ways to prove a set identity: (1) establish set inclusion in each direction or (2) verify the identity (or its dual) by using already proven identities.

Cartesian Product

There is one final operation we will define using elements of $\mathcal{P}(S)$.

2.35 Definition Let A and B be subsets of S. The **Cartesian product (cross product)** of A and B, denoted by $A \times B$, is defined by: $A \times B = \{(x, y) \mid x \in A$ and $y \in B\}$. ☐

Thus, the Cartesian product of two sets A and B is the set of all ordered pairs whose first component comes from A and whose second comes from B. The cross product is not a binary operation on $\mathcal{P}(S)$. Although it acts on an ordered pair of members of $\mathcal{P}(S)$ and gives a unique result, the resulting set is not, in general, a subset of S, that is, it is not a member of $\mathcal{P}(S)$. The closure property for a binary operation fails to hold.

Because we will often be interested in the cross product of a set with itself, we will abbreviate $A \times A$ as A^2; in general, we use A^n to mean the set of all ordered n-tuples $(x_1, x_2, \ldots, x_n)$ of elements of A.

2.36 Practice Let $A = \{1, 2\}$ and $B = \{3, 4\}$.

(a) Find $A \times B$.
(b) Find $B \times A$.
(c) Find A^2.
(d) Find A^3. ☐

Countable and Uncountable Sets

In a finite set S, we can always designate one element as the first member, s_1, another element as the second member, s_2, and so forth. If there are k elements in the set, then these can be listed in the order we have selected:

$$s_1, s_2, \ldots, s_k$$

This list represents the entire set.

If the set is infinite, we may still be able to select a first element s_1, a second element s_2, and so forth, so that the list

$$s_1, s_2, s_3, \ldots$$

represents all elements of the set. Such an infinite set is said to be **denumerable.** Both finite and denumerable sets are **countable** sets because we can count, or enumerate, all of their elements. Being countable does not always mean that we can give a value for the total number of elements in the set, but that we can say, "Here is a first one," "Here is a second one," and so on through the set. There are, however, infinite sets that are **uncountable.** In an uncountable set, there is no way to count out the elements and get the whole set in the process. Before we prove that uncountable sets exist, let's look at some denumerable (countably infinite) sets.

2.37 Example   The set $\mathbb{N}$ is denumerable.

To prove denumerability, we need only exhibit a counting scheme. For the set $\mathbb{N}$ of nonnegative integers, it is clear that

$$0, 1, 2, 3, \ldots$$

is an enumeration that will eventually include every member of the set.

□

2.38 Practice   Prove that the set of even positive integers is denumerable.   □

2.39 Example   The set $\mathbb{Q}^+$ of positive rational numbers is denumerable.

We assume that each positive rational number is written as a fraction of positive integers. We can write all such fractions having the numerator 1 in one row, all those having the numerator 2 in a second row, and so on:

1/1	1/2	1/3	1/4	1/5	· · ·
2/1	2/2	2/3	2/4	2/5	· · ·
3/1	3/2	3/3	3/4	3/5	· · ·
4/1	4/2	4/3	4/4	4/5	· · ·

To show that the set of all fractions in this array is denumerable, we will thread an arrow through the entire array, beginning with 1/1; following the arrow gives an enumeration of the set. Thus the fraction 1/3 is the fourth member in this enumeration:

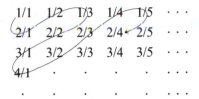

Therefore the set represented by the array is denumerable. To obtain an enumeration of $\mathbb{Q}^+$, we use the enumeration of the above set but eliminate any fractions not in lowest terms. This avoids the problem of listing both 1/2 and 2/4, for example, which represent the same positive rational. The enumeration of $\mathbb{Q}^+$ thus begins with

1/1, 2/1, 1/2, 1/3, 3/1, 4/1, . . .

For example, we have eliminated 2/2, which reduces to 1/1. □

2.40 Practice   What is the 10th fraction in the above enumeration of fractions? □

Now let's show that there is an infinite set that is not denumerable. The proof technique that seems appropriate to prove that set *A* does *not* have property *B* is to assume that *A* does have property *B* and look for a contradiction. The proof below is a very famous proof by contradiction known as **Cantor's diagonalization method,** after Georg Cantor, the "father of set theory."

2.41 Example   We will show that the set of all the real numbers between 0 and 1 is uncountable. We will write such numbers in decimal form; thus any member of the set can be written as

$0.d_1d_2d_3$ . . .

A number such as 0.24999999 . . . can be written in alternative form as 0.2500000 . . . . In order to avoid writing the same element twice, we will choose (arbitrarily) to always use the former representation and not the latter. Now let us assume that our set is countable. Therefore some enumeration of the set exists. We can depict an enumeration of the set as follows, where d_{ij} is the jth decimal digit in the ith number in the enumeration:

$$0.d_{11}d_{12}d_{13} \ldots$$
$$0.d_{21}d_{22}d_{23} \ldots$$
$$0.d_{31}d_{32}d_{33} \cdots$$

$$\begin{matrix} \cdot & \cdot & \cdot & \cdot \\ \cdot & \cdot & \cdot & \cdot \\ \cdot & \cdot & \cdot & \cdot \end{matrix}$$

We now construct a real number $p = 0.p_1p_2p_3 \ldots$ as follows: p_i is always chosen to be 5 if $d_{ii} \neq 5$ and 6 if $d_{ii} = 5$. Thus p is a real number between 0 and 1. For instance, if the enumeration begins with

0.342134 . . .

0.257001 . . .

0.546122 . . .

0.716525 . . .

then $d_{11} = 3$, $d_{22} = 5$, $d_{33} = 6$, and $d_{44} = 5$, so $p_1 = 5$, $p_2 = 6$, $p_3 = 5$, and $p_4 = 6$. Thus p begins with 0.5656 . . . .

If we compare p with the enumeration of the set, p differs from the first number at the first decimal digit, from the second number at the second decimal digit, from the third number at the third decimal digit, and so on. Therefore p does not agree with any of the representations in the enumeration. Furthermore, because p contains no 0s to the right of the decimal, it is not the alternative representation of any of the numbers in the enumeration. Therefore p is a different real number between 0 and 1 than any in the enumeration, yet the enumeration was supposed to include all members of the set. Here, then, is the contradiction, and the set of all real numbers between 0 and 1 is indeed uncountable. $\square$

Why is the proof in Example 2.41 called a "diagonalization" method?

Although it is interesting and perhaps surprising to learn that there are uncountable sets, we will usually be concerned with countable sets.

✔ Checklist

Definitions

equal sets (*p. 72*)
empty set (*p. 73*)
null set (*p. 73*)
subset (*p. 73*)
proper subset (*p. 73*)

power set (*p. 74*)
ordered pair (*p. 75*)
binary operation (*p. 76*)
well-defined operation (*p. 76*)
closed set under an operation (*p. 76*)
unary operation (*p. 76*)
universal set (*p. 77*)
union of sets (*p. 78*)
intersection of sets (*p. 78*)
complement of a set (*p. 78*)
set difference (*p. 78*)
disjoint sets (*p. 78*)
dual of a set identity (*p. 81*)
Cartesian product (cross product) of sets (*p. 81*)
denumerable set (*p. 82*)
countable set (*p. 82*)
uncountable set (*p. 82*)
Cantor's diagonalization method (*p. 83*)

Techniques

Describe sets by a list of elements and by a defining property.

Prove that one set is a subset of another.

Find the power set of a set.

Check whether something is a binary or unary operation on a set.

Form new sets by taking the union, intersection, complement, and cross product of sets.

Prove set identities by (1) showing set inclusion in each direction or (2) using the basic set identities.

Demonstrate the denumerability of a denumerable set.

Use the Cantor diagonalization method to prove certain sets uncountable.

Main Ideas

Sets and how they can be related (equal, subset, etc.) and combined (union, intersection, etc.).

Notation for standard sets.

The power set of a set with n elements, which has 2^n elements.

Binary and unary operations on sets.

Basic set identities exist (in dual pairs) and can be used to prove other set identities; once an identity is proved in this manner, its dual is also true.

Countable sets can be enumerated, and uncountable sets exist.

Exercises Section 2.1

★ 1. Let $S = \{2, 5, 17, 27\}$. Which of the following are true?
   (a) $5 \in S$
   (b) $2 + 5 \in S$
   (c) $\varnothing \in S$
   (d) $S \in S$

2. Let $B = \{x \mid x \in \mathbb{Q} \text{ and } -1 < x < 2\}$. Which of the following are true?
   (a) $0 \in B$
   (b) $-1 \in B$
   (c) $-0.84 \in B$
   (d) $\sqrt{2} \in B$

3. List the elements of each set.
   ★ (a) $\{x \mid x \in \mathbb{N} \text{ and } x^2 < 25\}$
   ★ (b) $\{x \mid x = y + 3 \text{ for } y \in \{0, 4, 7\}\}$
   (c) $\{x \mid x \text{ is one of the first three U.S. presidents}\}$
   (d) $\{x \mid x \in \mathbb{R} \text{ and } x^2 = -1\}$
   (e) $\{x \mid x \in \mathbb{R} \text{ and } x^2 - 2x - 3 = 0\}$
   (f) $\{x \mid x \in \mathbb{Z} \text{ and } |x| < 4\}$ ($|x|$ denotes the absolute value function)

4. Give the defining property of each set.
   (a) $\{1, 2, 3, 4, 5\}$
   (b) $\{1, 3, 5, 7, 9, 11, \ldots\}$
   (c) $\{\text{Melchior, Gaspar, Balthazar}\}$
   (d) $\{0, 1, 10, 11, 100, 101, 110, 111, 1000, \ldots\}$

★ 5. Given the description of a set A as $A = \{2, 4, 8, \ldots\}$, do you think $16 \in A$? (Check the answer given for this problem.)

6. Let

$$A = \{x \mid x \in \mathbb{N} \text{ and } 1 < x < 50\}$$
$$B = \{x \mid x \in \mathbb{R} \text{ and } 1 < x < 50\}$$
$$C = \{x \mid x \in \mathbb{Z} \text{ and } |x| \geq 25\}$$

Which of the following statements are true?
   (a) $A \subseteq B$
   (b) $17 \in A$
   (c) $A \subseteq C$
   (d) $-40 \in C$
   (e) $\sqrt{3} \in B$

(f) $\{0, 1, 2\} \subseteq A$

(g) $\varnothing \in B$

(h) $\{x \mid x \in \mathbb{Z} \text{ and } x^2 > 625\} \subseteq C$

7. Let

$$R = \{1, 3, \pi, 4.1, 9, 10\}$$
$$S = \{\{1\}, 3, 9, 10\}$$
$$T = \{1, 3, \pi\}$$
$$U = \{\{1, 3, \pi\}, 1\}$$

Which of the following are true? For those that are not, why not?

⋆(a) $S \subseteq R$ ⋆(b) $1 \in R$

⋆(c) $1 \in S$ ⋆(d) $1 \subseteq U$

⋆(e) $\{1\} \subseteq T$ ⋆(f) $\{1\} \subseteq S$

(g) $T \subset R$ (h) $\{1\} \in S$

(i) $\varnothing \subseteq S$ (j) $T \subseteq U$

(k) $T \in U$ (l) $T \notin R$

(m) $T \subseteq R$ (n) $S \subseteq \{1, 3, 9, 10\}$

8. Which of the following are true, where A, B, and C represent arbitrary sets?

⋆(a) If $A \subseteq B$ and $B \subseteq A$, then $A = B$ ⋆(b) $\{\varnothing\} = \varnothing$

⋆(c) $\{\varnothing\} = \{0\}$ ⋆(d) $\varnothing \in \{\varnothing\}$

⋆(e) $\varnothing \subseteq A$ ⋆(f) $\varnothing \in A$

(g) $\{\varnothing\} = \{\{\varnothing\}\}$ (h) If $A \subset B$ and $B \subseteq C$, then $A \subset C$

(i) If $A \neq B$ and $B \neq C$, then $A \neq C$ (j) If $A \in B$ and $B \nsubseteq C$, then $A \notin C$

9. Program QUAD finds and prints solutions to quadratic equations of the form $ax^2 + bx + c = 0$. Program EVEN lists all the even integers from $-2n$ to $2n$. Let Q denote the set of values output by QUAD and E denote the set of values output by EVEN.

(a) Show that for $a = 1$, $b = -2$, $c = -24$, and $n = 50$, $Q \subseteq E$.

(b) Show that for the same values of a, b, and c, but a value for n of 2, $Q \nsubseteq E$.

10. Find $\mathscr{P}(S)$ for $S = \{a, b\}$.

11. Find $\mathscr{P}(S)$ for $S = \{1, 2, 3, 4\}$. How many elements do you expect this set to have?

12. (a) Find $\mathscr{P}(S)$ for $S = \{\varnothing, \{\varnothing\}, \{\varnothing, \{\varnothing\}\}\}$.

(b) Find $\mathscr{P}(\mathscr{P}(S))$ for $S = \{a, b\}$.

13. Solve for x and y.

(a) $\{x, x + 2\} = \{5, 3\}$

(b) $(2x, y) = (16, 7)$

(c) $(2x - y, x + y) = (-2, 5)$

14. (a) Recall that ordered pairs must have the property that $(x, y) = (u, v)$ if and only if $x = u$ and $y = v$. Prove that $\{\{x\}, \{x, y\}\} = \{\{u\}, \{u, v\}\}$ if and only if $x = u$ and $y = v$. Therefore, although we know that $(x, y) \neq \{x, y\}$, we can define the ordered pair (x, y) as the set $\{\{x\}, \{x, y\}\}$.

(b) Show by an example that we cannot define the ordered triple (x, y, z) as the set $\{\{x\}, \{x, y\}, \{x, y, z\}\}$.

15. Which of the following are binary or unary operations on the given sets? For those that are not, why not?

⋆ (a) $x \circ y = x + 1$; $S = \mathbb{N}$

⋆ (b) $x \circ y = x + y - 1$; $S = \mathbb{N}$

⋆

(c) $x \circ y = \begin{cases} x - 1 & \text{if } x \text{ is odd} \\ x & \text{if } x \text{ is even} \end{cases}$ $S = \mathbb{Z}$

(d) $x^{\#} = \ln x$; $S = \mathbb{R}$

(e) $x^{\#} = x^2$; $S = \mathbb{Z}$

(f)

$\circ$	1	2	3
1	1	2	3
2	2	3	4
3	3	4	5

$S = \{1, 2, 3\}$

(g) $x \circ y = $ the taller of x and y; $S = $ set of all people living in Arkansas

(h) $x \circ y = $ that person, x or y, whose name appears first in an alphabetical sort; $S = $ set of 10 people with different names

(i) $x \circ y = \begin{cases} 1/x & \text{if } x \text{ is positive} \\ 1/(-x) & \text{if } x \text{ is negative} \end{cases}$ $S = \mathbb{R}$

(j) $x \circ y = xy$; $S = $ set of all finite-length strings of symbols from the set $\{p, q, r\}$

16. How many different binary operations can be defined on a set with n elements? (Hint: think about filling in a table.)

17. We have written binary operations in **infix** notation, where the operation symbol appears between the two operands, as in $A + B$. Evaluation of a complicated arithmetic expression is more efficient when the operations are written in **postfix** notation, where the operation symbol appears after the two operands, as in $AB+$. Many compilers change expressions in a computer program from infix to postfix form. One way to produce an equivalent postfix expression from an infix expression is to write the infix expression with a full set of parentheses, move each operator to replace its corresponding right parenthesis, and then eliminate all left parentheses. Thus,

$$A * B + C$$

becomes, when fully parenthesized,

$$((A * B) + C)$$

and the postfix notation is

$$AB * C +$$

Rewrite each of the following in postfix notation:
(a) $(A + B) * (C - D)$
(b) $A ** B - C * D$
(c) $A * C + B/(C + D * B)$

18. Let

$$A = \{p, q, r, s\}$$
$$B = \{r, t, v\}$$
$$C = \{p, s, t, u\}$$

be subsets of $S = \{p, q, r, s, t, u, v, w\}$. Find
(a) $B \cap C$ (b) $A \cup C$
(c) C' (d) $A \cap B \cap C$
(e) $B - C$ (f) $(A \cup B)'$
(g) $A \times B$ (h) $(A \cup B) \cap C'$

19. Let

$$A = \{2, 4, 5, 6, 8\}$$
$$B = \{1, 4, 5, 9\}$$
$$C = \{x \mid x \in \mathbb{Z} \text{ and } 2 \le x < 5\}$$

be subsets of $S = \{0, 1, 2, 3, 4, 5, 6, 7, 8, 9\}$. Find:
★(a) $A \cup B$ ★(b) $A \cap B$
★(c) $A \cap C$ (d) $B \cup C$
(e) $A - B$ (f) A'
(g) $A \cap A'$ ★(h) $(A \cap B)'$
(i) $C - B$ (j) $(C \cap B) \cup A'$
(k) $(B - A)' \cap (A - B)$ (l) $(C' \cup B)'$
(m) $B \times C$

20. Consider the following subsets of $\mathbb{Z}$:

$$A = \{x \mid x = 3y \text{ for } y \in \mathbb{Z} \text{ and } y \ge 4\}$$
$$B = \{x \mid x = 2y \text{ for } y \in \mathbb{Z}\}$$
$$C = \{x \mid x \in \mathbb{Z} \text{ and } |x| \le 10\}$$

Describe each of the following sets in terms of A, B, and C and set operations:
(a) set of all odd integers
(b) $\{-10, -8, -6, -4, -2, 0, 2, 4, 6, 8, 10\}$
(c) $\{x \mid x = 6y \text{ for } y \in \mathbb{Z} \text{ and } y \ge 2\}$
(d) $\{-9, -7, -5, -3, -1, 1, 3, 5, 7, 9\}$
(e) $\{x \mid x = 2y + 1 \text{ for } y \in \mathbb{Z} \text{ and } y \ge 5\} \cup \{x \mid x = 2y - 1 \text{ for } y \in \mathbb{Z} \text{ and } y \le -5\}$

21. Which of the following are true, where A and B represent arbitrary sets?

  ★(a) $A \cup A = A$

  ★(c) $(A \cap B)' = A' \cap B'$

  ★(e) $A - B = (B - A)'$

  (g) If $A \cap B = \varnothing$, then $A \subset B$

  (i) $\varnothing \times A = \varnothing$

  (b) $B \cap B = B$

  (d) $(A')' = A$

  (f) $(A - B) \cap (B - A) = \varnothing$

  (h) $B \times A = A \times B$

  (j) $\varnothing \cap \{\varnothing\} = \varnothing$

22. For each of the following statements, find conditions on sets A and B to make the statement true:

  ★(a) $A \cup B = A$

  (c) $A \cup \varnothing = \varnothing$

  (e) $A \cup B \subseteq A \cap B$

  (b) $A \cap B = A$

  (d) $B - A = \varnothing$

23. A binary operation on sets called the **symmetric difference** is defined by $A \oplus B = (A - B) \cup (B - A)$.

  (a) Draw a Venn diagram to illustrate $A \oplus B$.

  (b) For $A = \{3, 5, 7, 9\}$ and $B = \{2, 3, 4, 5, 6\}$, what is $A \oplus B$?

  (c) Prove that $A \oplus B = (A \cup B) - (A \cap B)$ for arbitrary sets A and B.

  (d) For an arbitrary set A, what is $A \oplus A$? What is $\varnothing \oplus A$?

  (e) Prove that $A \oplus B = B \oplus A$ for arbitrary sets A and B.

  (f) For any sets A, B, and C, prove that $(A \oplus B) \oplus C = A \oplus (B \oplus C)$.

24. The operations of set union and set intersection can be defined as n-ary operations.

  (a) Give a definition similar to Definition 2.23 for $A_1 \cup A_2 \cup \cdots \cup A_n$ where n is any positive integer.

  (b) Give a recursive definition for $A_1 \cup A_2 \cup \cdots \cup A_n$ for any positive integer n.

25. If S is any finite set, we denote the number of elements in S by $|S|$. Then for A and B finite sets, $|A \cup B| = |A| + |B| - |A \cap B|$ (see part (a) below). This equation, which can be extended by induction to the case $|A_1 \cup A_2 \cup \cdots \cup A_n|$, is called the **Principle of Inclusion and Exclusion.**

  (a) Show that for A and B finite sets, $|A \cup B| = |A| + |B| - |A \cap B|$.

  (b) Show that for finite sets A, B, and C,

$$|A \cup B \cup C| = |A| + |B| + |C| - |A \cap B|$$
$$- |A \cap C| - |B \cap C| + |A \cap B \cap C|$$

  (Hint: write $A \cup B \cup C$ as $A \cup (B \cup C)$.)

  (c) Thirty-five students in a programming class did the same programming assignment on the same machine. These students fell into the following categories: 19 wrote programs that ran in ≤ 2 minutes; 19 wrote programs that produced an output of ≤ 15

pages; and 15 wrote programs with ≤ 5 variable names. Also, 9 students wrote programs that ran in ≤ 2 minutes and produced an output of ≤ 15 pages; 7 students wrote programs that produced an output of ≤ 15 pages and used ≤ 5 variable names; 6 students wrote programs that ran in ≤ 2 minutes and used ≤ 5 variable names. How many students wrote programs that ran in ≤ 2 minutes, produced ≤ 15 pages of output, and used ≤ 5 variable names?

26. Verify the basic set identities on page 79 by showing set inclusion in each direction. (We have already done 3a and 4a.)

27. A and B are subsets of a set S. Prove the following set identities by showing set inclusion in each direction:
    (a) $(A \cup B)' = A' \cap B'$ ⎫
    (b) $(A \cap B)' = A' \cup B'$ ⎬ DeMorgan's Laws
    (c) $A \cup (B \cap A) = A$
    (d) $(A \cap B')' \cup B = A' \cup B$
    (e) $(A \cap B) \cup (A \cap B') = A$

28. Prove that for subsets $A_1, A_2, \ldots, A_n$ of a set S, the following identities hold, where $n \geq 2$. (See Exercise 27.)
    (a) $(A_1 \cup A_2 \cup \cdots \cup A_n)' = A_1' \cap A_2' \cap \cdots \cap A_n'$
    (b) $(A_1 \cap A_2 \cap \cdots \cap A_n)' = A_1' \cup A_2' \cup \cdots \cup A_n'$

29. (a) A, B, and C are subsets of a set S. Prove the following set identities using the basic set identities listed in this section:
    ★(1) $(A \cup B) \cap (A \cup B') = A$
    (2) $[((A \cap C) \cap B) \cup ((A \cap C) \cap B')] \cup (A \cap C)' = S$
    (3) $(A \cup C) \cap [(A \cap B) \cup (C' \cap B)] = A \cap B$
    (b) State the dual of each of the above identities.

30. The operations of set union and set intersection can be extended to apply to an infinite family of sets. We may describe the family as the collection of all sets A_i, where i takes on any of the values of a fixed set I. Here, I is called the **index set** for the family. The union of the family, $\bigcup_{i \in I} A_i$, is defined by $\bigcup_{i \in I} A_i = \{x \mid x$ is a member of some $A_i\}$. The intersection of the family, $\bigcap_{i \in I} A_i$, is defined by $\bigcap_{i \in I} A_i = \{x \mid x$ is a member of each $A_i\}$.
    (a) Let $I = \{1, 2, 3, \ldots\}$, and for each $i \in I$, let A_i be the set of real numbers in the interval $(-1/i, 1/i)$. What is $\bigcup_{i \in I} A_i$? What is $\bigcap_{i \in I} A_i$?
    (b) Let $I = \{1, 2, 3, \ldots\}$, and for each $i \in I$, let A_i be the set of real numbers in the interval $[-1/i, 1/i]$. What is $\bigcup_{i \in I} A_i$? What is $\bigcap_{i \in I} A_i$?

★ 31. Prove that the set of odd positive integers is denumerable.

32. Prove that the set $\mathbb{Z}$ of all integers is denumerable.

33. Prove that the set of all finite-length strings of the letter "a" is denumerable.

34. Prove that the set $\mathbb{Z} \times \mathbb{Z}$ is denumerable.

35. Use Cantor's diagonalization method to show that the set of all infinite sequences of positive integers is not countable.

36. Explain why any subset of a countable set is countable.

37. Explain why the union of any two denumerable sets is denumerable.

38. Sets can have sets as elements (see Exercise 7, for example). Let B be the set defined as follows:

$$B = \{S \mid S \text{ is a set and } S \notin S\}$$

Argue that both $B \in B$ and $B \notin B$ are true. This contradiction is called **Russell's paradox,** after the famous philosopher and mathematician Bertrand Russell, who stated it in 1901. (A carefully constructed axiomatization of set theory puts some restrictions on what can be called a set. All ordinary sets are still sets, but peculiar sets that get us into trouble, like B above, seem to be avoided.)

SECTION 2.2   COUNTING

Combinatorics is that branch of mathematics that deals with counting. Often we want to know how many members there are in some finite set. This seemingly trivial question can be difficult to answer. We have already answered two "how many" questions—how many rows are there in a truth table with n statement letters, and how many subsets are there in a set with n elements? (Actually, as we've noted, these can be thought of as the same question.) Counting questions are important whenever we have finite resources (How much storage does a particular computer program require? How many users can a given computer configuration support?), or whenever we are interested in efficiency (How many computations does a particular algorithm involve? See the discussion on analysis of algorithms in Section 1.5). Counting is also the basis for probability and statistics.

The Multiplication Principle

Many counting problems can be solved by applying the **Multiplication Principle.** Before stating this principle, let's consider the tree of Figure 2.3. Remember we used it in Chapter 1 to count the number of rows in a truth table with n statement letters. There are two possible outcomes when choosing the truth value of the first statement letter, two outcomes when choosing the truth value of the second statement letter, and so on. The tree structure illustrates that the total number of possible outcomes is the *product* of the number of outcomes for each step. This type of reasoning is behind the Multiplication Principle.

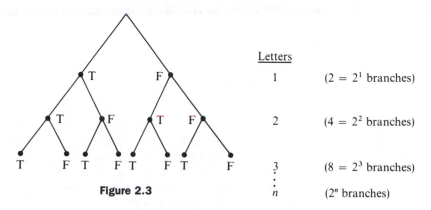

	Letters	
	1	$(2 = 2^1$ branches$)$
	2	$(4 = 2^2$ branches$)$
	3	$(8 = 2^3$ branches$)$
	n	$(2^n$ branches$)$

Figure 2.3

2.42 Definition

Multiplication Principle: If there are n_1 possible outcomes for a first event and then n_2 possible outcomes for a second event, there are $n_1 \cdot n_2$ possible outcomes for the sequence of the two events.   □

The Multiplication Principle can be extended by induction to apply to a sequence of any finite number of events.

The Multiplication Principle is useful whenever we want to count the total number of possible outcomes for a task that can be broken down into a sequence of successive subtasks.

2.43 Example

The last part of your telephone number contains 4 digits. How many 4-digit numbers are there?

We can think of the number of 4-digit numbers as the total number of possible outcomes for the sequence of subtasks of choosing the first digit, then the next digit, then the third, and finally the fourth digit. The first digit can be any one of the 10 digits from 0 to 9, so there are 10 possible outcomes for the first subtask. Likewise, there are 10 different possibilities for the second digit, and for the third and for the fourth. Using the Multiplication Principle, we multiply the number of outcomes for each subtask in the sequence. Therefore there are $10 \cdot 10 \cdot 10 \cdot 10 = 10,000$ different numbers.   □

The number of possible outcomes for successive events after the first event is affected if the same element cannot be used again, that is, if repetitions are not allowed.

2.44 Example   How many 4-digit numbers are there if the same digit cannot be used twice?

Again we have the sequence of subtasks of selecting the 4 digits, but no repetitions are allowed. There are 10 choices for the first digit, only 9 choices for the second because we can't use what we used before, and so on. There are $10 \cdot 9 \cdot 8 \cdot 7 = 5040$ different numbers. □

2.45 Example   (a) How many ways are there to choose 3 officers from a club of 25 people?

(b) How many ways are there to choose 3 officers from a club of 25 people if someone can hold more than one office?

In (a), there are three successive subtasks with no repetitions. The first subtask, choosing the first officer, has 25 possible outcomes. The second subtask has 24 outcomes, the third 23 outcomes. The total number of outcomes is $25 \cdot 24 \cdot 23 = 13,800$. In (b), the same three subtasks are done in succession, but repetitions are allowed. The total number of outcomes is $25 \cdot 25 \cdot 25 = 15,625$. □

2.46 Practice   If a man has 4 suits, 8 shirts, and 5 ties, how many outfits can he put together? □

The Addition Principle

Suppose we want to select a dessert from 3 pies and 4 cakes. In how many ways can this be done? There are two events here, one with 3 outcomes (choosing a pie) and one with 4 outcomes (choosing a cake). However, we are not doing a sequence of two events here, since we are only getting one dessert, which must be chosen from the two disjoint sets of possibilities. The number of different outcomes is the total number of choices we have, $3 + 4 = 7$. This illustrates the **Addition Principle.**

2.47 Definition   **Addition Principle:** If A and B are disjoint events with n_1 and n_2 possible outcomes, respectively, then the total number of possible outcomes for event A or B is $n_1 + n_2$. □

The Addition Principle can be extended by induction to the case of any finite number of disjoint events.

2.48 Example   A customer wants to purchase a vehicle from a dealer. The dealer has 23 autos and 14 trucks in stock. How many selections does the customer have?

The customer wants to choose a car or truck. These are disjoint events; choosing an auto has 23 outcomes, and choosing a truck has 14 outcomes. By the Addition Principle, choosing a vehicle has $23 + 14 = 37$ outcomes. $\square$

Notice the requirement that the outcomes for events A and B be disjoint sets. Thus if a customer wanted to purchase a vehicle from a dealer who had 23 autos, 14 trucks, and 17 red vehicles in stock, we could not conclude that the customer had $23 + 14 + 17$ choices!

The Addition Principle is useful whenever we want to count the total number of possible outcomes for a task that can be broken down into disjoint cases. Frequently the Addition Principle is used in conjunction with the Multiplication Principle.

2.49 Example How many 4-digit numbers begin with a 4 or a 5?

We can consider the two disjoint cases—numbers that begin with 4 and numbers that begin with 5. Counting the numbers that begin with 4, there is 1 outcome for the subtask of choosing the first digit, then 10 possible outcomes for the subtasks of choosing each of the other three digits. Hence by the Multiplication Principle there are $1 \cdot 10 \cdot 10 \cdot 10 = 1000$ ways to get a 4-digit number beginning with 4. The same reasoning shows that there are 1000 ways to get a 4-digit number beginning with 5. By the Addition Principle, there are $1000 + 1000 = 2000$ total possible outcomes. $\square$

Often a counting problem can be solved in more than one way. Although the possibility of a second solution might seem confusing, it provides an excellent way to check our work—if two different ways of looking at the problem produce the same answer, it increases our confidence that we have analyzed the problem correctly.

2.50 Example Consider the problem of Example 2.49 again. We can avoid using the Addition Principle by thinking of the problem as four successive subtasks, where the first subtask, choosing the first digit, has 2 possible outcomes—choosing a 4 or choosing a 5. Then there are $2 \cdot 10 \cdot 10 \cdot 10 = 2000$ possible outcomes. $\square$

2.51 Practice If a woman has 7 blouses, 5 skirts, and 9 dresses, how many different outfits does she have? $\square$

✔ Checklist

Definitions

combinatorics (*p. 92*)
Multiplication Principle (*p. 93*)
Addition Principle (*p. 94*)

Techniques

Use the Multiplication Principle and the Addition Principle for counting the number of objects in a finite set.

Main Ideas

The Multiplication Principle is used to count the number of possible outcomes for a sequence of subtasks, whereas the Addition Principle is used to count the number of possible outcomes for disjoint cases.

Exercises Section 2.2

★ 1. A frozen yogurt shop allows you to choose one flavor (vanilla, strawberry, lemon, cherry, or peach), one topping (chocolate shavings, crushed toffee, or crushed peanut brittle), and one condiment (whipped cream or shredded coconut). How many different desserts are possible?

★ 2. In Exercise 1, how many dessert choices do you have if you are allergic to strawberries and chocolate?

3. A video game on a microcomputer is begun by choosing from three menus. The first menu (number of players) has 4 selections, the second menu (level of play) has 8, and the third menu (speed) has 6. In how many configurations can the game be played?

4. A multiple choice exam has 20 questions, each with 4 possible answers, and 10 additional questions, each with 5 possible answers. How many different answer sheets are possible?

5. A user's password on a large computer system consists of 3 letters followed by 2 digits. How many different passwords are possible?

6. On the computer system of Exercise 5, how many passwords are possible if uppercase and lowercase letters can be distinguished?

★ 7. A telephone conference call is being placed from Central City to Booneville by way of Cloverdale. There are 45 trunk lines from Central City to Cloverdale, and 13 from Cloverdale to Booneville. How many different ways can the call be placed?

8. How many Social Security numbers are possible?

★ 9. A palindrome is a word that reads the same forwards and backwards. How many 5-letter, one-word, English language palindromes are possible?

10. How many 3-digit numbers less than 600 can be made using the digits 8, 6, 4, and 2?

⋆ 11. An identifier in BASIC must be either a single letter or a letter followed by a single digit. How many identifiers are possible?

12. A president and vice-president must be chosen for the executive committee of an organization. There are 17 volunteers from the Eastern Division, and 24 volunteers from the Western Division. If both officers must come from the same division, in how many ways can the officers be selected?

⋆ 13. How many 8-length binary strings are there? (Each character is either the digit 0 or the digit 1.)

14. How many 8-length binary strings begin and end with 0?

⋆ 15. How many 8-length binary strings begin or end with 0?

16. How many 8-length binary strings have 1 as the second digit?

17. How many 8-length binary strings begin with 111?

18. How many 8-length binary strings contain exactly one 0?

19. How many 8-length binary strings begin with 10 or have a 0 as the third digit?

20. How many 8-length binary strings are palindromes? (See Exercise 9.)

In Exercises 21 through 28, a hand consists of 1 card drawn from a standard 52-card deck with flowers on the back, and 1 card drawn from a standard 52-card deck with birds on the back.

⋆ 21. How many different hands are possible?

22. How many hands consist of a pair of aces?

⋆ 23. How many hands consist of two of a kind?

24. How many hands contain exactly one king?

25. How many hands have a face value of 5 (aces count as 1).

26. How many hands have a face value of less than 5?

27. How many hands do not contain any face cards?

28. How many hands contain at least one face card?

29. Prove by induction that if there are m events, $m \geq 2$, with possible outcomes of $n_1, n_2, \ldots n_m$, respectively, then there are $n_1 \cdot n_2 \cdots n_m$ possible outcomes for the sequence of these events.

30. How is the Addition Principle related to the Principle of Inclusion and Exclusion? (See Exercise 25, Section 2.1.)

SECTION 2.3   PERMUTATIONS AND COMBINATIONS

Permutations

In Section 2.2, we discussed the problem of counting all of the 4-digit numbers with no repeated digits (Example 2.44). In this problem, the number 1259 is not the same as the number 2951 because the order of the 4 digits is important. An ordered arrangement of objects is called a **permutation.** Determining the number of 4-digit numbers having no repeated digits can be viewed as counting the number of permutations, or arrangements, there are of 4 distinct objects chosen from a set of 10 objects (the digits). The answer, found by using the Multiplication Principle, is $10 \cdot 9 \cdot 8 \cdot 7$. In general, the number of permutations of r distinct objects chosen from n objects is denoted by $P(n, r)$. Therefore the solution to the problem of the 4-digit number without repeated digits can be expressed as $P(10, 4)$.

A formula for $P(n, r)$ can be written using the factorial function. For a positive integer n, **n factorial** is defined as $n(n - 1)(n - 2) \cdots 1$ and denoted by $n!$; also, 0! is defined to have the value 1. From the definition of $n!$, we see that

$$n! = n(n - 1)!$$

and that for $r < n$,

$$\frac{n!}{(n - r)!} = \frac{n(n - 1) \cdots (n - r + 1)(n - r)!}{(n - r)!}$$

$$= n(n - 1) \cdots (n - r + 1)$$

Using the factorial function,

$$P(10, 4) = 10 \cdot 9 \cdot 8 \cdot 7$$

$$= \frac{10 \cdot 9 \cdot 8 \cdot 7 \cdot 6 \cdot 5 \cdot 4 \cdot 3 \cdot 2 \cdot 1}{6 \cdot 5 \cdot 4 \cdot 3 \cdot 2 \cdot 1} = \frac{10!}{6!} = \frac{10!}{(10 - 4)!}$$

In general, $P(n, r)$ is given by the formula

$$P(n, r) = \frac{n!}{(n - r)!}$$

As a special case, the number of permutations of all n objects is $P(n, n) = n!/0! = n!$

2.52 Example   The value of $P(7, 3)$ is

$$\frac{7!}{(7 - 3)!} = \frac{7!}{4!} = \frac{7 \cdot 6 \cdot 5 \cdot 4 \cdot 3 \cdot 2 \cdot 1}{4 \cdot 3 \cdot 2 \cdot 1} = 7 \cdot 6 \cdot 5 = 210$$

$\square$

2.53 Example The number of permutations of 3 objects, say *a*, *b*, and *c*, is given by $P(3, 3) = 3! = 3 \cdot 2 \cdot 1 = 6$. The six permutations of *a*, *b*, and *c* are

$$abc \quad acb \quad bac \quad bca \quad cab \quad cba \qquad \Box$$

2.54 Example How many 3-letter words (not necessarily meaningful) can be formed from the word "compiler" if no letters can be repeated? Here the arrangement of letters matters, and we want to know the number of permutations of 3 distinct objects taken from 8 objects. The answer is $P(8, 3) = 8!/5! = 336$. $\qquad \Box$

Note that we could have solved Example 2.54 just by using the Multiplication Principle—there are 8 choices for the first letter, 7 for the second letter, and 6 for the third letter, so the answer is $8 \cdot 7 \cdot 6 = 336$. $P(n, r)$ simply gives us a new way to think about the problem, as well as a compact notation.

2.55 Practice In how many ways can a president and vice-president be selected from a group of 20 people? $\qquad \Box$

2.56 Practice In how many ways can 6 people be seated in a row of 6 chairs? $\qquad \Box$

Counting problems can have counting problems as subtasks.

2.57 Example A library has 4 books on operating systems, 7 on programming, and 3 on data structures. Let's see how many ways these books can be arranged on a shelf, given that all books on the same subject must be together. We can think of this problem as a sequence of subtasks. First we consider the subtask of arranging the 3 subjects. There are 3! outcomes to this subtask, that is, 3! different orderings of subject matter. The next subtasks are arranging the books on operating systems (4! outcomes), then arranging the books on programming (7! outcomes), and finally arranging the books on data structures (3! outcomes). Thus, by the Multiplication Principle, the final number of arrangements of all the books is $(3!)(4!)(7!)(3!) = 4,354,560$. $\qquad \Box$

Combinations

Sometimes we want to select *r* objects from a set of *n* objects, but we don't care how they are arranged. Then we are counting the number of **combinations** of *r* distinct objects chosen from *n* objects, denoted by $C(n, r)$. For each such combination, there are *r*! ways to permute the *r* objects. By the Multiplication Principle, the number of permutations of *r* distinct objects chosen from *n* objects is the product of the number of ways to choose the objects, $C(n, r)$, multiplied by the number of ways to arrange the objects chosen, $r!$. Thus,

$$C(n, r) \cdot r! = P(n, r)$$

or

$$C(n, r) = \frac{P(n, r)}{r!} = \frac{n!}{r!(n - r)!}$$

2.58 Example   The value of $C(7, 3)$ is

$$\frac{7!}{3!(7 - 3)!} = \frac{7!}{3!4!} = \frac{7 \cdot 6 \cdot 5 \cdot 4 \cdot 3 \cdot 2 \cdot 1}{3 \cdot 2 \cdot 1 \cdot 4 \cdot 3 \cdot 2 \cdot 1}$$
$$= \frac{7 \cdot 6 \cdot 5}{3 \cdot 2 \cdot 1} = 7 \cdot 5 = 35$$

From Example 2.52, the value of $P(7, 3)$ is 210, and $C(7, 3) \cdot (3!) = 35(6) = 210 = P(7, 3)$.   □

2.59 Example   How many 5-card poker hands can be dealt from a 52-card deck? Here order does not matter; $C(52, 5) = 52!/(5!47!) = 2{,}598{,}960$.   □

Unlike earlier problems, the answer to Example 2.59 cannot be easily obtained by applying the Multiplication Principle. Thus, $C(n, r)$ gives us a way to solve new problems.

2.60 Practice   In how many ways can a committee of 3 be chosen from a group of 12 people?   □

In the solution of counting problems, $C(n, r)$ can be used in conjunction with the Multiplication Principle or the Addition Principle.

2.61 Example   A committee of 8 students is to be selected from a class consisting of 19 freshmen and 34 sophomores.
(a) In how many ways can 3 freshmen and 5 sophomores be selected?
(b) In how many ways can a committee with exactly 1 freshman be selected?
(c) In how many ways can a committee with at most 1 freshman be selected?
(d) In how many ways can a committee with at least 1 freshman be selected?

For part (a), we have a sequence of two subtasks, selecting freshmen and selecting sophomores. The Multiplication Principle should be used. Each of the two subtasks is a combinations problem, since order does not matter. There are $C(19, 3)$ ways to choose the freshmen and $C(34, 5)$

ways to choose the sophomores. Hence, the answer is

$$C(19, 3) \cdot C(34, 5) = \frac{19!}{3!16!} \cdot \frac{34!}{5!29!} = (969)(278{,}256)$$

For part (b), we again have a sequence of subtasks: selecting the single freshman and then selecting the rest of the committee from among the sophomores. There are $C(19, 1)$ ways to select the single freshman, and $C(34, 7)$ ways to select the remaining 7 members from the sophomores. By the Multiplication Principle, the answer is

$$C(19, 1) \cdot C(34, 7) = \frac{19!}{1!(19 - 1)!} \cdot \frac{34!}{7!(34 - 7)!} = 19(5{,}379{,}616)$$

For part (c), we get at most 1 freshman by having exactly 1 freshman or by having 0 freshmen. Because these are disjoint events, we will use the Addition Principle. The number of ways to select exactly 1 freshman is the answer to part (b). The number of ways to select 0 freshmen is the same as the number of ways to select the entire 8-member committee from among the 34 sophomores, $C(34, 8)$. Thus the answer is

$$C(19, 1) \cdot C(34, 7) + C(34, 8) = \text{some big number}$$

We can attack part (d) in several ways. One way is to use the Addition Principle, thinking of the disjoint possibilities as: exactly 1 freshman, exactly 2 freshmen, . . . , exactly 8 freshmen. We could compute each of these numbers and then add them. However, it is easier to do the problem by counting all the ways the committee of 8 can be selected from the total pool of 53 people and then eliminate (subtract) the number of committees with 0 freshmen (all sophomores). Thus the answer is

$$C(53, 8) - C(34, 8) \qquad\qquad\qquad \square$$

✔ Checklist

Definitions

permutation (*p. 98*)
n factorial (*p. 98*)
combination (*p. 99*)

Techniques

Find the number of permutations of *r* objects chosen from *n* objects.

Find the number of combinations of *r* objects chosen from *n* objects.

Use permutations and combinations in conjunction with the Multiplication Principle and the Addition Principle.

Main Ideas

There are formulas to help us count the number of ways to arrange objects from a set and the number of ways to select objects from a set.

Exercises Section 2.3

1. Compute the value of the following expressions:
   ⋆ (a) $P(7, 2)$ (b) $P(8, 5)$
   (c) $P(6, 4)$ (d) $P(n, 1)$
   (e) $P(n, n - 1)$

2. How many batting orders are possible for a 9-man baseball team?

3. The 14 teams in the local Little League are listed in the newspaper. How many listings are possible?

4. How many permutations of the letters in "computer" are there? How many of these end in a vowel?

⋆ 5. How many distinct permutations of the letters in "error" are there? (Remember, the various Rs cannot be distinguished from one another.)

6. In how many ways can 6 people be seated in a circle of 6 chairs? (Only relative positions in the circle can be distinguished.)

⋆ 7. In how many ways can first, second, and third prize in a pie-baking contest be given to 15 contestants?

8. (a) Stock designations on an exchange are limited to 3 letters. How many different designations are there?
   (b) How many different designations are there if letters cannot be repeated?

9. In how many different ways can you seat 11 men and 8 women in a row if the men all sit together and the women all sit together?

10. In how many different ways can you seat 11 men and 8 women in a row if no 2 women are to sit together?

11. Compute the value of the following expressions:
    ⋆ (a) $C(10, 7)$ (b) $C(9, 2)$ (c) $C(8, 6)$ (d) $C(n, n - 1)$

12. What is the significance of $C(n, n)$? Compute its value. What is the significance of $C(n, 1)$? Compute its value.

⋆ 13. Quality control wants to test 25 microprocessor chips from the 300 manufactured each day. In how many ways can this be done?

14. A soccer team carries 18 men on the roster; 11 men make a team. In how many ways can the team be chosen?

★ 15. In how many ways can a jury of 5 men and 7 women be selected from a panel of 17 men and 23 women?

16. In how many ways can a librarian select 4 novels and 3 plays from a collection of 21 novels and 11 plays?

Exercises 17 through 20 deal with the following situation: Of a company's personnel, 7 work in design, 14 in manufacturing, 4 in testing, 5 in sales, 2 in accounting, and 3 in marketing. A committee of 6 people is to be formed to meet with upper management.

17. In how many ways can the committee be formed if there is to be 1 member from each department?

18. In how many ways can the committee be formed if there must be exactly 2 members from manufacturing?

19. In how many ways can the committee be formed if the accounting department is not to be represented, and marketing is to have exactly 1 representative?

20. In how many ways can the committee be formed if manufacturing is to have at least 2 representatives?

Exercises 21 through 26 concern a 5-card hand from a standard 52-card deck.

★ 21. How many hands consist of 3 spades and 2 hearts?

22. How many hands consist of all diamonds?

★ 23. How many hands consist of all the same suit?

24. How many hands consist of all face cards?

25. How many hands contain three of a kind?

26. How many hands contain a full house (three of a kind and a pair)?

For Exercises 27 through 30, a set of 4 coins is selected from a box containing 5 dimes and 7 quarters.

★ 27. Find the number of sets of 4 coins.

28. Find the number of sets in which 2 are dimes and 2 are quarters.

29. Find the number of sets composed of all dimes or all quarters.

30. Find the number of sets with 3 or more quarters.

Exercises 31 through 34 concern a computer network with 60 switching nodes.

31. The network is designed to withstand the failure of any 2 nodes. In how many ways can such a failure occur?

32. In how many ways can 1 or 2 nodes fail?

33. If 1 node has failed, in how many ways can 7 nodes be selected without encountering the failed node?

34. If 2 nodes have failed, in how many ways can 7 nodes be selected to include exactly 1 failed node?

In Exercises 35 through 38, a congressional committee of 3 is to be chosen from a set of 5 Democrats, 3 Republicans, and 4 independents.

★ 35. In how many ways can the committee be chosen?

36. In how many ways can the committee be chosen if it must include at least 1 independent?

★ 37. In how many ways can the committee be chosen if it cannot include both Democrats and Republicans?

38. In how many ways can the committee be chosen if it must have at least 1 Democrat and at least 1 Republican?

In Exercises 39 through 42, a hostess wishes to invite 6 dinner guests from a list of 14 friends.

39. In how many ways can she choose her guests?

40. In how many ways can she choose her guests if 6 of them are boring and 6 of them are interesting, and she wants to have at least 1 of each?

41. In how many ways can she choose her guests if 2 of her friends dislike each other and neither will come if the other is present?

42. In how many ways can she choose her guests if 2 of her friends are very fond of each other and one won't come without the other?

43. Prove that $C(n, r) = C(n, n - r)$.

44. Prove that $C(n, k)$ satisfies the recurrence relation
$$C(n, k) = C(n - 1, k) + C(n - 1, k - 1) \qquad \text{for } n \geq k > 0$$

45. Prove that
$$C(k, k) + C(k + 1, k) + \cdots + C(n, k)$$
$$= C(n + 1, k + 1) \qquad \text{for } n \geq k > 0$$

(Hint: use Exercise 44.)

SECTION 2.4   THE BINOMIAL THEOREM

The Theorem and Its Proof

The expression $C(n, r)$ is also called a **binomial coefficient** because $C(n, r)$ for various values of r occurs in the expansion of the binomial $(a + b)^n$, where a and b stand for real numbers. In the familiar case of $(a + b)^2$,

$$(a + b)^2 = a^2 + 2ab + b^2$$

and the coefficients of these three terms, 1, 2, and 1, are also the values of $C(2, 0)$, $C(2, 1)$, and $C(2, 2)$, respectively.

2.62 Practice   Compute $(a + b)^3$; show that the coefficients of the terms, written in order of descending powers of a, are given by $C(3, 0)$, $C(3, 1)$, $C(3, 2)$, and $C(3, 3)$, respectively.   □

The cases $(a + b)^2$ and $(a + b)^3$ suggest that we can write $(a + b)^n$ for any positive integer n as follows:

$$(a + b)^n = C(n, 0)a^n b^0 + C(n, 1)a^{n-1}b^1$$
$$+ C(n, 2)a^{n-2}b^2 + \cdots + C(n, k)a^{n-k}b^k + \cdots \qquad (1)$$
$$+ C(n, n - 1)a^1 b^{n-1} + C(n, n)a^0 b^n$$

Equation (1) is called the **binomial theorem.** Because it is stated ''for any positive integer n,'' it seems most appropriate to prove the binomial theorem by using induction on n.

For the basis step, $n = 1$, equation (1) becomes

$$(a + b)^1 = C(1, 0)a^1 b^0 + C(1, 1)a^0 b^1$$
$$= a + b \qquad (\text{because } C(1, 0) = C(1, 1) = 1)$$

so the basis step is satisfied.

As the inductive hypothesis, we assume

$$(a + b)^k = C(k, 0)a^k b^0 + C(k, 1)a^{k-1}b^1$$
$$+ \cdots + C(k, k - 1)a^1 b^{k-1} + C(k, k)a^0 b^k$$

Now consider

$$(a + b)^{k+1} = (a + b)^k(a + b) = (a + b)^k a + (a + b)^k b$$
$$= [C(k, 0)a^k b^0 + C(k, 1)a^{k-1}b^1 + \cdots$$
$$+ C(k, k - 1)a^1 b^{k-1} + C(k, k)a^0 b^k]a$$
$$+ [C(k, 0)a^k b^0 + C(k, 1)a^{k-1}b^1$$
$$+ \cdots + C(k, k - 1)a^1 b^{k-1} + C(k, k)a^0 b^k]b$$
$$(\text{by the inductive hypothesis})$$
$$= C(k, 0)a^{k+1}b^0 + C(k, 1)a^k b^1 + \cdots + C(k, k - 1)a^2 b^{k-1}$$
$$+ C(k, k)a^1 b^k + C(k, 0)a^k b^1 + C(k, 1)a^{k-1}b^2 + \cdots$$
$$+ C(k, k - 1)a^1 b^k + C(k, k)a^0 b^{k+1}$$

To complete the proof, do Practice 2.63.

2.63 Practice   Complete the induction proof. You will need the identity $C(k + 1, m) = C(k, m) + C(k, m - 1)$. (See Exercise 44, Section 2.3.)   □

Another, less algebraic, way to prove the binomial theorem involves a counting argument, hence it is called a **combinatorial proof.** Writing $(a + b)^n$ as $(a + b)(a + b) \cdots (a + b)$ (n factors), we know that the answer (using the distributive law of numbers) is the sum of all values obtained by multiplying each term in a factor by a term from every other factor. For example, using b as the term from k factors and a as the term from the remaining $n - k$ factors produces the expression $a^{n-k}b^k$. Using b from a different set of k factors and a from the $n - k$ remaining factors also produces $a^{n-k}b^k$. How many such terms are there? There are $C(n, k)$ different ways to select k factors from which to use b; hence there are $C(n, k)$ such terms. After adding these terms together, the coefficient of $a^{n-k}b^k$ is $C(n, k)$. As k ranges from 0 to n, the result of summing the terms is the binomial theorem.

Applying the Binomial Theorem

2.64 Example   Using the binomial theorem, we can write out the expansion of $(x - 3)^4$ as follows:

$$(x - 3)^4 = C(4, 0)x^4(-3)^0 + C(4, 1)x^3(-3)^1 + C(4, 2)x^2(-3)^2$$
$$+ C(4, 3)x^1(-3)^3 + C(4, 4)x^0(-3)^4$$
$$= x^4 + 4x^3(-3) + 6x^2(9) + 4x(-27) + 81$$
$$= x^4 - 12x^3 + 54x^2 - 108x + 81 \qquad \square$$

2.65 Practice   Expand $(x + 1)^5$ using the binomial theorem. $\qquad \square$

The binomial theorem tells us that the $(k + 1)$st term in the expansion of $(a + b)^n$ is $C(n, k)a^{n-k}b^k$. This allows us to find individual terms in the expansion without computing the entire expression.

2.66 Practice   What is the fifth term in the expansion of $(x + y)^7$? $\qquad \square$

By using various values for a and b in the binomial theorem, certain identities can be obtained.

2.67 Example   Let $a = b = 1$ in the binomial theorem. Then

$$(1 + 1)^n = C(n, 0) + C(n, 1) + \cdots + C(n, k) + \cdots + C(n, n)$$

or

$$2^n = C(n, 0) + C(n, 1) + \cdots + C(n, k) + \cdots + C(n, n) \qquad (2)$$

Actually, equation (2) can be proved on its own using a combinatorial proof. The number $C(n, k)$ is the number of ways to select k items from a set of n items and can be thought of as the number of k-element subsets of an n-element set. The right side of equation (2) therefore represents the total number of all the subsets (of all sizes) of an n-element set. But we already know that the number of such subsets is 2^n.   □

✔ Checklist

Definitions

> binomial coefficient (*p. 104*)
> binomial theorem (*p. 105*)
> combinatorial proof (*p. 106*)

Techniques

> Use the binomial theorem to expand a binomial or to find a particular term in the expansion.

Main Ideas

> The binomial theorem provides a formula for expanding a binomial without multiplying it out.

Exercises Section 2.4

In Exercises 1 through 8, expand the expression using the binomial theorem.

★ 1. $(a + b)^4$ 2. $(x + y)^6$

★ 3. $(a + 2)^5$ 4. $(a - 4)^4$

★ 5. $(2x + 3y)^3$ 6. $(3x - 1)^5$

   7. $(2p - 3q)^4$ 8. $(3x + \frac{1}{2})^5$

9. Find the fourth term in the expansion of $(a + b)^{10}$.

10. Find the seventh term in the expansion of $(x - y)^{12}$.

★ 11. Find the sixth term in the expansion of $(2x - 3)^9$.

12. Find the fifth term in the expansion of $(3a + 2b)^7$.

★ 13. Find the last term in the expansion of $(x - 3y)^8$.

14. Find the last term in the expansion of $(ab + 3x)^6$.

★ 15. Find the third term in the expansion of $(4x - 2y)^5$.

16. Find the fourth term in the expansion of $(3x - \frac{1}{2})^8$.

17. Use the binomial theorem (more than once) to expand $(a + b + c)^3$.

18. Expand $(1 + 0.1)^5$ in order to compute $(1.1)^5$.

★ 19. What is the coefficient of x^3y^4 in the expansion of $(2x - y + 5)^8$?

20. What is the coefficient of $x^5y^2z^2$ in the expansion of $(x + y + 2z)^9$?

21. Use the binomial theorem to prove that
$$C(n, 0) - C(n, 1) + C(n, 2) - \cdots + (-1)^n C(n, n) = 0$$

22. Use the binomial theorem to prove that
$$C(n, 0) + C(n, 1)2 + C(n, 2)2^2 + \cdots + C(n, n)2^n = 3^n$$

23. (a) Expand $(1 + x)^n$.

(b) Differentiate both sides of the equation from part (a) with respect to x to obtain
$$n(1 + x)^{n-1} = C(n, 1) + 2C(n, 2)x + 3C(n, 3)x^2$$
$$+ \cdots + nC(n, n)x^{n-1}$$

(c) Prove that
$$C(n, 1) + 2C(n, 2) + 3C(n, 3) + \cdots + nC(n, n) = n2^{n-1}$$

(d) Prove that
$$C(n, 1) - 2C(n, 2) + 3C(n, 3) - 4C(n, 4)$$
$$+ \cdots + (-1)^{n-1}nC(n, n) = 0$$

24. (a) Prove that
$$\frac{2^{n+1} - 1}{n + 1} = C(n, 0) + \frac{1}{2}C(n, 1) + \frac{1}{3}C(n, 2)$$
$$+ \cdots + \frac{1}{n + 1} C(n, n)$$

(b) Prove that
$$\frac{1}{n + 1} = C(n, 0) - \frac{1}{2}C(n, 1) + \frac{1}{3}C(n, 2)$$
$$+ \cdots + (-1)^n \frac{1}{n + 1} C(n, n)$$

(Hint: integrate both sides of the equation from part (a) of Exercise 23.)

25. Pascal's triangle is created by letting row 0 consist of the digit 1 and row 1 consist of two 1s. Each successive row then has 1s at the far left and right with each element in the middle the sum of the two nearest numbers in the row above it (see Figure 2.4).

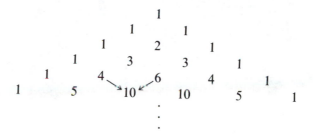

Figure 2.4

Prove that the coefficients in the expansion of $(a + b)^n$ are given by row n of Pascal's triangle. (Hint: see Exercise 44, Section 2.3.)

Chapter 3

Relations, Functions, and Matrices

Elements within a set or elements in different sets often have a special connection with one another that can be described as a *relation*. We will study relations in Section 3.1. There we see that binary relations, relations between pairs of elements, can have various properties. One type of binary relation with certain properties is called a partial ordering; elements related by a partial ordering can be represented graphically. Another type of binary relation is an equivalence relation; elements related by an equivalence relation can be grouped into classes.

A function is a special kind of binary relation. Functions as well as relations describe a number of real-world situations. Functions can also have special properties, as discussed in Section 3.2.

Matrices are considered in Section 3.3, where we develop an arithmetic on matrices. You are probably already familiar with the roles of matrices in representing data or in solving systems of linear equations, but they have a number of other applications. In the rest of this book, we shall use matrices as representations of graphs, as examples of mathematical structures, and as tools in coding theory.

SECTION 3.1   RELATIONS

Binary Relations

If we learn that two people, Henrietta and Horace, are related, we understand that there is some family connection between them—that (Henrietta, Horace) stands out from other ordered pairs of people because there is a relationship (cousins, sister and brother, or whatever) that Henrietta and Horace satisfy. The mathematical analog is to distinguish certain ordered pairs of objects from other ordered pairs because the components of the distinguished pairs satisfy some relationship that the components of the other pairs do not.

3.1 Example   Let $S = \{1, 2\}$ and $T = \{2, 3\}$, then $S \times T = \{(1, 2), (1, 3), (2, 2), (2, 3)\}$. If we are interested in the relationship of equality, then $(2, 2)$ is the only distinguished element of $S \times T$, that is, the only ordered pair whose components are equal. If we are interested in the relationship of one number being less than another, we would choose $(1, 2)$, $(1, 3)$, and $(2, 3)$ as the distinguished ordered pairs of $S \times T$.   □

In Example 3.1, we could pick out the distinguished ordered pairs (x, y) by saying that $x = y$ or that $x < y$. Similarly, the notation $x \rho y$ indicates that the ordered pair (x, y) satisfies a relation ρ. The relation ρ may be defined by some verbal description or simply by listing the distinguished pairs that ρ selects.

3.2 Example   Let $S = \{1, 2\}$ and $T = \{2, 3, 4\}$. On the set $S \times T = \{(1, 2), (1, 3), (1, 4), (2, 2), (2, 3), (2, 4)\}$, a relation ρ can be defined by $x \rho y$ if and only if $x = \frac{1}{2}y$, abbreviated $x \rho y \leftrightarrow x = \frac{1}{2}y$. Thus $(1, 2)$ and $(2, 4)$ satisfy ρ. Alternatively, the same ρ could be defined by saying that $\{(1, 2), (2, 4)\}$ is the set of ordered pairs satisfying ρ.   □

We have been talking about binary relations, relations between two objects. In Example 3.2, one way to define the binary relation ρ is to specify a subset of $S \times T$. Formally, a binary relation is actually a subset of $S \times T$.

3.3 Definition   Given sets S and T, a **binary relation** ρ on $S \times T$ is a subset of $S \times T$.   □

Generally, the set of ordered pairs that is a binary relation will be defined by describing the relation rather than by listing the ordered pairs.

3.4 Example   Let $S = \{1, 2\}$ and $T = \{2, 3, 4\}$. Let ρ be given by the description $x \rho y \leftrightarrow x + y$ is odd. Then $(1, 2) \in \rho$, $(1, 4) \in \rho$, and $(2, 3) \in \rho$.   □

3.5 Example   Let $S = \{1, 2\}$ and $T = \{2, 3, 4\}$. If ρ is defined on $S \times T$ by $\rho = \{(2, 3), (2, 4)\}$, then $2 \rho 3$ and $2 \rho 4$ but not, for instance, $1 \rho 4$. Here ρ seems to have no obvious verbal description.   □

3.6 Practice   For each of the following binary relations ρ on $N \times N$, decide which of the given ordered pairs belong to ρ:

(a) $x \rho y \leftrightarrow x = y + 1$; $(2, 2), (2, 3), (3, 3), (3, 2)$
(b) $x \rho y \leftrightarrow x$ divides y; $(2, 4), (2, 5), (2, 6)$

(c) $x \rho y \leftrightarrow x$ is odd; (2, 3), (3, 4), (4, 5), (5, 6)

(d) $x \rho y \leftrightarrow x > y^2$; (1, 2), (2, 1), (5, 2), (6, 4), (4, 3)   □

We can define *n*-ary relations by generalizing the definition for binary relations.

3.7 Practice   Complete the following definition: Given sets $S_1, S_2, \ldots, S_n$, an **n-ary relation** on $S_1 \times S_2 \times \cdots \times S_n$ is ———.   □

As a special case, a **unary relation** ρ on a set S is just a particular subset of S. An element $x \in S$ satisfies ρ if and only if x belongs to the subset.

Relations are also called **predicates,** so there are unary, binary, and *n*-ary predicates. We encountered predicates in Chapter 1, where we used notation such as $P(x)$ to denote that x satisfied a unary predicate (relation) P or $Q(x, y)$ to denote that (x, y) satisfied a binary predicate (relation) Q. Thus, if P is a unary relation or predicate, then $P(x)$ and $x \in P$ mean the same thing; if Q is a binary relation or predicate, then $Q(x, y)$, $(x, y) \in Q$, and $x Q y$ all mean the same thing.

Often we will be interested in a binary relation on a set S, meaning the relation is a subset of S^2. In general, an *n*-ary relation on S is a subset of S^n, a set of ordered *n*-tuples of members of S.

Operations on Relations

Suppose B is the set of all binary relations on a given set S. We can define two binary operations and one unary operation on B. Let ρ and σ belong to B. We define a new binary relation, $\rho + \sigma$, on S (that is, a new member of B) by

$x(\rho + \sigma)y \leftrightarrow x \rho y$ or $x \sigma y$

A second new binary relation, $\rho \cdot \sigma$, on S is defined by

$x(\rho \cdot \sigma)y \leftrightarrow x \rho y$ and $x \sigma y$

And finally, a new binary relation, ρ', on S is given by

$x \rho' y \leftrightarrow$ not $x \rho y$

3.8 Practice   Since ρ and σ above are actually subsets of S^2, express the new relations $\rho + \sigma$, $\rho \cdot \sigma$, and ρ' in terms of these subsets.   □

3.9 Practice   Let ρ and σ be binary relations on $\mathbb{N}$ defined by $x \rho y \leftrightarrow x = y$ and $x \sigma y \leftrightarrow x < y$.

Give verbal descriptions for (a), (b), and (c).

(a) What is the relation $\rho + \sigma$?

(b) What is the relation ρ'?

(c) What is the relation σ'?

(d) What is the relation $\rho \cdot \sigma$? (Give a set description.)   □

The following facts about the operations $+$, $\cdot$, and $'$ on relations are immediate consequences of Practice 3.8 and the basic set identities found in Section 2.1:

1a. $\rho + \sigma = \sigma + \rho$ 1b. $\rho \cdot \sigma = \sigma \cdot \rho$

2a. $(\rho + \sigma) + \gamma = \rho +$ 2b. $(\rho \cdot \sigma) \cdot \gamma = \rho \cdot (\sigma \cdot \gamma)$
    $(\sigma + \gamma)$

3a. $\rho + (\sigma \cdot \gamma) = (\rho + \sigma) \cdot$ 3b. $\rho \cdot (\sigma + \gamma) = (\rho \cdot \sigma) +$
    $(\rho + \gamma)$ $(\rho \cdot \gamma)$

4a. $\rho + \varnothing = \rho$ 4b. $\rho \cdot S^2 = \rho$

5a. $\rho + \rho' = S^2$ 5b. $\rho \cdot \rho' = \varnothing$

Properties of Relations

A binary relation on a set S may have certain properties. For example, the relation ρ of equality on S, $(x, y) \in \rho \leftrightarrow x = y$, has three properties: (1) for any $x \in S$, $x = x$, or $(x, x) \in \rho$; (2) for any $x, y \in S$, if $x = y$ then $y = x$, or $(x, y) \in \rho \rightarrow (y, x) \in \rho$; and (3) for any $x, y, z \in S$, if $x = y$ and $y = z$, then $x = z$, or $((x, y) \in \rho$ and $(y, z) \in \rho) \rightarrow (x, z) \in \rho$. These three properties make the equality relation reflexive, symmetric, and transitive.

3.10 Definition

Let ρ be a binary relation on a set S. Then

ρ is **reflexive** means: $x \in S \rightarrow (x, x) \in \rho$

ρ is **symmetric** means: $(x, y \in S \wedge (x, y) \in \rho) \rightarrow (y, x) \in \rho$

ρ is **transitive** means: $(x, y, z \in S \wedge (x, y) \in \rho \wedge (y, z) \in \rho) \rightarrow$
                                   $(x, z) \in \rho$ □

3.11 Example

Consider the relation $\leq$ on the set $\mathbb{N}$. This relation is reflexive because for any nonnegative integer x, $x \leq x$. It is also a transitive relation because for any nonnegative integers x, y, and z, $x \leq y$ and $y \leq z$ implies $x \leq z$. However, $\leq$ is not symmetric; $3 \leq 4$ does not imply $4 \leq 3$. In fact, for any $x, y \in \mathbb{N}$, if both $x \leq y$ and $y \leq x$, then $y = x$. This characteristic is described by saying that $\leq$ is antisymmetric.   □

3.12 Definition

Let ρ be a binary relation on a set S. Then ρ is **antisymmetric** means

$$(x, y \in S \wedge (x, y) \in \rho \wedge (y, x) \in \rho) \rightarrow x = y$$ □

3.13 Example

Let $S = \mathscr{P}(\mathbb{N})$. Define a binary relation ρ on S by $A \rho B \leftrightarrow A \subseteq B$. Then ρ is reflexive because every set is a subset of itself. Also, ρ is transitive because if A is a subset of B and B is a subset of C, then A is a subset of C. Finally, ρ is antisymmetric because if A is a subset of B and B is a subset of A, then A and B are equal sets. $\square$

The properties of symmetry and antisymmetry for binary relations are not precisely opposites. The equality relation on a set S is both symmetric and antisymmetric. However, the equality relation on S, or a subset of this relation, is the only relation having both these properties. To illustrate, suppose ρ is a symmetric and antisymmetric relation on S, and let $(x, y) \in \rho$. By symmetry, it follows that $(y, x) \in \rho$. But by antisymmetry, $x = y$. Thus, only equal elements can be related.

3.14 Practice   Test the following binary relations on the given sets S for reflexivity, symmetry, antisymmetry, and transitivity:

(a) $S = \mathbb{N}$; $x \rho y \leftrightarrow x + y$ is even
(b) $S = \mathbb{N}$; $x \rho y \leftrightarrow x$ divides y
(c) $S = $ set of all lines in the plane; $x \rho y \leftrightarrow x$ is parallel to y or x coincides with y
(d) $S = \mathbb{N}$; $x \rho y \leftrightarrow x = y^2$
(e) $S = \{0, 1\}$; $x \rho y \leftrightarrow x = y^2$
(f) $S = \{x \mid x$ is a person living in Peoria$\}$; $x \rho y \leftrightarrow x$ is older than y
(g) $S = \{x \mid x$ is a student in your class$\}$; $x \rho y \leftrightarrow x$ sits in the same row as y
(h) $S = \{1, 2, 3\}$; $\rho = \{(1, 1), (2, 2), (3, 3), (1, 2), (2, 1)\}$ $\square$

For the rest of this section we will concentrate on two types of binary relations that are characterized by which properties (reflexivity, symmetry, antisymmetry, and transitivity) they satisfy.

Partial Orderings

3.15 Definition   A binary relation on a set S that is reflexive, antisymmetric, and transitive is called a **partial ordering** on S. $\square$

From the previous examples and Practice 3.14, we have the following instances of partial orderings:

on $\mathbb{N}$: $x \rho y \leftrightarrow x \leq y$
on $\mathcal{P}(\mathbb{N})$: $A \rho B \leftrightarrow A \subseteq B$
on $\mathbb{N}$: $x \rho y \leftrightarrow x$ divides y
on $\{0, 1\}$: $x \rho y \leftrightarrow x = y^2$

If ρ is a partial ordering on S, then the ordered pair (S, ρ) is called a **partially ordered set**. We will denote an arbitrary, partially ordered set by $(S, \leq)$ (in any particular case, $\leq$ has some definite meaning such as "less than or equal to," "is a subset of," "divides," and so on).

Let $(S, \leq)$ be a partially ordered set, and let $A \subseteq S$. Then $\leq$ is a set of ordered pairs of elements of S, some of which may be ordered pairs of elements of A. If we select from $\leq$ the ordered pairs of elements of A, this new set is called the **restriction** of $\leq$ to A and is a partial ordering on A. (Do you see why the three required properties still hold?) For instance, once we know the relation "x divides y" is a partial ordering on $\mathbb{N}$, we automatically know that "x divides y" is a partial ordering on $\{1, 2, 3, 6, 12, 18\}$.

We want to introduce some terminology about partially ordered sets. Let $(S, \leq)$ be a partially ordered set. If $x \leq y$, then either $x = y$ or $x \neq y$. If $x \leq y$ but $x \neq y$, we write $x < y$ and say that x is a **predecessor** of y or y is a **successor** of x. A given y may have many predecessors, but if $x < y$ and there is no z with $x < z < y$, then x is an **immediate predecessor** of y.

3.16 Practice   Consider the relation "x divides y" on $\{1, 2, 3, 6, 12, 18\}$.

(a) Write the ordered pairs (x, y) of this relation.
(b) Write all the predecessors of 6.
(c) Write all the immediate predecessors of 6.   □

We can graph the partially ordered set $(S, \leq)$ if S is finite. Each of the elements of S is represented by a dot, called a **node**, or **vertex**, of the graph. If x is an immediate predecessor of y, then the node for y is placed above the node for x, and the two nodes are connected by a straight line segment.

3.17 Example   Consider $\mathcal{P}(\{1, 2\})$ under the relation of set inclusion. This is a partially ordered set. (We already know that $(\mathcal{P}(\mathbb{N}), \subseteq)$ is a partially ordered set.) The elements of $\mathcal{P}(\{1, 2\})$ are $\emptyset$, $\{1\}$, $\{2\}$, and $\{1, 2\}$. The binary relation $\subseteq$ consists of the following ordered pairs:

$$(\emptyset, \emptyset) \quad (\{1\}, \{1\}) \quad (\{2\}, \{2\}) \quad (\{1, 2\}, \{1, 2\}) \quad (\emptyset, \{1\}) \quad (\emptyset, \{2\})$$
$$(\emptyset, \{1, 2\}) \quad (\{1\}, \{1, 2\}) \quad (\{2\}, \{1, 2\})$$

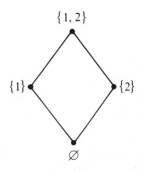

Figure 3.1

The graph of this partially ordered set appears in Figure 3.1. Notice that although $\emptyset$ is not an immediate predecessor of $\{1, 2\}$, it is a predecessor of $\{1, 2\}$ (shown on the graph by the chain of upward line segments connecting $\emptyset$ with $\{1, 2\}$). □

3.18 Practice Draw the graph for the relation "x divides y" on $\{1, 2, 3, 6, 12, 18\}$. □

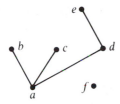

Figure 3.2

The graph of a partially ordered set (also called its **Hasse diagram**) conveys all the information about the partial ordering. We can reconstruct the set of ordered pairs making up the partial ordering just by looking at the graph. Thus, given the graph in Figure 3.2 of a partial ordering $\leq$ on a set $\{a, b, c, d, e, f\}$, we can conclude that $\leq$ is the set

$$\{(a, a), (b, b), (c, c), (d, d), (e, e), (f, f), (a, b), (a, c), (a, d),$$
$$(a, e), (d, e)\}$$

Two elements of S may be unrelated in a partial ordering of S. In Example 3.17, $\{1\}$ and $\{2\}$ are unrelated; so are 2 and 3, and 12 and 18 in Practice 3.18. In Figure 3.2, f is not related to any other element. A partial ordering in which every element of the set is related to every other element is called a **total ordering**, or **chain**. The graph for a total ordering looks like Figure 3.3. The relation $\leq$ on $\mathbb{N}$ is a total ordering.

Again, let $(S, \leq)$ be a partially ordered set. If there is a $y \in S$ with $y \leq x$ for all $x \in S$, then y is a **least element** of the partially ordered set. A least element, if it exists, is unique. If y and z are both least elements, then $y \leq z$ because y is least and $z \leq y$ because z is least; by antisymmetry, $y = z$. An element $y \in S$ is **minimal** if there is no $x \in S$ with $x < y$. Similar definitions apply for greatest element and maximal elements.

Figure 3.3

3.19 Practice Define **greatest element** and **maximal element** in a partially ordered set $(S, \leq)$. □

3.20 Example In the partially ordered set of Practice 3.18, 1 is both least and minimal. Twelve and eighteen are both maximal, but there is no greatest element. □

A least element is always minimal, and a greatest element is always maximal, but the converses are not true (see Example 3.20). In a totally ordered set, however, a minimal element is a least element, and a maximal element is a greatest element.

3.21 Practice Draw the graph for a partially ordered set with four elements in which there are two minimal elements but no least element and two maximal elements but no greatest element, and in which each element is related to two other elements. □

Partial orderings satisfy the properties of reflexivity, antisymmetry, and transitivity. Another type of binary relation, which we'll study next, satisfies a different set of properties.

Equivalence Relations

3.22
Definition

A binary relation on a set S that is reflexive, symmetric, and transitive is called an **equivalence relation** on S.   □

We have already come upon the following examples of equivalence relations:

on any set S: $x \rho y \leftrightarrow x = y$
on $\mathbb{N}$: $x \rho y \leftrightarrow x + y$ is even
on the set of all lines in the plane: $x \rho y \leftrightarrow x$ is parallel to y or coincides with y
on $\{0, 1\}$: $x \rho y \leftrightarrow x = y^2$
on $\{x \mid x$ is a student in your class$\}$: $x \rho y \leftrightarrow x$ sits in the same row as y
on $\{1, 2, 3\}$: $\rho = \{(1, 1), (2, 2), (3, 3), (1, 2), (2, 1)\}$

We can illustrate an important feature of an equivalence relation on a set by looking at $S = \{x \mid x$ is a student in your class$\}$, $x \rho y \leftrightarrow x$ sits in the same row as y. Let's group together all those students in set S who are related to one another. We come up with Figure 3.4. We have partitioned the set S into subsets in such a way that everyone in the class belongs to one and only one subset.

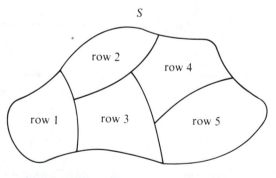

Figure 3.4

3.23
Definition

A **partition** of a set S is a collection of nonempty disjoint subsets of S whose union equals S.   □

Any equivalence relation, as we will see, partitions the set on which it is defined. The subsets making up the partition, often called the **blocks** of the partition, are formed by grouping together related elements, as in the above case.

For ρ an equivalence relation on a set S and $x \in S$, we let $[x]$ denote the set of all members of S related to x, called the **equivalence class** of x. Thus,

$$[x] = \{y \mid y \in S \wedge x \, \rho \, y\}$$

3.24 Example   In the case where $x \, \rho \, y \leftrightarrow$ "x sits in the same row as y," suppose that John, Chuck, Jose, Judy, and Ted all sit in row 3. Then [John] = {John, Chuck, Jose, Judy, Ted}. Also [John] = [Ted] = [Judy], and so on. There can be more than one name for a given equivalence class.   □

For ρ an equivalence relation on S, the distinct equivalence classes of members of S form a partition of S. To satisfy the definition of a partition, we must show that (1) the union of these distinct classes equals S and (2) the distinct classes are disjoint. To show that the union of the classes equals S is easy, since it is essentially a set equality; we prove it by showing set inclusion in each direction. Every equivalence class is a subset of S, so the union of classes is also contained in S. For the opposite direction, let $x \in S$. Then $x \, \rho \, x$ (reflexivity of ρ); thus, $x \in [x]$, and every member of S belongs to some equivalence class, hence to the union of classes.

Now let $[x]$ and $[z]$ be two distinct classes; that is, $[x] \neq [z]$. We need to show that $[x] \cap [z] = \varnothing$. We will do a proof by contradiction. Therefore, we assume that $[x] \cap [z] \neq \varnothing$ and thus that there is a $y \in S$ such that $y \in [x] \cap [z]$.

$y \in [x] \cap [z]$	(assumption)
$y \in [x], y \in [z]$	(definition of $\cap$)
$x \, \rho \, y, z \, \rho \, y$	(definition of $[x]$ and $[z]$)
$x \, \rho \, y, y \, \rho \, z$	(symmetry of ρ)
$x \, \rho \, z$	(transitivity of ρ)

Now we can show that $[x] = [z]$; we prove set inclusion in each direction. Let

$$q \in [z] \qquad ([z] \neq \varnothing)$$

Then

$z \, \rho \, q$	(definition of $[z]$)
$x \, \rho \, z$	(from above)
$x \, \rho \, q$	(transitivity of ρ)
$q \in [x]$	(definition of $[x]$)
$[z] \subseteq [x]$	(definition of $\subseteq$)
$[x] \subseteq [z]$	(Practice 3.25 below)
$[x] = [z]$	($[z] \subseteq [x]$ and $[x] \subseteq [z]$)

Therefore, $[x] = [z]$, which is a contradiction because $[x]$ and $[z]$ are distinct. Thus, our assumption was wrong; $[x] \cap [z] = \emptyset$, and distinct equivalence classes are disjoint.

3.25 Practice   For the above argument, supply the proof that $[x] \subseteq [z]$.   □

We have shown that an equivalence relation on a set determines a partition. The converse is also true. Given a partition of a set S, we define a relation ρ by $x \, \rho \, y \leftrightarrow$ "x is in the same subset of the partition as y."

3.26 Practice   Show that ρ, as defined above, is an equivalence relation on S; that is, show that ρ is reflexive, symmetric, and transitive.   □

We have proved the following about equivalence relations.

3.27 Theorem   An equivalence relation on a set S determines a partition of S, and a partition of a set S determines an equivalence relation on S.   □

3.28 Example   The equivalence relation on $\mathbb{N}$ given by

$$x \, \rho \, y \leftrightarrow x + y \text{ is even}$$

partitions $\mathbb{N}$ into two equivalence classes. If x is an even number, then for any even number y, $x + y$ is even, and $y \in [x]$. All even numbers form one class. If x is an odd number and y is any odd number, $x + y$ is even and $y \in [x]$. All odd numbers form the second class. The partition can be pictured as in Figure 3.5. Notice again that an equivalence class may have more than one name, or representative. In this example, $[2] = [8] = [1048]$, and so on, $[1] = [17] = [947]$, and so on.   □

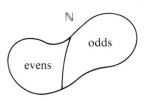

Figure 3.5

3.29 Practice   For each of the following equivalence relations, describe the corresponding equivalence classes:

    (a) On the set of all lines in the plane, $x \, \rho \, y \leftrightarrow x$ is parallel to y or x coincides with y.

    (b) On the set $\mathbb{N}$, $x \, \rho \, y \leftrightarrow x = y$.

    (c) On $\{1, 2, 3\}$, $\rho = \{(1, 1), (2, 2), (3, 3), (1, 2), (2, 1)\}$.   □

Partitioning a set into equivalence classes is helpful because it is often convenient to treat the classes themselves as entities. We will conclude this section with two examples where this is the case (other examples occur throughout the rest of the text).

3.30 Example   Let $S = \{a/b \,|\, a, b \in \mathbb{Z}, b \neq 0\}$. S is therefore the set of all fractions. Two fractions such as $\frac{1}{2}$ and $\frac{2}{4}$ are said to be equivalent. Formally, a/b is equivalent to c/d, denoted by $a/b \sim c/d$, if and only if $ad = bc$. We will show that the binary relation $\sim$ on S is an equivalence relation. First, $a/b \sim a/b$ because $ab = ba$. Also, if $a/b \sim c/d$, then $ad = bc$, or $cb = da$ and $c/d \sim a/b$. Hence, $\sim$ is reflexive and symmetric. To show that $\sim$ is transitive, let $a/b \sim c/d$ and $c/d \sim e/f$. Then $ad = bc$ and $cf = de$. Multiplying the first equation by f and the second by b, we get $adf = bcf$ and $bcf = bde$. Therefore, $adf = bde$, or $af = be$. Thus, $a/b \sim e/f$, and $\sim$ is transitive. Some sample equivalence classes of S formed by this equivalence relation are

$$\left[\frac{1}{2}\right] = \left\{ \cdots, \frac{-3}{-6}, \frac{-2}{-4}, \frac{-1}{-2}, \frac{1}{2}, \frac{2}{4}, \frac{3}{6}, \cdots \right\}$$

$$\left[\frac{3}{10}\right] = \left\{ \cdots, \frac{-9}{-30}, \frac{-6}{-20}, \frac{-3}{-10}, \frac{3}{10}, \frac{6}{20}, \frac{9}{30}, \cdots \right\}$$

The set $\mathbb{Q}$ of rational numbers can be regarded as the set of all equivalence classes of S. A single rational number, such as $[\frac{1}{2}]$, has many fractions representing it, although we customarily use the reduced fractional representation. When we add two rational numbers, such as $[\frac{1}{2}] + [\frac{3}{10}]$, we look for representatives from the classes having the same denominator and add those representatives. Our answer is the class to which the resulting sum belongs, and we usually name the class by using a reduced fraction. Thus, to add $[\frac{1}{2}] + [\frac{3}{10}]$, we represent $[\frac{1}{2}]$ by $\frac{5}{10}$ and $[\frac{3}{10}]$ by $\frac{3}{10}$. The sum of $\frac{5}{10}$ and $\frac{3}{10}$ is $\frac{8}{10}$, and $[\frac{8}{10}]$ is customarily named $[\frac{4}{5}]$. This procedure is so familiar that it is generally written as $\frac{1}{2} + \frac{3}{10} = \frac{4}{5}$; nonetheless, classes of fractions are being manipulated by means of representatives. $\quad\square$

3.31 Example   We will define a binary relation of **congruence modulo 4** on the set $\mathbb{Z}$ of integers. An integer x is congruent modulo 4 to y, symbolized by $x \equiv_4 y$, or $x \equiv y \pmod 4$, if $x - y$ is an integral multiple of 4. Congruence modulo 4 is an equivalence relation on $\mathbb{Z}$. (Can you prove this?) To construct the equivalence classes, note that $[0]$, for example, will contain all integers differing from 0 by a multiple of 4, such as 4, 8, -12, and so on. The distinct equivalence classes are

$$[0] = \{\ldots, -8, -4, 0, 4, 8, \ldots\}$$
$$[1] = \{\ldots, -7, -3, 1, 5, 9, \ldots\}$$

$[2] = \{. . . , -6, -2, 2, 6, 10, . . .\}$

$[3] = \{. . . , -5, -1, 3, 7, 11, . . .\}$

There is nothing special about the choice of 4 here, and we can give a definition for congruence modulo n for any positive integer n. This binary relation is always an equivalence relation. Later we will look at situations in which the resulting classes themselves are the entities. And, as in Example 3.30, the classes will be manipulated by means of representatives. We will also see how manipulating these classes relates to the arithmetic performed by a computer.   ☐

3.32 Practice   State the definition of **congruence modulo** n for an arbitrary positive integer n.   ☐

3.33 Practice   What are the equivalence classes corresponding to the relation of congruence modulo 5 on $\mathbb{Z}$?   ☐

✔ Checklist

Definitions

binary relation (*p. 112*)
n-ary relation (*p. 113*)
unary relation (*p. 113*)
predicate (*p. 113*)
reflexive relation (*p. 114*)
symmetric relation (*p. 114*)
transitive relation (*p. 114*)
antisymmetric relation (*p. 114*)
partial ordering (*p. 115*)
partially ordered set (*p. 115*)
restriction of a partial ordering (*p. 116*)
predecessor in a partial ordering (*p. 116*)
successor in a partial ordering (*p. 116*)
immediate predecessor in a partial ordering (*p. 116*)
node (vertex) (*p. 116*)
Hasse diagram (*p. 117*)
total ordering (chain) (*p.* 117)
least element (*p. 117*)
minimal element (*p. 117*)
greatest element (*p. 117*)
maximal element (*p. 117*)
equivalence relation (*p. 118*)
partition (*p. 118*)
block (*p. 118*)

equivalence class (*p. 119*)
congruence modulo *n* (*p. 122*)

Techniques

Test an ordered pair for membership in a binary relation.

Test a binary relation for reflexivity, symmetry, antisymmetry, and transitivity.

Graph a partially ordered set.

Find least, minimal, greatest, or maximal elements in a partially ordered set.

Find the equivalence classes associated with an equivalence relation.

Main Ideas

A binary relation on a set S, formally a subset of $S \times S$; the distinctive relationship satisfied by the relation's members often has a verbal description as well.

Operations on binary relations on a set $(+, \cdot, ')$.

Partially ordered sets and their graphs.

An equivalence relation on a set and the associated equivalence classes, which may themselves be treated as entities. An equivalence relation on a set S always determines a partition of S, and conversely, a partition of a set S always defines an equivalence relation on S.

Exercises Section 3.1

★ 1. For each of the following binary relations ρ on $\mathbb{N}$, decide which of the given ordered pairs belong to ρ:

   (a) $x \rho y \leftrightarrow x + y < 7$; $(1, 3), (2, 5), (3, 3), (4, 4)$
   (b) $x \rho y \leftrightarrow x = y + 2$; $(0, 2), (4, 2), (6, 3), (5, 3)$
   (c) $x \rho y \leftrightarrow 2x + 3y = 10$; $(5, 0), (2, 2), (3, 1), (1, 3)$
   (d) $x \rho y \leftrightarrow y$ is a perfect square; $(1, 1), (4, 2), (3, 9), (25, 5)$

2. Decide which of the given items satisfy the relation.

   (a) ρ a binary relation on $\mathbb{Z}$, $x \rho y \leftrightarrow x = -y$;
   $(1, -1), (2, 2), (-3, 3), (-4, -4)$
   (b) ρ a unary relation on $\mathbb{N}$, $x \in \rho \leftrightarrow x$ is prime;
   $19, 21, 33, 41$
   (c) ρ a ternary relation on $\mathbb{N}$, $(x, y, z) \in \rho \leftrightarrow x^2 + y^2 = z^2$;
   $(1, 1, 2), (3, 4, 5), (0, 5, 5), (8, 6, 10)$
   (d) ρ a binary relation on $\mathbb{Q}$, $x \rho y \leftrightarrow x \leq 1/y$;
   $(1, 2), (-3, -5), (-4, \frac{1}{2}), (\frac{1}{2}, \frac{1}{3})$

3. For each of the following binary relations on $\mathbb{R}$, draw a figure to show the region of the plane it describes:

★(a) $x \rho y \leftrightarrow y \leq 2$

(b) $x \rho y \leftrightarrow x = y - 1$

(c) $x \rho y \leftrightarrow x^2 + y^2 \leq 25$

(d) $x \rho y \leftrightarrow x \geq y$

4. For each of the following figures, give the binary relation on $\mathbb{R}$ that describes the shaded area:

(a)

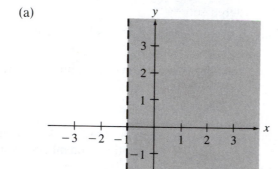

Figure 3.6

(b)

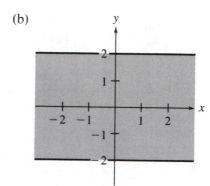

Figure 3.7

(c)

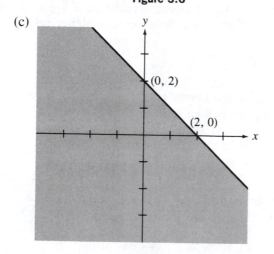

Figure 3.8

(d)

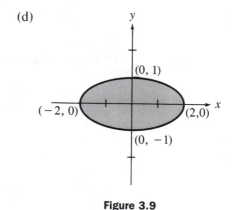

Figure 3.9

★ 5. Let ρ and σ be binary relations on $\mathbb{N}$ defined by $x \rho y \leftrightarrow x$ divides y, $x \sigma y \leftrightarrow 5x \leq y$. Decide which of the given ordered pairs satisfy the following relations:

    (a) $\rho + \sigma$; (2, 6), (3, 17), (2, 1), (0, 0)

    (b) $\rho \cdot \sigma$; (3, 6), (1, 2), (2, 12)

    (c) ρ'; (1, 5), (2, 8), (3, 15)

    (d) σ'; (1, 1), (2, 10), (4, 8)

6. Let $S = \{0, 1, 2, 4, 6\}$. Test the following binary relations on S for reflexivity, symmetry, antisymmetry, and transitivity:

    (a) $\rho = \{(0, 0), (1, 1), (2, 2), (4, 4), (6, 6), (0, 1), (1, 2), (2, 4), (4, 6)\}$

    (b) $\rho = \{(0, 1), (1, 0), (2, 4), (4, 2), (4, 6), (6, 4)\}$

    (c) $\rho = \{(0, 1), (1, 2), (0, 2), (2, 0), (2, 1), (1, 0), (0, 0), (1, 1),$
        $(2, 2)\}$

    (d) $\rho = \{(0, 0), (1, 1), (2, 2), (4, 4), (6, 6), (4, 6), (6, 4)\}$

    (e) $\rho = \varnothing$

7. Test the following binary relations on the given sets S for reflexivity, symmetry, antisymmetry, and transitivity:

    ★ (a) $S = \mathbb{Q}$
        $x \rho y \leftrightarrow |x| \leq |y|$

    ★ (b) $S = \mathbb{Z}$
        $x \rho y \leftrightarrow x - y$ is an integral multiple of 3

    ★ (c) $S = \mathbb{N}$
        $x \rho y \leftrightarrow x \cdot y$ is even

    (d) $S = \mathbb{N}$
        $x \rho y \leftrightarrow x$ is odd

    (e) $S =$ set of all squares in the plane
        $S_1 \rho S_2 \leftrightarrow$ length of side of $S_1 =$ length of side of S_2

    (f) $S =$ set of all finite-length strings of characters
        $x \rho y \leftrightarrow$ number of characters in $x =$ number of characters in y

    (g) $S =$ set of all people in the United States
        $x \rho y \leftrightarrow x$ is the brother of y

8. Which of the binary relations of Exercise 7 are equivalence relations? For each equivalence relation, describe the associated equivalence classes.

9. For each case below, think of a set S and a binary relation ρ on S (different from any in the examples or problems) satisfying the given conditions.

    (a) ρ is reflexive and symmetric but not transitive.

    (b) ρ is reflexive and transitive but not symmetric.

    (c) ρ is not reflexive or symmetric but is transitive.

    (d) ρ is reflexive but neither symmetric nor transitive.

10. Let ρ and σ be binary relations on a set S.

    (a) If ρ and σ are reflexive, is $\rho + \sigma$ reflexive? Is $\rho \cdot \sigma$ reflexive?

    (b) If ρ and σ are symmetric, is $\rho + \sigma$ symmetric? Is $\rho \cdot \sigma$ symmetric?

    (c) If ρ and σ are antisymmetric, is $\rho + \sigma$ antisymmetric? Is $\rho \cdot \sigma$ antisymmetric?

    (d) If ρ and σ are transitive, is $\rho + \sigma$ transitive? Is $\rho \cdot \sigma$ transitive?

11. Two additional properties of a binary relation ρ are defined as follows:

ρ is **irreflexive** means: $x \in S \rightarrow (x, x) \notin \rho$

ρ is **asymmetric** means: $(x, y \in S \wedge (x, y) \in \rho) \rightarrow (y, x) \notin \rho$

★ (a) Give an example of a binary relation ρ on set $S = \{1, 2, 3\}$ that is neither reflexive nor irreflexive.

(b) Give an example of a binary relation ρ on set $S = \{1, 2, 3\}$ that is neither symmetric nor asymmetric.

(c) Prove that if ρ is an asymmetric relation on a set S, then ρ is irreflexive.

(d) Prove that if ρ is an irreflexive and transitive relation on a set S, then ρ is asymmetric.

(e) Prove that if ρ is a nonempty, symmetric, and transitive relation on a set S, then ρ is not irreflexive.

12. Let ρ be a relation on a set S. For $A \subseteq S$, define

$$\#A = \{x \in S \mid x \rho y \text{ for all } y \in A\}$$
$$A\# = \{x \in S \mid y \rho x \text{ for all } y \in A\}$$

(a) Prove that if ρ is symmetric, then $\#A = A\#$.

(b) Prove that if $A \subseteq B$, then $\#B \subseteq \#A$ and $B\# \subseteq A\#$.

(c) Prove that $A \subseteq (\#A)\#$.

(d) Prove that $A \subseteq \#(A\#)$.

13. Graph the following partial orderings:

(a) $S = \{a, b, c\}$
$\rho = \{(a, a), (b, b), (c, c), (a, b), (b, c), (a, c)\}$

(b) $S = \{a, b, c, d\}$
$\rho = \{(a, a), (b, b), (c, c), (d, d), (a, b), (a, c)\}$

(c) $S = \{\varnothing, \{a\}, \{a, b\}, \{c\}, \{a, c\}, \{b\}\}$
$A \rho B \leftrightarrow A \subseteq B$

14. For Exercise 13, name any least elements, minimal elements, greatest elements, or maximal elements.

15. Graph the partial ordering "x divides y" on the set $\{2, 3, 5, 7, 21, 42, 105, 210\}$. Name any least elements, minimal elements, greatest elements, or maximal elements. Name a totally ordered subset with four elements.

★ 16. Graph each of the two partially ordered sets.

(a) $S = \{1, 2, 3, 5, 6, 10, 15, 30\}$
$x \rho y \leftrightarrow x$ divides y

(b) $S = \mathscr{P}(\{1, 2, 3\})$
$A \rho B \leftrightarrow A \subseteq B$

What do you notice about the structure of these two graphs?

17. For each graph of a partial ordering, list the ordered pairs that belong to the relation.

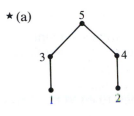

Figure 3.10

Figure 3.11

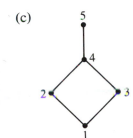

Figure 3.12

Figure 3.13

18. Let (S, ρ) and (T, σ) be two partially ordered sets. Define a relation μ on $S \times T$ by $(s_1, t_1) \, \mu \, (s_2, t_2) \leftrightarrow s_1 \, \rho \, s_2$ and $t_1 \, \sigma \, t_2$. Show that μ is a partial ordering on $S \times T$.

19. (a) Let $(S, \leq)$ be a partially ordered set and define a new binary relation $\geq$ on S by $x \geq y \leftrightarrow y \leq x$. Show that $(S, \geq)$ is a partially ordered set, called the **dual** of $(S, \leq)$.

    (b) If $(S, \leq)$ is a finite, partially ordered set with the graph shown in Figure 3.13, draw the graph of $(S, \geq)$.

    (c) Let $(S, \leq)$ be a totally ordered set and let $X = \{(x, x) \, | \, x \in S\}$. Show that the set difference $\geq \; - \; X$ equals the set $\leq'$.

20. A computer program is to be written that will generate a dictionary, or the index for a book. We will assume a maximum length of n characters per word. Thus, we are given a set S of words of length at most n, and we want to produce a linear list of these words arranged in alphabetical order. There is a natural total ordering $\leq$ on alphabetic characters ($a < b, b < c$, etc.), and we will assume our words contain only alphabetic characters. We want to define a total ordering $\leq$ on S that will arrange the members of S alphabetically. The idea is to compare two words X and Y character by character, passing over equal characters. If at any point the X-character alphabetically precedes the corresponding Y-character, then X precedes Y; if all characters in X are equal to the corresponding Y-characters but we run out of characters in X before characters in Y, then X precedes Y. Otherwise, Y precedes X.

    Formally, let $X = (x_1, x_2, \ldots, x_j)$ and $Y = (y_1, y_2, \ldots, y_k)$ be members of S with $j \leq k$. Let β (for blank) be a new symbol, and fill out X with $k - j$ blanks on the right. X can now be written $(x_1, x_2, \ldots, x_k)$. Let β precede any alphabetic character. Then $X \leq Y$ if

$$x_1 \neq y_1 \quad \text{and} \quad x_1 \leq y_1$$

or

$$x_1 = y_1, x_2 = y_2, \ldots, x_m = y_m \quad (m \leq k)$$
$$x_{m+1} \neq y_{m+1} \quad \text{and} \quad x_{m+1} \leq y_{m+1}$$

Otherwise, $Y \leq X$.

Note that because the ordering $\leq$ on alphabetic characters is a total ordering, if $Y \leq X$ by "otherwise," then there exists $m \leq k$ such that $x_1 = y_1, x_2 = y_2, \ldots, x_m = y_m, x_{m+1} \neq y_{m+1}$ and $y_{m+1} \leq x_{m+1}$.

(a) Show that $\leq$ on S as defined above is a total ordering.
(b) Apply the total ordering described to the words "boo," "bug," "be," "bah," and "bugg." Note why each word precedes the next.

★ 21. Exercise 20 discusses a total ordering on a set of words of length at most n that will produce a linear list in alphabetical order. Suppose we want to generate a list of all the distinct words in a text (for example, a compiler must create a symbol table of variable names). As in Exercise 20, we will assume that the words contain only alphabetic characters because there is a natural precedence relation already existing ($a < b$, $b < c$, etc.). If numeric or special characters are involved, they must be assigned a precedence relation with alphabetic characters (the collating sequence must be determined). If we list words alphabetically, it is a fairly quick procedure deciding whether a word currently being processed is new, but to fit the new word into place, all successive words must be moved one unit down the line. If the words are listed in the order in which they are processed, new words are simply tacked onto the end and no rearranging is necessary, but each word being processed has to be compared with each member of the list to determine if it is new. Thus, both logical linear lists have disadvantages.

We will describe a listing process called a **binary tree search** that usually allows a quick determination of whether a word is new and requires no juggling to fit it into place if it is, thus combining the advantages of both methods above. Suppose we want to process the phrase "when in the course of human events." The first word in the text is used to label the first node of a graph.

when

•

Figure 3.14

Once a node is labeled, it drops down a left and right arc, putting two unlabeled nodes below the one just labeled. When the next word in the text is processed, it is compared with the first node. When the word being processed alphabetically precedes the label of a node, the left arc is taken, and when the word alphabetically follows the label, the right arc is taken. The word becomes the label of the first unlabeled node it reaches. (If the word equals a node label, the next word in the text is processed.) This procedure continues for the entire text. Thus,

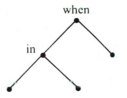

Figure 3.15

then

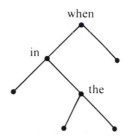

Figure 3.16

then

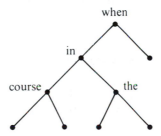

Figure 3.17

until finally

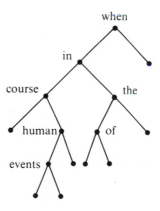

Figure 3.18

By traversing the nodes of this graph in the proper order (described by always processing the left nodes below a node first, then the node, then the right nodes below it), an alphabetical listing "course, events, human, in, of, the, when" is produced.

(a) The type of graph shown in Figure 3.18 is called a **tree**. Turned upside down, it could be viewed as the graph of a partial ordering $\leq$. What would be the least element? Is there a greatest element? Which of the following would belong to $\leq$: (in, of), (the, of), (in, events), (course, of)?

Here the tree structure contains more information than the partial ordering, because we are interested not only in whether a word w_1 precedes a word w_2 but also whether w_2 is to the left or right of w_1.

(b) Use the binary tree search to graph "Old King Cole was a merry old soul." Considering the graph (upside down) as a partial ordering, name the maximal elements.

22. The alphabetical ordering defined in Exercise 20 can be applied to words of any finite length. If we define $A*$ to be the set of all finite-length "words" (strings of characters, not necessarily meaningful) from the English alphabet, then the alphabetical ordering on $A*$ has all words composed only of the letter "a" preceding all other words. Thus, all the words in the infinite list

    a, aa, aaa, aaaa, . . .

precede words such as "b" or "aaaaaaab." Therefore this list does not enumerate $A*$, because we can never count up to any words with any characters other than "a." However, the set $A*$ is denumerable. Prove this by ordering $A*$ by length (all words of length 1 precede all words of length 2, etc.) and then alphabetically ordering words of the same length.

★ 23. (a) For the equivalence relation $\rho = \{(a, a), (b, b), (c, c), (a, c), (c, a)\}$, what is the set $[a]$? Does it have any other names?

(b) For the equivalence relation $\rho = \{(1, 1), (2, 2), (1, 2), (2, 1), (1, 3), (3, 1), (3, 2), (2, 3), (3, 3), (4, 4), (5, 5), (4, 5), (5, 4)\}$, what is the set $[3]$? What is the set $[4]$?

(c) For the equivalence relation of congruence modulo 2 on the set $\mathbb{Z}$, what is the set $[1]$?

(d) For the equivalence relation of congruence modulo 5 on the set $\mathbb{Z}$, what is the set $[-3]$?

24. (a) Given the partition $\{1, 2\}$ and $\{3, 4\}$ of the set $S = \{1, 2, 3, 4\}$, list the ordered pairs in the corresponding equivalence relation.

(b) Given the partition $\{a, b, c\}$ and $\{d, e\}$ of the set $S = \{a, b, c, d, e\}$, list the ordered pairs in the corresponding equivalence relation.

25. (a) Compute the total number of possible partitions of a 3-element set.
    (b) Compute the total number of possible partitions of a 4-element set.

26. Given two partitions π_1 and π_2 of a set S, π_1 is a **refinement** of π_2 if each block of π_1 is a subset of a block of π_2. Show that refinement is a partial ordering on the set of all partitions of S.

27. Let S be the set of all propositions (unquantified statements) with n statement letters (see Section 1.1). Let ρ be a binary relation on S defined by $P \rho Q \leftrightarrow$ "$P \leftrightarrow Q$ is a tautology." Show that ρ is an equivalence relation on S and describe the resulting equivalence classes.

28. (a) Show that the EQUIVALENCE statement in the FORTRAN language defines an equivalence relation on variable names. Describe the resulting equivalence classes.
    (b) Show that the DEFINED attribute of the DECLARE statement in PL/I defines an equivalence relation on variable names. Describe the resulting equivalence classes.

SECTION 3.2   FUNCTIONS

In this section we discuss functions, which are essentially special cases of binary relations. This view of a function is a rather sophisticated one, however, and we will work up to it gradually.

The Definition

"Function" is a common enough word even in nontechnical contexts. A newspaper may have an article on how starting salaries for this year's college graduates have increased over those for last year's graduates. The article might say something like "The salary increase varies depending upon the degree program" or "The salary increase is a function of the degree program." It may illustrate this functional relationship with a graph like Figure 3.19. The graph shows that each degree program has some figure showing the salary increase associated with it, that no degree program has more than one figure associated with it, and that both the physical sciences and the liberal arts have the same figure, 3%.

Of course, we also use mathematical functions in algebra and calculus. The equation $g(x) = x^3$ expresses a functional relationship between values for x and corresponding values that result when x is replaced in

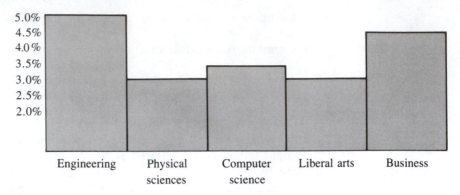

Figure 3.19

the equation by its values. Thus an x-value of 2 has the number $2^3 = 8$ associated with it. (This number is expressed as $g(2) = 8$.) Similarly, $g(1) = 1^3 = 1$, $g(-1) = (-1)^3 = -1$, and so on. For each x-value, the corresponding $g(x)$-value is unique. If we were to graph this function on a rectangular coordinate system, the points $(2, 8)$, $(1, 1)$, and $(-1, -1)$ would be points on the graph. If we allow x to take on any real-number value, the resulting graph is the continuous curve shown in Figure 3.20.

The function in the salary increase example could be described as follows. We set the stage by the diagram in Figure 3.21, which indicates that the function always starts with a given degree program and that a particular salary increase is associated with that degree program. The association itself is described by the set of ordered pairs {(engineering, 5.0%), (physical sciences, 3.0%), (computer science, 3.5%), (liberal arts, 3.0%), (business, 4.5%)}.

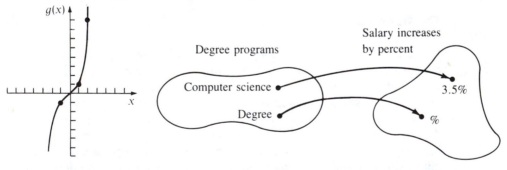

Figure 3.20　　　　　　　　　**Figure 3.21**

For the algebraic example $g(x) = x^3$, Figure 3.22 shows that the function always starts with a given real number and that a second real number is associated with it. The association itself is described by $\{(x, g(x)) \mid g(x) = x^3\}$, or simply $g(x) = x^3$. This set includes $(2, 8)$, $(1, 1)$,

Real numbers   Real numbers

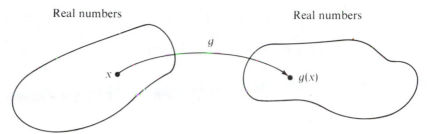

Figure 3.22

$(-1, -1)$, but because it is an infinite set, we cannot list all its members; we have to describe them.

From the above examples, we can conclude that there are three parts to a function: (1) a set of starting values, (2) a set from which associated values come, and (3) the association itself. The set of starting values is called the **domain** of the function, and the set from which associated values come is called the **codomain** of the function. Thus, the picture for an arbitrary function f is shown in Figure 3.23. Here f is a function from S to T, symbolized $f : S \to T$. The association itself is a set of ordered pairs, each of the form (s, t) where $s \in S$, $t \in T$, and t is the value from T that the function associates with the value s from S; $t = f(s)$. Hence, the association is a subset of $S \times T$ (a binary relation on $S \times T$). But the important property of this relation is that every member of S must have one and only one T-value associated with it, so every $s \in S$ will appear exactly once as the first component of an (s, t) pair. (This property does not prevent a given T-value from appearing more than once.)

Domain S Codomain T

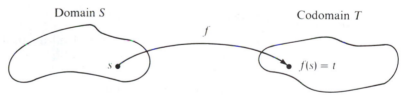

Figure 3.23

We are now ready for a formal definition of a function.

3.34
Definition

Let S and T be sets. A **function (mapping)** $f : S \to T$ is a subset of $S \times T$ where each member of S appears exactly once as the first component of an ordered pair. S is the **domain** and T the **codomain** of the function. If (s, t) belongs to the function, then t is denoted by $f(s)$; t is the **image** of s under f, s is a **preimage** of t under f, and f is said to map s to t. For $A \subseteq S$, $f(A)$ denotes $\{f(a) \mid a \in A\}$.   □

We have talked a lot about values from the sets S and T, but as our salary increase example shows, these values are not necessarily numbers, nor is the association itself necessarily described by an equation.

3.35 Practice   Which of the following are functions from the domain to the codomain indicated? For those that are not, why not?

(a) $f : S \to T$ where $S = T = \{1, 2, 3\}$, $f = \{(1, 1), (2, 3), (3, 1), (2, 1)\}$

(b) $g : \mathbb{Z} \to \mathbb{N}$ where g is defined by $g(x) = |x|$ (the absolute value of x)

(c) $h : \mathbb{N} \to \mathbb{N}$ where h is defined by $h(x) = x - 4$

(d) $f : S \to T$ where S is the set of all people in your hometown, T is the set of all Social Security numbers, and f associates with each person that person's Social Security number

(e) $g : S \to T$ where $S = \{1972, 1973, 1974, 1975\}$, $T = \{\$20,000, \$30,000, \$40,000, \$50,000, \$60,000\}$, and g is defined by the graph in Figure 3.24

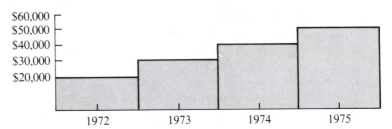

Profits of the American Earthworm Corp.

Figure 3.24

(f) $h : S \to T$ where S is the set of all quadratic polynomials in x with integer coefficients, $T = \mathbb{N}$, and h is defined by $h(ax^2 + bx + c) = b + c$

(g) $f : \mathbb{R} \to \mathbb{R}$ where f is defined by $f(x) = 4x - 1$

(h) $g : \mathbb{N} \to \mathbb{N}$ where g is defined by

$$g(x) = \begin{cases} x + 3 & \text{if } x \geq 5 \\ x & \text{if } x \leq 5 \end{cases}$$ □

3.36 Practice   (a) For $f : \mathbb{Z} \to \mathbb{Z}$ given by $f(x) = x^2$, what is the image of -4?

(b) What are the preimages of 9?   □

3.37 Example   The sequences given by recursive definitions in Section 1.5 are functions whose domains are the positive integers.   □

3.38 Example   In Section 2.1 we defined a binary operation ∘ on a set S as associating a unique member of S, $x \circ y$, with every (x, y) pair of elements of S. Thus, any binary operation on S is a function: $S \times S \to S$.   □

Let's return to our earlier example of $g : \mathbb{R} \to \mathbb{R}$ where g is defined by $g(x) = x^3$. It is common in algebra and calculus to say "consider the function $g(x) = x^3$," implying that the equation *is* the function. Technically, the equation only describes how to compute associated values. The function $h : \mathbb{R} \to \mathbb{R}$ given by $h(x) = x^3 - 3x + 3(x + 5) - 15$ is the same function as g since it contains the same ordered pairs. However, the equation is different in that it says to do different things to any given x-value. On the other hand, the function $f : \mathbb{Z} \to \mathbb{R}$ given by $f(x) = x^3$ is not the same function as g. The domain has been changed, which changes the set of ordered pairs. The graph of $f(x)$ would consist of discrete (separated) points (see Figure 3.25).

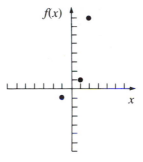

Figure 3.25

Most of the functions in which we are interested will have this latter feature. In a digital computer, information is processed in a series of distinct (discrete) steps. Even in situations where one quantity varies continuously with another, we approximate by taking data at discrete, small intervals, much as the graph of $g(x)$ (Figure 3.20) is approximated by the graph of $f(x)$ (Figure 3.25).

Finally, let's look at the function $k : \mathbb{R} \to \mathbb{C}$ given by $k(x) = x^3$. The equation and domain is the same as for $g(x)$; the codomain has been enlarged, but this does not affect the ordered pairs. Is this considered the same function as $g(x)$? It is not, but to see why we need a further definition.

Special Functions

Onto Functions

Let $f : S \to T$ be an arbitrary function with domain S and codomain T (see Figure 3.26). Part of the definition of a function is that every member of

S has an image under f and that all the images are members of T; the set R of all such images is called the **range** of the function f. Thus, $R = \{f(s) \mid s \in S\}$, or $R = f(S)$. Clearly, $R \subseteq T$; the range R is shaded in Figure 3.27. If it should happen that $R = T$, that is, that the range coincides with the codomain, then the function is called an onto function.

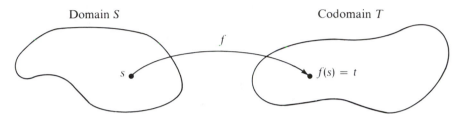

Figure 3.26

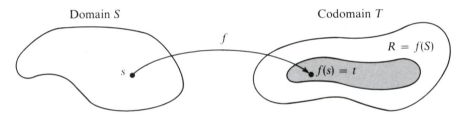

Figure 3.27

**3.39
Definition**

A function $f : S \to T$ is an **onto** or **surjective** function if the range of f equals the codomain of f.   □

In every function with range R and codomain T, $R \subseteq T$. To prove that a given function is onto, we must show that $T \subseteq R$; then $R = T$. We must therefore show that an arbitrary member of the codomain is a member of the range, that is, is the image of some member of the domain. On the other hand, if we can produce one member of the codomain that is not the image of any member of the domain, then we have proved that the function is not onto.

3.40 Example

The function $g : \mathbb{R} \to \mathbb{R}$ defined by $g(x) = x^3$ is an onto function. To prove that $g(x)$ is onto, let r be an arbitrary real number, and let $x = \sqrt[3]{r}$. Then x is a real number, so x belongs to the domain of g and $g(x) = (\sqrt[3]{r})^3 = r$. Hence, any member of the codomain is the image under g of a member of the domain. The function $k : \mathbb{R} \to \mathbb{C}$ given by $k(x) = x^3$ is not surjective. There are many complex numbers (i, for example) that cannot be obtained by cubing a real number. Thus, g and k are not equal functions. *For two functions to be equal, they must have the same domain and the same codomain and consist of the same ordered pairs.*   □

3.41 Example   Let $f : \mathbb{Q} \to \mathbb{Q}$ be defined by $f(x) = 3x + 2$. To test whether f is onto, let $q \in \mathbb{Q}$. We want an $x \in \mathbb{Q}$ such that $f(x) = 3x + 2 = q$. When we solve this equation for x, we find that $x = (q - 2)/3$ is the only possible value and is indeed a member of $\mathbb{Q}$. Thus, q is the image of a member of $\mathbb{Q}$ under f, and f is onto. However, the function $h : \mathbb{Z} \to \mathbb{Q}$ defined by $h(x) = 3x + 2$ is not onto because there are many values $q \in \mathbb{Q}$, for example 0, for which the equation $3x + 2 = q$ has no integer solution.   □

3.42 Practice   Which of the functions found in Practice 3.35 are onto functions?   □

One-to-One Functions

The definition of a function guarantees a unique image for every member of the domain. A given member of the range may have more than one preimage, however. In our very first example of a function (salary increases) both physical sciences and liberal arts were preimages of 3%. This function was not one-to-one.

**3.43
Definition**   A function $f : S \to T$ is **one-to-one**, or **injective**, if no member of T is the image under f of two distinct elements of S.   □

To prove that a function is one-to-one, we often assume that there are elements s_1 and s_2 of S with $f(s_1) = f(s_2)$, and then show that $s_1 = s_2$. To prove a function is not injective, we produce an element in the range with two preimages in the domain.

3.44 Example   The function $g : \mathbb{R} \to \mathbb{R}$ defined by $g(x) = x^3$ is one-to-one because if x and y are real numbers with $g(x) = g(y)$, then $x^3 = y^3$ and $x = y$. The function $f : \mathbb{R} \to \mathbb{R}$ given by $f(x) = x^2$ is not injective because, for example, $f(2) = f(-2) = 4$. However, the function $h : \mathbb{N} \to \mathbb{N}$ given by $h(x) = x^2$ is injective since if x and y are nonnegative integers with $h(x) = h(y)$, then $x^2 = y^2$; because x and y are both nonnegative, $x = y$.   □

3.45 Practice   Which of the functions found in Practice 3.35 are one-to-one functions?   □

Bijections

**3.46
Definition**   A function $f : S \to T$ is **bijective** if it is both one-to-one and onto.   □

3.47 Example   The function $g : \mathbb{R} \to \mathbb{R}$ given by $g(x) = x^3$ is a bijection. The function in part (g) of Practice 3.35 is a bijection. The function $f : \mathbb{R} \to \mathbb{R}$ given by $f(x) = x^2$ is not a bijection (not one-to-one) and neither is the function $k : \mathbb{R} \to \mathbb{C}$ given by $k(x) = x^3$ (not onto).   □

3.48
Definition   A set S is **equivalent** to a set T if there exists a bijection $f : S \rightarrow T$. Two sets that are equivalent have the same **cardinality**.   □

If S is equivalent to T, then all the members of S and T are paired off by f in a one-to-one correspondence. If S and T are finite sets, this pairing off can only happen when S and T are the same size. With infinite sets, the idea of size gets a bit fuzzy, because we can sometimes prove that a given set is equivalent to what *seems* to be a smaller set. The correct extension of the idea of size to encompass infinite sets is cardinality.

3.49 Practice   Describe a bijection $f : \mathbb{Z} \rightarrow \mathbb{N}$, thus showing that $\mathbb{Z}$ is equivalent to $\mathbb{N}$ ($\mathbb{Z}$ and $\mathbb{N}$ have the same cardinality) even though $\mathbb{N} \subset \mathbb{Z}$.   □

If we have found a bijection between a set S and $\mathbb{N}$, we have established a one-to-one correspondence between the members of S and the nonnegative integers. We can name the members of S according to this correspondence, writing s_0 for the value of S associated with 0, s_1 for the value of S associated with 1, and so on. Then the list

$$s_0, s_1, s_2, \ldots$$

includes all of the members of S. Since this list constitutes an enumeration of S, S is a denumerable set. Conversely, if S is denumerable, then a listing of the members of S exists and can be used to define a bijection between S and $\mathbb{N}$. Therefore a set is denumerable if and only if it is equivalent to $\mathbb{N}$.

For finite sets, we know that if S has n elements, then $\mathcal{P}(S)$ has 2^n elements. Of course, $2^n > n$, and we cannot find a bijection between a set with n elements and a set with 2^n elements. Therefore S and $\mathcal{P}(S)$ are not equivalent. This result is also true for infinite sets.

3.50 Theorem   **Cantor's Theorem**: For any set S, S and $\mathcal{P}(S)$ are not equivalent.

Proof: We will do a proof by contradiction and assume that S and $\mathcal{P}(S)$ are equivalent. Let f be the bijection between S and $\mathcal{P}(S)$. For any member s of S, $f(s)$ is a member of $\mathcal{P}(S)$, so that $f(s)$ is a set containing some members of S, possibly containing s itself. Now we define a set $X = \{x \in S \mid x \notin f(x)\}$. Because X is a subset of S, it is an element of $\mathcal{P}(S)$ and therefore must be equal to $f(y)$ for some $y \in S$. Then y either is or is not a member of X. If $y \in X$, then by the definition of X, $y \notin f(y)$, but since $f(y) = X$, then $y \notin X$. On the other hand, if $y \notin X$, then since $X = f(y)$, $y \notin f(y)$, and by the definition of X, $y \in X$. In either case, there is a contradiction, and our original assumption is incorrect. Therefore S and $\mathcal{P}(S)$ are not equivalent.   □

The proof of Cantor's Theorem depends upon the nature of set X, which was carefully constructed to provide the crucial contradiction. In this sense, the proof is similar to the diagonalization method (Example 2.41) used to prove the existence of an uncountable set. Indeed, the existence of an uncountable set can be shown directly from Cantor's Theorem.

3.51 Example   The set $\mathbb{N}$ is, of course, a denumerable set. By Cantor's Theorem, the set $\mathcal{P}(\mathbb{N})$ is not equivalent to $\mathbb{N}$ and is therefore not a denumerable set, although it is clearly infinite. $\square$

Composition of Functions

Suppose that f and g are functions with $f:S \to T$ and $g:T \to U$. Then for any $s \in S$, $f(s)$ is a member of T, which is also the domain of g. Thus, the function g can be applied to $f(s)$. The result is $g(f(s))$, a member of U; see Figure 3.28. In effect, taking an arbitrary member s of S, applying the function f, and then applying the function g to $f(s)$ is the same as associating a unique member of U with s. In short, we have created a function: $S \to U$, called the composition function of f and g and denoted by $g \circ f$.

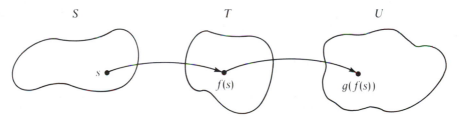

Figure 3.28

3.52 Definition   Let $f:S \to T$, and $g:T \to U$. Then the **composition function, $g \circ f$**, is a function from S to U defined by $(g \circ f)(s) = g(f(s))$. $\square$

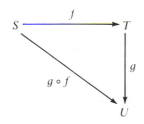

Figure 3.29

Note that the function $g \circ f$ is applied right to left; function f is applied first and then function g.

The diagram in Figure 3.29 illustrates the definition of the composition function. The corners indicate the domains and codomains of the three functions. The diagram says that, starting with an element of S, if we follow path $g \circ f$ or path f and then path g, we get to the same element in U. Diagrams illustrating that alternate paths produce the same effect are called **commutative diagrams**.

3.53 Practice   Let $f:\mathbb{R} \to \mathbb{R}$ be defined by $f(x) = x^2$. Let $g:\mathbb{R} \to \mathbb{R}$ be the truncation function; that is, $g(x)$ equals the integral part of x—any digits to the right of the decimal point are dropped.

   (a) What is the value of $(g \circ f)(2.3)$?
   (b) What is the value of $(f \circ g)(2.3)$?   □

   From Practice 3.53, we see that order is important in function composition, which should not be surprising. If we were to write a computer program to carry out function composition, we would generally require an assignment statement to compute each function. Reversing the order of composition would essentially reverse the order of two program statements, which almost always changes the program results.
   Function composition preserves the property of being onto or of being one-to-one. Again, let $f:S \to T$ and $g:T \to U$, but also suppose that both f and g are onto functions. Then the composition function $g \circ f$ is also onto. Recall that $g \circ f:S \to U$, so we must pick an arbitrary $u \in U$ and show that it has a preimage under $g \circ f$ in S. Because g is surjective, there exists $t \in T$ such that $g(t) = u$. And because f is surjective, there exists $s \in S$ such that $f(s) = t$. Then $(g \circ f)(s) = g(f(s)) = g(t) = u$, and $g \circ f$ is an onto function.

3.54 Practice   Let $f:S \to T$ and $g:T \to U$, and assume that both f and g are one-to-one functions. Prove that $g \circ f$ is a one-to-one function. (Hint: assume that $(g \circ f)(s_1) = (g \circ f)(s_2)$.)   □

   We have now proved the theorem below.

3.55 Theorem   The composition of two bijections is a bijection.   □

Permutation Functions

   Theorem 3.55 pertains to the composition function *when it exists*, that is, when the domains and codomains permit the creation of a composition function. Of course, this is no problem if only one set is involved, that is, if we are only discussing functions that map a set A into A.

3.56 Definition   For a given set A, $S_A = \{f \mid f:A \to A$ and f is a bijection$\}$. S_A is thus the set of all bijections of set A into (and therefore onto) itself; such functions are called **permutations** of A.   □

   Theorem 3.55 says that if we perform function composition on two members of S_A, we get a (unique) member of S_A; thus, function composition is a binary operation on the set S_A.

In Section 2.3 we described a permutation of objects in a set as being an ordered arrangement of those objects. Similarly, permutation functions represent ordered arrangements of the objects in the domain. If $A = \{1, 2, 3, 4\}$, one permutation function of A, call it f, is given by $f = \{(1, 2), (2, 3), (3, 1), (4, 4)\}$. We can also describe function f in array form by listing the elements of the domain in a row and, directly beneath, the images of these elements under f. Thus,

$$f = \begin{pmatrix} 1 & 2 & 3 & 4 \\ 2 & 3 & 1 & 4 \end{pmatrix}$$

The bottom row is an ordered arrangement of the objects in the top row. A shorter way to describe the permutation f is to use cycle notation and write $f = (1, 2, 3)$—understood to mean that f maps each element listed to the one on its right, the last element listed to the first, and any element of the domain not listed to itself. Therefore, the cycles $(1, 2, 3)$, $(2, 3, 1)$, and $(3, 1, 2)$ all represent f.

3.57 Practice   (a) Let $A = \{1, 2, 3, 4, 5\}$, and let $f \in S_A$ be given in array form by

$$f = \begin{pmatrix} 1 & 2 & 3 & 4 & 5 \\ 4 & 2 & 3 & 5 & 1 \end{pmatrix}$$

Write f in cycle form.
(b) Let $A = \{1, 2, 3, 4, 5\}$, and let $g \in S_A$ be given in cycle form by $g = (2, 4, 5, 3)$. Write g in array form.   □

If f and g are members of S_A for some set A, then $g \circ f \in S_A$, and the action of $g \circ f$ on any member of A is determined by applying function f and then function g. If f and g are cycles, $g \circ f$ is still computed the same way. If $A = \{1, 2, 3, 4\}$ and $f, g \in S_A$ are given by $f = (1, 2, 3)$ and $g = (2, 3)$, then $g \circ f = (2, 3) \circ (1, 2, 3)$. For element 1 of A, $1 \rightarrow 2$ under f and $2 \rightarrow 3$ under g, so $1 \rightarrow 3$ under $g \circ f$. Similarly, $2 \rightarrow 3$ under f and $3 \rightarrow 2$ under g, so $2 \rightarrow 2$ under $g \circ f$. Testing what happens to 3 and 4, we conclude that $g \circ f = (1, 3)$. But if we compute $f \circ g$, we get $(1, 2)$. (Function composition on S_A is, like subtraction, a binary operation where the order of the elements matters.)

3.58 Practice   Let $A = \{1, 2, 3, 4, 5\}$, and let f and g belong to S_A. Compute $g \circ f$ for the following:

(a) $f = (5, 2, 3)$; $g = (3, 4, 1)$. Write the answer in cycle form.
(b) $f = (1, 2, 3, 4)$; $g = (3, 2, 4, 5)$. Write the answer in array form.   □

If A is an infinite set, not every permutation of A can be written as a cycle. Even when A is a finite set, not every permutation of A can be written as a cycle. The permutation $g \circ f$ of Practice 3.58(b) is not a cycle. However, a permutation on a finite set can always be written as the composition of two or more cycles with no common elements. The permutation

$$\begin{pmatrix} 1 & 2 & 3 & 4 & 5 \\ 4 & 2 & 5 & 1 & 3 \end{pmatrix}$$

is $(1, 4) \circ (3, 5)$ or $(3, 5) \circ (1, 4)$. (When there are no common elements, the order of the cycles does not matter.)

3.59 Practice   Write

$$\begin{pmatrix} 1 & 2 & 3 & 4 & 5 & 6 \\ 2 & 4 & 5 & 1 & 3 & 6 \end{pmatrix}$$

as a composition of cycles.   □

Inverse Functions

There is one more important property of bijective functions. Let $f : S \to T$ be a bijection. Because f is onto, every $t \in T$ has a preimage in S. And because f is one-to-one, that preimage is unique. We could associate with each element t of T a unique member of S, namely, that $s \in S$ such that $f(s) = t$. This association describes a function g, $g : T \to S$. The picture for f and g is given in Figure 3.30. The domains and codomains of g and f are such that we can form both $g \circ f : S \to S$ and $f \circ g : T \to T$. If $s \in S$, then $(g \circ f)(s) = g(f(s)) = g(t) = s$. Thus, $g \circ f$ maps each element of S to itself. A function on a set S that maps each element to itself is called the **identity function** on S, denoted by i_S. Hence, $g \circ f = i_S$.

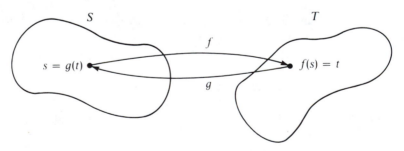

Figure 3.30

3.60 Practice   Show that $f \circ g = i_T$.   □

We have now seen that if f is a bijection, $f : S \to T$, then there is a function $g : T \to S$ with $g \circ f = i_S$ and $f \circ g = i_T$. The converse is also true. To prove the converse, suppose $f : S \to T$ and there exists $g : T \to S$ with $g \circ f = i_S$ and $f \circ g = i_T$. We can prove that f is a bijection. To show that f is onto, let $t \in T$. Then $t = i_T(t) = (f \circ g)(t) = f(g(t))$. Because $g : T \to S$, $g(t) \in S$, and $g(t)$ is the preimage under f of t. To show that f is one-to-one, suppose $f(s_1) = f(s_2)$. Then $g(f(s_1)) = g(f(s_2))$ and $(g \circ f)(s_1) = (g \circ f)(s_2)$, implying $i_S(s_1) = i_S(s_2)$, or $s_1 = s_2$. Thus, f is a bijection.

3.61
Definition

Let f be a function, $f : S \to T$. If there exists a function $g : T \to S$ such that $g \circ f = i_S$ and $f \circ g = i_T$, then g is called the **inverse function** of f, denoted by f^{-1}.   □

We have proved the following theorem.

3.62 Theorem

Let $f : S \to T$. Then f is a bijection if and only if f^{-1} exists.   □

Actually, we have been a bit sneaky in talking about *the* inverse function of f. What we have shown is that if f is a bijection, this is equivalent to the existence of *an* inverse function. But it is easy to see that there is only one such inverse function. *When you want to prove that something is unique, the standard technique is to assume that there are two such things and then obtain a contradiction.* Thus, suppose f has two inverse functions, f_1^{-1} and f_2^{-1} (existence of either means that f is a bijection). Both f_1^{-1} and f_2^{-1} are functions from T to S; if they are not the same function, then they must act differently somewhere. Assume that there is a $t \in T$ such that $f_1^{-1}(t) \neq f_2^{-1}(t)$. Because f is one-to-one, it follows that $f(f_1^{-1}(t)) \neq f(f_2^{-1}(t))$, or $(f \circ f_1^{-1})(t) \neq (f \circ f_2^{-1})(t)$. But both $f \circ f_1^{-1}$ and $f \circ f_2^{-1}$ are i_T, so $t \neq t$, which is a contradiction. We are therefore justified in speaking of f^{-1} as *the* inverse function of f. If f is a bijection so that f^{-1} exists, then f is the inverse function for f^{-1}; therefore, f^{-1} is also a bijection.

3.63 Practice

$f : \mathbb{R} \to \mathbb{R}$ given by $f(x) = 3x + 4$ is a bijection. Describe f^{-1}.   □

A Final Remark

The definition of a function includes functions of more than one variable. We can have a function $f : S_1 \times S_2 \times \cdots \times S_n \to T$ that associates with each ordered n-tuple of elements $(s_1, s_2, \ldots, s_n)$, $s_i \in S_i$, a unique element of T. Recall from Section 2.1 that S^n is the notation for the set of all ordered n-tuples of elements of S.

3.64 Example

$f : \mathbb{Z}^2 \to \mathbb{N}$ given by $f(x, y) = x^2 + 2y^2$ maps the ordered pair of integers $(-2, 3)$ to the nonnegative integer 22. This function is not one-to-one because $(2, 3)$ also maps to 22.   □

3.65 Example   Let P be a proposition (an unquantified statement) with n statement letters (see Section 1.1). The truth table for P defines a function mapping $\{T, F\}^n \rightarrow \{T, F\}$. This function will be onto unless P is a tautology or a contradiction.   □

✔ Checklist

Definitions

domain (*p. 133*)
codomain (*p. 133*)
function (mapping) (*p. 133*)
image (*p. 133*)
preimage (*p. 133*)
range (*p. 136*)
onto (surjective) function (*p. 136*)
one-to-one (injective) function (*p. 137*)
bijection (*p. 137*)
equivalent sets (*p. 138*)
cardinality of sets (*p. 138*)
Cantor's Theorem (*p. 138*)
composition function (*p. 139*)
commutative diagram (*p. 139*)
permutation function (*p. 140*)
identity function (*p. 142*)
inverse function (*p. 143*)

Techniques

Test whether a given relation is a function.

Test a function for being one-to-one or onto.

Find the image of an element under functon composition.

Write permutations of a set in array or cycle form.

Main ideas

Functions, especially bijections.

Composition of functions, which preserves bijectiveness.

Permutations on a set and their composition.

The inverse function of a bijection is itself a bijection.

Exercises Section 3.2

1. Let $S = \{0, 2, 4, 6\}$ and $T = \{1, 3, 5, 7\}$. Determine whether each of the following sets of ordered pairs is a function with domain S and codomain T. If so, is it one-to-one? Is it onto?
   (a) $\{(0, 2), (2, 4), (4, 6), (6, 0)\}$
   (b) $\{(6, 3), (2, 1), (0, 3), (4, 5)\}$
   (c) $\{(2, 3), (4, 7), (0, 1), (6, 5)\}$
   (d) $\{(2, 1), (4, 5), (6, 3)\}$
   (e) $\{(6, 1), (0, 3), (4, 1), (0, 7), (2, 5)\}$

2. For any bijections in Exercise 1, describe the inverse function.

★ 3. Which of the following are functions from the domain to the codomain given? Which functions are one-to-one? Which functions are onto? Describe the inverse function for any bijective function.
   (a) $f : \mathbb{Z} \to \mathbb{N}$ where f is defined by $f(x) = x^2 + 1$
   (b) $g : \mathbb{N} \to \mathbb{Q}$ where g is defined by $g(x) = 1/x$
   (c) $h : \mathbb{Z} \times \mathbb{N} \to \mathbb{Q}$ where h is defined by $h(z, n) = z/(n + 1)$
   (d) $f : \{1, 2, 3\} \to \{p, q, r\}$ where $f = \{(1, q), (2, r), (3, p)\}$
   (e) $g : \mathbb{N} \to \mathbb{N}$ where g is given by $g(x) = 2^x$
   (f) $h : \mathbb{R}^2 \to \mathbb{R}^2$ where h is defined by $h(x, y) = (y + 1, x + 1)$

4. Show that each of the following is a function from S to T. Which are surjective (onto)? Which are injective (one-to-one)?
   (a) $K : S \to T$ where S is a set of 16 data records, T is the set of integers from 1 to 16 (representing 16 five-position data fields), and K is given by the following FORTRAN program:

```
        DIMENSION NUM(16)
        K1 = 16
        K2 = 1
        DO 100 I = 1, K1
        K = (I − 1)/K2 + K2
        READ 1000,   (NUM(J), J = 1, K)
    100 PRINT 2000, K
              .
              .
              .
   1000 FORMAT (16I5)
   2000 FORMAT (I6)
        END
```

   (b) The program is the same as part (a) except S contains 15 records and line 2 of the program is replaced by K1 = 15.
   (c) The program is the same as part (a) except line 3 of the program is replaced by K2 = 2.

5. Let $S = \{a, b, c, d\}$ and $T = \{x, y, z\}$.

(a) Give an example of a function from S to T that is neither onto nor one-to-one.

(b) Give an example of a function from S to T that is onto but not one-to-one.

(c) Can you find a function from S to T that is one-to-one?

6. Let f be a function, $f: S \to T$.
   (a) Show that for all subsets A and B of S, $f(A \cap B) \subseteq f(A) \cap f(B)$.
   (b) Show that $f(A \cap B) = f(A) \cap f(B)$ for all subsets A and B of S if and only if f is one-to-one.

★7. (a) Let $S = \{a, b, c\}$ and $T = \{p, q\}$. How many functions f are there such that $f: S \to T$? How many of these are onto functions?
   (b) Let S be a set with m elements and T be a set with n elements. How many functions f are there such that $f: S \to T$?
   (c) Let S and T be sets with m and n elements, respectively, and assume $m \le n$. How many injective functions are there from S to T?

8. (a) Let f be a function, $f: S \to S$, where S is a set with n elements. Show that f is one-to-one if and only if f is onto.
   (b) Find an infinite set S and a function $f: S \to S$ such that f is one-to-one but not onto.
   (c) Find an infinite set S and a function $f: S \to S$ such that f is onto but not one-to-one.

★9. By the definition of a function f from S to T, f is a subset of $S \times T$ where the image of every $s \in S$ under f is uniquely determined as the second component of the ordered pair (s, t) in f. Now consider any binary relation ρ on $S \times T$. The relation ρ is a subset of $S \times T$ in which some elements of S may not appear at all as first components of an ordered pair and some may appear more than once. We can view ρ as a **nondeterministic function** from a subset of S to T. An $s \in S$ not appearing as the first component of an ordered pair represents an element outside the domain of ρ. For an $s \in S$ appearing once or more as a first component, ρ can select for the image of s any one of the corresponding second components.

Let $S = \{1, 2, 3\}$, $T = \{a, b, c\}$, and $U = \{m, n, o, p\}$. Let ρ be a binary relation on $S \times T$ and σ be a binary relation on $T \times U$ defined by

$$\rho = \{(1, a), (1, b), (2, b), (2, c), (3, c)\}$$
$$\sigma = \{(a, m), (a, o), (a, p), (b, n), (b, p), (c, o)\}$$

Thinking of ρ and σ as nondeterministic functions from S to T and T to U, respectively, we can form the composition $\sigma \circ \rho$, a nondeterministic function from S to U.

(a) What is the set of possible images of 1 under $\sigma \circ \rho$?

(b) What is the set of possible images of 2 under $\sigma \circ \rho$? Of 3?

10. Let $S = \{1, 2, 3, 4\}$, $T = \{1, 2, 3, 4, 5, 6\}$, and $U = \{6, 7, 8, 9, 10\}$. Also, let $f = \{(1, 2), (2, 4), (3, 3), (4, 6)\}$ be a function from S to T, and let $g = \{(1, 7), (2, 6), (3, 9), (4, 7), (5, 8), (6, 9)\}$ be a function from T to U. Write the ordered pairs in the function $g \circ f$.

★ 11. Let $f:\mathbb{N} \to \mathbb{N}$ be defined by $f(x) = x + 1$. Let $g:\mathbb{N} \to \mathbb{N}$ be defined by $g(x) = 3x$.
   (a) What is $(g \circ f)(5)$?   (b) What is $(f \circ g)(5)$?
   (c) What is $(g \circ f)(x)$?   (d) What is $(f \circ g)(x)$?
   (e) What is $(f \circ f)(x)$?   (f) What is $(g \circ g)(x)$?

12. (a) Let $f:\mathbb{R} \to \mathbb{Z}$ be defined by $f(x) = $ greatest integer $\leq x$. Let $g:\mathbb{Z} \to \mathbb{N}$ be defined by $g(x) = x^2$. What is $(g \circ f)(-4.7)$?
   (b) Let f map the set of books into the integers where f assigns to each book the page number of the last page. Let $g:\mathbb{Z} \to \mathbb{Z}$ be given by $g(x) = 2x$. What is $(g \circ f)$ (this book)?
   (c) Let f map strings of alphabetic characters and blank spaces into strings of alphabetic consonants where f takes any string and removes all vowels and all blanks. Let g map strings of alphabetic consonants into integers where g maps a string into the number of characters it contains. What is $(g \circ f)$ (abraham lincoln)?

★ 13. Let $A = \{1, 2, 3, 4, 5\}$. Write each of the following permutations on A in cycle form:
   (a) $f = \begin{pmatrix} 1 & 2 & 3 & 4 & 5 \\ 3 & 1 & 5 & 4 & 2 \end{pmatrix}$
   (b) $f = \{(1, 4), (2, 5), (3, 2), (4, 3), (5, 1)\}$

14. Let $A = \{a, b, c, d\}$. Write each of the following permutations on A in array form:
   (a) $f = \{(a, c), (b, b), (c, d), (d, a)\}$
   (b) $f = (c, a, b, d)$
   (c) $f = (d, b, a)$
   (d) $f = (a, b) \circ (b, d) \circ (c, a)$

★ 15. (a) Let $A = \{1, 2, 3\}$. Write the elements of S_A, the set of all permutations on A.
   (b) Let $A = \{1, 2, \dots, n\}$. How many elements are in the set S_A?

16. Let A be any set and let S_A be the set of all permutations of A. Let $f, g, h \in S_A$. Prove that the functions $h \circ (g \circ f)$ and $(h \circ g) \circ f$ are equal, thereby showing that we can write $h \circ g \circ f$ without parentheses to indicate grouping.

★ 17. Find the composition of the following cycles representing permutations on $A = \{1, 2, 3, 4, 5, 6, 7, 8\}$. Write your answer as a single cycle or as a composition of cycles with no common elements.
   (a) $(1, 3, 4) \circ (5, 1, 2)$
   (b) $(2, 7, 8) \circ (1, 2, 4, 6, 8)$

(c) $(1, 3, 4) \circ (5, 6) \circ (2, 3, 5) \circ (6, 1)$ (By Exercise 16, we can omit parentheses indicating grouping.)

(d) $(2, 7, 1, 3) \circ (2, 8, 7, 5) \circ (4, 2, 1, 8)$

18. The "pushdown store," or "stack," is a storage device operating much like a set of plates stacked on a spring in a cafeteria. All storage locations are initially empty. An item of data is added to the top of the stack by a "push" instruction, which pushes any previously stored items further down in the stack. Only the topmost item on the stack is accessible at any moment, and it is fetched and removed from the stack by a "pop" instruction.

    Let's consider strings of integers that are an even number of characters in length; half of the characters are positive integers, and the other half are zeros. We process these strings through a pushdown store as follows: as we read from left to right, the push instruction is applied to any nonzero integer, and a zero causes the pop instruction to be applied to the stack, thus printing the "popped" integer. Thus, processing the string 12030040 results in an output of 2314, and processing 12304000 results in an output of 3421. (A string such as 10020340 cannot be handled by this procedure, because we cannot pop two integers from a stack containing only one integer.) Both 2314 and 3421 could be thought of as permutations,

    $$\begin{pmatrix} 1 & 2 & 3 & 4 \\ 2 & 3 & 1 & 4 \end{pmatrix} \quad \text{and} \quad \begin{pmatrix} 1 & 2 & 3 & 4 \\ 3 & 4 & 2 & 1 \end{pmatrix}$$

    respectively, on the set $A = \{1, 2, 3, 4\}$.

    (a) What permutation of $A = \{1, 2, 3, 4\}$ is generated by applying this procedure to the string 12003400?

    (b) Name a permutation of $A = \{1, 2, 3, 4\}$ that cannot be generated from any string where the digits 1, 2, 3, and 4 appear in order, no matter where the zeros are placed.

★ 19. For each of the following bijections $f:\mathbb{R} \to \mathbb{R}$, find f^{-1}.

    (a) $f(x) = 2x$

    (b) $f(x) = x^3$

    (c) $f(x) = (x + 4)/3$

20. (a) Let f be a function, $f:S \to T$. If there exists a function $g:T \to S$ such that $g \circ f = i_S$, then g is called a **left inverse** of f. Show that f has a left inverse if and only if f is one-to-one.

    (b) Let f be a function, $f:S \to T$. If there exists a function $g:T \to S$ such that $f \circ g = i_T$, then g is called a **right inverse** of f. Show that f has a right inverse if and only if f is onto.

    (c) Let $f:\mathbb{N} \to \mathbb{N}$ be given by $f(x) = 3x$. Then f is one-to-one. Find two different left inverse functions for f.

21. Let $f:S \to T$ and $g:T \to U$ be functions.

    (a) Prove that if $g \circ f$ is one-to-one, so is f.

    (b) Prove that if $g \circ f$ is onto, so is g.

(c) Find an example where $g \circ f$ is one-to-one but g is not one-to-one.

(d) Find an example where $g \circ f$ is onto but f is not onto.

22. Let f and g be bijections, $f : S \to T$ and $g : T \to U$. Then f^{-1} and g^{-1} exist. Also, $g \circ f$ is a bijection from S to U. Show that $(g \circ f)^{-1} = f^{-1} \circ g^{-1}$.

23. Let $\mathscr{C}$ be a collection of sets, and define a binary relation ρ on $\mathscr{C}$ as follows: for $S, T \in \mathscr{C}$, $S \rho T \leftrightarrow$ "S is equivalent to T." Show that ρ is an equivalence relation on $\mathscr{C}$.

24. Let f be a function, $f : S \to T$.

(a) Define a binary relation ρ on S by $x \rho y \leftrightarrow f(x) = f(y)$. Prove that ρ is an equivalence relation.

(b) For $S = T = \mathbb{Z}$ and $f(x) = 3x^2$, what is [4] under the equivalence relation of part (a)?

SECTION 3.3   MATRICES

Terminology

A **matrix** is a rectangular arrangement of data, usually numbers. Thus $\mathbf{A}$ is a matrix where

$$\mathbf{A} = \begin{bmatrix} 1 & 0 & 4 \\ 3 & -6 & 8 \end{bmatrix}$$

Here $\mathbf{A}$ has two rows and three columns. The **dimensions** of the matrix are the number of rows and columns; thus $\mathbf{A}$ is a 2×3 matrix.

Elements of a matrix $\mathbf{A}$ are denoted by a_{ij}, where i is the row number of the element in the matrix and j is the column number. In the above matrix $\mathbf{A}$, $a_{23} = 8$ because 8 is the element in row 2, column 3 of $\mathbf{A}$.

3.66 Example   Let

$$\mathbf{B} = \begin{bmatrix} 1 & 5 & 7 & 0 \\ -5 & 1 & 9 & 4 \\ 3 & 0 & -4 & 2 \end{bmatrix}$$

Then $\mathbf{B}$ is a 3×4 matrix with $b_{14} = 0$, $b_{22} = 1$, and $b_{31} = 3$.   □

3.67 Practice   In the matrix

$$\mathbf{A} = \begin{bmatrix} 1 & 4 & -6 & 8 \\ 3 & 0 & 1 & -7 \end{bmatrix}$$

what is a_{23}? What is a_{24}? What is a_{13}?   □

In a matrix, the arrangement of the entries is significant. Therefore, for two matrices to be **equal** they must have the same dimensions and the same entries in each location.

3.68 Example   Let

$$\mathbf{X} = \begin{bmatrix} x & 4 \\ 1 & y \\ z & 0 \end{bmatrix} \qquad \mathbf{Y} = \begin{bmatrix} 3 & 4 \\ 1 & 6 \\ 2 & w \end{bmatrix}$$

If $\mathbf{X} = \mathbf{Y}$, then $x = 3$, $y = 6$, $z = 2$, and $w = 0$.   □

We will often be interested in square matrices, in which the number of rows equals the number of columns. If $\mathbf{A}$ is an $n \times n$ square matrix, then the elements $a_{11}, a_{22}, \ldots , a_{nn}$ form the **main diagonal** of the matrix. If the corresponding elements match when we think of folding the matrix along the main diagonal and flipping one half over onto the other, then the matrix is symmetric about the main diagonal. In a **symmetric** matrix, $a_{ij} = a_{ji}$.

3.69 Example   The square 3×3 matrix

$$\mathbf{A} = \begin{bmatrix} 1 & 5 & 7 \\ 5 & 0 & 2 \\ 7 & 2 & 6 \end{bmatrix}$$

is symmetric. The upper triangular part, above the main diagonal, is a reflection of the lower triangular part. Note that $a_{21} = a_{12} = 5$.   □

Matrices are useful to represent any sort of data that logically fall into tabular form.

3.70 Example   Average temperatures in three different cities for each month can be neatly summarized in a 3×12 matrix.

City	J	F	M	A	M	J	J	A	S	O	N	D
P	23	26	38	47	58	71	78	77	69	55	39	33
Q	14	21	33	38	44	57	61	59	49	38	25	21
R	35	46	54	67	78	86	91	94	89	75	62	51

Month   □

A more general way to represent arrangements of data is the **array**. Arrays are n-dimensional arrangements of data, where n can be any positive integer. If $n = 1$, then the data are arranged in a single line, which is therefore a list or finite sequence of data items. If $n = 2$, the array is a matrix. If $n = 3$, we can picture layers of two-dimensional matrices. For $n > 3$, we can formally deal with the array elements, but we can't really visualize the arrangement. The array data structure is available in many high-level programming languages; generally, the number of elements expected in each dimension of the array must be declared in the program. The array **X** of Example 3.68, for instance, would be declared as a 3×2 array—a two-dimensional array (matrix) with three elements in one dimension and two in the other (that is, three rows and two columns).

Matrix Operations

We will define four arithmetic operations on matrices whose entries are numerical. The first involves multiplication of a matrix by a **scalar**, a single number. We simply multiply each entry of the matrix by this number and obtain a matrix with the same dimensions as the original matrix. This operation is called **scalar multiplication**.

3.71 Example   The result of multiplying matrix

$$A = \begin{bmatrix} 1 & 4 & 5 \\ 6 & -3 & 2 \end{bmatrix}$$

by the scalar $r = 3$ is

$$3A = \begin{bmatrix} 3 & 12 & 15 \\ 18 & -9 & 6 \end{bmatrix}$$

□

Addition of two matrices **A** and **B** is only defined when **A** and **B** have the same dimensions; then it is simply a matter of adding the corresponding elements. Formally, if **A** and **B** are both $n \times m$ matrices, then $C = A + B$ is an $n \times m$ matrix with entries

$$c_{ij} = a_{ij} + b_{ij}$$

3.72 Example   For

$$A = \begin{bmatrix} 1 & 3 & 6 \\ 2 & 0 & 4 \\ -4 & 5 & 1 \end{bmatrix} \qquad B = \begin{bmatrix} 0 & -2 & 8 \\ 1 & 5 & 2 \\ 2 & 3 & 3 \end{bmatrix}$$

the matrix $A + B$ is

$$A + B = \begin{bmatrix} 1 & 1 & 14 \\ 3 & 5 & 6 \\ -2 & 8 & 4 \end{bmatrix}$$

□

3.73 Practice   For $r = 2$,

$$\mathbf{A} = \begin{bmatrix} 1 & 7 \\ -3 & 4 \\ 5 & 6 \end{bmatrix} \qquad \mathbf{B} = \begin{bmatrix} 4 & 0 \\ 9 & 2 \\ -1 & 4 \end{bmatrix}$$

find $r\mathbf{A} + \mathbf{B}$.   □

Subtraction of matrices is defined by $\mathbf{A} - \mathbf{B} = \mathbf{A} + (-1)\mathbf{B}$.

A **zero matrix** is a matrix all of whose entries are 0. If we add an $n \times m$ zero matrix, denoted by $\mathbf{0}$, to any $n \times m$ matrix $\mathbf{A}$, the result is matrix $\mathbf{A}$. We can symbolize this by the matrix equation

$$\mathbf{0} + \mathbf{A} = \mathbf{A}$$

This equation is true because of a similar equation that holds for all the individual numerical entries, $0 + a_{ij} = a_{ij}$. Other matrix equations are also true because of similar equations that hold for the individual entries.

3.74 Example   If $\mathbf{A}$ and $\mathbf{B}$ are $n \times m$ matrices and r and s are scalars, the following matrix equations are true.

$$\mathbf{0} + \mathbf{A} = \mathbf{A}$$
$$\mathbf{A} + \mathbf{B} = \mathbf{B} + \mathbf{A}$$
$$(\mathbf{A} + \mathbf{B}) + \mathbf{C} = \mathbf{A} + (\mathbf{B} + \mathbf{C})$$
$$r(\mathbf{A} + \mathbf{B}) = r\mathbf{A} + r\mathbf{B}$$
$$(r + s)\mathbf{A} = r\mathbf{A} + s\mathbf{A}$$
$$r(s\mathbf{A}) = (rs)\mathbf{A}$$

To prove that $\mathbf{A} + \mathbf{B} = \mathbf{B} + \mathbf{A}$, for instance, it is sufficient to note that $a_{ij} + b_{ij} = b_{ij} + a_{ij}$ for each entry in matrices $\mathbf{A}$ and $\mathbf{B}$.   □

One might expect that in **multiplication** of matrices individual elements are simply multiplied. The definition is not that simple, however. The definition of matrix multiplication is based on the use of matrices in mathematics to represent functions called linear transformations, which map points in the real-number plane to points in the real-number plane. Although we won't use matrices in this way, we will use the standard definition for matrix multiplication.

To compute $\mathbf{A}$ times $\mathbf{B}$, $\mathbf{A} \cdot \mathbf{B}$, the number of columns in $\mathbf{A}$ must equal the number of rows in $\mathbf{B}$. Thus we can compute $\mathbf{A} \cdot \mathbf{B}$ if $\mathbf{A}$ is an $n \times m$ matrix and $\mathbf{B}$ is an $m \times p$ matrix. The result is an $n \times p$ matrix. An entry in row i, column j of $\mathbf{A} \cdot \mathbf{B}$ is obtained by multiplying elements in row i of $\mathbf{A}$ by the corresponding elements in column j of $\mathbf{B}$ and adding the results. Formally, $\mathbf{A} \cdot \mathbf{B} = \mathbf{C}$, where

$$c_{ij} = \sum_{k=1}^{m} a_{ik}b_{kj}$$

3.75 Example   Let

$$A = \begin{bmatrix} 2 & 4 & 3 \\ 4 & -1 & 2 \end{bmatrix} \qquad B = \begin{bmatrix} 5 & 3 \\ 2 & 2 \\ 6 & 5 \end{bmatrix}$$

A is a 2 × 3 matrix and **B** is a 3 × 2 matrix, so the product **A · B** exists and is a 2 × 2 matrix **C**. To find element c_{11}, we multiply corresponding elements of row 1 of **A** and column 1 of **B** and add the results:

$$2(5) + 4(2) + 3(6) = 10 + 8 + 18 = 36$$

$$\begin{bmatrix} 2 & 4 & 3 \\ 4 & -1 & 2 \end{bmatrix} \begin{bmatrix} 5 & 3 \\ 2 & 2 \\ 6 & 5 \end{bmatrix} = \begin{bmatrix} 36 & - \\ - & - \end{bmatrix}$$

Element c_{12} is obtained by multiplying corresponding elements of row 1 of **A** and column 2 of **B** and adding the results:

$$\begin{bmatrix} 2 & 4 & 3 \\ 4 & -1 & 2 \end{bmatrix} \begin{bmatrix} 5 & 3 \\ 2 & 2 \\ 6 & 5 \end{bmatrix} = \begin{bmatrix} 36 & 29 \\ - & - \end{bmatrix}$$

The complete product is

$$\begin{bmatrix} 2 & 4 & 3 \\ 4 & -1 & 2 \end{bmatrix} \begin{bmatrix} 5 & 3 \\ 2 & 2 \\ 6 & 5 \end{bmatrix} = \begin{bmatrix} 36 & 29 \\ 30 & 20 \end{bmatrix}$$

□

3.76 Practice   Compute **A · B** and **B · A** for

$$A = \begin{bmatrix} 1 & 4 \\ 6 & -2 \end{bmatrix} \qquad B = \begin{bmatrix} 3 & 6 \\ 3 & 4 \end{bmatrix}$$

□

From Practice 3.76, we see that even if **A** and **B** have dimensions so that both **A · B** and **B · A** are defined, **A · B** need not equal **B · A**. There are, however, several matrix equations involving multiplication that are true.

3.77 Example   Where **A**, **B**, and **C** are matrices of appropriate dimensions, the following matrix equations are true:

$$A(B \cdot C) = (A \cdot B)C$$
$$A(B + C) = A \cdot B + A \cdot C$$
$$(A + B)C = A \cdot C + B \cdot C$$

Verifying these equations for matrices of particular dimensions is simple, if tedious.   □

The $n \times n$ matrix with 1s along the main diagonal and 0s elsewhere is called the **identity matrix**, denoted by **I**. If we multiply **I** times any $n \times n$ matrix **A**, we get **A** as the result. The equation

$$\mathbf{I} \cdot \mathbf{A} = \mathbf{A} \cdot \mathbf{I} = \mathbf{A}$$

holds.

3.78 Practice   Let

$$\mathbf{I} = \begin{bmatrix} 1 & 0 \\ 0 & 1 \end{bmatrix} \qquad \mathbf{A} = \begin{bmatrix} a_{11} & a_{12} \\ a_{21} & a_{22} \end{bmatrix}$$

Verify that $\mathbf{I} \cdot \mathbf{A} = \mathbf{A} \cdot \mathbf{I} = \mathbf{A}$.   ☐

✔ Checklist

Definitions

matrix (*p. 149*)   scalar multiplication (*p. 151*)
dimensions of a matrix (*p. 149*)   addition of matrices (*p. 151*)
equal matrices (*p. 150*)   subtraction of matrices (*p. 152*)
main diagonal (*p. 150*)   zero matrix (*p. 152*)
symmetric matrix (*p. 150*)   multiplication of matrices (*p. 152*)
array (*p. 150*)   identity matrix (*p. 154*)
scalar (*p. 151*)

Techniques

Add, subtract, multiply, and perform scalar multiplication on matrices.

Main Ideas

Matrices are rectangular arrangements of data that are used to represent information in tabular form.

Matrices have their own arithmetic, with operations of addition, subtraction, multiplication, and scalar multiplication.

Exercises Section 3.3

★ 1. For the matrix

$$\mathbf{A} = \begin{bmatrix} 1 & 2 \\ 3 & 0 \\ -4 & 1 \end{bmatrix}$$

what is a_{12}? What is a_{31}?

2. Find x and y if

$$\begin{bmatrix} 1 & 3 \\ x & x+y \end{bmatrix} = \begin{bmatrix} 1 & 3 \\ 2 & 6 \end{bmatrix}$$

★ 3. Find x, y, z, and w if

$$\begin{bmatrix} x+y & 2x-3y \\ z-w & z+2w \end{bmatrix} = \begin{bmatrix} 4 & -7 \\ -6 & 6 \end{bmatrix}$$

4. If **A** is a symmetric matrix, find u, v, and w:

$$\mathbf{A} = \begin{bmatrix} 2 & w & u \\ 7 & 0 & v \\ 1 & -3 & 4 \end{bmatrix}$$

In Exercises 5 through 18, if $r = 3$, $s = -2$,

$$\mathbf{A} = \begin{bmatrix} 2 & 1 \\ -1 & 0 \\ 3 & 4 \end{bmatrix} \qquad \mathbf{B} = \begin{bmatrix} 4 & 1 & 2 \\ 6 & -1 & 5 \\ 1 & 3 & 2 \end{bmatrix}$$

$$\mathbf{C} = \begin{bmatrix} 2 & 4 \\ 6 & -1 \end{bmatrix} \qquad \mathbf{D} = \begin{bmatrix} 4 & -6 \\ 1 & 3 \\ 2 & -1 \end{bmatrix}$$

compute the following (if possible):

★ 5. $\mathbf{A} + \mathbf{D}$ 6. $\mathbf{A} - \mathbf{D}$

   7. $r\mathbf{B}$ 8. $s\mathbf{C}$

★ 9. $\mathbf{A} + r\mathbf{D}$ 10. $\mathbf{B} - r\mathbf{C}$

  11. $r(\mathbf{A} + \mathbf{D})$ 12. $r(s\mathbf{C})$

★13. $\mathbf{B} \cdot \mathbf{D}$ 14. $\mathbf{D} \cdot \mathbf{C}$

  15. $\mathbf{A} \cdot \mathbf{C}$ 16. $\mathbf{C} \cdot \mathbf{A}$

★17. $\mathbf{C}^2 = \mathbf{C} \cdot \mathbf{C}$ 18. $\mathbf{B} \cdot \mathbf{A} + \mathbf{D}$

For Exercises 19 through 22, let

$$\mathbf{A} = \begin{bmatrix} 3 & -1 \\ 2 & 5 \end{bmatrix} \qquad \mathbf{B} = \begin{bmatrix} 4 & 1 \\ 2 & -1 \end{bmatrix} \qquad \mathbf{C} = \begin{bmatrix} 6 & -5 \\ 2 & -2 \end{bmatrix}$$

19. Compute $\mathbf{A} \cdot \mathbf{B}$ and $\mathbf{B} \cdot \mathbf{A}$.

20. Compute $\mathbf{A}(\mathbf{B} \cdot \mathbf{C})$ and $(\mathbf{A} \cdot \mathbf{B})\mathbf{C}$.

★ 21. Compute $\mathbf{A}(\mathbf{B} + \mathbf{C})$ and $\mathbf{A} \cdot \mathbf{B} + \mathbf{A} \cdot \mathbf{C}$.

22. Compute $(\mathbf{A} + \mathbf{B})\mathbf{C}$ and $\mathbf{A} \cdot \mathbf{C} + \mathbf{B} \cdot \mathbf{C}$.

23. If

$$\mathbf{A} = \begin{bmatrix} 2 & 3 \\ 4 & 1 \end{bmatrix} \qquad \mathbf{B} = \begin{bmatrix} x & 3 \\ y & 2 \end{bmatrix}$$

find x and y if $\mathbf{A} \cdot \mathbf{B} = \mathbf{B} \cdot \mathbf{A}$.

24. Prove that $\mathbf{I}^2 = \mathbf{I}$ for any identity matrix $\mathbf{I}$.

25. An $n \times n$ matrix **A** is **invertible** if there exists an $n \times n$ matrix **B** such that $\mathbf{A} \cdot \mathbf{B} = \mathbf{B} \cdot \mathbf{A} = \mathbf{I}$; **B** is called the **inverse** of **A**.
    ★(a) Show that for

    $$\mathbf{A} = \begin{bmatrix} 1 & 3 \\ 2 & 2 \end{bmatrix} \qquad \mathbf{B} = \begin{bmatrix} -\frac{1}{2} & \frac{3}{4} \\ \frac{1}{2} & -\frac{1}{4} \end{bmatrix}$$

    $\mathbf{A} \cdot \mathbf{B} = \mathbf{B} \cdot \mathbf{A} = \mathbf{I}.$
    ★(b) Show that

    $$\mathbf{A} = \begin{bmatrix} 1 & 2 \\ 2 & 4 \end{bmatrix}$$

    is not invertible.
    (c) Show that

    $$\mathbf{A} = \begin{bmatrix} a_{11} & a_{12} \\ a_{21} & a_{22} \end{bmatrix}$$

    is invertible with inverse

    $$\mathbf{B} = \frac{1}{a_{11}a_{22} - a_{12}a_{21}} \begin{bmatrix} a_{22} & -a_{12} \\ -a_{21} & a_{11} \end{bmatrix}$$

    if and only if $a_{11}a_{22} - a_{12}a_{21} \neq 0.$

26. The **transpose** of a matrix **A**, $\mathbf{A}^T$, is obtained by interchanging its rows and columns. Thus, if we denote the element in row i, column j of **A** by $\mathbf{A}(i, j)$, then $\mathbf{A}^T(i, j) = \mathbf{A}(j, i)$.
    (a) Find $\mathbf{A}^T$ for

    $$\mathbf{A} = \begin{bmatrix} 1 & 3 & 4 \\ 6 & -2 & 1 \end{bmatrix}$$

    (b) Prove that if **A** is a square matrix, then **A** is symmetric if and only if $\mathbf{A}^T = \mathbf{A}$.
    (c) Prove that $(\mathbf{A} + \mathbf{B})^T = \mathbf{A}^T + \mathbf{B}^T$.
    (d) Prove that $(\mathbf{A} \cdot \mathbf{B})^T = \mathbf{B}^T \cdot \mathbf{A}^T$.

★ 27. Find two 2×2 matrices **A** and **B** such that $\mathbf{A} \cdot \mathbf{B} = \mathbf{0}$ but $\mathbf{A} \neq \mathbf{0}$ and $\mathbf{B} \neq \mathbf{0}$.

28. Find three 2×2 matrices **A**, **B**, and **C** such that $\mathbf{A} \cdot \mathbf{C} = \mathbf{B} \cdot \mathbf{C}$ but $\mathbf{A} \neq \mathbf{B}$.

29. Write a computer program to multiply two $n \times n$ matrices; assume $n \leq 10$.

Chapter 4

Graphs and Trees

A graph is a visual representation of certain data items and the connections between some of these items. A tree is a particular type of graph where the connections between data items are not circular. A surprising number of real-world situations—organization charts, road maps, transportation and communication networks, and so forth—can be represented as graphs or trees. Later we shall see other uses of graphs to represent logic networks, finite-state machines, and formal-language derivations. Graph theory is an extensive topic; Section 4.1 presents some of the considerable terminology of graph theory.

To represent a graph in computer memory, data must be arranged in a way that preserves all the information contained in the visual graph. Two approaches to representing graphs within a computer, adjacency matrices and adjacency lists, are discussed in Section 4.2.

Much research has been done on finding efficient solutions to certain kinds of graph problems. Such questions as whether there is a path through a graph that uses each arc, or edge, of the graph exactly once and how to find the shortest path between two nodes of a graph have algorithmic solutions. These are discussed in Section 4.3.

Finally, algorithms exist that allow *traversal* of the graph—a systematic visitation of all nodes of the graph—in various ways. These algorithms are discussed in Section 4.4.

SECTION 4.1   GRAPH TERMINOLOGY AND APPLICATIONS

Graphs

One way to while away the hours on an airplane trip is to look at the literature in the seat pockets. This material almost always includes a map showing the routes of the airline you are flying, such as in Figure 4.1. All

157

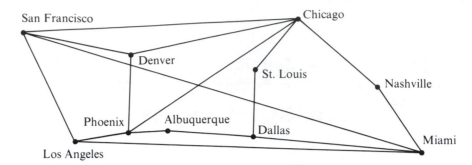

Figure 4.1

of this route information could be expressed in paragraph form; for example, there is a direct route between Chicago and Nashville but not between St. Louis and Nashville. However, the paragraph would be rather long and involved, and we would not be able to assimilate the information as quickly and clearly as we can from the map. There are many cases where ''a picture is worth a thousand words.''

Figure 4.1 is a graph. We have also talked about graphs of functions on rectangular coordinate systems. There are many types of graphs—bar graphs, picture graphs, and pie (or circle) graphs, for instance (see Figure 4.2). Each type is a visual representation of data. But our definition of a graph is more restrictive; according to it, only the airline map is a graph.

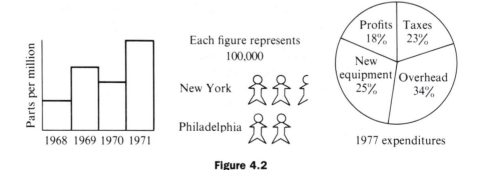

Figure 4.2

4.1 Definition   A **graph** is an ordered triple (N, A, f) where

N = a nonempty set of **nodes,** or **vertices**

A = a set of **arcs,** or **edges**

f is a function associating with each arc a an unordered pair x–y of nodes called the **endpoints** of a.   □

Our graphs will always have a finite number of nodes and arcs.

4.2 *Example*   The set of nodes in the airline map is {Chicago, Nashville, Miami, Dallas, St. Louis, Albuquerque, Phoenix, Denver, San Francisco, Los Angeles}. There are 16 arcs; for example, Phoenix–Albuquerque is an arc (here we are naming an arc by its endpoints), Albuquerque–Dallas is an arc, and so on.   □

4.3 *Example*   Four more graphs are shown in Figure 4.3.   □

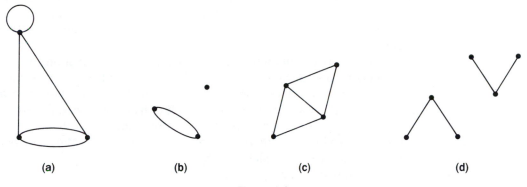

(a)   (b)   (c)   (d)

Figure 4.3

Graphs that at first glance appear different may really be the same according to the definition of a graph. In particular, the shape of an arc is not specified, so that ⟶ and ⌁ are the same graph. Notice that arcs can intersect at points that are not nodes of the graph.

4.4 *Practice*   According to the definition of a graph, which graph in Figure 4.4 is not the same graph as the one shown in Figure 4.3b?   □

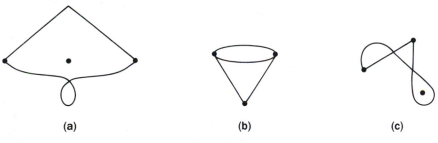

(a)   (b)   (c)

Figure 4.4

4.5 *Practice*   Sketch a graph having nodes {1, 2, 3, 4, 5}, arcs {a_1, a_2, a_3, a_4, a_5, a_6}, and function $f(a_1) = 1\text{--}2$, $f(a_2) = 1\text{--}3$, $f(a_3) = 3\text{--}4$, $f(a_4) = 3\text{--}4$, $f(a_5) = 4\text{--}5$, $f(a_6) = 5\text{--}5$.   □

We will need some graph theory terminology. Two nodes in a graph are **adjacent** if they are the endpoints of an arc. In the graph of Practice 4.5, 4 and 5 are adjacent nodes, 3 and 5 are not, and 5 is adjacent to itself. A **loop** in a graph is an arc with endpoints $n–n$ for some node n; an example in Practice 4.5 is arc a_6 with endpoints 5–5. A graph with no loops is **loop-free.** Two arcs with the same endpoints are **parallel arcs**; thus arcs a_3 and a_4 in Practice 4.5 are parallel. A **simple graph** is one with no loops or parallel arcs. Figures 4.3c and 4.3d are simple graphs; 4.3a and 4.3b are not. An **isolated node** is adjacent to no other node, as in Figure 4.3b. The **degree** of a node is the number of arc ends at that node. In the graph of Practice 4.5, node 2 has degree 1, and nodes 3, 4, and 5 all have degree 3.

The next two definitions concern the nature of the function f relating arcs to endpoints. Because f is a function, each arc has a unique pair of endpoints. If f is a one-to-one function, then there is at most one arc associated with a pair of endpoints; such graphs have no parallel arcs. In a **complete** graph, any two distinct nodes are adjacent. In this case, f is almost an onto function (there does not have to be a loop at every node). Figure 4.3a shows a complete graph.

A **path** from node n_0 to node n_k is a sequence

$$n_0, a_0, n_1, a_1, \ldots, n_{k-1}, a_{k-1}, n_k$$

of nodes and arcs where for each i, the endpoints of arc a_i are $n_i–n_{i+1}$. In the graph of Practice 4.5, a path from node 2 to node 4 is as follows: 2, a_1, 1, a_2, 3, a_4, 4. The **length** of a path is the number of arcs it contains. A graph is **connected** if there is a path from any node to any other node. The graphs in Figures 4.3a and 4.3c are connected; those in Figures 4.3b and 4.3d are not. A **cycle** in a graph is a path from some node n_0 back to n_0 where no arc appears more than once, n_0 is the only node appearing more than once, and n_0 occurs only at the ends. (Nodes and arcs may be repeated in a path but not, except for node n_0, in a cycle.) In the graph of Practice 4.5, 3, a_3, 4, a_4, 3 is a cycle. A graph with no cycles is **acyclic.** Any acyclic graph is simple, but not conversely.

Trees

A **tree** is an acyclic, connected graph. Figure 4.5 pictures two trees. Perversely, computer scientists like to draw trees with the *root* at the top. A tree can also be defined recursively: a single node is a tree. For trees with more than one node, a single node r is the root of the tree, and the remaining nodes are divided into disjoint trees with roots $r_1, r_2, \ldots, r_t$, with a single arc between r and r_i, $1 \le i \le t$. The nodes $r_1, r_2, \ldots, r_t$ are **children** of r, and r is a **parent** of $r_1, r_2, \ldots, r_t$.

Because a tree is a connected graph, there is a path from the root to any other node in the tree; because the tree is acyclic, that path is

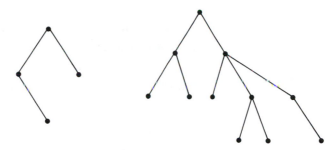

Figure 4.5

unique. The **depth** of any node in a tree is the length of the path from the root to the node; the root itself has depth 0. The **height** of the tree is the maximum depth of any node in the tree; in other words, it is the length of the longest path from the root to any node. A node with no children is called a **leaf** of the tree; all nonleaves are **internal nodes.** A **forest** is an acyclic graph (not necessarily connected); thus a forest is a disjoint collection of trees. Figure 4.5 shows a forest.

Binary **trees,** where each node has at most two children, are of particular interest. In a binary tree, each child of a node is designated as either the **left child** or the **right child.** A **full binary tree** occurs when all internal nodes have two children, and all leaves are at the same depth. Figure 4.6 shows a binary tree of height 4, and Figure 4.7 shows a full binary tree of height 3.

Figure 4.6

Figure 4.7

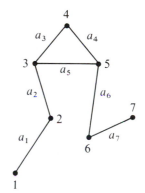

Figure 4.8

4.6 Practice   Answer the following questions for the graph shown in Figure 4.8:

    (a) Is the graph simple?
    (b) Complete?
    (c) Connected?
    (d) Find two paths from 3 to 6.

(e) Find a cycle.

(f) Name one arc whose removal will make the graph a tree.

(g) Name one arc whose removal will make the graph not connected.

□

4.7 Practice   Answer the following questions about the binary tree shown in Figure 4.9:

(a) What is the root?

(b) What is the height?

(c) What is the left child of node 2?

(d) What is the depth of node 5?

□

Figure 4.9

Trees are fertile ground (no pun intended) for proofs by induction, either on the number of nodes or arcs or on the height. The next theorem, for example, which seems to hold for all pictures of trees that we can draw, is proved by induction.

4.8 Theorem   A tree with n nodes has $n - 1$ arcs.

Proof: We use induction on n, $n \geq 1$. For $n = 1$, the tree consists of a single node and no arcs, so the number of arcs is 1 less than the number of nodes. Assume that any tree with k nodes has $k - 1$ arcs, and consider a tree with $k + 1$ nodes. Let x be a leaf of the tree (a leaf must exist since the tree is finite). Then x has a unique parent y. Remove from the tree the node x and the single arc a connecting x and y. The remaining graph is still a tree and has k nodes. Therefore, by the inductive hypothesis, it has $k - 1$ arcs, and the original graph, containing arc a, had $(k - 1) + 1 = k$ arcs.   □

Directed Graphs

We might want the arcs of a graph to begin at one node and end at another, in which case we would use a directed graph.

4.9 Definition   A **directed graph** is an ordered triple (N, A, f) where

N = a set of nodes

A = a set of arcs

f is a function associating with each arc a an ordered pair (x, y) of nodes where x is the **initial point** and y is the **terminal point** of a.

□

In a directed graph, then, there is a direction associated with each arc. The concept of a path extends as we might expect: a **path** from node n_0 to node n_k is a sequence $n_0, a_0, n_1, a_1, \ldots, n_{k-1}, a_{k-1}, n_k$ where for each i, n_i is the initial point of arc a_i and n_{i+1} is the terminal point of a_i. If a path exists from node n_0 to node n_k, then n_k is **reachable** from n_0. The definition of a cycle also carries over to directed graphs.

4.10 Example   In the directed graph of Figure 4.10, there are many paths from node 1 to node 3: 1, a_4, 3 and 1, a_1, 2, a_2, 2, a_2, 2, a_3, 3 are two possibilities. Node 3 is certainly reachable from 1. Node 1, however, is not reachable from any other node. The cycles in this graph are the loop a_2 and the path 3, a_5, 4, a_6, 3.   □

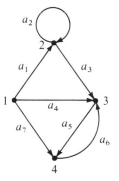

Figure 4.10

Besides imposing direction on the arcs of a graph, we may want to modify the basic definition of a graph in other ways. We often want the nodes of a graph to carry identifying information, like the names of the cities in the map of airline routes. This would be a **labeled graph.** We may want to use a **weighted graph,** where each arc has some numerical value, or *weight,* associated with it. For example, we might want to indicate the distances of the various routes in the airline map.

Applications

Although the idea of a graph is very simple, an amazing number of situations have relationships between items that lend themselves to graphical representation. Not surprisingly, there are many graphs in this book. We have used them to visualize partially ordered finite sets. The commutative diagram illustrating composition of functions, Figure 3.29, is a directed graph. Chapter 6 will introduce logic networks and represent them essentially as directed graphs. Directed graphs will also be used to describe finite-state machines in Chapter 9, and the derivations of words in certain formal languages will be shown as trees in Chapter 12 (these are the parse trees generated by a compiler while analyzing a computer program).

Exercise 21 of Section 3.1 describes the organization of data into a binary tree structure. By using these trees, a collection of records can be efficiently searched to locate a particular record or to determine that a record is not in the collection. Examples of such a search would be checking for a volume in a library, for a patient's medical record in a hospital, or for an individual's credit record at the bank.

We saw that the airline route map was a graph. A representation of any network of transportation routes (a road map, for example), communication lines (as in a computer network), or product or service distribution routes such as natural gas pipelines or water mains is a graph. The chemical structure of a molecule is represented graphically. A family tree is a graph, although if there were intermarriages, it would not be a tree in the technical sense. (Information obtained from a family tree is not only interesting, it is useful for research in medical genetics.) The organization chart indicating who reports to whom in a large company is usually a tree. The underlying structure in a program flowchart is a directed graph; so is that of an electrical circuit. Directed graphs are used to optimize completion of a complex task, perhaps the manufacture of a product, through procedures called PERT (program evaluation and review technique) or critical-path analysis.

4.11 Practice   Draw the underlying graph in each of the following cases:

(a) Figure 4.11 is a road map for part of Arizona.
(b) Figure 4.12 is a representation of an ozone molecule with three oxygen atoms.

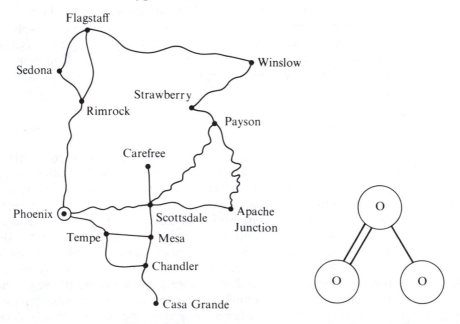

Figure 4.11   Figure 4.12

(c) Figure 4.13 is the flowchart for a computer program that reads a sequence of nonnegative integers, prints those integers greater than 7, and stops when the input is 0. (This graph will be directed.)   □

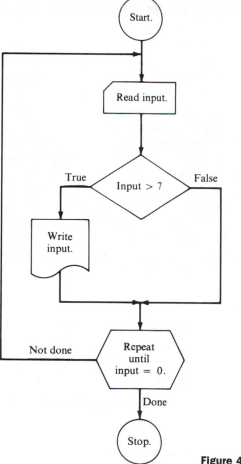

Figure 4.13

4.12 Example   Algebraic expressions involving binary operations can be represented by labeled binary trees. The leaves are labeled as operands, and the internal nodes are labeled as binary operations. For any internal node, the binary operation of its label is performed on the expressions associated with its left and right subtrees. Thus the binary tree in Figure 4.14 represents the algebraic expression $(2 + x) - (y * 3)$.   □

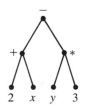

Figure 4.14

4.13 Practice   What is the expression tree for $(2 + 3) * 5$?   □

✔ Checklist

Definitions

graph (*p. 158*)   children nodes (*p. 160*)
node (vertex) (*p. 158*)   parent node (*p. 160*)
arc (edge) (*p. 158*)   depth of a node (*p. 161*)
endpoints (*p. 158*)   height of a tree (*p. 161*)
adjacent nodes (*p. 160*)   leaf (*p. 161*)
loop (*p. 160*)   internal node (*p. 161*)
loop-free graph (*p. 160*)   forest (*p. 161*)
parallel arcs (*p. 160*)   binary tree (*p. 161*)
simple graph (*p. 160*)   left child (*p. 161*)
isolated node (*p. 160*)   right child (*p. 161*)
degree of a node (*p. 160*)   full binary tree (*p. 161*)
complete graph (*p. 160*)   directed graph (*p. 162*)
path (*p. 160*)   initial point (*p. 162*)
length of a path (*p. 160*)   terminal point (*p. 162*)
connected graph (*p. 160*)   reachable node (*p. 163*)
cycle (*p. 160*)   labeled graph (*p. 163*)
acyclic graph (*p. 160*)   weighted graph (*p. 163*)
tree (*p. 160*)

Techniques

Use graph terminology.

Construct expression trees.

Main Ideas

Graphical representation of diverse situations.

Exercises Section 4.1

★ 1. According to the definition of a graph, which graph of Figure 4.15 is not the same as the others?

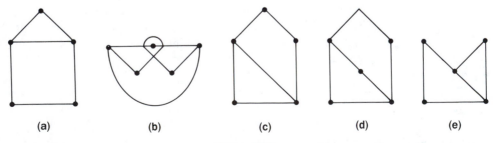

(a)   (b)   (c)   (d)   (e)

Figure 4.15

★ 2. Sketch a picture of each of the following graphs:
   (a) Five nodes, one adjacent to all the others, no others adjacent. Is this a tree?
   (b) A simple graph with three nodes, each of degree 2.
   (c) Four nodes, with cycles of length 1, 2, 3, and 4.

★ 3. (a) Draw a simple, complete graph with four nodes.
   (b) Draw a simple, complete graph with five nodes.
   (c) Draw a simple, complete graph with six nodes.
   (d) Conjecture how many arcs there are in a simple, complete graph with n nodes.
   (e) Prove your conjecture.

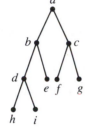

Figure 4.16

4. Answer the following questions about the graph of Figure 4.16:
   (a) Is this a binary tree?
   (b) Is it a full binary tree?
   (c) What is the parent of e?
   (d) What is the right child of e?
   (e) What is the depth of g?
   (f) What is the height of the tree?

★ 5. Use the directed graph in Figure 4.17 to answer the following questions:

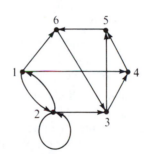

Figure 4.17

   (a) Which nodes are reachable from node 3?
   (b) What is the length of the shortest path from node 3 to node 6?
   (c) Give a path from node 1 to node 6 of length 8.

In Exercises 6 through 9, draw the expression tree.

6. $((x - 2) * 3) + (5 + 4)$

★ 7. $((2 * x - 3 * y) + 4 * z) + 1$

8. $1 - [2 - (3 - (4 - 5))]$

9. $((6 \div 2) * 4) + ((1 + x) * (5 + 3))$

10. Prove that a binary tree has at most 2^d nodes at depth d.

11. (a) Draw a full binary tree of height 2. How many nodes does it have?

(b) Draw a full binary tree of height 3. How many nodes does it have?

(c) Conjecture how many nodes there are in a full binary tree of height h.

(d) Prove your conjecture. (Hint: use Exercise 10.)

In Exercises 12 through 14, trees are considered to be the same (not distinct) if they have the same parent–child structure. Thus the two trees in Figure 4.18 are not distinct because each represents a single root, located at r, with four children.

12. Show that there are two distinct trees with 3 nodes.

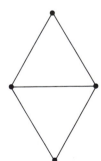

★ 13. Show that there are four distinct trees with 4 nodes.

14. Show that there are nine distinct trees with 5 nodes.

●★ 15. Prove that a simple graph is a tree if and only if there is a unique path between any two nodes.

Figure 4.18

16. Prove that a full binary tree with x internal nodes has $x + 1$ leaves.

17. Prove that the number of leaves in any binary tree is 1 more than the number of nodes with two children.

18. Let G be a simple graph. Prove that G is a tree if and only if G is connected and if the removal of any single arc from G makes G unconnected.

19. Let G be a simple graph. Prove that G is a tree if and only if G is connected and the addition of one arc to G results in a graph with exactly one cycle.

20. A **planar** graph is a graph that can be drawn so that its arcs intersect only at nodes. Figure 4.19 shows a connected planar graph; it divides the plane into two enclosed regions. **Euler's Formula** says that in any connected planar graph with n nodes, a arcs, and r enclosed regions,

$$n - a + r = 1$$

Verify Euler's Formula for the graph in Figure 4.19.

21. Prove Euler's Formula (see Exercise 20). (Hint: use induction on a, beginning with $a = 0$; consider two ways to add one new arc to a graph.)

Figure 4.19

SECTION 4.2   COMPUTER REPRESENTATIONS OF GRAPHS

We have said that the major advantage of a graph is its visual representation of information. However, for storage and manipulation within a computer, this information must be represented in other ways. We will consider two data structures for representing graphs—adjacency matrices and adjacency lists.

Adjacency Matrix

Suppose a graph has m nodes, numbered $n_1, n_2, \ldots, n_m$. We can form an $m \times m$ matrix where entry i,j is the number of arcs between nodes n_i and n_j. This matrix is called the **adjacency matrix A** of the graph. Thus,

$$a_{ij} = p \qquad \text{where there are } p \text{ arcs between } n_i \text{ and } n_j$$

4.14 Example

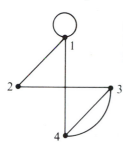

Figure 4.20

The adjacency matrix for the graph in Figure 4.20 is a 4×4 matrix. Entry 1,1 is a 1 due to the loop at node 1. All other elements on the main diagonal are 0. Entry 2,1 (second row, first column) is a 1 because there is one arc between node 2 and node 1, which also means that entry 1,2 is a 1. So far we have

$$\mathbf{A} = \begin{bmatrix} 1 & 1 & - & - \\ 1 & 0 & - & - \\ - & - & 0 & - \\ - & - & - & 0 \end{bmatrix}$$

□

4.15 Practice

Complete the adjacency matrix for Figure 4.20. □

The adjacency matrix in Practice 4.15 is symmetric, which will be true for the adjacency matrix of any undirected graph—if there are p arcs between n_i and n_j, there are certainly p arcs between n_j and n_i. The symmetry of the matrix means that only elements on or below the main diagonal need to be stored. Therefore, all the information contained in the graph in Figure 4.20 is contained in the array below, and the graph could be reconstructed from this array.

An arc between nodes n_i and n_j is a path of length 1. Let's compute the product of an adjacency matrix $\mathbf{A}$ with itself. Recalling the definition of matrix multiplication from Section 3.3, entry i,j in $\mathbf{A}^2$, $\mathbf{A}^2 (i, j)$, is given by

$$\mathbf{A}^2(i, j) = \sum_{k=1}^{m} a_{ik}a_{kj} = a_{i1}a_{1j} + a_{i2}a_{2j} + \cdots + a_{im}a_{mj}$$

If a term in this sum, such as $a_{i2}a_{2j}$, is 0, then either $a_{i2} = 0$ or $a_{2j} = 0$, and there is either no path of length 1 from n_i to n_2 or none from n_2 to n_j. Thus, there are no paths of length 2 from n_i to n_j through n_2. If $a_{i2}a_{2j}$ is not 0, then $a_{i2} = p$ and $a_{2j} = q$, where p and q are positive integers.

Therefore there are p paths of length 1 from n_i to n_2, and q paths of length 1 from n_2 to n_j. This result gives pq possible paths of length 2 from n_i to n_j through n_2. The sum of all such terms gives all possible paths of length 2 from n_i to n_j. Thus, $\mathbf{A}^2$ gives the number of all possible paths of length 2 between nodes of the graph.

4.16 Practice   Compute $\mathbf{A}^2$ for the graph of Figure 4.20. Can you find five distinct paths of length 2 from node 4 to node 4 on the graph?   □

4.17 Practice   (a) What information would be contained in entry i,j of matrix $\mathbf{A}^n$?

(b) What type of proof is called for to prove this formally?   □

In a directed graph, the adjacency matrix $\mathbf{A}$ reflects the direction of the arcs. For a directed matrix,

$$a_{ij} = p \qquad \text{where there are } p \text{ arcs from } n_i \text{ to } n_j$$

An adjacency matrix for a directed graph will not necessarily be symmetric, since an arc from n_i to n_j does not imply an arc from n_j to n_i. Again, by the same argument as before, the entries of $\mathbf{A}^n$ give the number of paths of length n from one node to another.

We may want to know whether one node in a directed graph can be reached from another. In a flowchart, for example, any statement(s) corresponding to a node that is not reachable from the start node could be eliminated from the program. For node n_j to be reached from node n_i, there must be a path of some length from n_i to n_j. If we form the adjacency matrix $\mathbf{A}$ for the graph and compute $\mathbf{A}^2$, $\mathbf{A}^3$, $\mathbf{A}^4$, . . . , any path would eventually show up as a positive i,j entry in one of these matrices. But we cannot compute an infinite string of matrices. Fortunately, the length of a path that we should look for is limited. If there are m nodes in the graph, we can write a path with at most $m - 1$ arcs (and m nodes) before a node repeats itself. In a longer path, any section of path between repeating nodes is a cycle that can be eliminated, thereby shortening the path.

Thus, if a path exists from n_i to n_j, there will be a path from n_i to n_j of length at most $m - 1$. We need only compute $\mathbf{A}$, $\mathbf{A}^2$, . . . , $\mathbf{A}^{m-1}$ to locate such a path. Alternatively, we can define a new matrix $\mathbf{R}$ where

$$\mathbf{R} = \mathbf{A} + \mathbf{A}^2 + \cdots + \mathbf{A}^{m-1}$$

Then n_j is reachable from n_i if and only if entry i,j in $\mathbf{R}$ is positive.

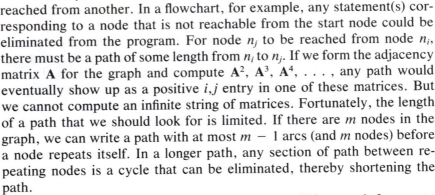

Figure 4.21

4.18 Example   Consider the directed graph of Figure 4.21. The adjacency matrix is

$$\mathbf{A} = \begin{bmatrix} 1 & 1 & 0 & 0 \\ 0 & 0 & 1 & 1 \\ 0 & 1 & 0 & 0 \\ 0 & 0 & 1 & 0 \end{bmatrix}$$

   □

4.19 Practice   Compute the matrix **R** for the directed graph of Figure 4.21.   □

    In a simple weighted graph, the entries in the adjacency matrix can indicate the weight of an arc by the appropriate number rather than just indicating the presence of an arc by the number 1.

Adjacency List

Many graphs, far from being complete graphs, have relatively few arcs. Such graphs have sparse adjacency matrices; that is, the adjacency matrices contain many zeros. Yet if the graph has m nodes, it still requires m^2 data items to represent the adjacency matrix, even if many of these items are 0. Any algorithm or procedure in which every arc in the graph must be examined would require looking at all m^2 items in the matrix, since there is no way of knowing which entries are nonzero without examining them. In other words, for a given node n_i, to find all of the arcs of which n_i is an endpoint requires scanning the entire ith row of the adjacency matrix, a total of m items.

    A more efficient representation of a graph with relatively few arcs, is obtained by storing only the nonzero entries of the adjacency matrix. This representation consists of a list, for each node, of all the nodes adjacent to it. Pointers are used to get us from one item in the list to the next. Such an arrangement is called a **linked list.** There is an array of m pointers, one for each node, to get each list started. This **adjacency list** representation, although it requires extra storage for the pointers, may still be more efficient than an adjacency matrix. One disadvantage of the adjacency list representation is that to find out if a certain node n_j is adjacent to n_i may require scanning all of n_i's adjacency list, whereas we could access element i,j of the adjacency matrix directly.

4.20 Example   The adjacency list for the graph of Figure 4.20 contains a 4-element array of pointers, one for each node. The pointer for each node points to an adjacent node, which points to another adjacent node, and so forth. The adjacency list structure is shown in Figure 4.22.

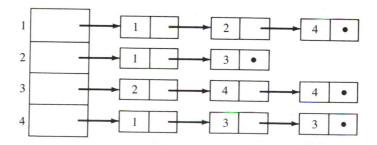

Figure 4.22

In the figure the dot indicates a **nil pointer,** meaning that there is nothing more to be pointed to or that the end of the list has been reached. We have dealt with parallel arcs by listing a given node more than once on the adjacency list for n_i if there is more than one arc between n_i and that node.   □

4.21 Practice   Draw the adjacency list representation for the graph shown in Figure 4.23.
                                                                                   □

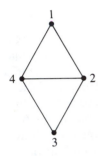

In an undirected graph, each arc is represented twice. If n_j is on the adjacency list of n_i, then n_i is also on the adjacency list of n_j. The adjacency list representation for a directed graph puts n_j on the list for n_i if there is an arc from n_i to n_j; n_i would not necessarily be on the adjacency list for n_j. For a labeled graph or a weighted graph, additional data items can be stored with the node name in the adjacency list.

Figure 4.23

4.22 Example   Figure 4.24a shows a weighted directed graph. The adjacency list representation for this graph is shown in Figure 4.24b. For each record in the list, the first data item is the node, the second is the weight of the arc to that node, and the third is the pointer. Note that entry 4 in the array of startup pointers is nil because there are no arcs that begin at node 4.
                                                                                   □

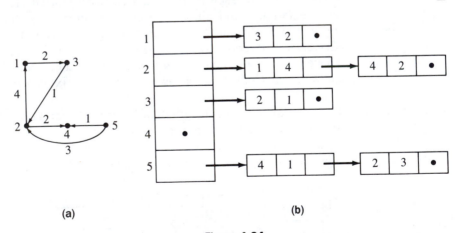

(a)   (b)

Figure 4.24

In a programming language that does not support pointers, we can still achieve the effect of an adjacency list by using a multicolumn array (or an array of records), where one column contains the nodes and another column contains the array index of the next node on the adjacency list—

a "pseudopointer." The disadvantage of this approach is that the maximum amount of storage space that might be needed must be initially set aside for the array; new space cannot be dynamically created as the program executes.

4.23 Example   The array–pointer representation of the graph of Figure 4.24a is shown in Figure 4.25. A nil pointer is indicated by an array index of 0.

In this array row 2, representing node 2, has a pointer to index 7. At index 7 of the array, we find node 1 with weight 4, representing the arc of weight 4 from node 2 to node 1. The pointer to index 8 says that the adjacency list for node 2 has more entries. At index 8, we learn that there is an arc from 2 to 4 of weight 2, and that this completes the adjacency list for node 2.   □

	Node	Weight	Pointer
1			6
2			7
3			9
4			0
5			10
6	3	2	0
7	1	4	8
8	4	2	0
9	2	1	0
10	4	1	11
11	2	3	0

Figure 4.25

Binary Tree Representation

Because a tree is also a graph, all the representations we have discussed for graphs in general can also be used for trees. Binary trees, however, have special characteristics that allow particularly simple representations. For example, the important facts about any node in a binary tree are the names of its left and right children. Thus we can use a two-column array (or an array of records) where the data for each node is the left and right child of that node. An additional item to store that would be useful if we need to backtrack up the tree is the parent of a node as well as its children.

4.24 Example

For the binary tree shown in Figure 4.26, the left child–right child representation is given in Figure 4.27. Zeros again indicate nil pointers. □

Figure 4.26

	Left child	Right child
1	2	3
2	4	5
3	0	6
4	0	0
5	0	0
6	0	0

Figure 4.27

4.25 Practice

Give the left child–right child representation of the binary tree in Figure 4.28. □

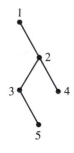

Figure 4.28

Directed Graphs and Binary Relations

A directed graph with no parallel arcs defines a binary relation ρ on the set of nodes. Thus, $n_i \rho n_j$ if there is an arc from n_i to n_j. Conversely, if ρ is a binary relation on a set N, we define a corresponding directed graph with no parallel arcs with N the set of nodes and an arc from n_i to n_j whenever $n_i \rho n_j$. Thus, there is a one-to-one correspondence between binary relations and directed graphs with no parallel arcs. If ρ is a reflexive relation, then $n_i \rho n_i$ for every node n_i, and the corresponding graph will have loops at every node, with 1s along the main diagonal of its adjacency matrix. If ρ is symmetric, then $n_i \rho n_j$ implies $n_j \rho n_i$, and the adjacency matrix of the graph is symmetric. When we used graphs to represent partial orderings in Section 3.1, we assumed an upward direction for each arc and eliminated the loops at the nodes. We also eliminated an arc from node n_i to n_j if there was a longer path from n_i to n_j.

4.26 Practice

Explain why the associated binary relation ρ is not antisymmetric for the directed graph in Figure 4.10. □

✔ **Checklist**

Definitions

> adjacency matrix (*p. 169*)
> linked list (*p. 171*)
> adjacency list (*p. 171*)
> nil pointer (*p. 172*)

Techniques

> Construct adjacency matrices and adjacency lists for graphs and directed graphs.
>
> Find which nodes can be reached from which in a directed graph.

Main Ideas

> Matrix representation of graphs.
>
> Linked list representation of graphs.
>
> One-to-one correspondence between binary relations and directed graphs with no parallel arcs.

Exercises Section 4.2

For Exercises 1 through 6, write the adjacency matrix for the graph in the specified figure.

★ 1.

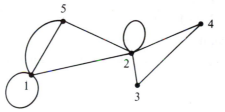

Figure 4.29

2.

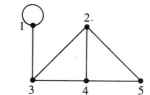

Figure 4.30

3.

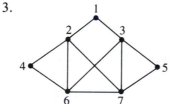

Figure 4.31

4.

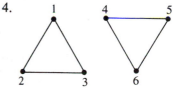

Figure 4.32

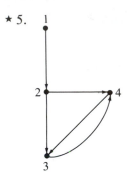

★ 5.

Figure 4.33

6.

Figure 4.34

7. For the graph in Figure 4.31, compute A^2 to find the number of paths of length 2 from node 2 to node 6.

8. For the graph of Figure 4.29, count the number of paths of length 3 from node 5 to node 2. Then check by computing A^3.

★ 9. For the directed graph of Figure 4.34, compute the matrix R to show that node 1 cannot be reached from node 3.

10. For the directed graph of Figure 4.17, Section 4.1, compute the matrix R to show that nodes 1 and 2 cannot be reached from node 3.

11. The adjacency matrix for an undirected graph is given in *lower triangular form* by

$$\begin{bmatrix} 2 & & & \\ 1 & 0 & & \\ 0 & 1 & 1 & \\ 0 & 1 & 2 & 0 \end{bmatrix}$$

Draw the graph.

12. The adjacency matrix for a directed graph is given by

$$\begin{bmatrix} 0 & 1 & 1 & 0 & 0 \\ 0 & 0 & 0 & 0 & 0 \\ 0 & 0 & 1 & 1 & 0 \\ 0 & 0 & 1 & 0 & 2 \\ 1 & 0 & 0 & 0 & 0 \end{bmatrix}$$

Draw the graph.

For Exercises 13 through 18, draw the adjacency list representation for the indicated graph.

13. Figure 4.29

14. Figure 4.30

★ 15. Figure 4.31

16. Figure 4.32

17. Figure 4.33

18. Figure 4.34

Figure 4.35

19. Refer to the graph in Figure 4.35.
    (a) Draw the adjacency list representation.
    (b) How many storage locations are required for the adjacency list? (A pointer takes one storage location.)
    (c) How many storage locations would be required in an adjacency matrix for this graph?

20. Draw the adjacency list representation for the weighted directed graph of Figure 4.36.

★ 21. For the directed graph of Figure 4.34, construct the array–pointer representation.

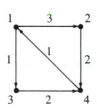

Figure 4.36

22. For the weighted directed graph of Figure 4.36, construct the array–pointer representation.

23. Draw the undirected graph represented by the adjacency list in Figure 4.37.

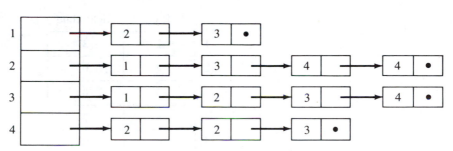

Figure 4.37

24. Draw the directed graph represented by the adjacency list in Figure 4.38.

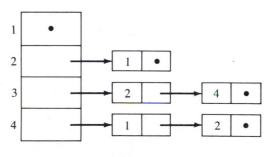

Figure 4.38

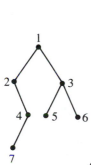

Figure 4.39

★ 25. Write the left child–right child representation for the binary tree in Figure 4.39.

26. Write the left child–right child representation for the binary tree in Figure 4.40.

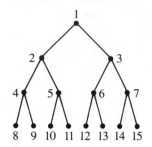

Figure 4.40

★ 27. Draw the binary tree represented by the left child–right child representation of Figure 4.41.

	Left child	Right child
1	2	3
2	4	0
3	5	0
4	6	7
5	0	0
6	0	0
7	0	0

Figure 4.41

28. Draw the binary tree represented by the left child–right child representation of Figure 4.42.

	Left child	Right child
1	2	0
2	3	4
3	0	0
4	5	6
5	0	0
6	0	0

Figure 4.42

29. Figure 4.43 represents a binary tree in which the left child and parent of each node are given. Draw the binary tree.

	Left child	Parent
1	2	0
2	4	1
3	0	1
4	0	2
5	0	2
6	0	3

Figure 4.43

30. Figure 4.44 represents a tree (not necessarily binary), where for each node, the leftmost child and the closest right sibling of that child are given. Draw the tree.

	Left child	Right sibling
1	2	0
2	5	3
3	0	4
4	8	0
5	0	6
6	0	7
7	0	0
8	0	0

Figure 4.44

★ 31. List the ordered pairs in the binary relation ρ defined by the directed graph of Figure 4.45.

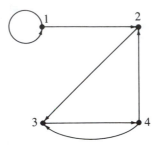

Figure 4.45

32. Let ρ be a binary relation defined on the set $\{0, \pm 1, \pm 2, \pm 4, \pm 16\}$ by $x \rho y \leftrightarrow y = x^2$. Draw the associated directed graph.

SECTION 4.3   GRAPH ALGORITHMS

Because graphs have so many applications, there is a great deal of interest in finding efficient ways to work with them, especially to carry out tasks such as traveling along the arcs of a graph or visiting the various nodes. In this section, we consider three graph problems—all dealing with certain kinds of paths in a given graph—and their solutions.

The Euler Path Problem

The Euler path problem originated many years ago. Swiss mathematician Leonhard Euler (pronounced "oiler") (1707–1783) was intrigued by a puzzle popular among the townfolk of Königsberg (an East Prussian city now called Kaliningrad, which is in the Soviet Union). The river flowing through the city branched around an island. Various bridges crossed the river as shown in Figure 4.46. The puzzle was to decide whether a person could walk through the city crossing each bridge only once. It is possible, in theory, to answer the question by listing all possible routes, so some dedicated Königsberger could have solved this particular puzzle. Euler's idea was to represent the situation as a graph (see Figure 4.47) where the bridges are arcs and the land masses are nodes. He then solved the general question of when an Euler path exists in any graph.

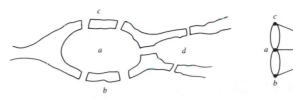

Figure 4.46   **Figure 4.47**

**4.27
Definition**

An **Euler path** in a graph G is a path that uses each arc of G exactly once.

□

4.28 Practice

(a)

(b)

Figure 4.48

Do Euler paths exist for either graph in Figure 4.48? (Use trial and error to answer. This is the old children's game of whether you can trace the whole graph without lifting your pencil and without retracing any arcs.)

□

For this discussion we will assume that all graphs are connected, since an Euler path generally cannot exist otherwise. Whether an Euler path exists in a given graph hinges on the degrees of the nodes. A node is **even** if its degree is even and **odd** if its degree is odd. It turns out that every graph has an even number of odd nodes. To see this, choose any graph and let N be the number of odd nodes in it, $N(1)$ the number of nodes of degree 1, $N(2)$ the number of nodes of degree 2, and so on. Then the sum S of the degrees of all the nodes of the graph is

$$S = 1 \cdot N(1) + 2 \cdot N(2) + 3 \cdot N(3) + \cdots + k \cdot N(k) \tag{1}$$

for some k. This sum is, in fact, a count of the total number of arc ends in the graph. Because the number of arc ends is twice the number of arcs, S is an even number. We will reorganize equation (1) to group together terms representing odd nodes and terms representing even nodes:

$$S = \underbrace{2 \cdot N(2) + 4 \cdot N(4) + \cdots + 2m \cdot N(2m)}_{\text{even nodes}}$$
$$+ \underbrace{1 \cdot N(1) + 3 \cdot N(3) + \cdots + (2n + 1) \cdot N(2n + 1)}_{\text{odd nodes}}$$

The sum of the terms representing even nodes is an even number. If we subtract it from both sides of the equation, we get a new equation

$$S' = 1 \cdot N(1) + 3 \cdot N(3) + \cdots + (2n + 1) \cdot N(2n + 1) \tag{2}$$

where S' is an even number. Now if we rewrite equation (2) as

$$S' = \underbrace{1 + 1 + \cdots + 1}_{N(1) \text{ terms}} + \underbrace{3 + 3 + \cdots + 3}_{N(3) \text{ terms}} + \cdots$$
$$+ \underbrace{(2n + 1) + (2n + 1) + \cdots + (2n + 1)}_{N(2n + 1) \text{ terms}}$$

we see that there are N terms in the sum and that each term is an odd number. For the sum of N odd numbers to be even, N must be even. (Can you prove this?) We have proved Theorem 4.29.

4.29 Theorem   The number of odd nodes in any graph is even.   □

Now suppose a graph has an odd node n of degree $2k + 1$ and that an Euler path exists in the graph but does not start at n. Then for each arc we use to enter n, there is a new arc to use leaving n until we have used k pairs of arcs. The next time we enter n, there is no new arc on which to leave. Thus, even if our path does not begin at n, it must end at n. Or equivalently, the path either begins or ends at n. Therefore, if there are more than two odd nodes in the graph, there can be no path. Thus, there are two possible cases where an Euler path may exist—on a graph with no odd nodes or on one with two odd nodes.

Consider the graph with no odd nodes. Pick any node n and begin an Euler path. Whenever you enter a node, you will always have another arc on which to exit until you get back to n. If you have used up every arc of the graph, you are done. If not, there is some node n' of your path with unused arcs. Construct an Euler path beginning and ending at n', much as you did the previous section of path, using all new arcs. Attach this cycle as a side trip on the original path. If you have now used up every arc of the graph, you are done. If not, continue this process until every arc has been covered.

If there are exactly two odd nodes, an Euler path can be started beginning at one odd node and ending at the other. If the path has not covered all of the arcs, extra circuits can be patched in as in the previous case.

We now have a complete solution to the Euler path problem.

4.30 Theorem   An Euler path exists in a connected graph if and only if there are no odd nodes or two odd nodes. For the case of no odd nodes, the path can begin at any node and will end there; for the case of two odd nodes, the path must begin at one odd node and end at the other.   □

4.31 Practice   Using Theorem 4.30, work Practice 4.28 again.   □

4.32 Practice   Is the Königsberg walk possible?   □

Theorem 4.30 is actually an algorithm to determine if an Euler path exists on an arbitrary connected graph. To make it look more like an algorithm, we'll rewrite it in pseudocode. The essence of the algorithm is to count the number of nodes adjacent to each node and to determine whether this is an odd or an even number. If there are too many odd numbers, an Euler path does not exist.

In the following algorithm, the input is a connected graph G, represented by an $n \times n$ adjacency matrix $\mathbf{A}$; the output indicates whether an Euler path exists. Also, *odd* is a variable that keeps track of the number of odd nodes so far found in the graph. The degree of a node, *degree*, is found by adding the numbers in that node's row of the adjacency matrix, because this gives the number of nodes adjacent to that one node.

Algorithm EulerPath

1. $odd \leftarrow 0$
2. $i \leftarrow 1$
3. while $odd \leq 2$ and $i \leq n$ do
    begin
4.   *degree* $\leftarrow 0$
5.   for $j = 1$ to n do
      begin
6.   *degree* $\leftarrow$ *degree* $+ a_{ij}$
      end
7.   if *degree* is odd, then $odd \leftarrow odd + 1$
8. $i \leftarrow i + 1$
    end while
9. if $odd > 2$
   then write ''No Euler path exists''
   else write ''Yes, Euler path exists''

4.33 Example   The adjacency matrix for the graph of Figure 4.48a is shown below.

$$\begin{bmatrix} 0 & 2 & 1 & 0 & 0 \\ 2 & 0 & 1 & 0 & 0 \\ 1 & 1 & 0 & 1 & 1 \\ 0 & 0 & 1 & 0 & 2 \\ 0 & 0 & 1 & 2 & 0 \end{bmatrix}$$

When the algorithm first reaches step 5, *odd* and *degree* are both 0, and i and j are both 1. At step 6, the value of a_{11}, 0, is added to *degree*. Step 6 is within a loop; at the next execution of step 6, the value of a_{12}, 2, is added to *degree*, making 2 the new value of *degree*. At the next execution of step 6, the value of a_{13}, 1, is added to *degree*, making 3 the new value of *degree*. This value remains unchanged as the remaining entries in row 1 of the matrix are examined. At step 7, the value of *degree* is examined and found to be an odd number. Thus the value of *odd* is upgraded from 0 to 1. Neither the bounds on *odd* nor the bounds on the array size have been exceeded, so the while loop executes again, this time for row 2 of the array. Once again, *degree* is found to be odd, so the value of *odd* gets changed to 2. When the while loop is executed for row 3 of the array, the value of *degree* is even (4), so *odd* does not change, and the while loop is executed again, for $i = 4$. Row 4 again produces an odd value for *degree,* so *odd* is raised to 3. This terminates the while loop, and at step 9 the bad news is told that there is no Euler path.   □

4.34 Practice   Write the adjacency matrix for the Königsberg walk problem and trace the execution of algorithm *EulerPath*.   □

Let us try to analyze algorithm *EulerPath*. The important operation being done here is an examination of the elements of the adjacency matrix. In the worst case, almost the entire matrix must be examined before the number of nodes of odd degree can be finally determined. (We don't have to check the last row because there must be an even number of odd nodes; if the number of odd nodes exceeds 2, we will know that before checking the last row.) We can see that examining the entire matrix requires n^2 operations. Or, using the Multiplication Principle (see Section 2.2), the task of choosing a row of the matrix has n possible outcomes, and the task of choosing a column in the row also has n possible outcomes; thus there are $n \cdot n = n^2$ elements to examine. The algorithm itself reflects this counting argument: in the worst case, the while loop at line 3 must be executed n times (choosing a row). Within each execution of the while loop, the loop at line 5 requires n executions (choosing a column within that row). Even if we modify the algorithm so that the last row is not examined, there are still $(n - 1)n$ elements to examine—roughly the same as n^2 when n is large.

The term *order of magnitude* is used to denote functions that are roughly the same size.

4.35 Definition   Let f and g be two functions from $\mathbb{N} \to \mathbb{N}$. Then f and g are of the same **order of magnitude** if (1) there exist positive constants c_1 and n_1 such that $f(n) \leq c_1 g(n)$ for all $n \geq n_1$ and (2) there exist positive constants c_2 and n_2 such that $g(n) \leq c_2 f(n)$ for all $n \geq n_2$. We write $f = O(g)$ or $g = O(f)$ to denote functions of the same order of magnitude.*   □

4.36 Example   Let $f(n) = 3n^2$ and $g(n) = 200n^2 + 140n + 7$. Then f and g are of the same order of magnitude because $3n^2 \leq 200n^2 + 140n + 7$ for all n, and $200n^2 + 140n + 7 \leq 100(3n^2)$ for all $n \geq 2$. Both f and g are of the same order of magnitude as n^2. A polynomial function is the order of its term of highest degree. The functions $f(n) = n$ and $g(n) = n^2$ are not of the same order because we cannot force $n^2 \leq cn$ (this implies $n \leq c$) for all

---

*Many books define order of magnitude and use the notation $f = O(g)$ somewhat differently, letting $f = O(g)$ if $f(n) \leq cg(n)$ for $n \geq$ some n_0 and c a positive constant.

n greater than or equal to some n_0, no matter how large a value we choose for c. □

4.37 Practice Find specific values for n_1, n_2, c_1, and c_2 to show that

$$f(n) = 3n^2 + 4 \quad \text{and} \quad g(n) = n^2 + n + 3$$

are of the same order of magnitude. □

Using the notation for order of magnitude, we can say that the number of operations required by algorithm *EulerPath* in the worst case is $O(n^2)$.

If we represented the graph G by an adjacency list rather than an adjacency matrix, then the corresponding version of the algorithm would have to count the length of the adjacency list for each node and keep track of how many are of odd length. There would be n adjacency lists to examine, just as there were n rows of the adjacency matrix to examine, but the length of each adjacency list might be shorter than n, the length of a row of the matrix. It is possible to reduce the order of magnitude below n^2 if the number of arcs in the graph is small, but the worst case is still $O(n^2)$.

Another famous mathematician, William Rowan Hamilton (1805–1865), posed a problem in graph theory very much like Euler's. He asked how to tell whether a graph has a **Hamiltonian circuit,** a cycle using every node of the graph.

4.38 Practice Do Hamiltonian circuits exist for the graphs of Figure 4.48? (Use trial and error to answer.) □

Although we have a simple, efficient algorithm to determine if an Euler path exists on an arbitrary graph, no efficient procedure (trial and error is not efficient!) has been found for determining if a Hamiltonian circuit exists in an arbitrary graph. We will see in Chapter 11 why it is unlikely that such a procedure will ever be found.

However, in certain types of graphs, we can easily determine if a Hamiltonian circuit exists. In particular, a complete graph with $n > 2$ has a Hamiltonian circuit because no matter what node you are in, there is always an arc to travel to any unused node and finally an arc to return to the starting point.

Suppose we are dealing with a weighted graph. If a Hamiltonian circuit exists for the graph, can we find one with minimum weight? This problem is even more difficult than the Hamiltonian circuit problem. It is sometimes called the "traveling salesman" problem, in honor of the salespeople who must visit every city in their region once, return home, and minimize the distance traveled. (The Euler path problem is sometimes called the "highway inspector" problem where the arcs are thought of as roads to be traveled.)*

The Shortest-Path Problem

One minimum-distance path problem that can be solved efficiently is described next. Assume that we have a simple, weighted, connected graph where the weights are positive. Then a path exists between any two nodes x and y. How do we find the path with minimum weight? Because weight often represents distance, this problem has come to be known as the "shortest-path" problem. This problem is an important one to solve for computer or communications networks, where information at one node must be routed to another node in the most efficient way possible. It is also a useful problem to solve for product distribution, where stock in one city must be transported to another city.

The shortest-path algorithm works as follows. We want to find the minimum-length path from a given node x to a given node y. We build a set (we'll call it *IN*) that initially contains only x but grows as the algorithm proceeds. At any given time *IN* contains every node whose shortest path from x, using only nodes in *IN*, has so far been determined. For every node z outside *IN*, we keep track of the shortest distance $d(z)$ from x to that node, using a path whose only non-*IN* node is z. We also keep track of the node adjacent to z on this path, $s(z)$.

How do we let *IN* grow, that is, which node should be moved into *IN* next? We pick the non-*IN* node with the smallest distance d. Once we add that node, call it p, to *IN*, then we have to recompute d for all the remaining non-*IN* nodes, because there may be a shorter path from x going through p than there was before p belonged to *IN*. If there is a shorter path, we must also update $s(z)$ so that p is now shown to be the node adjacent to z on the current shortest path. As soon as y is moved into *IN*, *IN* stops growing. The current value of $d(y)$ is the length of the shortest path, and its nodes are found by looking at y, $s(y)$, $s(s(y))$, and so forth, until we have traced the path back to x.

A pseudocode form of the algorithm follows. The input is the adjacency matrix for a simple connected graph G with positive weights, and nodes x and y; the output is the shortest path between x and y and the length of that path. We actually assume a modified adjacency matrix **A**, where a_{ij} is the weight of the arc between i and j if one exists and a_{ij} has the value ∞ if no arc exists (here the symbol ∞ denotes a number larger than any weight in the graph).

---

*The traveling salesman problem for visiting all 48 capitals of the contiguous states has been solved by a mathematician at Bell Laboratories: a total of 10,628 miles is required ("Here's One for the Road—and Then Some," *Discover*, July 1985, pp. 13–16).

Algorithm ShortestPath

1. $IN \leftarrow \{x\}$
2. for all nodes z not in IN do
   begin
   $d(z) \leftarrow a_{xz}$
   $s(z) \leftarrow x$
   end
3. while $y \notin IN$ do
   begin
4. $p \leftarrow$ node z not in IN with $d(z)$ a minimum
5. $IN \leftarrow IN \cup \{p\}$
6. for all nodes z not in IN do
   begin
7. *old-distance* $\leftarrow d(z)$
8. $d(z) \leftarrow$ minimum of $(d(z), d(p) + a_{pz})$
9. if $d(z) \neq$ *old-distance*, then $s(z) \leftarrow p$
   end
   end while
10. write "The length of the shortest path is:" $d(y)$
11. write "The path is:" $y, s(y), s(s(y)), \ldots, x$

4.39 Example   Consider the graph in Figure 4.49 and the corresponding adjacency matrix shown in Figure 4.50.

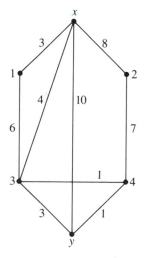

$$\begin{array}{c c c c c c c} & x & 1 & 2 & 3 & 4 & y \\ x & \infty & 3 & 8 & 4 & \infty & 10 \\ 1 & 3 & \infty & \infty & 6 & \infty & \infty \\ 2 & 8 & \infty & \infty & \infty & 7 & \infty \\ 3 & 4 & 6 & \infty & \infty & 1 & 3 \\ 4 & \infty & \infty & 7 & 1 & \infty & 1 \\ y & 10 & \infty & \infty & 3 & 1 & \infty \end{array}$$

Figure 4.49   **Figure 4.50**

We will trace algorithm *ShortestPath* on this graph. At the end of steps 1 and 2, we have initialized *IN* and all the direct distances from x

to other nodes:

$$IN = \{x\}$$

	1	2	3	4	y
d:	3	8	4	∞	10
s:	x	x	x	x	x

We now enter the while loop. At step 4 we search through the *d*-values for the node of minimum distance that is not in *IN*; this turns out to be node 1, with $d(1) = 3$. We throw node 1 into *IN* (step 5), and recompute all the *d*-values for the remaining nodes: 2, 3, 4, and *y* (the loop at steps 6 to 9).

$$p = 1$$
$$IN = \{x, 1\}$$
$$d(2) = \min(8, 3 + a_{12}) = \min(8, \infty) = 8$$
$$d(3) = \min(4, 3 + a_{13}) = \min(4, 6) = 4$$
$$d(4) = \min(\infty, 3 + a_{14}) = \min(\infty, \infty) = \infty$$
$$d(y) = \min(10, 3 + a_{1y}) = \min(10, \infty) = 10$$

There were no changes in the *d*-values, so step 9 produced no changes in the *s*-values (there were no shorter paths from *x* by going through node 1 than by going directly from *x*).

The second pass through the while loop at step 4 produces the following:

$$p = 3 \quad \text{(3 has the smallest } d\text{-value of 2, 3, 4, or } y)$$
$$IN = \{x, 1, 3\}$$
$$d(2) = \min(8, 4 + \infty) = 8$$
$$d(4) = \min(\infty, 4 + 1) = 5, \text{ a change, so update } s(4)$$
$$d(y) = \min(10, 4 + 3) = 7, \text{ a change, so update } s(y)$$

	1	2	3	4	y
d:	3	8	4	5	7
s:	x	x	x	3	3

Shorter paths from *x* to the two nodes 4 and *y* are found by going through 3.

On the next pass,

$$p = 4$$
$$IN = \{x, 1, 3, 4\}$$
$$d(2) = \min(8, 5 + 7) = 8$$
$$d(y) = \min(7, 5 + 1) = 6, \text{ a change, update } s(y)$$

$$
\begin{array}{c|ccccc}
 & 1 & 2 & 3 & 4 & y \\
\hline
d: & 3 & 8 & 4 & 5 & 6 \\
s: & x & x & x & 3 & 4
\end{array}
$$

Processing the while loop again, we get

$p = y$

$IN = \{x, 1, 3, 4, y\}$

$d(2) = \min(8, 6 + \infty) = 8$

$$
\begin{array}{c|ccccc}
 & 1 & 2 & 3 & 4 & y \\
\hline
d: & 3 & 8 & 4 & 5 & 6 \\
s: & x & x & x & 3 & 4
\end{array}
$$

Now that y is part of IN, the while loop terminates. The path length is $d(y) = 6$, and the path goes through y, $s(y) = 4$, $s(4) = 3$, and $s(3) = x$. Thus the path consists of nodes x, 3, 4, and y. (The algorithm gives us these nodes in reverse order.) By looking at the graph in Figure 4.49 and checking all the possibilities, we can see that this is the shortest path.

□

Algorithm *ShortestPath* terminates when y is put into IN, even though there may be other nodes in the graph not yet in IN (such as node 2 in Example 4.39). How do we know that a still shorter path cannot be found through one of these nodes? If we continue processing until all nodes have been included in IN, the d-values then represent the shortest paths from x to any node, using all of the values in IN, that is, the shortest path using any nodes of the graph. But new nodes are brought into IN in order of increasing d-values. A node z brought into IN later than y has as its shortest path from x one that is at least as long as the d-value of y when y was brought into IN. Therefore there cannot be a shorter path from x to y via z because there is not even a shorter path just between x and z.

4.40 Practice Trace algorithm *ShortestPath* on the graph shown in Figure 4.51. Show the values for p and IN and the d-values and s-values for each pass through the while loop. Write out the shortest path and its length.   □

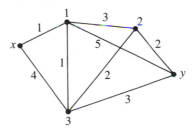

Figure 4.51

At step 4 of algorithm *ShortestPath*, more than one node *p* may give a minimum *d*-value, in which case *p* can be selected arbitrarily. There may also be more than one shortest path between *x* and *y* in a graph.

Algorithm *ShortestPath* also works for directed graphs, if the adjacency matrix is in the appropriate form. It also works for unconnected graphs; if *x* and *y* are not in the same component, then *d(y)* will remain ∞ throughout. After *y* has been brought into *IN*, the algorithm will terminate, and this value of ∞ for *d(y)* will indicate that no path exists between *x* and *y*.

We may think of algorithm *ShortestPath* as being a "nearsighted" algorithm. It cannot see the entire graph at once to pick out overall shortest paths; it only picks out shortest paths relative to the set *IN* at each step. Such an algorithm is called a **greedy algorithm**—always taking the approach of doing what seems best based on its limited immediate knowledge. In this case, what seems best at the time turns out to be best overall.

How efficient is the shortest-path algorithm? Most of the work seems to take place within the loop at step 6. Here the algorithm checks all *n* nodes to determine which nodes *z* are not in *IN*, and recomputes *d(z)* for those nodes, possibly also changing *s(z)*. The necessary quantities *d(z)*, *d(p)*, and a_{pz} for a given *z* are directly available. Therefore the loop at step 6 requires *O(n)* operations. In addition, at step 4 the algorithm determines *p*, which can also be done in *O(n)* operations by checking all *n* nodes. With the additional small amount of work at step 5, each execution of the while loop takes *O(n)* operations. In the worst case, *y* is the last node brought into *IN*, and the while loop at step 3 will be executed *n* − 1 times. Therefore the total number of operations in steps 3 through 9 is *O(n(n* − 1)) = *O(n²)*. The initialization (steps 1 and 2) and steps 10 and 11 together take *O(n)* operations, so the algorithm requires *O(n + n²) = O(n²)* operations in the worst case.

What if we keep *IN* (or rather the complement of *IN*) as some sort of linked list, so that at steps 4 and 6 all the nodes of the graph do not have to be examined to see which are not in *IN*? Surely this would make the algorithm more efficient. Note that the number of nodes not in *IN* is initially *n* − 1, and that number decreases by 1 for each pass through the while loop. At steps 4 and 6 the algorithm thus has to perform *n* − 1 operations on the first pass, then *n* − 2, then *n* − 3, and so on. But

$$(n - 1) + (n - 2) + \cdots + 1 = \frac{(n - 1)n}{2} \qquad \text{(prove this by induction)}$$

$$= O(n^2)$$

Thus the worst-case situation still requires *O(n²)* operations.

The Minimal Spanning Tree Problem

A problem encountered in designing networks is how to connect all the nodes efficiently, where nodes can be computers, telephones, warehouses, and so on. A minimal spanning tree may provide an economical solution.

4.41
Definition

A **spanning tree** for a graph is a tree whose set of nodes coincides with the set of nodes for the graph and whose arcs are (some of) the arcs of the graph.   □

A spanning tree thus connects all the nodes of a graph with no excess arcs (no cycles). There is an algorithm for constructing a **minimal spanning tree,** a spanning tree with minimal weight, for a given weighted, connected graph.

The algorithm proceeds very much like the shortest-path algorithm. There is a set *IN*, which initially contains one arbitrary node. For every node *z* not in *IN*, we keep track of the shortest distance $d(z)$ between *z* and any node in *IN*. We successively add nodes to *IN*, where the next node added is one that is not in *IN* and whose distance $d(z)$ is minimal. The arc having this minimal distance is then made part of the spanning tree. Because there may be ties here, the minimal spanning tree of a graph is not unique. The algorithm terminates when all nodes of the graph are in *IN*. We won't write out the algorithm (which, like the shortest-path algorithm, requires $O(n^2)$ operations in the worst case and is a greedy algorithm), but will simply illustrate it with an example.

4.42   Example

We will find a minimal spanning tree for the graph of Figure 4.49. We let node 1 be the arbitrary initial node in *IN*. Next we consider all the nodes adjacent to any node in *IN*, that is, all nodes adjacent to 1, and select the closest one, which is node *x*. Now $IN = \{1, x\}$, and we consider all nodes not in *IN* that are adjacent to either 1 or *x*. The closest such node is 3, 4 units away from *x*. For $IN = \{1, x, 3\}$, the next closest node is node 4, 1 unit away from 3. The remaining nodes are added in the order *y* and then 2. Figure 4.52 shows the minimal spanning tree.   □

4.43   Practice

Find a minimal spanning tree for the graph of Figure 4.51.   □

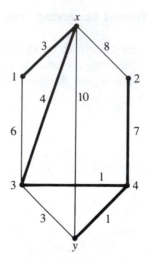

Figure 4.52

✔ Checklist

Definitions

> Euler path (*p. 181*)
> even node (*p. 181*)
> odd node (*p. 181*)
> order of magnitude (*p. 184*)
> Hamiltonian circuit (*p. 185*)
> greedy algorithm (*p. 190*)
> spanning tree (*p. 191*)
> minimal spanning tree (*p. 191*)

Techniques

> Determine whether an Euler path exists in a graph (use algorithm *EulerPath*).
>
> Find a shortest path from *x* to *y* in a graph (use algorithm *ShortestPath*).
>
> Find a minimal spanning tree for a graph.

Main Ideas

> Simple criterion for determining existence of Euler paths in a graph; no simple criterion for Hamiltonian circuits.
>
> Order of magnitude as a gauge of algorithm efficiency.

Algorithms of $O(n^2)$ in the worst case to determine existence of an Euler path, to find a shortest path between two nodes, or to find a minimal spanning tree in a graph with n nodes.

Exercises Section 4.3

For Exercises 1 through 4, determine whether the graph in the specified figure has an Euler path.

★ 1.

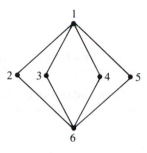

Figure 4.53

2.

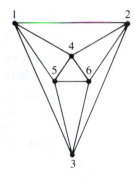

Figure 4.54

3.

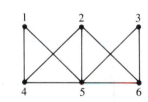

Figure 4.55

4.

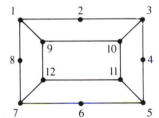

Figure 4.56

★ 5. Draw the adjacency matrix for the graph of Figure 4.53. In applying algorithm *EulerPath*, what is the value of *odd* after the second pass through the while loop at step 3?

6. Draw the adjacency matrix for the graph of Figure 4.55. In applying algorithm *EulerPath*, what is the value of *odd* after the fourth pass through the while loop at step 3?

7. Prove that $f(n) = 140n + 25$ and $g(n) = n - 7$ are of the same order of magnitude.

8. Prove that $n^{1/2}$ and n are not of the same order of magnitude.

9. Write a computer program to implement the Euler path algorithm.

10. Consider a connected graph with $2n$ odd vertices, $n \geq 2$. By Theorem 4.30, an Euler path does not exist for this graph.

(a) What is the minimum number of disjoint Euler paths, each traveling some of the arcs of the graph, necessary to travel each arc exactly once?

(b) Show that the minimum number is sufficient.

For Exercises 11 through 14, decide (by trial and error) whether Hamiltonian circuits exist.

★ 11. Figure 4.53

12. Figure 4.54

13. Figure 4.55

14. Figure 4.56

★ 15. Prove that a Hamiltonian circuit always exists in a connected graph where every node has degree 2.

16. (a) Is it possible to walk in and out of each room in the house shown in Figure 4.57 so that each door of the house is used exactly once?

(b) Why or why not?

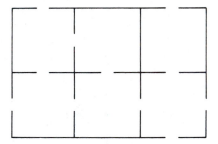

Figure 4.57

For Exercises 17 through 20, use the graph of Figure 4.58. Apply algorithm *ShortestPath* for the pairs of nodes given; show the values for *p* and *IN* and the *d*-values and *s*-values for each pass through the while loop. Write out the shortest path and its length.

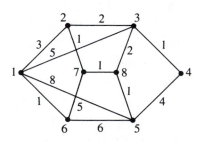

Figure 4.58

★ 17. 2, 5   18. 3, 6

  19. 1, 5   20. 4, 7

For Exercises 21 and 22, use the directed graph of Figure 4.59. Apply algorithm *ShortestPath* to the nodes given; show the values for *p* and *IN* and the *d*-values and *s*-values for each pass through the while loop. Write out the shortest path and its length.

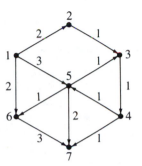

Figure 4.59

★ 21. From 1 to 7   22. From 3 to 1

★ 23. Give an example to show that the shortest-path algorithm does not work when there are negative weights.

  24. Give an example to show that adding the node closest to *IN* at each step, as is done in the minimal spanning tree algorithm, will not guarantee a shortest path.

  25. Write a computer program to implement the shortest-path algorithm.

  26. Modify algorithm *ShortestPath* so that it finds the shortest paths from *x* to all other nodes in the graph.

For Exercises 27 through 30, find a minimal spanning tree for the graph in the specified figure.

  27. Figure 4.58   28.

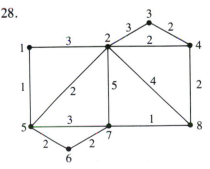

Figure 4.60

★ 29.

30.

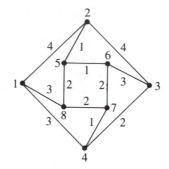

Figure 4.61

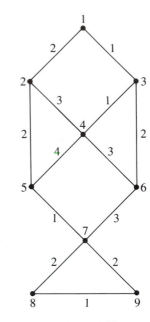

Figure 4.62

31. Write a computer program to find a minimal spanning tree for a graph.

32. Adding new nodes and arcs to a graph may result in a spanning tree for the new graph that has less weight than a spanning tree for the original graph. (The new spanning tree could represent a minimal-cost network for communications between a group of cities obtained by adding a switchboard in a location outside any of the cities.)
    (a) Find a spanning tree of minimum weight for the labeled graph of Figure 4.63. What is its weight?

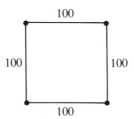

Figure 4.63

    (b) Put a node in the center of the square. Add new arcs from the center to the corners. Find a spanning tree for the new graph, and compute its (approximate) weight.

SECTION 4.4   ALGORITHMS FOR TRAVERSING GRAPHS

The last section concerned various kinds of paths through a graph under certain constraints—not retracing arcs, not passing through a node more than once, or minimizing the distance traveled. In this section we deal with a simpler problem—we only want to list all the nodes of a connected graph in some particular order. This means we must visit each node at least once, but we can visit it more than once if we don't add it to the list again. We can also retrace arcs on the graph if necessary, and clearly this would in general be necessary if we were to visit each node in a tree. This process is called *traversing* a graph, so we will present several algorithms for **graph traversal.**

Depth-First Search and Breadth-First Search

In the first two algorithms for graph traversal, we begin at an arbitrary node *a* of the graph, mark it visited, and write it in the list. In the **depth-first search** procedure, we then strike out on a path away from *a*, proceeding as far as possible until there are no more unvisited nodes on that path. We then back up the path, sending out more paths at each node, if possible, as we retreat back to *a*. In the **breadth-first search** we first fan out from node *a* to nodes adjacent to *a*, then we fan out from those nodes, and so on. To illustrate the difference between these two approaches, Figure 4.64 shows the first few nodes visited (marked by circles) in a graph using depth-first search, and Figure 4.65 shows the first few nodes visited in the same graph using breadth-first search.

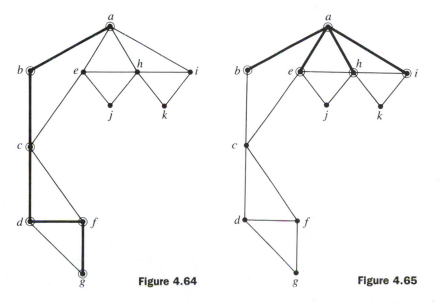

Figure 4.64   Figure 4.65

Depth-First Search

The easiest way to describe the depth-first search algorithm is to use recursion, where the algorithm invokes itself in the course of its execution. In the following algorithm, the input is a connected graph G and a specified node a; the output is a list of all nodes in G in depth-first order:

Algorithm DepthFirst(G, a)

1. mark a visited
2. write a in the list
3. for each n adjacent to a do
       begin
4.   if n not visited then *DepthFirst(G, n)*
       end

The recursion occurs at line 4, where the algorithm is invoked with a new node specified as the starting point. We have not indicated here how to mark visited nodes or how to find nodes n adjacent to a at step 3. These details depend on whether the graph G is represented as an adjacency matrix or an adjacency list.

4.44 Example   We will apply depth-first search to the graph of Figure 4.64, where a is the initially specified node. We first mark that we have visited a (it's helpful in tracing the execution of the algorithm to circle a visited node), and then we write a in the list (steps 1 and 2). Next (step 3) we search the nodes adjacent to a for an unvisited node. We have a choice of unvisited nodes adjacent to a (b, e, h, and i); let us select node b. (Just so we all get the same answers, let's agree to choose the node that is alphabetically first when we have a choice; in practice, the choice would probably be determined by how the vertices were stored in an adjacency matrix or an adjacency list.) At step 4 we invoke the depth-first search algorithm beginning with node b.

    This means we go back to step 1 of the algorithm, where the specified node is now b rather than a. Thus at steps 1 and 2, we mark b visited and write it in the list. At step 3 we search through nodes adjacent to b to find an unmarked node. Node a is adjacent to b, but it is marked. Node c will do, and we invoke the depth-first search algorithm beginning with node c.

    Node c is marked and added to the list, and we look for unmarked nodes adjacent to c. By our alphabetical convention, we select node d. Continuing in this fashion, we then visit nodes f and g. When we get to node g, we have reached a dead end, because there are no unvisited adjacent nodes. Thus the loop at step 3 of the instance of the algorithm for node g is complete. (The graph at this point looks like Figure 4.64.)

    We are therefore done with the algorithm for node g, but node g

was (one of) the unmarked nodes adjacent to node f, and we are still in the loop at step 3 for the instance of the algorithm for node f. As it happens, g was the only unvisited node when we were processing f, therefore we complete the loop at step 3 and thus the algorithm for node f. Similarly, backing up to node d, the algorithm finds no other adjacent unmarked nodes, and it backs up again to the instance of the algorithm for node c. Thus, after processing node d and everything that came after it until the dead end, we are still in the loop at step 3 for the algorithm applied to node c. We look for other adjacent, unmarked nodes and find one—node e. Therefore we apply depth-first search to node e, which leads to nodes h, i, and k before another dead end is reached. Backing up, we have a final new path to try from node h, and that leads to node j. The complete list of the nodes, in the order in which they would be written out, is

$$a \quad b \quad c \quad d \quad f \quad g \quad e \quad h \quad i \quad k \quad j$$ □

Example 4.44 makes the depth-first search process sound very complex, but it is much easier to carry out than to write down, as you will see in Practice 4.45.

4.45 Practice Write the nodes in a depth-first search of the graph in Figure 4.66. Begin with node a. □

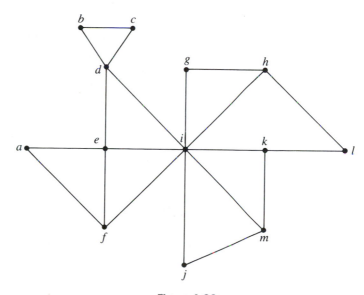

Figure 4.66

Breadth-First Search

In the breadth-first search, nodes are visited by fanning out from the current node, almost like the concentric circles of ripples in a pond. We apply the breadth-first search algorithm to a connected graph G with a specified starting node a; the objective is to output a list of nodes in breadth-first order. To write the algorithm neatly, we will invent an imaginary node called *before-list*. Its purpose is to locate the start of the list in the sense that when we are currently at *before-list,* there is a next node in the list and that node is the starting node a.

In the following algorithm, the input is a connected graph G and a specified node a; the output is a list of all nodes in G in breadth-first order:

Algorithm BreadthFirst(G, a)

1. *current-node* ← *before-list*
2. while a next node x in the list after *current-node* exists do
   begin
3.   *current-node* ← x
4.   for each n adjacent to *current-node* do
   begin
5.   if n not visited then visit n and write n in the list
   end for loop
   end while loop

4.46 Example   Let's do a breadth-first search of the graph of Figure 4.65 (this is the same graph on which we did the depth-first search in Example 4.44). We begin by positioning ourselves before the output list (step 1). When we first reach the while loop at step 2, there is a next node in the list; that node is a, so at step 3 node a becomes the current node. At step 4, we begin to search through all the nodes adjacent to the current node, looking for unvisited nodes to visit and add to the list (step 5). We may have a choice of nodes to visit here; as before, and purely as a convention, we will agree to visit the nodes in alphabetical order. Thus, after completing the loop at step 4 for the first time, we have visited b, e, h, and i, in that order, and added them to the list. The graph at this point looks like Figure 4.65, and the current list of nodes is

> a b e h i

For the next iteration of the while loop at step 2, there is a next node in the list after the current node a, namely b, which then (step 3) becomes the current node. We repeat the for loop at step 4 on nodes adjacent to b. The only previously unvisited node here is c, which gets added to the list. The list is now

> a b e h i c

Performing the while loop again, we advance to the next node in the list, node *e*. A search of the nodes adjacent to node *e* produces one new node, node *j*. The graph now looks like Figure 4.67, and the list is

a   b   e   h   i   c   j

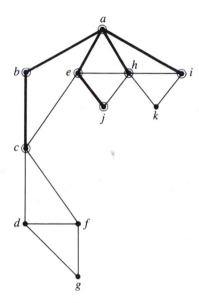

Figure 4.67

When the current node advances to *h*, we pick up one new node in the list, node *k*. When the current node advances to *i*, no new nodes are added to the list.

The current node advances to *c*, and searching the nodes adjacent to *c* turns up two new nodes, *d* and *f*, which are added to the list. The list now is

a   b   e   h   i   c   j   k   d   f

When the current node is *j*, and then *k*, no new nodes are added to the list. When the current node is *d*, a new node *g* is found, and the list is

a   b   e   h   i   c   j   k   d   f   g

As the current node advances to *f* and then *g*, no new nodes are found. When we complete the while loop for current node *g*, there are no next nodes in the list, so the while loop and thus the algorithm terminate. The final output list is

a   b   e   h   i   c   j   k   d   f   g   □

Like the depth-first search, the breadth-first search is not difficult to execute; one must simply keep track (by circles) of which nodes have been visited and (by using a pointer, which might be your index finger) which node is the current node.

4.47 Practice   Write the nodes in a breadth-first search of the graph in Figure 4.66, beginning with node a.   □

Tree Traversal

Because depth-first search and breadth-first search apply to any graph, they also apply to trees. However, there are three other traversal algorithms that are useful for trees—**preorder, inorder,** and **postorder.**

In these traversal methods, it is helpful to use the recursive definition of a tree, where the root of a tree has branches down to roots of subtrees (see Figure 4.68). We will therefore assume that a tree T has a root r; any subtrees are labeled left to right as $T_1, T_2, \ldots, T_t$. Because we are using a recursive definition of a tree, it will be easy to state the tree traversal algorithms in recursive form.

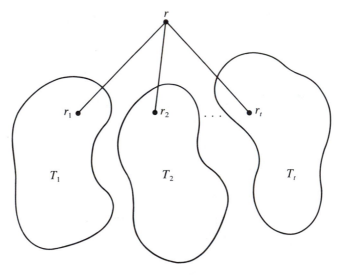

Figure 4.68

The terms *preorder, inorder,* and *postorder* refer to when the root of a tree is visited compared with when the subtree nodes are visited. In preorder traversal, the root of the tree is visited first and then the subtrees are processed left to right, each in preorder.

In the following algorithm, the input is a tree T with, from left to right, subtrees $T_1, T_2, \ldots, T_t$; the output is a list of nodes in preorder:

Algorithm Preorder(T)

1. write r
2. for $i = 1$ to t do *Preorder(T_i)*

In inorder traversal, the left subtree is processed by an inorder traversal, then the root is visited, and then the remaining subtrees are processed from left to right, each in inorder. If the tree is a binary tree, the result is that the root is visited between processing of the two subtrees.

In the following algorithm, the input is a tree T with, from left to right, subtrees $T_1, T_2, \ldots, T_t$; the output is a list of nodes in inorder:

Algorithm Inorder(T)

1. *Inorder(T_1)*
2. write r
3. for $i = 2$ to t do *Inorder(T_i)*

Finally, in postorder traversal, the root is visited last, after all subtrees have been processed from left to right in postorder.

In the following algorithm, the input is a tree T with, from left to right, subtrees $T_1, T_2, \ldots, T_t$; the output is a list of nodes in postorder:

Algorithm Postorder(T)

1. for $i = 1$ to t do *Postorder(T_i)*
2. write r

4.48 Example Consider the tree shown in Figure 4.69.

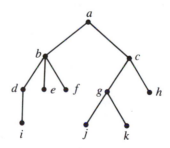

Figure 4.69

For a preorder traversal, we first write the root a and then do a preorder traversal on the left subtree, rooted at b. The preorder traversal of this subtree allows us to write out b and then proceed to a preorder traversal of the left subtree of b, which is rooted at d. We write out d and then do a preorder traversal of the left subtree of d, which is rooted at i. After we write out i and attempt to do a preorder traversal of the subtrees of i, we find no subtrees. Therefore we are done with the preorder traversal of the subtree rooted at i, and we back up to consider any other subtrees of d; there are none. Backing up to b, we find other subtrees of b. Processing these left to right, we write out e and then f. We are now done with all of the subtrees of b, and we back up to node a to look for subtrees farther to the right, finding one rooted at c. We write out c, move to its leftmost subtree rooted at g, and write out g. Processing the subtrees of g, we write out j and k; then backing up to c, we process its remaining subtree and write out h. Node c has no other subtrees; backing up to a, a has no other subtrees, and we are done. The list of nodes in preorder traversal is

$$a \quad b \quad d \quad i \quad e \quad f \quad c \quad g \quad j \quad k \quad h \qquad \square$$

4.49 Example　To do an inorder traversal of the tree in Figure 4.69, we process left subtrees first. This leads us down to node i, which has no subtrees. Therefore we write out i and back up to d. We have processed the left subtree of d, so we can write out d. Since node d has no further subtrees, we back up to b. We have processed the left subtree of b, so we write out b and then process its remaining subtrees, writing out e and f. We back up to a, write it out, and then process the right subtree of a. This leads to nodes j, g, k, c, and h, in that order, and we are done. Thus the inorder list of nodes is

$$i \quad d \quad b \quad e \quad f \quad a \quad j \quad g \quad k \quad c \quad h \qquad \square$$

4.50 Example　The tree of Figure 4.69 produces the following list of nodes from a post-order traversal:

$$i \quad d \quad e \quad f \quad b \quad j \quad k \quad g \quad h \quad c \quad a \qquad \square$$

4.51 Practice　Do a preorder, inorder, and postorder traversal of the tree in Figure 4.70.
$$\square$$

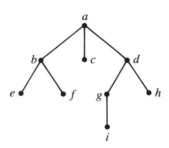

Figure 4.70

4.52 Example   Recall from Example 4.12 that algebraic expressions can be represented as binary trees. If we do an inorder traversal of the expression tree, we retrieve the original algebraic expression. For the expression tree of Figure 4.71, for example, an inorder traversal gives the expression

$$(2 + x) * 4$$

where the parentheses are added as we complete the processing of a subtree. This form of an algebraic expression, where the operation symbol appears between the two operands, is called **infix** notation. Parentheses are necessary here to determine the order of operations. Without parentheses, the expression becomes $2 + x * 4$, which is also an infix expression but is not what is intended.

A preorder traversal of Figure 4.71 gives the expression

$$* \quad + \quad 2 \quad x \quad 4$$

Here the operation symbol precedes its operands. This form of an expression is called **prefix** notation or **Polish** notation. The expression can be translated into infix form as follows:

Figure 4.71 $* \quad + \quad 2 \quad x \quad 4 \rightarrow * \quad (2 + x) \quad 4 \rightarrow (2 + x) * 4$

A postorder traversal gives the expression

$$2 \quad x \quad + \quad 4 \quad *$$

where the operation symbol follows its operands. This form of an expression is called **postfix** notation or **reverse Polish** notation. The expression can be translated into infix form as follows:

$$2 \quad x \quad + \quad 4 \quad * \rightarrow (2 + x) \quad 4 \quad * \rightarrow (2 + x) * 4$$

Neither prefix nor postfix form requires parentheses to avoid ambiguity. These notations are therefore more efficient, if less familiar, representational forms for algebraic expressions than infix notation; compilers often change algebraic expressions in computer programs from infix to postfix notation for more efficient processing.   □

4.53 Practice   Write the expression tree for

$$a + (b * c - d)$$

and write the expression in prefix and postfix notation.   □

✔ **Checklist**

Definitions

graph traversal (*p. 197*)
depth-first search (*p. 197*)
breadth-first search (*p. 197*)

preorder traversal (*p. 202*)
inorder traversal (*p. 202*)
postorder traversal (*p. 202*)
infix notation (*p. 205*)
prefix notation (*p. 205*)
Polish notation (*p. 205*)
postfix notation (*p. 205*)
reverse Polish notation (*p. 205*)

Techniques

Conduct a depth-first search and a breadth-first search of a graph.

Conduct a preorder, inorder, and postorder traversal of a tree.

Main Ideas

Algorithms to visit the nodes of a graph systematically.

Exercises Section 4.4

For Exercises 1 through 6, write the nodes in a depth-first search of the graph in Figure 4.72, beginning with the given node.

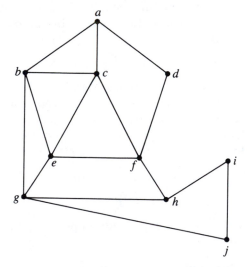

Figure 4.72

★ 1. *a*   2. *c*   3. *d*   4. *g*   ★ 5. *e*   6. *h*

For Exercises 7 through 10, write the nodes in a depth-first search of the graph in Figure 4.73, beginning with the given node.

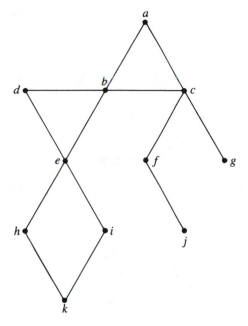

Figure 4.73

★ 7. *a*   8. *e*   ★ 9. *f*   10. *h*

For Exercises 11 through 16, write the nodes in a breadth-first search of the graph in Figure 4.72, beginning with the given node.

★ 11. *a*   12. *c*   13. *d*   14. *g*   15. *e*   16. *h*

For Exercises 17 through 20, write the nodes in a breadth-first search of the graph in Figure 4.73, beginning with the given node.

★ 17. *a*   18. *e*   19. *f*   20. *h*

For Exercises 21 through 26, write the list of nodes from a preorder traversal, an inorder traversal, and a postorder traversal of the specified graph.

★ 21.   22.

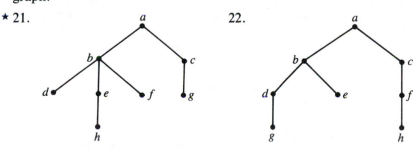

Figure 4.74   Figure 4.75

23.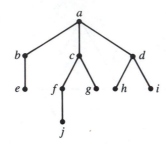

Figure 4.76

24.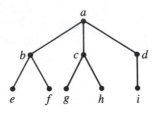

Figure 4.77

★ 25.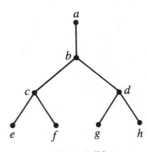

Figure 4.78

26.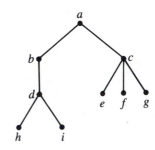

Figure 4.79

★ 27. Write in prefix and postfix notation: $\frac{3}{4} + (2 - y)$

28. Write in prefix and postfix notation: $(x * y + 3/z) * 4$

29. Write in infix and postfix notation: $-$ $*$ $+$ 2 3 $*$ 6 x 7

30. Write in infix and postfix notation: $-$ $+$ $-$ x y z w

★ 31. Write in prefix and infix notation: 4 7 x $-$ $*$ z $+$

32. Write in prefix and infix notation: x 2 w $+$ y z $*$ $-$ $/$

33. Given a **binary search tree** (the result of doing a binary tree search; see Exercise 21, Section 3.1), an inorder tree traversal produces an alphabetical listing of the tree nodes. Construct a binary search tree for "To be or not to be, that is the question," and then do an inorder traversal.

34. Construct a binary search tree (see Exercise 33) for "In the high and far off times the Elephant, O Best Beloved, had no trunk," and then do an inorder traversal.

★ 35. Find an example of a tree whose infix and postfix traversals yield the same list of nodes.

36. Find two different trees that have the same list of nodes under a preorder traversal.

37. Find a way to traverse a tree in *level order,* that is, so that all nodes at the same depth are listed, from left to right, for increasing depth. (Hint: we already have a way to do this.)

Chapter 5

Structures and Simulations

You may wonder what the title of this book means. (Does anyone ever read the title of a textbook?) What is a mathematical structure? This phrase might conjure up an image of highly intricate architecture or maybe the building that houses the computer center!

In this chapter we will explore what a mathematical structure is. We can liken a mathematical structure to a human skeleton. We can think of the skeleton as the basic structure of the human body. People may be thin or fat, short or tall, black or white, and so on, but stripped down to skeletons they all look pretty much alike. Although the outward appearances differ, the inward structure, the shape and arrangement of the bones, is the same. Similarly, mathematical structures represent the underlying sameness in situations that may appear different outwardly. The structures are defined to model, or simulate, the common characteristics. This simulation of common characteristics encountered in diverse situations we will call Simulation I. In Section 5.1, we define two mathematical structures: the *Boolean algebra* structure and the *monoid* structure.

Once we have defined some structures, we can talk about another kind of simulation. In some cases, one instance of a particular structure can do the job of (simulate) another instance of that structure. This sort of simulation, Simulation II, can be used to classify occurrences of structures and recognize those occurrences that are really the same. These ideas are discussed in Section 5.2.

SECTION 5.1   STRUCTURES—SIMULATION I

The Boolean Algebra Structure

We have just said that mathematical structures are defined to simulate characteristics common to diverse situations. Let's review three situations we have already encountered.

In Section 1.1 we studied unquantified statements of logic. We used the symbols $\vee$ and $\wedge$ to stand for disjunction and conjunction, respectively, and A' to denote the negation of a statement A. Equality between statements meant that the statements were equivalent (they had the same truth values). Zero stood for any contradiction, a statement with truth values always false, and 1 stood for any tautology, a statement with truth values always true. Recall the list of tautologies:

1a. $A \vee B = B \vee A$ 1b. $A \wedge B = B \wedge A$

2a. $(A \vee B) \vee C =$ 2b. $(A \wedge B) \wedge C =$
     $A \vee (B \vee C)$ $A \wedge (B \wedge C)$

3a. $A \vee (B \wedge C) =$ 3b. $A \wedge (B \vee C) =$
     $(A \vee B) \wedge (A \vee C)$ $(A \wedge B) \vee (A \wedge C)$

4a. $A \vee 0 = A$ 4b. $A \wedge 1 = A$

5a. $A \vee A' = 1$ 5b. $A \wedge A' = 0$

In Section 2.1, we studied set identities among the subsets of a set S. Here $\cup$ and $\cap$ denoted the union and intersection of sets, respectively, A' was the complement of a set A, and $\varnothing$ was the empty set. Recall the list of set identities:

1a. $A \cup B = B \cup A$ 1b. $A \cap B = B \cap A$

2a. $(A \cup B) \cup C =$ 2b. $(A \cap B) \cap C =$
     $A \cup (B \cup C)$ $A \cap (B \cap C)$

3a. $A \cup (B \cap C) =$ 3b. $A \cap (B \cup C) =$
     $(A \cup B) \cap (A \cup C)$ $(A \cap B) \cup (A \cap C)$

4a. $A \cup \varnothing = A$ 4b. $A \cap S = A$

5a. $A \cup A' = S$ 5b. $A \cap A' = \varnothing$

Finally, in Section 3.1, for ρ and σ members of the set of all binary relations on a given set S, we defined binary relations $\rho + \sigma$, $\rho \cdot \sigma$, and ρ'. Then

1a. $\rho + \sigma = \sigma + \rho$ 1b. $\rho \cdot \sigma = \sigma \cdot \rho$

2a. $(\rho + \sigma) + \gamma =$ 2b. $(\rho \cdot \sigma) \cdot \gamma =$
     $\rho + (\sigma + \gamma)$ $\rho \cdot (\sigma \cdot \gamma)$

3a. $\rho + (\sigma \cdot \gamma) =$ 3b. $\rho \cdot (\sigma + \gamma) =$
     $(\rho + \sigma) \cdot (\rho + \gamma)$ $(\rho \cdot \sigma) + (\rho \cdot \gamma)$

4a. $\rho + \varnothing = \rho$ 4b. $\rho \cdot S^2 = \rho$

5a. $\rho + \rho' = S^2$ 5b. $\rho \cdot \rho' = \varnothing$

These three lists of properties are similar. The disjunction of statements, the union of sets, the sum of binary relations—all seem to play the same roles in their respective environments. So do the conjunction of statements, the intersection of sets, and the product of binary relations. A contradiction seems to correspond to the empty set, and a tautology to S for sets or S^2 for binary relations.

Now suppose we try to characterize the similarities in these three cases formally. In each case we are talking about items from a set: a set of statements, a set of subsets of a set S, or a set of binary relations on a set S. In each case we have two binary operations and one unary operation on the members of the set: disjunction/conjunction/negation, union/intersection/complementation, and sum/product/complementation, respectively. In each case there are two distinguished elements of the set: $0/1$, $\varnothing/S$, and $\varnothing/S^2$, respectively. Finally, there are the 10 properties that hold in each case. Whenever all these features are present, we say that we have a Boolean algebra.

Definition

5.1 Practice

Complete the following definition: A **Boolean algebra** is a set B on which are defined two binary operations, $+$ and $\cdot$, and one unary operation, $'$, and in which there are two distinct elements, 0 and 1, such that the following properties hold for all $x, y, z \in B$:

1a. $x + y = y + x$	1b. _____ (commutative properties)
2a. _____	2b. _____ (associative properties)
3a. _____	3b. _____ (distributive properties)
4a. _____	4b. $x \cdot 1 = x$ (identities)
5a. _____	5b. _____ (complements)

□

Your definition should look like the one given below.

5.2 Definition

A **Boolean algebra** is a set B on which are defined two binary operations, $+$ and $\cdot$, and one unary operation, $'$, and in which there are two distinct elements, 0 and 1, such that the following properties hold for all $x, y, z \in B$:

1a. $x + y = y + x$	1b. $x \cdot y = y \cdot x$ (commutative properties)
2a. $(x + y) + z =$ $x + (y + z)$	2b. $(x \cdot y) \cdot z =$ $x \cdot (y \cdot z)$ (associative properties)
3a. $x + (y \cdot z) =$ $(x + y) \cdot (x + z)$	3b. $x \cdot (y + z) =$ $x \cdot y + x \cdot z$ (distributive properties)
4a. $x + 0 = x$	4b. $x \cdot 1 = x$ (identities)
5a. $x + x' = 1$	5b. $x \cdot x' = 0$ (complements) □

What, then, is the Boolean algebra structure? It is a formalization that simulates, or models, the three cases we have considered (and perhaps others as well). There is a subtle philosophical distinction between the formalization itself, the *idea* of the Boolean algebra structure, and any instance of the formalization, such as the three cases with which we have been working. Nevertheless, we will often use the term "Boolean algebra" to describe both the idea and its occurrences. This usage should not be confusing. We often have a mental idea ("chair," for example), and whenever we encounter a concrete example of the idea, we also call it by our word for the idea.

The formalization helps us focus on the essential features common to all examples of Boolean algebras, and we can use these features— these facts from the definition of a Boolean algebra—to prove other facts about Boolean algebras. Then these new facts, once proved in general, hold in any particular instance of a Boolean algebra. To use our analogy, if we ascertain that in a typical human skeleton "the thighbone is connected to the kneebone," then we don't need to reconfirm this in every person we meet. This type of generalization is typical of modern, or abstract, mathematics.

We denote a Boolean algebra by $[B, +, \cdot, ', 0, 1]$.

5.3 Example   Let $B = \{0, 1\}$ and define binary operations $+$ and $\cdot$ on B by $x + y = \max(x, y)$, $x \cdot y = \min(x, y)$. Then we can illustrate the operations of $+$ and $\cdot$ by the tables below.

+	0	1		·	0	1
0	0	1		0	0	0
1	1	1		1	0	1

A unary operation $'$ can be defined by means of a table, as follows, instead of by a verbal description.

$'$	
0	1
1	0

Thus $0' = 1$ and $1' = 0$. Then $[B, +, \cdot, ', 0, 1]$ is a Boolean algebra. We can verify the 10 properties by checking all possible cases. Thus, for property 2b, the associativity of $\cdot$, we show that

$$(0 \cdot 0) \cdot 0 = 0 \cdot (0 \cdot 0) = 0$$
$$(0 \cdot 0) \cdot 1 = 0 \cdot (0 \cdot 1) = 0$$
$$(0 \cdot 1) \cdot 0 = 0 \cdot (1 \cdot 0) = 0$$
$$(0 \cdot 1) \cdot 1 = 0 \cdot (1 \cdot 1) = 0$$
$$(1 \cdot 0) \cdot 0 = 1 \cdot (0 \cdot 0) = 0$$
$$(1 \cdot 0) \cdot 1 = 1 \cdot (0 \cdot 1) = 0$$
$$(1 \cdot 1) \cdot 0 = 1 \cdot (1 \cdot 0) = 0$$
$$(1 \cdot 1) \cdot 1 = 1 \cdot (1 \cdot 1) = 1$$

For property 4a, we show that

$$0 + 0 = 0$$
$$1 + 0 = 1$$
□

Properties

There are many other properties that hold in any Boolean algebra. We can prove these additional properties by using the properties in the definition.

5.4 Example

The **idempotent property,** $x + x = x,$ holds in any Boolean algebra. Thus,

$$
\begin{aligned}
x + x &= (x + x) \cdot 1 &\text{(4b)}\\
&= (x + x) \cdot (x + x') &\text{(5a)}\\
&= x + (x \cdot x') &\text{(3a)}\\
&= x + 0 &\text{(5b)}\\
&= x &\text{(4a)}
\end{aligned}
$$
□

As with set identities, each property in the definition of a Boolean algebra has its **dual** as part of the definition, where the dual is obtained by interchanging $+$ and $\cdot$, and 1 and 0. Therefore, every time a new property P about Boolean algebras is proved, each step in that proof can be replaced by the dual of that step. The result is a proof of the dual of P. Thus, once we have proved P, we know that the dual of P also holds.

5.5 Example

The dual of the property in Example 5.4, $x \cdot x = x$, is true in any Boolean algebra. □

5.6 Practice

(a) What does the idempotent property of Example 5.4 become in the context of statement logic?

(b) What does it become in the context of sets? □

Once a property about Boolean algebra is proved, we can use it to prove new properties.

5.7 Practice   (a) Prove that the property $x + 1 = 1$ holds in any Boolean algebra. Give a reason for each step.

(b) What is the dual property?   □

More properties of Boolean algebras appear in Exercise 4 at the end of this section.

For x an element of a Boolean algebra B, the element x' is called the **complement** of x. The complement of x satisfies

$$x + x' = 1 \quad \text{and} \quad x \cdot x' = 0$$

Indeed, x' is the unique element with these two properties. To prove this, suppose x_1 is an element of B with

$$x + x_1 = 1 \quad \text{and} \quad x \cdot x_1 = 0$$

Then

$$
\begin{aligned}
x' &= x' + 0 & \text{(4a)} \\
&= x' + (x \cdot x_1) & (x \cdot x_1 = 0) \\
&= (x' + x) \cdot (x' + x_1) & \text{(3a)} \\
&= (x + x') \cdot (x' + x_1) & \text{(1a)} \\
&= 1 \cdot (x' + x_1) & (x + x' = 1) \\
&= (x' + x_1) \cdot 1 & \text{(1b)} \\
&= x' + x_1 & \text{(4b)}
\end{aligned}
$$

A similar argument shows that

$$x_1 = x_1 + x'$$

Then

$$x' = x' + x_1 = x_1 + x' = x_1$$

Thus $x' = x_1$ and x' is unique.

We need to examine what *uniqueness* means in the context of statement logic. We just concluded that x' is unique because any element with the two properties of the complement of x is equal to x'. If the Boolean algebra being discussed is statement logic, an equal sign denotes that two statements are equivalent, that is, have the same truth values. However, equal statements need not look exactly alike. For example, for statement letters A and B, the statements $A \vee (B \vee A)$ and $B \vee A$ are equal. (Write the truth tables for each.) Statements $A \wedge A'$ and $B \wedge (A \wedge A')$ are also equal even though $A \wedge A'$ does not involve B. The truth table for $A \wedge A'$

can be written with a B column and four rows by simply ignoring the truth value of B. Then the statements $A \wedge A'$ and $B \wedge (A \wedge A')$ clearly have the same truth values. Thus x' is unique in its truth table, but may be written in a number of forms.

The following theorem summarizes our observations.

5.8 Theorem   For any x in a Boolean algebra, if an element x_1 exists such that

$$x + x_1 = 1 \quad \text{and} \quad x \cdot x_1 = 0$$

then $x_1 = x'$.   □

5.9 Practice   Prove that $0' = 1$ and $1' = 0$. (Hint: $1' = 0$ will follow by duality from $0' = 1$. To show $0' = 1$, use Theorem 5.8.)   □

There are many ways to define a Boolean algebra. Indeed, in Definition 5.2 we could have omitted the associative properties, since these can be derived from the remaining properties of the definition. In Chapter 6 we will apply Boolean algebra to computer logic; Boolean algebra is also used to build finite-state machines (Chapter 10).

Development of the Monoid Structure

A Boolean algebra is our first example of a mathematical structure. From this example we can generalize about the possible components in the definition of any structure. First, there will be one or more sets of objects, with perhaps some elements of the set or sets singled out because they are special in some way. There may be binary or unary operations defined on the sets as in a Boolean algebra; in other cases, functions may be defined by using the sets as domains and codomains. Finally, there will be a list of properties that the elements of the set(s) obey under the operations or functions.

We defined the Boolean algebra structure to simulate three particular mathematical situations—statement logic, subsets of a set, and binary relations on a set. For the remainder of this section, we will consider four more mathematical situations, explore their properties, and then define another structure simulating these four cases.

For the first mathematical situation, let's look at 2×2 matrices where the matrix entries are integers. We will denote the set of such matrices by $M_2(\mathbb{Z})$. Now what happens when we multiply such matrices? According to the definition of matrix multiplication from Section 3.3, for

$$\begin{bmatrix} a_{11} & a_{12} \\ a_{21} & a_{22} \end{bmatrix} \quad \text{and} \quad \begin{bmatrix} b_{11} & b_{12} \\ b_{21} & b_{22} \end{bmatrix}$$

in $M_2(\mathbb{Z})$,

$$\begin{bmatrix} a_{11} & a_{12} \\ a_{21} & a_{22} \end{bmatrix} \begin{bmatrix} b_{11} & b_{12} \\ b_{21} & b_{22} \end{bmatrix} = \begin{bmatrix} a_{11}b_{11} + a_{12}b_{21} & a_{11}b_{12} + a_{12}b_{22} \\ a_{21}b_{11} + a_{22}b_{21} & a_{21}b_{12} + a_{22}b_{22} \end{bmatrix}$$

The resulting product is a unique member of $M_2(\mathbb{Z})$ (note that it has integer entries), so matrix multiplication is a binary operation on $M_2(\mathbb{Z})$. If the matrices **X**, **Y**, and **Z** belong to $M_2(\mathbb{Z})$, then it can be shown (Exercise 11) that

$$(\mathbf{X} \cdot \mathbf{Y}) \cdot \mathbf{Z} = \mathbf{X} \cdot (\mathbf{Y} \cdot \mathbf{Z})$$

This fact is called the *associative property* of multiplication on $M_2(\mathbb{Z})$. (In general, associativity of a binary operation means that the grouping of elements does not affect the answer.) Also, the 2×2 *identity matrix* **I**, where

$$\mathbf{I} = \begin{bmatrix} 1 & 0 \\ 0 & 1 \end{bmatrix}$$

belongs to $M_2(\mathbb{Z})$. Again from Section 3.3, we know that for any $\mathbf{X} \in M_2(\mathbb{Z})$,

$$\mathbf{X} \cdot \mathbf{I} = \mathbf{I} \cdot \mathbf{X} = \mathbf{X}$$

To summarize, matrix multiplication is a binary operation on $M_2(\mathbb{Z})$, the multiplication is associative, and there is an identity matrix.

In the second mathematical situation we want to consider, we will limit ourselves to 2×2 matrices where the elements on the main diagonal are integers and the two other elements are zero. We will denote this set of matrices by $M_2^D(\mathbb{Z})$. Thus a typical element of $M_2^D(\mathbb{Z})$ is a matrix of the form

$$\begin{bmatrix} a_{11} & 0 \\ 0 & a_{22} \end{bmatrix}$$

where a_{11} and a_{22} are integers.

5.10 Practice Prove that matrix multiplication is a binary operation on $M_2^D(\mathbb{Z})$.   □

Notice that $M_2^D(\mathbb{Z}) \subseteq M_2(\mathbb{Z})$, which allows us to conclude immediately that multiplication on $M_2^D(\mathbb{Z})$ is associative. If **X**, **Y**, and **Z** belong to $M_2^D(\mathbb{Z})$, they also belong to $M_2(\mathbb{Z})$, and we already know that the equation $(\mathbf{X} \cdot \mathbf{Y}) \cdot \mathbf{Z} = \mathbf{X} \cdot (\mathbf{Y} \cdot \mathbf{Z})$ is true in $M_2(\mathbb{Z})$. Also, the identity matrix **I** belongs to $M_2^D(\mathbb{Z})$, and for any **X** in $M_2^D(\mathbb{Z})$, $\mathbf{X} \cdot \mathbf{I} = \mathbf{I} \cdot \mathbf{X} = \mathbf{X}$ because this equation is true for any **X** in $M_2(\mathbb{Z})$. Therefore, $M_2^D(\mathbb{Z})$ under matrix multiplication has all of the properties that $M_2(\mathbb{Z})$ has under multiplication.

But there is another property as well. For **X** and **Y** any members of $M_2^D(\mathbb{Z})$,

$$\mathbf{X} \cdot \mathbf{Y} = \mathbf{Y} \cdot \mathbf{X}$$

This fact is called the *commutative property* of multiplication on $M_2^D(\mathbb{Z})$. (In general, commutativity of a binary operation means that the order of the elements does not affect the answer.) From Practice 3.76, we know that matrix multiplication on $M_2(\mathbb{Z})$ is not commutative.

5.11 Practice   Show that matrix multiplication on $M_2^D(\mathbb{Z})$ is commutative.   □

Our third mathematical case is a little less exotic. Here we will consider the set $\mathbb{R}$ of real numbers under ordinary addition. Certainly addition is a binary operation on $\mathbb{R}$. For any real numbers x, y, and z,

$$(x + y) + z = x + (y + z)$$

so that addition on $\mathbb{R}$ is associative. Next we look for an element of $\mathbb{R}$ that will have the same effect in the real numbers under addition that **I** does in $M_2(\mathbb{Z})$ under matrix multiplication. Recall that for all $\mathbf{X} \in M_2(\mathbb{Z})$,

$$\mathbf{X} \cdot \mathbf{I} = \mathbf{I} \cdot \mathbf{X} = \mathbf{X}$$

Now consider any $x \in \mathbb{R}$ for the operation of addition. An identity element i must be a real number such that for any $x \in \mathbb{R}$,

$$x + i = i + x = x$$

5.12 Practice   What is the identity element in the set $\mathbb{R}$ under addition?   □

5.13 Practice   Is addition on $\mathbb{R}$ commutative?   □

To summarize this case, addition is a binary operation on $\mathbb{R}$ that is associative and commutative, and an identity element exists. But again there is a new property. For every $x \in \mathbb{R}$, there is an element of $\mathbb{R}$ called the *inverse* of x, namely $-x$, such that

$$x + (-x) = (-x) + x = 0$$

5.14 Practice   Translate the inverse element property into a corresponding statement about $M_2^D(\mathbb{Z})$ under multiplication by completing the following: for each $\mathbf{X} \in M_2^D(\mathbb{Z})$, there is a matrix in $M_2^D(\mathbb{Z})$ called the inverse of **X**, denoted by $\mathbf{X}^{-1}$, such that

$$\underline{\hspace{2cm}} = \underline{\hspace{2cm}} = \underline{\hspace{2cm}}$$

(Remember that in $M_2^D(\mathbb{Z})$, the operation is multiplication and the identity is **I**.)   □

5.15 Example   Consider the matrix

$$\begin{bmatrix} 2 & 0 \\ 0 & 2 \end{bmatrix}$$

of $M_2^D(\mathbb{Z})$. If this element is to have an inverse matrix in $M_2^D(\mathbb{Z})$, it must be of the form

$$\begin{bmatrix} a_{11} & 0 \\ 0 & a_{22} \end{bmatrix}$$

and have the property that

$$\begin{bmatrix} 2 & 0 \\ 0 & 2 \end{bmatrix}\begin{bmatrix} a_{11} & 0 \\ 0 & a_{22} \end{bmatrix} = \begin{bmatrix} a_{11} & 0 \\ 0 & a_{22} \end{bmatrix}\begin{bmatrix} 2 & 0 \\ 0 & 2 \end{bmatrix} = \begin{bmatrix} 1 & 0 \\ 0 & 1 \end{bmatrix}$$

By the definition of matrix multiplication, the only matrix satisfying this equation has $a_{11} = a_{22} = \frac{1}{2}$, and it is not a member of $M_2^D(\mathbb{Z})$. Thus, unlike real numbers under addition, not every member of $M_2^D(\mathbb{Z})$ has an inverse element in $M_2^D(\mathbb{Z})$. This example can also be used to show that not every element of $M_2(\mathbb{Z})$ has an inverse element in $M_2(\mathbb{Z})$. □

In our final example, we look at the set $\mathbb{R}^+$ of positive real numbers under multiplication.

5.16 Practice   Consider $\mathbb{R}^+$ under multiplication.

    (a) Is multiplication a binary operation on $\mathbb{R}^+$?
    (b) Is multiplication of $\mathbb{R}^+$ associative? What equation must be satisfied?
    (c) Is multiplication on $\mathbb{R}^+$ commutative? What equation must be satisfied?
    (d) What equation would an identity element have to satisfy? Is there an identity element?
    (e) What equation expresses the inverse element property? Does every member of $\mathbb{R}^+$ have an inverse? □

5.17 Practice   If we change to $\mathbb{R}$ under multiplication, will your answers to the questions of Practice 5.16 change? □

Our final case, $\mathbb{R}^+$ under multiplication, has all of the properties of the previous case, $\mathbb{R}$ under addition.

Now, what kind of structure can we define that will simulate all four of these cases? All four cases involve one set and a binary operation on that set. Properties common to all four cases are associativity and the existence of an identity element.

5.18
Definition

$[S, \cdot]$ is a **monoid** if S is a nonempty set and $\cdot$ is a binary operation on S, and if the following are true:

   (a) The operation $\cdot$ is associative.

   (b) There is an element $i \in S$ (identity element) such that for any $s \in S$, $s \cdot i = i \cdot s = s$.   □

The monoid structure, like the Boolean algebra structure, is an abstraction of properties found in a number of different situations. In the definition of a monoid we denoted the binary operation by a multiplication symbol, but in any particular monoid this symbol has to be interpreted properly. In our four examples of monoids, the interpretations are matrix multiplication, matrix multiplication, real-number addition, and real-number multiplication, respectively.

Only our first example, $[M_2(\mathbb{Z}), \cdot]$, is a monoid and nothing more. $[M_2^D(\mathbb{Z}), \cdot]$ is not only a monoid but a commutative monoid. The last two cases, $[\mathbb{R}, +]$ and $[\mathbb{R}^+, \cdot]$, are not only monoids but also examples of a structure called a commutative group.

We defined the monoid structure in an attempt to model some arithmetic properties that we had observed in various contexts. In Chapter 7 we will look at several other structures that simulate arithmetic properties. Then we will design a structure to simulate a computer in Chapter 9 and consider a structure to simulate the most general sort of computational process in Chapter 11.

✔ Checklist

Definitions

   Boolean algebra (*p. 211*)
   idempotent property (of a Boolean algebra) (*p. 213*)
   dual (of a Boolean algebra property) (*p. 213*)
   complement (of a Boolean algebra element) (*p. 214*)
   monoid (*p. 219*)

Techniques

   Prove properties about Boolean algebras.

   Decide whether something is a Boolean algebra.

   Decide whether something is a monoid.

Main Ideas

   Mathematical structures as abstractions of common properties found in diverse situations (Simulation I).

Exercises Section 5.1

★1. Let $B = \{0, 1, a, a'\}$. $+$ and $\cdot$ are binary operations on B. $'$ is a unary operation on B defined by the table

$'$	
0	1
1	0
a	a'
a'	a

Suppose you know that $[B, +, \cdot, ', 0, 1]$ is a Boolean algebra. Making use of the properties that must hold in any Boolean algebra, fill in the following tables defining the binary operations $+$ and $\cdot$:

$+$	0	1	a	a'
0				
1				
a				
a'				

$\cdot$	0	1	a	a'
0				
1				
a				
a'				

2. Define two binary operations $+$ and $\cdot$ on the set $\mathbb{Z}$ of integers by $x + y = \max(x, y)$ and $x \cdot y = \min(x, y)$.
   (a) Show that the commutative, associative, and distributive properties of a Boolean algebra hold for these two operations on $\mathbb{Z}$.
   (b) Show that no matter what element of $\mathbb{Z}$ is chosen to be 0, the property $x + 0 = x$ of a Boolean algebra fails to hold.

3. Let S be the set $\{0, 1\}$. Then S^2 is the set of all ordered pairs of 0s and 1s; $S^2 = \{(0, 0), (0, 1), (1, 0), (1, 1)\}$. Consider the set B of all functions mapping S^2 to S. For example, one such function, $f(x, y)$, is given by

$$f(0, 0) = 0$$
$$f(0, 1) = 1$$
$$f(1, 0) = 1$$
$$f(1, 1) = 1$$

   (a) How many elements are in B?
   (b) For f_1 and f_2 members of B and $(x, y) \in S^2$, define

$$(f_1 + f_2)(x, y) = \max(f_1(x, y), f_2(x, y))$$
$$(f_1 \cdot f_2)(x, y) = \min(f_1(x, y), f_2(x, y))$$
$$f_1'(x, y) = \begin{cases} 1 & \text{if } f_1(x, y) = 0 \\ 0 & \text{if } f_1(x, y) = 1 \end{cases}$$

Suppose

$$f_1(0, 0) = 1 \qquad f_2(0, 0) = 1$$
$$f_1(0, 1) = 0 \qquad f_2(0, 1) = 1$$
$$f_1(1, 0) = 1 \qquad f_2(1, 0) = 0$$
$$f_1(1, 1) = 0 \qquad f_2(1, 1) = 0$$

What are the functions $f_1 + f_2$, $f_1 \cdot f_2$, and f_1'?

(c) Prove that $[B, +, \cdot, ', 0, 1]$ is a Boolean algebra where 0 and 1 are defined by

$$0(0, 0) = 0 \qquad 1(0, 0) = 1$$
$$0(0, 1) = 0 \qquad 1(0, 1) = 1$$
$$0(1, 0) = 0 \qquad 1(1, 0) = 1$$
$$0(1, 1) = 0 \qquad 1(1, 1) = 1$$

4. Prove the following properties of Boolean algebras. Give a reason for each step.

★ (a) $(x')' = x$

★ (b) $x + (x \cdot y) = x \qquad x \cdot (x + y) = x \qquad$ (absorption properties)

(c) $(x + y)' = x' \cdot y' \qquad (x \cdot y)' = x' + y' \qquad$ (DeMorgan's Laws)

(d) $x \cdot (y + (x \cdot z)) = (x \cdot y) + (x \cdot z)$
$x + (y \cdot (x + z)) = (x + y) \cdot (x + z) \qquad$ (modular properties)

(e) $(x + y) \cdot (x' + y) = y$
$(x \cdot y) + (x' \cdot y) = y$

(f) $(x + y) + (y \cdot x') = x + y$
$(x \cdot y) \cdot (y + x') = x \cdot y$

(g) $x + y' = x + (x' \cdot y + x \cdot y)'$

(h) $((x \cdot y) \cdot z) + (y \cdot z) = y \cdot z$

(i) $x + (x \cdot y + y) = (x + y) \cdot (y + 1)$

(j) $(x + y') \cdot z = ((x' + z') \cdot (y + z'))'$

5. A new binary operation $\oplus$ in a Boolean algebra is defined by

$$x \oplus y = x \cdot y' + y \cdot x'$$

Prove that

★ (a) $x \oplus y = y \oplus x$

(b) $x \oplus x = 0$

(c) $0 \oplus x = x$

(d) $1 \oplus x = x'$

6. Prove that in any Boolean algebra:

(a) If $x + y = 0$, then $x = 0$ and $y = 0$.

(b) $x = y$ if and only if $x \cdot y' + y \cdot x' = 0$.

7. (a) Find an example of a Boolean algebra with elements x, y, and z for which $x + y = x + z$ but $y \neq z$.

(b) Prove that in any Boolean algebra, if $x + y = x + z$ and $x' + y = x' + z$, then $y = z$.

8. Prove that the 0 element in any Boolean algebra is unique; prove that the 1 element in any Boolean algebra is unique.

9. A Boolean algebra may also be defined as a partially ordered set with certain additional properties. Let $(B, \leq)$ be a partially ordered set. For any $x, y \in B$, we define the **least upper bound** of x and y as an element z such that $x \leq z$, $y \leq z$ and if there is any element z^* with $x \leq z^*$ and $y \leq z^*$, then $z \leq z^*$. The **greatest lower bound** of x and y is an element w such that $w \leq x$, $w \leq y$, and if there is any element w^* with $w^* \leq x$ and $w^* \leq y$, then $w^* \leq w$. A **lattice** is a partially ordered set in which every two elements x and y have a least upper bound, denoted by $x + y$, and a greatest lower bound, denoted by $x \cdot y$.

★ (a) Prove that in any lattice:
   (i) $x \cdot y = x$ if and only if $x \leq y$
   (ii) $x + y = y$ if and only if $x \leq y$

(b) Prove that in any lattice:
   (i) $x + y = y + x$
   (ii) $x \cdot y = y \cdot x$
   (iii) $(x + y) + z = x + (y + z)$
   (iv) $(x \cdot y) \cdot z = x \cdot (y \cdot z)$

(c) A lattice L is **complemented** if there exists a least element 0 and a greatest element 1, and for every $x \in L$ there exists $x' \in L$ such that $x + x' = 1$ and $x \cdot x' = 0$. Prove that in a complemented lattice L,

$$x + 0 = x \qquad \text{and} \qquad x \cdot 1 = x$$

for all $x \in L$.

(d) A lattice L is **distributive** if

$$x + (y \cdot z) = (x + y) \cdot (x + z)$$

and

$$x \cdot (y + z) = (x \cdot y) + (x \cdot z)$$

for every $x, y, z \in L$. By parts (b) and (c), a complemented, distributive lattice is a Boolean algebra. Which of the Hasse diagrams of partially ordered sets in Figure 5.1 do not represent Boolean algebras? Why? (Hint: in a Boolean algebra, the complement of an element is unique.)

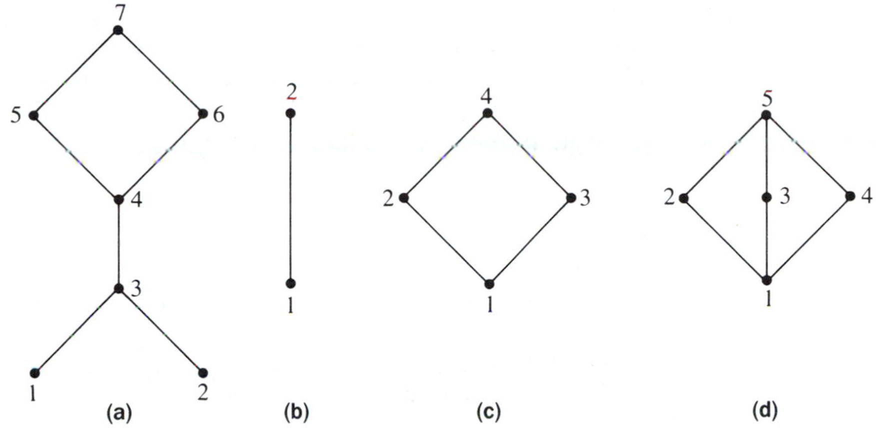

Figure 5.1

10. Prove that no Boolean algebra can have an odd number of elements. (Note that in the definition of a Boolean algebra, 0 and 1 are distinct elements of B, so B has at least two elements. Arrange the remaining elements of B so that each element is paired with its complement.)

11. Show that matrix multiplication on $M_2(\mathbb{Z})$ is associative.

12. Show that the set $M_2(\mathbb{Z})$ under matrix addition is a commutative monoid.

★ 13. (a) Write the equation that subtraction on $\mathbb{Z}$ must satisfy to be associative. Is subtraction on $\mathbb{Z}$ associative?
   (b) Write the equation that subtraction on $\mathbb{Z}$ must satisfy to be commutative. Is subtraction on $\mathbb{Z}$ commutative?

14. Each case below defines a binary operation, denoted by $\cdot$, on a given set. Which are associative? Which are commutative?

★ (a) on $\mathbb{Z}$: $x \cdot y = \begin{cases} x & \text{if } x \text{ is even} \\ x + 1 & \text{if } x \text{ is odd} \end{cases}$

★ (b) on $\mathbb{N}$: $x \cdot y = (x + y)^2$

   (c) on $\mathbb{R}^+$: $x \cdot y = x^4$

   (d) on $\mathbb{Q}$: $x \cdot y = \dfrac{xy}{2}$

   (e) on $\mathbb{R}^+$: $x \cdot y = \dfrac{1}{x + y}$

SECTION 5.2   MORPHISMS—SIMULATION II

In the preceding section we defined structures to simulate common properties found in various situations (Simulation I). Now we want to investigate what it might mean for one instance of a particular structure to simulate another instance of that structure (Simulation II). Such a "second-order simulation" occurs frequently when computers are used to solve problems. A real-world situation may be modeled by a continuous function (Simulation I), but the computer deals with a second function that is a discrete approximation of the first (Simulation II).

Isomorphisms

To illustrate isomorphisms, we will start with a very simple case.

5.19 *Example*   In one of the exercises in Chapter 3, you were asked to draw the graphs of each of two partially ordered sets:

    (a) $S = \{1, 2, 3, 5, 6, 10, 15, 30\}$; $x \, \rho \, y \leftrightarrow x$ divides y
    (b) $S = \mathcal{P}(\{1, 2, 3\})$; $A \, \rho \, B \leftrightarrow A \subseteq B$

The two graphs appear in Figure 5.2.

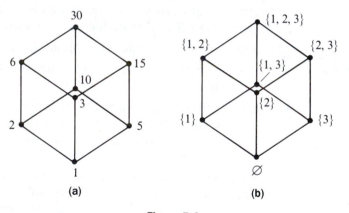

      (a)   (b)

Figure 5.2

     The striking thing about these two graphs is that if the node labels were removed, the graphs would be identical, even though they represent two different partially ordered sets. Let's define a one-to-one function f from the set of nodes in Figure 5.2a onto the set of nodes in Figure 5.2b:

$$f(1) = \varnothing \qquad f(6) = \{1, 2\}$$
$$f(2) = \{1\} \qquad f(10) = \{1, 3\}$$
$$f(3) = \{2\} \qquad f(15) = \{2, 3\}$$
$$f(5) = \{3\} \qquad f(30) = \{1, 2, 3\}$$

This function equates the two sets of nodes so that the relationships of Figure 5.2a are preserved in Figure 5.2b. For example, if x and y are nodes of Figure 5.2a with x an immediate predecessor of y, then in Figure 5.2b $f(x)$ is an immediate predecessor of $f(y)$. If in Figure 5.2a nodes x

and y are unrelated, then their images under f, $f(x)$ and $f(y)$, are unrelated in Figure 5.2b. The inverse function preserves the relationships of Figure 5.2b in Figure 5.2a.

Each graph simulates the other in the following sense: if we want to find a successor of node x in Figure 5.2a, for example, we can consult node $f(x)$ in Figure 5.2b, where we see that some node y' is a successor of $f(x)$. But y' is $f(y)$ for some node y in Figure 5.2a; we apply the inverse function to find y. Then y will be a successor of x in Figure 5.2a. In short, Figure 5.2b is a mirror image of Figure 5.2a. We obtain information about Figure 5.2a by "looking in the mirror," which amounts to applying f, working in Figure 5.2b, and then applying f^{-1}. In the same way, Figure 5.2a is a mirror image of Figure 5.2b, and we obtain information about Figure 5.2b by applying f^{-1}, working in Figure 5.2a, and then applying f.

The two partially ordered sets that these graphs represent are said to be *isomorphic*. □

A graph is not a very dynamic structure. We will look next at instances of a structure where a binary operation is available.

5.20 Example $[\mathbb{R}, +]$ and $[\mathbb{R}^+, \cdot]$ are both monoids. Can these two monoids be mirror images of each other? If so, there must be a one-to-one correspondence between the sets $\mathbb{R}$ and $\mathbb{R}^+$; we want a bijection $f : \mathbb{R} \to \mathbb{R}^+$. More than that, if $[\mathbb{R}^+, \cdot]$ is to act as a mirror for $[\mathbb{R}, +]$, then f has to preserve in $[\mathbb{R}^+, \cdot]$ the action of the operation $+$ in $\mathbb{R}$. Thus, if we want to find $x + y$ for x and y in $\mathbb{R}$, we should look in the mirror—take $f(x)$ and $f(y)$, which puts us in $[\mathbb{R}^+, \cdot]$, do the operation available there—that is, find $f(x) \cdot f(y)$—and then apply f^{-1}. This process should yield $x + y$. Thus for any x and y in $\mathbb{R}$, we want

$$x + y = f^{-1}(f(x) \cdot f(y))$$

or

$$f(x + y) = f(x) \cdot f(y)$$

This equation says the following: we want to be able to operate in $\mathbb{R}$ with $+$ and then map to $\mathbb{R}^+$, or to map to $\mathbb{R}^+$ and then operate there with $\cdot$; the result should be the same in either case. We can illustrate this with a commutative diagram. In Figure 5.3, $+$ is a binary operation taking members of $\mathbb{R} \times \mathbb{R}$ (ordered pairs of real numbers) to $\mathbb{R}$, $\cdot$ is a binary operation taking members of $\mathbb{R}^+ \times \mathbb{R}^+$ to $\mathbb{R}^+$, and $f \times f$ applies f to each component of an element of $\mathbb{R} \times \mathbb{R}$.

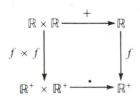

Figure 5.3

A function $f : \mathbb{R} \to \mathbb{R}^+$ that does the job is given by $f(x) = 2^x$. Then f is a one-to-one, onto function (a bijection). Also, for x and y in $\mathbb{R}$,

$$f(x + y) = 2^{x+y} = 2^x \cdot 2^y = f(x) \cdot f(y) \qquad \square$$

5.21 Practice   Use the function f defined in Example 5.20. Choose -4 and 7 as elements of $\mathbb{R}$.

(a) What is $f(-4 + 7)$? What paths in the commutative diagram of Figure 5.3 does this computation trace?
(b) What is $f(-4) \cdot f(7)$? What paths in the commutative diagram of Figure 5.3 does this computation trace? $\qquad \square$

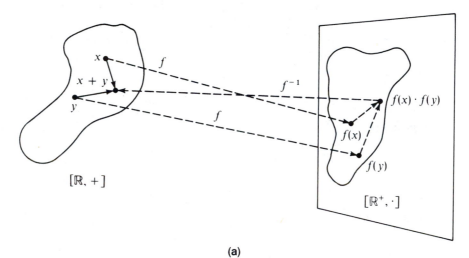

(a)

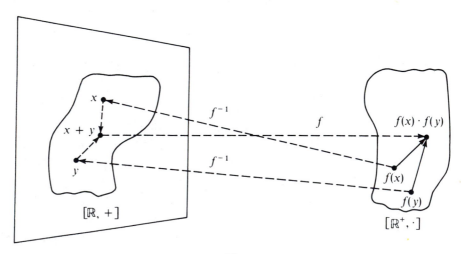

(b)

Figure 5.4

Figure 5.4a illustrates how $[\mathbb{R}^+, \cdot]$ simulates $[\mathbb{R}, +]$ in that $x + y$ in $[\mathbb{R}, +]$ can be done by mapping to $[\mathbb{R}^+, \cdot]$, operating there, and then taking f^{-1}. Figure 5.4b shows how $[\mathbb{R}, +]$ simulates $[\mathbb{R}^+, \cdot]$ in that $f(x) \cdot f(y)$ in $\mathbb{R}^+$ (remember that f is onto) can be done by using f^{-1} to get to $[\mathbb{R}, +]$, operating there, and then applying f. Each monoid simulates the other, and $[\mathbb{R}, +]$ and $[\mathbb{R}^+, \cdot]$ are said to be *isomorphic* monoids.

Two instances of a structure will be **isomorphic** whenever there is a bijection (called an **isomorphism**) between the elements of sets involved that preserves the effects of operations or functions on the sets. More specific definitions will be given for particular structures, but the general idea is that of the mirror pictures of Figure 5.4. Isomorphic structures are essentially the same since the elements are simply relabeled from one to the other. Therefore, we can use the idea of isomorphism to classify instances of a structure, lumping together those that are isomorphic.

Isomorphic Boolean Algebras

Now let's try to understand isomorphism for instances of a structure with which we are relatively familiar, a Boolean algebra. Suppose we have two Boolean algebras, $[B, +, \cdot, ', 0, 1]$ and $[b, \&, *, '', \emptyset, \mathcal{X}]$. This notation means that, for example, if x is in B, x' is the result of performing on x the unary operation defined in B, and if z is an element of b, z'' is the result of performing on z the unary operation defined in b. How would we define isomorphism between these two Boolean algebras? First, we would need a bijection f from B onto b. Then f must preserve in b the effects of the various operations in B. There are three operations, so we will need three equations. To preserve the operation $+$, we want to be able to operate in B and then map to b, or to map to b and operate there. (Think of the commutative diagram of Example 5.20.) Thus, for x and y in B, we require

$$f(x + y) = f(x) \ \& \ f(y)$$

5.22 Practice
(a) Write the equation requiring f to preserve the effect of the binary operation $\cdot$.

(b) Write the equation requiring f to preserve the effect of the unary operation $'$. □

Here is the definition of an isomorphism for Boolean algebras.

5.23 Definition
Let $[B, +, \cdot, ', 0, 1]$ and $[b, \&, *, '', \emptyset, \mathcal{X}]$ be Boolean algebras. A function $f: B \to b$ is an **isomorphism** from $[B, +, \cdot, ', 0, 1]$ to $[b, \&, *, '', \emptyset, \mathcal{X}]$ if

(a) f is a bijection
(b) $f(x + y) = f(x) \ \& \ f(y)$

(c) $f(x \cdot y) = f(x) * f(y)$
(d) $f(x') = (f(x))''$ □

5.24 Practice   Illustrate equations (b), (c), and (d) in Definition 5.23 by a commutative
diagram.   □

We already know (see the beginning of Section 5.1) that for any set
S, $\mathcal{P}(S)$ under the operations of union, intersection, and complementation
is a Boolean algebra. If we pick $S = \{1, 2\}$, then the elements of $\mathcal{P}(S)$ are
$\emptyset$, $\{1\}$, $\{2\}$, and $\{1, 2\}$. The operations are given by the tables

∪	$\emptyset$	$\{1, 2\}$	$\{1\}$	$\{2\}$
$\emptyset$	$\emptyset$	$\{1, 2\}$	$\{1\}$	$\{2\}$
$\{1, 2\}$	$\{1, 2\}$	$\{1, 2\}$	$\{1, 2\}$	$\{1, 2\}$
$\{1\}$	$\{1\}$	$\{1, 2\}$	$\{1\}$	$\{1, 2\}$
$\{2\}$	$\{2\}$	$\{1, 2\}$	$\{1, 2\}$	$\{2\}$

∩	$\emptyset$	$\{1, 2\}$	$\{1\}$	$\{2\}$
$\emptyset$	$\emptyset$	$\emptyset$	$\emptyset$	$\emptyset$
$\{1, 2\}$	$\emptyset$	$\{1, 2\}$	$\{1\}$	$\{2\}$
$\{1\}$	$\emptyset$	$\{1\}$	$\{1\}$	$\emptyset$
$\{2\}$	$\emptyset$	$\{2\}$	$\emptyset$	$\{2\}$

′	
$\emptyset$	$\{1, 2\}$
$\{1, 2\}$	$\emptyset$
$\{1\}$	$\{2\}$
$\{2\}$	$\{1\}$

In Exercise 1 of the last section, a Boolean algebra was defined on the
set $B = \{0, 1, a, a'\}$ where the tables defining the operations of $+$, $\cdot$, and
$'$ were

+	0	1	a	a'
0	0	1	a	a'
1	1	1	1	1
a	a	1	a	1
a'	a'	1	1	a'

·	0	1	a	a'
0	0	0	0	0
1	0	1	a	a'
a	0	a	a	0
a'	0	a'	0	a'

′	
0	1
1	0
a	a'
a'	a

We claim that the mapping $f : B \rightarrow \mathcal{P}(S)$ given by

$f(0) = \emptyset$
$f(1) = \{1, 2\}$
$f(a) = \{1\}$
$f(a') = \{2\}$

is an isomorphism. Certainly it is a bijection. For $x, y \in B$, we can verify each of the equations

$$f(x + y) = f(x) \cup f(y)$$
$$f(x \cdot y) = f(x) \cap f(y)$$
$$f(x') = (f(x))'$$

by examining all possible cases. Thus, for example,

$$f(a \cdot 1) = f(a) = \{1\} = \{1\} \cap \{1, 2\} = f(a) \cap f(1)$$

5.25 Practice   Verify the following equations:

(a) $f(0 + a) = f(0) \cup f(a)$
(b) $f(a + a') = f(a) \cup f(a')$
(c) $f(a \cdot a') = f(a) \cap f(a')$
(d) $f(1') = (f(1))'$ □

The remaining cases also hold. Even without testing all cases, it is pretty clear here that f is going to work because it merely relabels the entries in the tables for B so that they resemble the tables for $\mathcal{P}(S)$. In general, however, it may not be so easy to decide whether a given f is an isomorphism between two instances of a structure. Even harder to answer is the question of whether two given instances of a structure are isomorphic; we must either think up a function that works or show that no such function exists. One case where no such function exists is if the sets involved are not the same size; we cannot have a four-element Boolean algebra isomorphic to an eight-element Boolean algebra.

We just showed that a particular four-element Boolean algebra is isomorphic to $\mathcal{P}(\{1, 2\})$. It turns out that any finite Boolean algebra is isomorphic to the Boolean algebra of a power set. We will state this as a theorem, but we will not prove it.

5.26 Theorem   Let B be any Boolean algebra with n elements. Then $n = 2^m$ for some m, and B is isomorphic to $\mathcal{P}(\{1, 2, \ldots, m\})$.   □

Theorem 5.26 gives us two pieces of information. We know by Exercise 10 of the last section that a finite Boolean algebra must have an even number of elements; now we find that the even number must be a power of 2. Also we learn that finite Boolean algebras that are power sets are—in our lumping together of isomorphic things—really the only kinds of finite Boolean algebras. In a sense we have come full circle. We defined a Boolean algebra to represent many kinds of situations; now we find that (for the finite case) the situations, except for the labels of objects, are the same anyway!

An Example of a Homomorphism

When A is isomorphic to B, A and B can simulate each other. The success of the simulation depends on two properties of the isomorphism $f : A \rightarrow B$. First, f is a bijection (we use both the onto property and the existence of f^{-1}). Second, f preserves in B the effect of the operations in A (commutative diagrams hold). We get simulations of a less perfect nature by relaxing the bijection requirement on the function f. Before we look at an example, we need one more monoid to work with.

5.27 Example   The set $\mathbb{Z}$ of integers under multiplication is a monoid denoted by $[\mathbb{Z}, \cdot]$. The product of two integers exists and is a unique integer, so multiplication is a binary operation on $\mathbb{Z}$. Multiplication is associative because $(a \cdot b) \cdot c = a \cdot (b \cdot c)$ for all $a, b, c \in \mathbb{Z}$. The number $1 \in \mathbb{Z}$ is an identity element because $1 \cdot a = a \cdot 1 = a$ for all $a \in \mathbb{Z}$. $\square$

Now let's consider the monoids $[M_2^D(\mathbb{Z}), \cdot]$ and $[\mathbb{Z}, \cdot]$. Recall that $M_2^D(\mathbb{Z})$ is the set of all 2×2 diagonal matrices with integer entries. For

$$\begin{bmatrix} a_{11} & 0 \\ 0 & a_{22} \end{bmatrix}$$

in $M_2^D(\mathbb{Z})$, we define f by

$$f\left(\begin{bmatrix} a_{11} & 0 \\ 0 & a_{22} \end{bmatrix} \right) = a_{11}$$

Then $f : M_2^D(\mathbb{Z}) \rightarrow \mathbb{Z}$. Clearly, f is not one-to-one, so it is not a bijection. Now we claim that f preserves in $[\mathbb{Z}, \cdot]$ the effect of the operation in $M_2^D(\mathbb{Z})$:

$$f\left(\begin{bmatrix} a_{11} & 0 \\ 0 & a_{22} \end{bmatrix} \cdot \begin{bmatrix} b_{11} & 0 \\ 0 & b_{22} \end{bmatrix} \right) = f\left(\begin{bmatrix} a_{11} & 0 \\ 0 & a_{22} \end{bmatrix} \right) \cdot f\left(\begin{bmatrix} b_{11} & 0 \\ 0 & b_{22} \end{bmatrix} \right)$$

Notice that there are two different multiplications in this equation. The left side of the equation is matrix multiplication, the right, integer multiplication.

5.28 Practice   Verify the above equation. $\square$

The function f defined above is not an isomorphism because it is not a bijection; f is a **homomorphism** (it preserves the operation). Although a homomorphism in general need not be an onto function, this particular f is an onto mapping.

Because f is not one-to-one, f^{-1} does not exist. That means the mirror pictures don't work. A direct simulation of $[M_2^D(\mathbb{Z}), \cdot]$ by $[\mathbb{Z}, \cdot]$ is not possible using f. However, a type of simulation is still possible. We will define a relation ρ on $M_2^D(\mathbb{Z})$ by saying that two members of $M_2^D(\mathbb{Z})$ are related if they map under f to the same member of $\mathbb{Z}$. Thus,

$$\begin{bmatrix} a_{11} & 0 \\ 0 & a_{22} \end{bmatrix} \rho \begin{bmatrix} b_{11} & 0 \\ 0 & b_{22} \end{bmatrix} \leftrightarrow f\left(\begin{bmatrix} a_{11} & 0 \\ 0 & a_{22} \end{bmatrix}\right) = f\left(\begin{bmatrix} b_{11} & 0 \\ 0 & b_{22} \end{bmatrix}\right)$$

$$\leftrightarrow a_{11} = b_{11}$$

The relation ρ is an equivalence relation—reflexive, symmetric, and transitive—on $M_2^D(\mathbb{Z})$ (see Exercise 24, Section 3.2). Therefore, $M_2^D(\mathbb{Z})$ is partitioned into equivalence classes, and each class can be associated with an integer, namely the integer that is the 1,1 element of every member of the class.

We can use $[M_2^D(\mathbb{Z}), \cdot]$ to simulate $[\mathbb{Z}, \cdot]$ as follows. Let v and w belong to $\mathbb{Z}$; we want to compute $v \cdot w$ by simulation. The integer v is associated with a whole class of matrices in $M_2^D(\mathbb{Z})$, all those with 1,1 element v. Pick any member of this class, and name it $\mathbf{V}$. Similarly, w is associated with a class of matrices; pick any member $\mathbf{W}$ of this class. Compute $\mathbf{V} \cdot \mathbf{W}$ in $[M_2^D(\mathbb{Z}), \cdot]$. It will belong to a class of matrices; the integer in $\mathbb{Z}$ corresponding to this class is $v \cdot w$.

5.29 Example   Suppose we want to compute $3 \cdot 4$. One matrix from the class for 3 is

$$\begin{bmatrix} 3 & 0 \\ 0 & 5 \end{bmatrix}$$

and one matrix from the class for 4 is

$$\begin{bmatrix} 4 & 0 \\ 0 & -18 \end{bmatrix}$$

Then

$$\begin{bmatrix} 3 & 0 \\ 0 & 5 \end{bmatrix}\begin{bmatrix} 4 & 0 \\ 0 & -18 \end{bmatrix} = \begin{bmatrix} 12 & 0 \\ 0 & -90 \end{bmatrix}$$

The corresponding integer is 12.   □

So far we have f a homomorphism from $M_2^D(\mathbb{Z})$ onto $\mathbb{Z}$, and we have been able to simulate $[\mathbb{Z}, \cdot]$ by $[M_2^D(\mathbb{Z}), \cdot]$. The only difference between this case and Figure 5.4b is that we don't have f^{-1}, and we must choose preimages of v and w under f. Still, given that we can choose such preimages, all goes well. It is the other direction that is less than perfect. We want to simulate $[M_2^D(\mathbb{Z}), \cdot]$ by $[\mathbb{Z}, \cdot]$. We pick two matrices $\mathbf{X}$ and $\mathbf{Y}$ in $M_2^D(\mathbb{Z})$. We find $f(\mathbf{X})$ and $f(\mathbf{Y})$ and compute $f(\mathbf{X}) \cdot f(\mathbf{Y})$. So far so good. But here is the problem. When we try to go from $f(\mathbf{X}) \cdot f(\mathbf{Y})$, which is some integer z, back to $M_2^D(\mathbb{Z})$, we only know that z is associated with a class of matrices. Although we can choose a member of the class, we might not choose the one that is really $\mathbf{X} \cdot \mathbf{Y}$.

5.30 Example   Suppose we want to compute

$$\begin{bmatrix} 2 & 0 \\ 0 & -4 \end{bmatrix} \cdot \begin{bmatrix} -6 & 0 \\ 0 & 7 \end{bmatrix}$$

We find

$$f\left(\begin{bmatrix} 2 & 0 \\ 0 & -4 \end{bmatrix}\right) = 2 \qquad f\left(\begin{bmatrix} -6 & 0 \\ 0 & 7 \end{bmatrix}\right) = -6$$

and compute $(2)(-6) = -12$. The class associated with -12 is the set of all matrices with -12 in the 1,1 position. But only one member of this class, namely

$$\begin{bmatrix} -12 & 0 \\ 0 & -28 \end{bmatrix}$$

is the answer we want.   □

Thus, the simulation of $[M_2^D(\mathbb{Z}), \cdot]$ by $[\mathbb{Z}, \cdot]$ is imperfect; we can only obtain the answer to within an equivalence class. Actually, this result is not so surprising; $[M_2^D(\mathbb{Z}), \cdot]$ is somehow more complex than $[\mathbb{Z}, \cdot]$. It is possible to show that we can treat the classes of $M_2^D(\mathbb{Z})$ as objects and define an operation of multiplication on them. This new structure will be isomorphic to $[\mathbb{Z}, \cdot]$. Thus, in a way, a picture of $[\mathbb{Z}, \cdot]$ is somehow embedded within $[M_2^D(\mathbb{Z}), \cdot]$, and it makes sense that $[M_2^D(\mathbb{Z}), \cdot]$ simulates $[\mathbb{Z}, \cdot]$ but that $[\mathbb{Z}, \cdot]$ only imperfectly simulates $[M_2^D(\mathbb{Z}), \cdot]$.

What has been done with these two particular monoids will be done in a more general setting in Chapter 7. Meanwhile, let's summarize the ideas. Suppose A and B are two examples of a structure. Simulation must involve a function that preserves operations, a homomorphism. If there is in fact an isomorphism (a bijective homomorphism) from A onto B, then A simulates B and B simulates A; thus, we think of A and B as being essentially the same. If there is a homomorphism but not an isomorphism from A onto B (the function is not one-to-one), then A simulates B, but B only imperfectly simulates A; that is, B is isomorphic to a structure composed of classes of A.

We will also see in Chapter 7 that the nature of computer arithmetic is governed by these ideas of simulation and homomorphism.

✔ Checklist

Definition

        isomorphic instances of a structure (*p. 227*)
        isomorphism (*p. 227*)
        isomorphism for Boolean algebras (*p. 227*)
        homomorphism (*p. 230*)

Techniques

Write the equation meaning that a function f preserves an operation from one instance of a structure to another; verify or disprove such an equation.

Main Ideas

If there is an isomorphism (a bijection that preserves operations) from A to B, where A and B are instances of a structure, then A and B simulate each other (Simulation II), and, except for labels, A and B are the same.

All finite Boolean algebras are isomorphic to Boolean algebras that are power sets.

If there is a homomorphism (a function that preserves operations) from A onto B, where A and B are instances of a structure, then A simulates B and B imperfectly simulates A.

Exercises Section 5.2

★ 1. Example 5.20 defines an isomorphism from $[\mathbb{R}, +]$ to $[\mathbb{R}^+, \cdot]$.
   (a) Use $[\mathbb{R}^+, \cdot]$ to simulate the computation $-2 + 12$ in $[\mathbb{R}, +]$.
   (b) Use $[\mathbb{R}, +]$ to simulate the computation $4 \cdot 8$ in $[\mathbb{R}^+, \cdot]$.

2. In this section, an isomorphism from the Boolean algebra with set $B = \{0, 1, a, a'\}$ to the Boolean algebra with set $\mathcal{P}(\{1, 2\})$ was defined.
   (a) Use the Boolean algebra on $\mathcal{P}(\{1, 2\})$ to simulate the computation $1 \cdot a'$ in the Boolean algebra on B.
   (b) Use the Boolean algebra on $\mathcal{P}(\{1, 2\})$ to simulate the computation $(a)'$ in the Boolean algebra on B.
   (c) Use the Boolean algebra on B to simulate the computation $\{1\} \cup \{2\}$ in the Boolean algebra on $\mathcal{P}(\{1, 2\})$.
   (d) Use the Boolean algebra on B to simulate the computation $\{1\} \cap \{1, 2\}$ in the Boolean algebra on $\mathcal{P}(\{1, 2\})$.

★ 3. Let $(S, \leq)$ and $(S', \leq')$ be two partially ordered sets. $(S, \leq)$ is isomorphic to $(S', \leq')$ if there is a bijection $f: S \rightarrow S'$ such that for x, y in S, $x < y \rightarrow f(x) <' f(y)$ and $f(x) <' f(y) \rightarrow x < y$.
   (a) Show that there are exactly 2 nonisomorphic, partially ordered sets with 2 elements (use graphs).
   (b) Show that there are exactly 5 nonisomorphic, partially ordered sets with 3 elements.
   (c) How many nonisomorphic, partially ordered sets with 4 elements are there?

4. Find an example of two partially ordered sets $(S, \leq)$ and $(S', \leq')$ and

a bijection $f: S \to S'$ where, for x, y in S, $x < y \to f(x) <' f(y)$, but $f(x) <' f(y) \nrightarrow x < y$.

5. Let $S = \{0, 1\}$ and $\cdot$ be defined on S by

$\cdot$	0	1
0	1	0
1	0	1

Let $T = \{5, 7\}$ and $+$ be defined on T by

$+$	5	7
5	7	5
7	5	7

Then $[S, \cdot]$ and $[T, +]$ are both monoids.
(a) If a function f is an isomorphism from $[S, \cdot]$ to $[T, +]$, what two properties must f satisfy?
(b) Define a function f and prove it is an isomorphism from $[S, \cdot]$ to $[T, +]$.

6. On the set B of all functions mapping $\{0, 1\}^2$ to $\{0, 1\}$, we can define operations of $+$, $\cdot$, and $'$ by

$$(f_1 + f_2)(x, y) = \max(f_1(x, y), f_2(x, y))$$

$$(f_1 \cdot f_2)(x, y) = \min(f_1(x, y), f_2(x, y))$$

$$f_1'(x, y) = \begin{cases} 1 & \text{if } f_1(x, y) = 0 \\ 0 & \text{if } f_1(x, y) = 1 \end{cases}$$

Then, according to Exercise 3 of Section 5.1, $[B, +, \cdot, ', 0, 1]$ is a Boolean algebra of 16 elements. The following table assigns names to these 16 functions:

(x, y)	0	1	f_1	f_2	f_3	f_4	f_5	f_6	f_7	f_8	f_9	f_{10}	f_{11}	f_{12}	f_{13}	f_{14}
$(0, 0)$	0	1	1	1	1	1	1	1	0	0	0	1	0	0	0	0
$(0, 1)$	0	1	0	1	1	0	1	0	1	1	1	0	0	1	0	0
$(1, 0)$	0	1	1	0	1	0	0	1	1	1	0	0	1	0	1	0
$(1, 1)$	0	1	0	0	0	0	1	1	1	0	1	1	1	0	0	1

According to Theorem 5.26, this Boolean algebra is isomorphic to $[\mathcal{P}(\{1, 2, 3, 4\}), \cup, \cap, ', \emptyset, \{1, 2, 3, 4\}]$. Complete the following definition of an isomorphism from B to $\mathcal{P}(\{1, 2, 3, 4\})$:

$$0 \to \varnothing$$
$$1 \to \{1, 2, 3, 4\}$$
$$f_4 \to \{1\}$$
$$f_{12} \to \{2\}$$
$$f_{13} \to \{3\}$$
$$f_{14} \to \{4\}$$

★ 7. Suppose that $[B, +, \cdot, ', 0, 1]$ and $[b, \&, *, '', \varnothing, X]$ are isomorphic Boolean algebras and that f is an isomorphism from B to b.
   (a) Prove that $f(0) = \varnothing$.
   (b) Prove that $f(1) = X$.

 8. Show that the function $f : M_2(\mathbb{Z}) \to M_2^D(\mathbb{Z})$ given by

$$f\left(\begin{bmatrix} a_{11} & a_{12} \\ a_{21} & a_{22} \end{bmatrix}\right) = \begin{bmatrix} a_{11} & 0 \\ 0 & a_{22} \end{bmatrix}$$

is *not* a homomorphism from the monoid $[M_2(\mathbb{Z}), \cdot]$ to the monoid $[M_2^D(\mathbb{Z}), \cdot]$.

 9. (a) Let $\mathbb{R}^*$ denote the set of nonzero real numbers. Show that $[\mathbb{R}^*, \cdot]$ is a monoid.
   (b) Define a function $f : \mathbb{R}^* \to \mathbb{R}^+$ by $f(a) = |a|$, the absolute value function. Show that f is a homomorphism from $[\mathbb{R}^*, \cdot]$ to $[\mathbb{R}^+, \cdot]$.
   (c) Show that f is an onto function but is not one-to-one.
   (d) Define a relation ρ on $\mathbb{R}^*$ by $x \, \rho \, y \leftrightarrow f(x) = f(y)$. Then ρ is an equivalence relation on $\mathbb{R}^*$; describe the equivalence classes.
   (e) Use $[\mathbb{R}^*, \cdot]$ (and the classes of part (d)) to simulate the computation $3 \cdot 5$ in $[\mathbb{R}^+, \cdot]$.
   (f) Use $[\mathbb{R}^+, \cdot]$ to simulate the computation $(-2) \cdot 6$ in $[\mathbb{R}^*, \cdot]$ to within an equivalence class.

Chapter 6

Boolean Algebra and Computer Logic

This chapter establishes a relationship between the Boolean algebra structure and the wiring diagrams for the electronic circuits in computers, calculators, industrial control devices, telephone systems, and so forth.

Indeed, we will see in Section 6.1 that truth functions, expressions made up of variables and the operations of Boolean algebra, and these wiring diagrams are all related. As a result, we can effectively pass from one formulation to another and still preserve characteristic behavior with respect to truth values. We will also find that we can simplify wiring diagrams by using properties of Boolean algebras (as discussed in Chapter 5). In Section 6.2 we will look at two other procedures for simplifying wiring diagrams.

SECTION 6.1   LOGIC NETWORKS

Combinational Networks

Basic Logic Elements

The concept of a Boolean algebra was first formulated by George Boole, an English mathematician, around 1850 to model statement logic. In 1938 the American mathematician Claude Shannon perceived the parallel between statement logic and computer logic and realized that Boolean algebra could play a part in systematizing this new realm of electronics.

Recall that an unquantified statement is either true or false. Two statement letters, A and B, can be combined by one of two binary operations. If the operation is disjunction, for example, then $A \vee B$ is true or false as a function of the truth or falsity of A and B, according to the truth table in Figure 6.1. We can view A and B as inputs to a device that performs the disjunction operation. Thus, we have Figure 6.2.

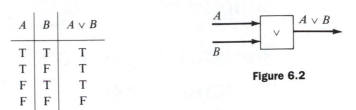

A	B	$A \vee B$
T	T	T
T	F	T
F	T	T
F	F	F

Figure 6.1

Figure 6.2

Remember that $\vee$ is an example of the Boolean algebra operation $+$ in the realm of statement logic. Any Boolean algebra operation can be associated with an electronic device. Figure 6.3 represents a device whose behavior parallels that of the Boolean operation $+$. In Figure 6.3 an input or output wire is assigned the value 1 or 0 (corresponding to T or F, respectively), depending on its voltage. There are two technical details to consider here. First, the voltages carried along wires x_1, x_2, and $x_1 + x_2$ may fluctuate somewhat. Consequently, we actually associate one range of voltage values with 1 and another range with 0. Second, we assume that the output of the device at a given instant is a function of its inputs at that instant; that is, we assume that the time required for the device to perform its actions is negligible. This assumption helps us avoid difficulties in timing when we connect devices together.

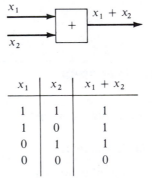

x_1	x_2	$x_1 + x_2$
1	1	1
1	0	1
0	1	1
0	0	0

Figure 6.3

Let's consider for a moment what might be inside the box in Figure 6.3. If we try to produce the required behavior for $x_1 + x_2$ using mechanical switches, the earliest technology used, we would wire a switch controlled by x_1 and one controlled by x_2 in parallel, so that the circuit would be broken only if both switches were open. Later technologies included vacuum tubes and then transistors. However, the actual physical implementation does not concern us here; we shall simply label our device to identify its behavior. We will have one device for each of the three Boolean operations. Each has a standard symbol. The **OR gate**, Figure 6.4a, behaves like the Boolean operation $+$. The **AND gate**, Figure 6.4b, represents the Boolean operation $\cdot$. Figure 6.4c shows an **inverter**, corresponding to the unary Boolean operation $'$. Because of the associativity property for $+$ and $\cdot$, the OR and AND gates can have more than two inputs.

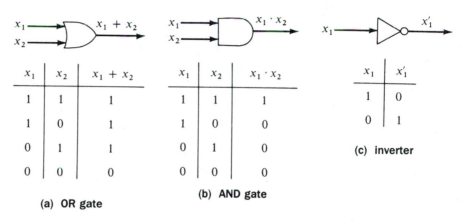

x_1	x_2	$x_1 + x_2$
1	1	1
1	0	1
0	1	1
0	0	0

(a) OR gate

x_1	x_2	$x_1 \cdot x_2$
1	1	1
1	0	0
0	1	0
0	0	0

(b) AND gate

x_1	x_1'
1	0
0	1

(c) inverter

Figure 6.4

Boolean Expressions

Each device in Figure 6.4 represents a single Boolean operation acting upon some variables; the corresponding truth table is also shown.

6.1 Definition   A **Boolean expression** in n variables, $x_1, x_2, \ldots, x_n$, is any finite string of symbols formed by applying the following rules:

1. $x_1, x_2, \ldots, x_n$ are Boolean expressions.
2. If P and Q are Boolean expressions, so are $(P + Q)$, $(P \cdot Q)$ and $(P)'$. □

(Definition 6.1 is another example of a recursive definition; rule 1 is the basis step, and rule 2 the inductive step.) When there is no chance of confusion, we can omit the parentheses introduced by rule 2. In addition, $\cdot$ takes precedence over $+$, so that $x_1 + x_2 \cdot x_3$ stands for $x_1 + (x_2 \cdot x_3)$; this convention also allows us to remove some parentheses. Finally, we will generally omit the symbol $\cdot$ and use juxtaposition, so that $x_1 \cdot x_2$ is written $x_1 x_2$.

6.2 Example x_3, $(x_1 + x_2)'x_3$, $(x_1 x_3 + x_4')x_2$, and $(x_1' x_2)'x_1$ are all Boolean expressions. □

Truth Functions

6.3 Definition   A **truth function** is a function f such that $f:\{0, 1\}^n \rightarrow \{0, 1\}$ for some integer $n \geq 1$.   □

The notation $\{0, 1\}^n$ denotes the set of all n-tuples of 0s and 1s. A truth function thus associates a value of 0 or 1 with each such n-tuple.

6.4 Example   The truth table for the Boolean operation $+$ describes a truth function f with $n = 2$. The domain of f is $\{(1, 1), (1, 0), (0, 1), (0, 0)\}$, and $f(1, 1) = 1$, $f(1, 0) = 1$, $f(0, 1) = 1$, $f(0, 0) = 0$. Similarly, the Boolean operation $\cdot$ describes a different truth function with $n = 2$, and the Boolean operation $'$ describes a truth function for $n = 1$.   □

6.5 Practice   (a) If we are writing a truth function $f:\{0, 1\}^n \rightarrow \{0, 1\}$ in tabular form (like a truth table), how many rows will the table have?

(b) How many different truth functions are there that take $\{0, 1\}^2 \rightarrow \{0, 1\}$?

(c) How many different truth functions are there that take $\{0, 1\}^n \rightarrow \{0, 1\}$?   □

Any Boolean expression defines a unique truth function, just as do the simple Boolean expressions $x_1 + x_2$, $x_1 x_2$, and x_1'.

6.6 Example   The Boolean expression $x_1 x_2' + x_3$ defines the truth function given by the table in Figure 6.5.   □

x_1	x_2	x_3	$x_1 x_2' + x_3$
1	1	1	1
1	1	0	0
1	0	1	1
1	0	0	1
0	1	1	1
0	1	0	0
0	0	1	1
0	0	0	0

Figure 6.5

Networks and Expressions

By combining AND gates, OR gates, and inverters, we can construct a logic network representing a given Boolean expression and producing the same truth function.

6.7 Example   The logic network for the Boolean expression $x_1x_2' + x_3$ is shown in Figure 6.6.   □

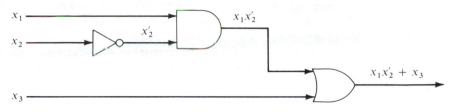

Figure 6.6

6.8 Practice   Design the logic network for the following Boolean expressions:

(a) $x_1 + x_2'$

(b) $x_1(x_2 + x_3)'$ □

Conversely, if we have a logic network, we can write a Boolean expression with the same truth function.

6.9 Example   A Boolean expression for the logic network in Figure 6.7 is $(x_1x_2 + x_3)'x_3$.   □

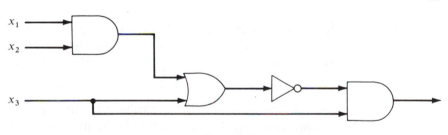

Figure 6.7

6.10 Practice   (a) Write a Boolean expression for the logic network in Figure 6.8.

(b) Write the truth function (in table form) for the network (and expression) of part (a).   □

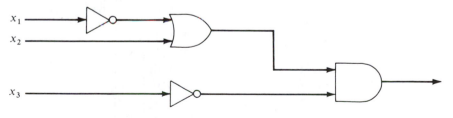

Figure 6.8

Logic networks constructed of AND gates, OR gates, and inverters are also called **combinational networks.** They have several features that we should note. First, input or output lines are not tied together except by passing through gates. The lines can be split, however, to serve as input to more than one device. There are no loops where the output of an element is part of the input to that same element. Finally, the output of a network is an instantaneous function of the input; there are no delay elements that capture and remember input signals. Notice also that the picture of any network is, in effect, a directed graph.

Canonical Form

Here is the situation so far (arrows indicate a procedure that we can carry out):

truth function $\leftarrow$ Boolean expression $\leftrightarrow$ logic network

We can write a unique truth function from either a network or an expression. Given an expression, we can find a network with the same truth function, and conversely. The last part of the puzzle concerns how to get from an arbitrary truth function to an expression (and hence a network) having that truth function. An algorithm to solve this problem is explained in the next example.

6.11 Example   Suppose we want to find a Boolean expression for the truth function f of Figure 6.9. There are four rows in the table, 1, 3, 4, and 7, for which f is 1. The basic form of our expression will be a sum of four terms

$$( ) + ( ) + ( ) + ( )$$

such that the first term has the value 1 for the input values of row 1, the second term has the value 1 for the input values of row 3, and so on. Thus, the entire expression has the value 1 for these inputs and no others—precisely what we want. (Other inputs cause each term in the sum, and hence the sum itself, to be 0.)

x_1	x_2	x_3	$f(x_1, x_2, x_3)$
1	1	1	1
1	1	0	0
1	0	1	1
1	0	0	1
0	1	1	0
0	1	0	0
0	0	1	1
0	0	0	0

Figure 6.9

Each individual term in the sum will be a product of the form $\alpha\beta\gamma$ where α is either x_1 or x_1', β is either x_2 or x_2', and γ is either x_3 or x_3'. If the input value of x_i, $i = 1, 2, 3$, in the row we are working on is 1, then x_i itself is used; if the input value of x_i in the row we are working on is 0, then x_i' is used. These values will force $\alpha\beta\gamma$ to be 1 for that row and 0 for all other rows. Thus, we have

row 1: $x_1x_2x_3$

row 3: $x_1x_2'x_3$

row 4: $x_1x_2'x_3'$

row 7: $x_1'x_2'x_3$

The final expression is

$$(x_1x_2x_3) + (x_1x_2'x_3) + (x_1x_2'x_3') + (x_1'x_2'x_3) \qquad \square$$

The procedure described in Example 6.11 always leads to an expression that is a sum of products, called the **canonical sum-of-products form**, or the **disjunctive normal form**, for the given truth function. The only case not covered by this procedure is when the function has a value of 0 everywhere. Then we use an expression such as

$$x_1x_1'$$

which is also a sum (one term) of products. Therefore, we can find a sum-of-products expression to represent any truth function. A pseudocode description of the algorithm is given below. For the following algorithm, the input is a truth function on n variables $x_1, x_2, \ldots, x_n$; the output is a Boolean expression in disjunctive normal form with the same truth function.

Algorithm Sum-of-Products

1. if truth function value is 0 for all rows of truth table
2. then single product is x_1x_1'
3. else
      begin
4.   while truth function value is 1 for any unprocessed row of truth table do
         begin
5.   consider next unprocessed row with truth value 1
6.   start a new product
7.   for $i = 1$ to n do
         begin
         if $x_i = 1$
         then put x_i in product
         else put x_i' in product
         end of for

8.   save product
9.   mark row as processed
      end of while
   end of else
10. expression = sum of all products

Because any expression has a corresponding network, any truth function has a logic network representation. Furthermore, the AND gate, OR gate, and inverter are the only devices needed to construct the network. Thus, we can build a network for any truth function with only three kinds of parts—and lots of wire! Later we will see that it is only necessary to stock one kind of part.

Given a truth function, the canonical sum-of-products form just described is one expression having this truth function, but it is not the only possible one. A method for obtaining a different expression for any truth function is given in Exercise 8 at the end of this section.

6.12 Example   The network for the canonical sum-of-products form of Example 6.11 is shown in Figure 6.10. We have drawn the inputs to each AND gate separately because it looks neater, but actually a single x_1, x_2, or x_3 input can be split as needed.   □

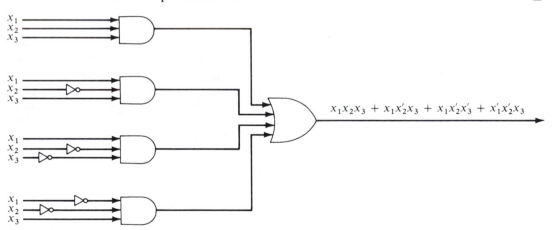

Figure 6.10

6.13 Practice   (a) Find the canonical sum-of-products form for the truth function of Figure 6.11.

(b) Draw the network for the expression of part (a).   □

Minimization

As already noted, a given truth function may be represented by more than one Boolean expression and hence by more than one logic network composed of AND gates, OR gates, and inverters.

x_1	x_2	x_3	$f(x_1, x_2, x_3)$
1	1	1	1
1	1	0	0
1	0	1	1
1	0	0	1
0	1	1	0
0	1	0	0
0	0	1	1
0	0	0	1

Figure 6.11

6.14 Example   The Boolean expression

$$x_1 x_3 + x_2'$$

has the truth function of Figure 6.11. The logic network corresponding to this expression is given by Figure 6.12. Compare this with your network in Practice 6.13b!   □

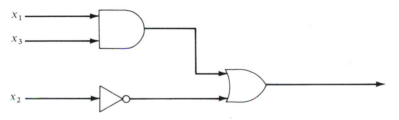

Figure 6.12

6.15 Definition   Two Boolean expressions are **equivalent** if they have the same truth function.   □

We know that

$$x_1 x_2 x_3 + x_1 x_2' x_3 + x_1 x_2' x_3' + x_1' x_2' x_3 + x_1' x_2' x_3'$$

and

$$x_1 x_3 + x_2'$$

for example, are equivalent Boolean expressions.

Clearly, equivalence of Boolean expressions is an equivalence relation on the set of all Boolean expressions in n variables. Each equivalence class is associated with a distinct truth function. Given a truth function, algorithm *Sum-of-Products* produces one particular member of the class associated with that function, namely, the canonical sum-of-products form. However, if we are trying to design the logic network for that function, we want to find a member of the class that is as simple as possible. We would rather build the network of Figure 6.12 than the one for Practice 6.13b.

How can we reduce a Boolean expression to an equivalent, simpler expression? We can use the properties of a Boolean algebra because they express the equivalence of Boolean expressions. If P is a Boolean expression containing the subexpression $x_1 + (x_2x_3)$, for example, and Q is the expression obtained from P by replacing $x_1 + (x_2x_3)$ with the equivalent expression $(x_1 + x_2)(x_1 + x_3)$, then P and Q are equivalent.

6.16 Example   Using the properties of Boolean algebra, we can reduce

$$x_1x_2x_3 + x_1x_2'x_3 + x_1x_2'x_3' + x_1'x_2'x_3 + x_1'x_2'x_3'$$

to

$$x_1x_3 + x_2'$$

as follows:

$$x_1x_2x_3 + x_1x_2'x_3 + x_1x_2'x_3' + x_1'x_2'x_3 + x_1'x_2'x_3'$$
$$= x_1x_2x_3 + x_1x_2'x_3 + x_1x_2'x_3 + x_1x_2'x_3' + x_1'x_2'x_3 + x_1'x_2'x_3'$$
$$= x_1x_3x_2 + x_1x_3x_2' + x_1x_2'x_3 + x_1x_2'x_3' + x_1'x_2'x_3 + x_1'x_2'x_3'$$
$$= x_1x_3(x_2 + x_2') + x_1x_2'(x_3 + x_3') + x_1'x_2'(x_3 + x_3')$$
$$= x_1x_3 \cdot 1 + x_1x_2' \cdot 1 + x_1'x_2' \cdot 1$$
$$= x_1x_3 + x_1x_2' + x_1'x_2'$$
$$= x_1x_3 + x_2'x_1 + x_2'x_1'$$
$$= x_1x_3 + x_2'(x_1 + x_1')$$
$$= x_1x_3 + x_2' \cdot 1$$
$$= x_1x_3 + x_2' \qquad\qquad \square$$

Unfortunately, one must be fairly clever to apply Boolean algebra properties to simplify an expression. In the next section we will discuss more systematic approaches to this minimization problem that require less ingenuity. For now, we should say a bit more about why we would want to minimize. When logic networks were built from separate gates and inverters, the cost of these elements was a considerable factor in the design, and it was desirable to have as few elements as possible. Now, however, most networks are built using integrated circuit technology, a development that began in the early 1960s. An integrated circuit is itself a logic network representing a certain truth function or functions, just as if some gates and inverters had been combined in the appropriate arrangement inside this package. These integrated circuits are then combined as needed to produce the desired result. Because the integrated circuits are extremely small and relatively inexpensive, it might seem pointless to bother minimizing a network. However, minimization is still important because the reliability of the final network is a function of the number of connections between the integrated circuit packages.

Moreover, the designers of integrated circuits are highly interested in the minimization problem. The square silicon chips in which integrated circuits are embedded may be no more than one-quarter inch on each side, yet they can contain the equivalent of half a million transistors for implementing truth functions. The wiring channels required to connect components on the chip may be so numerous that the wiring takes up more of the chip's "floor space" than the components themselves. Minimizing the number of components and the amount of wiring required to realize a desired truth function makes the chip less crowded and easier to design. Minimization also makes it possible to embed more functions in a single chip.

Programmable Logic Arrays

Instead of designing a custom chip to implement particular truth functions, a **PLA** (*programmable logic array*) can be used. A PLA is a chip that is already implanted with an array of AND gates and an array of OR gates, together with a rectangular grid of wiring channels, and some inverters. Once Boolean expressions in sum-of-products form have been determined for the truth functions, the required components in the PLA are activated. Although this chip is not very efficient and is practical only for smaller-scale circuit logic, the PLA can be mass-produced, and only a small amount of time (i.e., money) is then required to "program" it for the desired functions.

6.17 Example Figure 6.13a shows a PLA for the three inputs x_1, x_2, and x_3. There are four output lines, so four functions can be programmed in this PLA. When the PLA is programmed, the horizontal line going into an AND gate will pick up certain inputs, and the AND gate will form the product of these inputs. The vertical line going into an OR gate will, when programmed, allow the OR gate to form the sum of certain inputs. Figure 6.13b shows the same PLA programmed to produce the truth functions f_1—$x_1x_2x_3$ + $x_1x_2'x_3$ + $x_1x_2'x_3'$ + $x_1'x_2'x_3$—of Example 6.11 and f_2—$x_1x_2x_3$ + $x_1x_2'x_3$ + $x_1x_2'x_3'$ + $x_1'x_2'x_3$ + $x_1'x_2'x_3'$—of Practice 6.13; the dots represent activation points. □

A Useful Network

We can design a network that adds binary numbers, a basic operation that a computer must be able to perform. The rules for adding two binary digits are summarized in Figure 6.14.

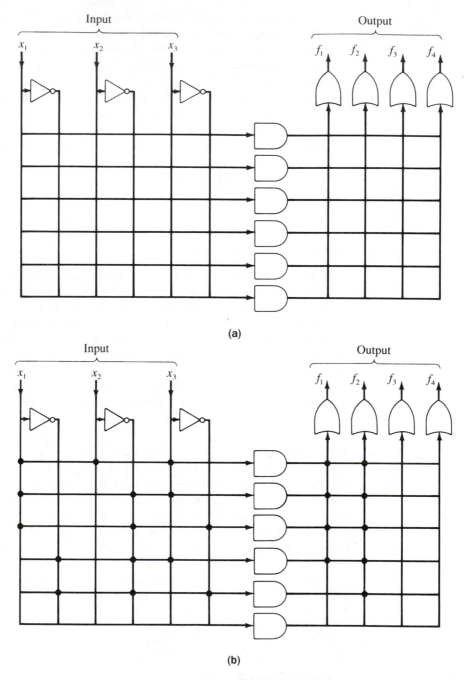

(a)

(b)

Figure 6.13

x_1	x_2	sum
1	1	10
1	0	1
0	1	1
0	0	0

(a)

x_1	x_2	s
1	1	0
1	0	1
0	1	1
0	0	0

x_1	x_2	c
1	1	1
1	0	0
0	1	0
0	0	0

(b)

Figure 6.14

We express the sum as a single sum digit s (the right-hand digit of the actual sum) together with a single carry digit c. This gives us the two truth functions of Figure 6.14b. The canonical sum-of-products form for each truth function is

$$s = x_1'x_2 + x_1x_2'$$

$$c = x_1x_2$$

An equivalent Boolean expression for s is

$$s = (x_1 + x_2)(x_1x_2)'$$

Figure 6.15a shows a network with inputs x_1 and x_2 and outputs s and c. This device, for reasons which will be clear shortly, is called a **half-adder.**

To add two n-digit binary numbers, we add column by column from the low-order to the high-order digits. The ith column (except for the very first column) has as input its two binary digits x_1 and x_2 plus the carry digit from the addition of column $i - 1$ to its right. Thus we need a device incorporating the previous carry digit as input. This can be accomplished by adding x_1 and x_2, using a half-adder, and then adding the previous carry digit c_{i-1} (using another half-adder) to the result. Again, a sum digit and final carry digit c_i are output where c_i is 1 if either half-adder produces

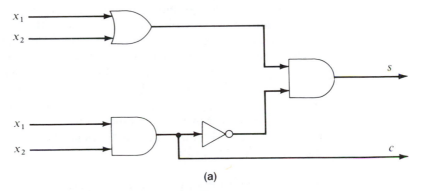

(a)

Figure 6.15

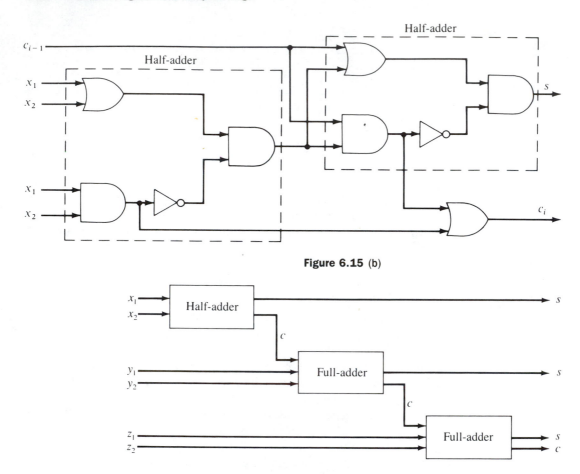

Figure 6.15 (b)

Figure 6.16

a 1 as its carry digit. The **full-adder** is shown in Figure 6.15b. The full-adder is thus composed of two half-adders and an additional OR gate.

To add two n-digit binary numbers, the two low-order digits, where there is no input carry digit, can be added with a half-adder. Then the carry signal must be propagated through $n - 1$ full-adders. Although we have assumed that gates output instantaneously, there is in fact a small time delay that can be appreciable for large n. Circuitry that speeds up the addition process is available for today's computers, although more time can be saved by a clever representation of the numbers to be added.

Figure 6.16 shows the modules required to add two 3-digit binary numbers $z_1y_1x_1$ and $z_2y_2x_2$. The resulting sum is found by reading the output digits from top to bottom.

6.18 Practice   Trace the operation of the circuit in Figure 6.16 as it adds 101 and 111.

□

Other Logic Elements

The basic elements used in integrated circuits are not really AND and OR gates and inverters, but NAND and NOR gates. Figure 6.17 shows the standard symbol for the **NAND gate** (the NOT AND gate) and its truth function. The NAND gate alone is sufficient to realize any truth function because networks using only NAND gates can do the job of inverters, OR gates, and AND gates. Figure 6.18 shows these networks.

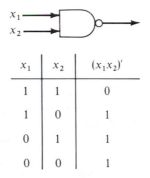

x_1	x_2	$(x_1 x_2)'$
1	1	0
1	0	1
0	1	1
0	0	1

Figure 6.17

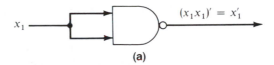

(a)

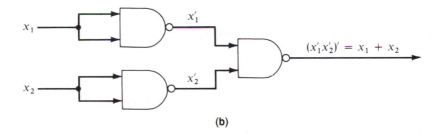

(b)

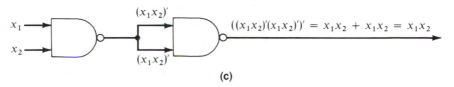

(c)

Figure 6.18

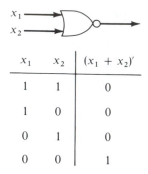

x_1	x_2	$(x_1 + x_2)'$
1	1	0
1	0	0
0	1	0
0	0	1

Figure 6.19

The **NOR gate** (the NOT OR gate) and its truth function appear in Figure 6.19. An exercise at the end of this section asks you to construct networks using only NOR gates for inverters, OR gates, and AND gates.

Although we can construct a NAND network for a truth function by replacing AND gates, OR gates, and inverters in the canonical form or a minimized form with the appropriate NAND networks, we can often obtain a simpler network by using the properties of NAND elements directly.

6.19 Practice   (a) Rewrite the network of Figure 6.12 with NAND elements by directly replacing the AND gate, OR gate, and inverter, as in Figure 6.18.

(b) Rewrite the Boolean expression $x_1x_3 + x_2'$ for Figure 6.12 using DeMorgan's Laws, and then construct a network using only two NAND elements.   □

Constructing Truth Functions

We know how to write a Boolean expression and construct a network from a given truth function. Often the truth function itself must first be deduced from the description of the actual problem.

6.20 Example   At a mail-order cosmetics firm, an automatic control device is used to supervise the packaging of orders. The firm sells lipstick, perfume, makeup, and nail polish. As a bonus item, shampoo is included with any order that includes perfume or any order that includes lipstick, makeup, and nail polish. How can we design the logic network that controls whether shampoo is packaged with an order?

The inputs to the network will represent the four items that can be ordered. We label these

$$x_1 = \text{lipstick}$$

$$x_2 = \text{perfume}$$

$$x_3 = \text{makeup}$$

$$x_4 = \text{nail polish}$$

The value of x_i will be 1 when that item is included in the order and 0 otherwise. The output from the network should be 1 if shampoo is to be packaged with the order and 0 otherwise. The truth table for the circuit appears in Figure 6.20a. The canonical sum-of-products form for this truth function is lengthy, but the expression $x_1 x_3 x_4 + x_2$ also represents the function. Figure 6.20b shows the logic network for this expression.   □

x_1	x_2	x_3	x_4	$f(x_1, x_2, x_3, x_4)$
1	1	1	1	1
1	1	1	0	1
1	1	0	1	1
1	1	0	0	1
1	0	1	1	1
1	0	1	0	0
1	0	0	1	0
1	0	0	0	0
0	1	1	1	1
0	1	1	0	1
0	1	0	1	1
0	1	0	0	1
0	0	1	1	0
0	0	1	0	0
0	0	0	1	0
0	0	0	0	0

(a)

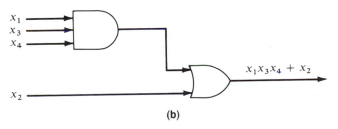

(b)

Figure 6.20

6.21 Practice   A hall light is controlled by two light switches, one at each end. Find (a) a truth function, (b) a Boolean expression, and (c) a logic network that allows the light to be switched on or off by either switch.   □

In some problems the corresponding truth functions have certain undefined values because certain combinations of input cannot occur (see Exercise 17 at the end of this section). Under these "don't-care" conditions, any value may be assigned to the output.

In a programming language where the Boolean operators AND, OR, and NOT are available, designing the logic of a computer program may consist in part of choosing appropriate truth functions and their corresponding Boolean expressions (see Exercise 11 of Section 1.1).

✔ Checklist

Definitions

OR gate (*p. 239*)
AND gate (*p. 239*)
inverter (*p. 239*)
Boolean expression (*p. 239*)
truth function (*p. 240*)
combinational network (*p. 241*)
canonical sum-of-products form (disjunctive normal form) (*p. 243*)
equivalent Boolean expressions (*p. 245*)
PLA (*p. 247*)
half-adder (*p. 249*)
full-adder (*p. 249*)
NAND gate (*p. 251*)
NOR gate (*p. 252*)

Techniques

Find the truth function corresponding to a given Boolean expression or logic network.

Construct a logic network with the same truth function as a given Boolean expression.

Write a Boolean expression with the same truth function as a given logic network.

Write the Boolean expression in canonical sum-of-products form for a given truth function.

Using only NAND gates, find a network having the same truth

function as a given network with AND gates, OR gates, and inverters.

Find a truth function satisfying the description of a particular problem.

Main Ideas

We can effectively convert information from any of the three forms below to any other form:

truth function ↔ Boolean expression ↔ logic network

A Boolean expression can sometimes be converted to a simpler, equivalent expression using the properties of Boolean algebra, thus producing a simpler network for a given truth function.

Exercises Section 6.1

1. Construct logic networks for the following Boolean expressions, using AND gates, OR gates, and inverters:
   ★ (a) $(x_1' + x_2)x_3$
   (b) $(x_1 + x_2)' + x_1'x_3$
   (c) $x_1'x_2 + (x_1x_2)'$
   (d) $(x_1 + x_2)'x_3 + x_3'$

2. Write a Boolean expression and a truth function for each of the logic networks shown in Figures 6.21 through 6.24.

(a)

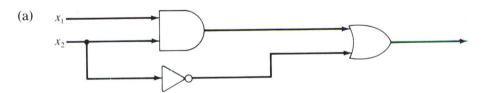

Figure 6.21

(b)

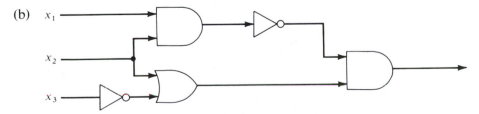

Figure 6.22

(c)

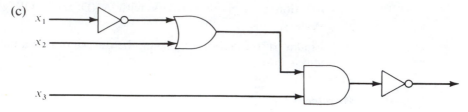

Figure 6.23

(d)

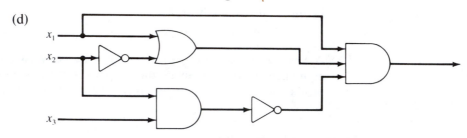

Figure 6.24

3. (a) Write the truth function for the Boolean operation

$$x \oplus y = xy' + yx'$$

(b) Draw the logic network for $x \oplus y$.
(c) Show that the network of Figure 6.25 also represents $x \oplus y$. Explain why the network illustrates that $\oplus$ is the **exclusive OR** operation.

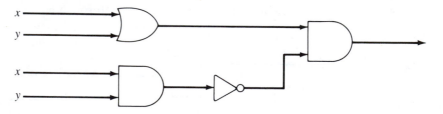

Figure 6.25

4. Find the canonical sum-of-products form for the truth functions in Figures 6.26 through 6.29.

(a)

x_1	x_2	$f(x_1, x_2)$
1	1	0
1	0	0
0	1	0
0	0	1

Figure 6.26

(b)

x_1	x_2	$f(x_1, x_2)$
1	1	1
1	0	0
0	1	1
0	0	0

Figure 6.27

(c)

x_1	x_2	x_3	$f(x_1, x_2, x_3)$
1	1	1	0
1	1	0	1
1	0	1	1
1	0	0	0
0	1	1	1
0	1	0	0
0	0	1	0
0	0	0	1

Figure 6.28

(d)

x_1	x_2	x_3	$f(x_1, x_2, x_3)$
1	1	1	0
1	1	0	0
1	0	1	1
1	0	0	1
0	1	1	0
0	1	0	1
0	0	1	0
0	0	0	0

Figure 6.29

★ 5. (a) Find the canonical sum-of-products form for the truth function in Figure 6.30.
   (b) Draw the logic network for the expression of part (a).
   (c) Use properties of a Boolean algebra to reduce the expression of part (a) to an equivalent expression whose network requires only two logic elements. Draw the network.

x_1	x_2	x_3	$f(x_1, x_2, x_3)$
1	1	1	0
1	1	0	1
1	0	1	0
1	0	0	1
0	1	1	0
0	1	0	0
0	0	1	0
0	0	0	0

Figure 6.30

x_1	x_2	x_3	$f(x_1, x_2, x_3)$
1	1	1	1
1	1	0	0
1	0	1	0
1	0	0	0
0	1	1	1
0	1	0	1
0	0	1	0
0	0	0	0

Figure 6.31

6. (a) Find the canonical sum-of-products form for the truth function in Figure 6.31.
   (b) Draw the logic network for the expression of part (a).
   (c) Use properties of a Boolean algebra to reduce the expression of part (a) to an equivalent expression whose network requires only three logic elements. Draw the network.

7. (a) Show that the two Boolean expressions

$$(x_1 + x_2)(x_1' + x_3)(x_2 + x_3)$$

and

$$(x_1 x_3) + (x_1' x_2)$$

are equivalent by writing the truth table for each.

(b) Write the canonical sum-of-products form equivalent to the two expressions of part (a).

(c) Use properties of a Boolean algebra to reduce one of the expressions of part (a) to the other.

★ 8. There is also a **canonical product-of-sums form (conjunctive normal form)** for any truth function. This expression has the form

$$( )( ) \cdots ( )$$

with each factor a sum of the form

$$\alpha + \beta + \cdots + \omega$$

where $\alpha = x_1$ or x_1', $\beta = x_2$ or x_2', and so on. Each factor is constructed to have a value of 0 for the input values of one of the rows of the truth function having value 0. Thus, the entire expression has value 0 for these inputs and no others. Find the canonical product-of-sums form for the truth functions of Exercise 4 above.

9. Figure 6.32 shows an unprogrammed PLA for three inputs, x_1, x_2, and x_3. Program this PLA to generate the truth functions f_1 and f_3 represented by

$$f_1: \quad x_1 x_2 x_3 + x_1' x_2 x_3' + x_1' x_2' x_3$$

$$f_3: \quad x_1 x_2' x_3' + x_1' x_2' x_3 + x_1' x_2' x_3'$$

10. The **2's complement** of an n-digit binary number p is an n-digit binary number q such that $p + q$ equals an n-digit representation of zero

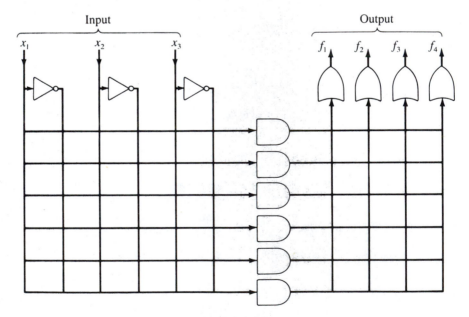

Figure 6.32

(any carry digit to column $n + 1$ is ignored). Thus 01110 is the 2's complement of 10010 because

$$
\begin{array}{r}
10010 \\
+\quad \underline{01110} \\
(1)00000
\end{array}
$$

The 2's complement idea can be used to represent negative integers in binary form. Given a binary number p, the 2's complement of p is found by scanning p from low-order to high-order digits (right to left). As long as digit i of p is 0, digit i of q is 0. When the first 1 of p is encountered, say at digit j, then digit j of q is 1, but for the remaining digits, $j < i \le n$, $q_i = p_i'$. For $p = 10010$, for instance, the rightmost 0 digit of p stays a 0 digit in q, and the first 1 digit stays a 1 digit. The remaining digits of q, however, are the reverse of the digits in p (see Figure 6.33).

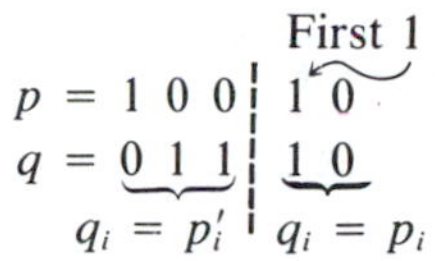

$$
\begin{array}{l}
\overset{\text{First 1}}{} \\
p = 1\ 0\ 0\ |\ \overset{\frown}{1\ 0} \\
q = \underbrace{0\ 1\ 1}\ |\ \underline{1\ 0} \\
\quad q_i = p_i'\ |\ q_i = p_i
\end{array}
$$

Figure 6.33

For each binary number p, find the 2's complement of p, namely q, and then calculate $p + q$.

(a)   1100

(b)   1001

(c)   011

11. For any digit x_i in a binary number p, let r_i be the corresponding digit in q, the 2's complement of p (see Exercise 10). The value of r_i depends on the value of x_i and also on the position of x_i relative to the first 1 digit in p. For the ith digit, let c_{i-1} denote a 0 if the digits p_j, $1 \le j \le i - 1$, are 0 and a 1 otherwise. A value c_i must be computed to move on to the next digit.

(a)   Give a truth function for r_i with inputs x_i and c_{i-1}. Give a truth function for c_i with inputs x_i and c_{i-1}.

(b)   Write Boolean expressions for the truth functions of part (a). Simplify as much as possible.

(c)   Design a circuit module to output r_i and c_i from inputs x_i and c_{i-1}.

(d)   Using the modules of part (c), design a circuit to find the 2's complement of a 3-digit binary number zyx. Trace the operation of the circuit in computing the 2's complement of 110.

12. (a)   Construct a network for the following expression using only NAND

elements. Replace the AND and OR gates and inverters with the appropriate NAND networks.

$$x_3'x_1 + x_2'x_1 + x_3'$$

(b) Use the properties of a Boolean algebra to reduce the expression of part (a) to one whose network would require only three NAND gates. Draw the network.

★ 13. Replace the network of Figure 6.34 with an equivalent network using one AND gate, one OR gate, and one inverter.

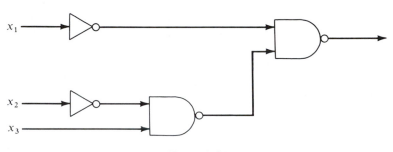

Figure 6.34

14. Using only NOR elements, construct networks that can replace (a) an inverter, (b) an OR gate, and (c) an AND gate.

★ 15. Explain why Exercises 20 and 21 of Section 1.1 prove that the NAND gate and the NOR gate, respectively, are sufficient to realize any truth function.

16. Find an equivalent network for the half-adder module that uses exactly five NAND gates.

★ 17. You have just been hired at Mercenary Motors. Your job is to design a logic network so that a car can be started only when the automatic transmission is in neutral or park, and the driver's seat belt is fastened. Find a truth function, a Boolean expression, and a logic network. (There is a don't-care condition to the truth function, since the car cannot be in both neutral and park.)

18. Mercenary Motors has expanded into the calculator business. You need to design the circuitry for the display readout on a new calculator. This design involves a two-step process.
(a) Any digit 0, 1, . . . , 9 put into the calculator is first converted to binary form. Figure 6.35 illustrates this conversion, which involves four separate networks, one each for x_1 to x_4. Each

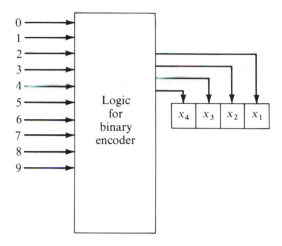

Figure 6.35

network has 10 inputs, but only 1 input can be on at any given moment. Write a Boolean expression and then draw a network for x_2.

(b) The binary form of the digit is then converted into a visual display by activating a pattern of 7 outputs arranged as shown in Figure 6.36a. To display the digit 3, for example, requires that y_1, y_2, y_3, y_5, and y_7 be on, as in Figure 6.36b. Thus, the second step of the process can be represented by Figure 6.37, which involves 7 separate networks, one each for y_1 to y_7, each with 4 inputs, x_1 to x_4. Write a truth function, a Boolean expression, and a network for y_5 and for y_6.

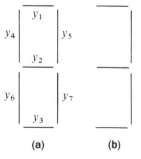

(a)   (b)

Figure 6.36

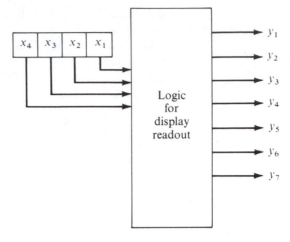

Figure 6.37

SECTION 6.2   MINIMIZATION

The Minimization Process

Remember from the last section that a given truth function is associated with an equivalence class of Boolean expressions. If we want to design a logic network for the function, our *ideal* would be to have a procedure that chooses a "simplest" Boolean expression from the class. What we consider simple will depend upon the technology employed in building the network, what kind of logic elements are available, and so on. At any rate, we probably want to minimize the total number of connections that must be made and the total number of logic elements used. (As we discuss minimization procedures, keep in mind that other factors may influence the economics of the situation. If a network is to be built only once, the time spent on minimization is costlier than building the network. But if the network is to be mass-produced, then the cost of minimization time may be worthwhile.)

We have had some experience in simplifying Boolean expressions by applying the properties of Boolean algebra. However, we had no procedure to use: we simply had to guess, attacking each problem individually. What we want now is a mechanical procedure that we can use without having to be clever or insightful. Unfortunately, we won't develop our ideal procedure. However, we already know how to select the canonical sum-of-products form from the equivalence class of expressions for a given truth function. In this section we will discuss two procedures to reduce a canonical sum-of-products form to a minimal sum-of-products form. Therefore, we can minimize within the framework of a sum-of-products form, and reduce, if not completely minimize, the number of elements and connections required.

6.22 Example The Boolean expression

$$x_1 x_2 x_3 + x_1' x_2 x_3 + x_1' x_2 x_3'$$

is in sum-of-products form. An equivalent minimal sum-of-products form is

$$x_2 x_3 + x_2 x_1'$$

Implementing a network for this form would require two AND gates, one OR gate, and an inverter. Using one of the distributive laws of Boolean algebra, this expression reduces to

$$x_2(x_3 + x_1')$$

which requires only one AND gate, one OR gate, and an inverter, but it is no longer in sum-of-products form. Thus, a minimal sum-of-products form may not be minimal in an absolute sense. □

There are two extremely useful equivalencies in minimizing a sum-of-products form. They are

$$x_1 x_2 + x_1' x_2 = x_2$$

and

$$x_1 + x_1' x_2 = x_1 + x_2$$

6.23 Practice Use properties of Boolean algebra to reduce the following:

(a) $x_1 x_2 + x_1' x_2$ to x_2
(b) $x_1 + x_1' x_2$ to $x_1 + x_2$ □

The equivalency $x_1 x_2 + x_1' x_2 = x_2$ means, for example, that the expression $x_1' x_2 x_3' x_4 + x_1' x_2' x_3' x_4$ reduces to $x_1' x_3' x_4$. Thus, when we have a sum of two products that differ in only one factor, we can eliminate that factor. However, the canonical sum-of-products form for a truth function of four variables, for example, might be quite long and require some searching to locate two product terms differing by only one factor. To help us in this search, we can use the *Karnaugh map*. The Karnaugh map is a visual representation of the truth function so that terms in the canonical sum-of-products form that differ by only one factor can be quickly matched.

The Karnaugh Map

Figure 6.38

In the canonical sum-of-products form for a truth function, we are interested in values of the input variables that produce outputs of 1. The Karnaugh map records the 1s of the function in an array that forces products of inputs differing by only one factor to be adjacent. The array form for a two-variable function is given in Figure 6.38. Notice that the square corresponding to x_1x_2, the upper left-hand square, is adjacent to squares $x_1'x_2$ and x_1x_2', which differ in one factor from x_1x_2; however, it is not adjacent to the $x_1'x_2'$ square, which differs in two factors from x_1x_2.

6.24 Example   The truth function of Figure 6.39 is represented by the Karnaugh map of Figure 6.40. At once we can observe 1s in two adjacent squares, so there are two terms in the canonical sum-of-products form differing by one variable; again from the map, we see that the variable that changes is x_1. It can be eliminated. We conclude that the function can be represented by x_2. Indeed, the canonical sum-of-products form for the function is $x_1x_2 + x_1'x_2$, which, by our basic reduction rule, reduces to x_2. However, we did not have to write the canonical form—we only had to look at the map.   □

6.25 Practice   Draw the Karnaugh map and use it to find a reduced expression for the function in Figure 6.41.   □

x_1	x_2	$f(x_1, x_2)$
1	1	1
1	0	0
0	1	1
0	0	0

Figure 6.39

Figure 6.40

x_1	x_2	$f(x_1, x_2)$
1	1	0
1	0	0
0	1	1
0	0	1

Figure 6.41

Maps for Three and Four Variables

The array forms for functions of three and four variables are shown in Figure 6.42. In these arrays, adjacent squares also differ by only one variable. However, in Figure 6.42a, the leftmost and rightmost squares in a row also differ by one variable, so we consider them adjacent. (They would in fact be adjacent if we wrapped the map around a cylinder and glued the left and right edges together.) In Figure 6.42b, the leftmost and rightmost squares in a row are adjacent (differ by exactly one variable), and also the top and bottom squares in a column are adjacent.

In three-variable maps, two adjacent squares marked with 1 allow one variable to be eliminated, and four adjacent squares marked with 1

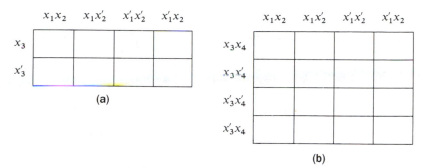

Figure 6.42

(either in a single row or arranged in a square) allow two variables to be eliminated.

6.26 Example   In the map of Figure 6.43, the squares that combine for a reduction are shown as a block. These four adjacent squares reduce to x_3 (eliminate the changing variables x_1 and x_2). The reduction uses our basic reduction rule more than once:

$$x_1x_2x_3 + x_1x_2'x_3 + x_1'x_2'x_3 + x_1'x_2x_3$$

$$= x_1x_3(x_2 + x_2') + x_1'x_3(x_2' + x_2)$$

$$= x_1x_3 + x_1'x_3$$

$$= x_3(x_1 + x_1')$$

$$= x_3 \qquad \square$$

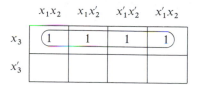

Figure 6.43

In four-variable maps, two adjacent squares marked with 1 allow one variable to be eliminated, four adjacent squares marked with 1 allow two variables to be eliminated, and eight adjacent squares marked with 1 allow three variables to be eliminated.

Figure 6.44 illustrates some ways in which two adjacent marked squares could occur. Figure 6.45 illustrates some ways in which four adjacent marked squares could occur, and Figure 6.46 shows possibilities for eight adjacent marked squares.

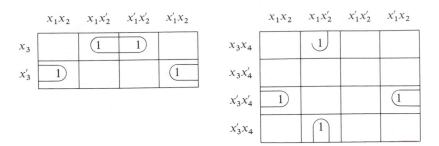

Figure 6.44

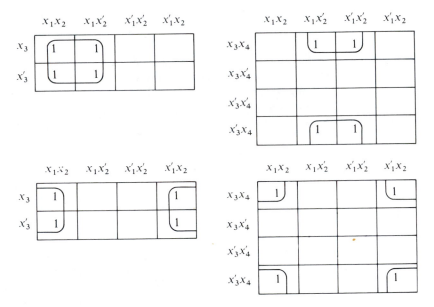

Figure 6.45

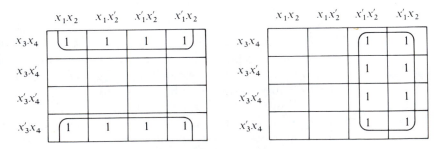

Figure 6.46

6.27 Example   In the map of Figure 6.47, the four outside corners reduce to $x_2 x_4$ and the inside square reduces to $x_2' x_4'$.   ☐

6.28 Practice   Find the two terms represented by the map in Figure 6.48.   ☐

	$x_1 x_2$	$x_1 x_2'$	$x_1' x_2'$	$x_1' x_2$
$x_3 x_4$	1			1
$x_3 x_4'$		1	1	
$x_3' x_4'$		1	1	
$x_3' x_4$	1			1

Figure 6.47

	$x_1 x_2$	$x_1 x_2'$	$x_1' x_2'$	$x_1' x_2$
$x_3 x_4$	1	1		
$x_3 x_4'$	1	1		
$x_3' x_4'$				1
$x_3' x_4$				1

Figure 6.48

Using the Karnaugh Map

How do we find a minimal sum-of-products form from a Karnaugh map (or from a truth function or a canonical sum-of-products form)? We must use every marked square of the map, and we want to include every marked square in the largest combination of marked squares possible since doing so provides maximum reduction of the expression. However, we cannot begin by simply looking for the largest blocks of marked squares on the map.

6.29 Example   In the Karnaugh map of Figure 6.49, if we simply looked for the largest block of marked squares, we would use the column of 1s and reduce it to $x_1' x_2'$. However, we would still have four marked squares unaccounted for. Each of these marked squares can be combined into a two-square block in only one way (see Figure 6.50), and each of these blocks has to be included. But when this is done, every square in the column of 1s is used, and the term $x_1' x_2'$ is superfluous. The minimal sum-of-products form for this map becomes

$$x_2' x_3 x_4 + x_1' x_3 x_4' + x_2' x_3' x_4' + x_1' x_3' x_4$$ ☐

	$x_1 x_2$	$x_1 x_2'$	$x_1' x_2'$	$x_1' x_2$
$x_3 x_4$		1	1	
$x_3 x_4'$			1	1
$x_3' x_4'$		1	1	
$x_3' x_4$			1	1

Figure 6.49

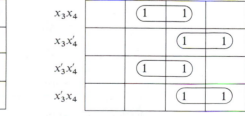

	$x_1 x_2$	$x_1 x_2'$	$x_1' x_2'$	$x_1' x_2$
$x_3 x_4$		1	1	
$x_3 x_4'$			1	1
$x_3' x_4'$		1	1	
$x_3' x_4$			1	1

Figure 6.50

To avoid the redundancy illustrated by Example 6.29, we analyze the map as follows. First, we form terms for those marked squares that cannot be combined with anything. Then we use the remaining marked squares to find those that can be combined only into two-square blocks and in only one way. Then among the unused marked squares, that is, those not already assigned to a block, we find those that can be combined only into four-square blocks and in only one way; then we look for any unused squares that go uniquely into eight-square blocks. At each step, if an unused marked square can go into more than one block, we do nothing with it. Finally, we take any unused marked squares that are left (for which there was a choice of blocks) and select blocks that include them in the most efficient manner. Actually, if there are many 1s in the map, thus allowing many different blockings, even this procedure may not lead to a minimal form (see Example 6.34).

6.30 Example   In Figure 6.51a, we have shown the only square that cannot be combined into a larger block. In Figure 6.51b, we have formed the unique two-square block for the $x_1 x_2' x_3'$ square and the unique two-square block for the $x_1' x_2' x_3$ square. All marked squares are covered. The minimal sum-of-products expression is

$$x_1 x_2 x_3 + x_2' x_3' + x_1' x_2'$$

Formally, the last two terms are obtained by expanding $x_1' x_2' x_3'$ into $x_1' x_2' x_3' + x_1' x_2' x_3'$ and then combining it with each of its neighbors.   □

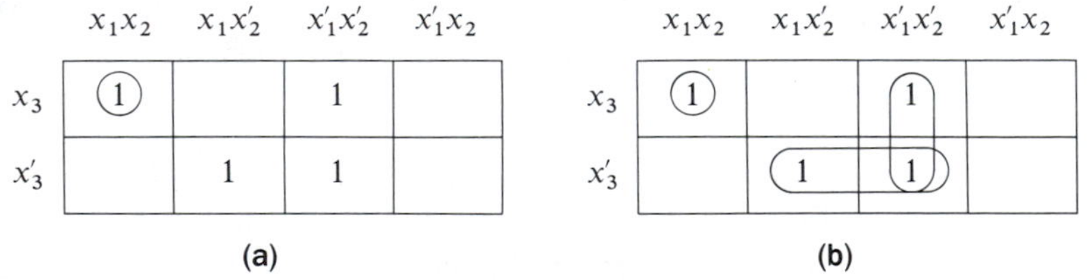

Figure 6.51

6.31 Example   Figure 6.52a shows the unique two-square blocks for the $x_1' x_2' x_3 x_4$ square and the $x_1 x_2' x_3' x_4'$ square. In Figure 6.52b the two unused squares have been combined into a unique four-square block. The minimal sum-of-products expression is

$$x_1 x_3 + x_2' x_3 x_4 + x_1 x_2' x_4'$$ □

6.32 Example   Figure 6.53a shows the unique two-square blocks. We can assign the remaining unused marked square to either of two different two-square blocks; these blocks are shown in Figure 6.53b. There are two minimal sum-of-products forms,

$$x_1 x_2' x_4' + x_1' x_2 x_3 + x_2' x_3 x_4'$$

and

$$x_1 x_2' x_4' + x_1' x_2 x_3 + x_1' x_3 x_4'$$

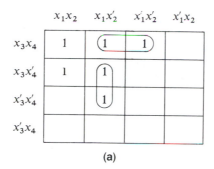

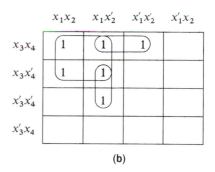

Figure 6.52

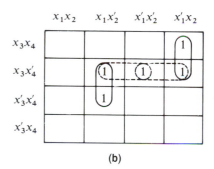

Figure 6.53

6.33 Example   Figure 6.54a shows the unique two-square and four-square blocks. The remaining two unused marked squares can be assigned to two-square blocks in two different ways, as shown in Figures 6.54b and 6.54c. Assigning them together to a single two-square block is more efficient, since it produces a sum-of-products form with three terms rather than four. The minimal sum-of-products expression is

$$x_1 x_3 + x_1' x_2 x_3' + x_2' x_3' x_4'$$

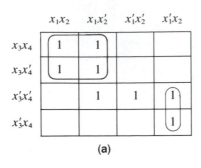
(a)

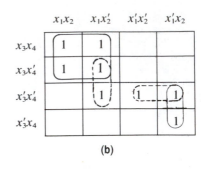
(b)

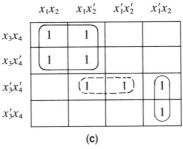
(c)

Figure 6.54

6.34 Example   Consider the map of Figure 6.55a. Here the two unique four-square blocks determined by the squares with * have been chosen. In Figure 6.55b, the remaining unmarked squares, for which there was a choice of blocks, are combined into blocks as efficiently as possible. The resulting sum-of-products form is

$$x_1x_3 + x_1'x_3' + x_3x_4 + x_1'x_2 + x_1x_2'x_4' \tag{1}$$

Yet in Figure 6.55c, choosing a different four-square block at the top leads to another sum-of-products form,

$$x_2x_3 + x_1'x_3' + x_3x_4 + x_1x_2'x_4'$$

which is simpler than (1).   □

6.35 Practice   Write the minimal sum-of-products expression for the map shown in Figure 6.56.   □

We have used Karnaugh maps for functions of two, three, and four variables. By using three-dimensional drawings, Karnaugh maps for functions of five, six, or even more variables can be constructed, but the visualization gets too complicated to be worthwhile.

If the Karnaugh map corresponds to a function with don't-care conditions, then the don't-care squares on the map can be left blank or assigned the value 1, whichever aids the minimization process.

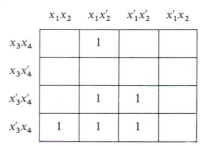

	x_1x_2	x_1x_2'	$x_1'x_2'$	$x_1'x_2$
x_3x_4	1	1	1	1
x_3x_4'	1	*1		1
$x_3'x_4'$		1	*1	1
$x_3'x_4$			1	1

(a)

	x_1x_2	x_1x_2'	$x_1'x_2'$	$x_1'x_2$
x_3x_4	1	1	1	1
x_3x_4'	1	*1		1
$x_3'x_4'$		1	*1	1
$x_3'x_4$			1	1

(b)

	x_1x_2	x_1x_2'	$x_1'x_2'$	$x_1'x_2$
x_3x_4	1	1	1	1
x_3x_4'	1	1		1
$x_3'x_4'$		1	1	1
$x_3'x_4$			1	1

(c)

Figure 6.55

	x_1x_2	x_1x_2'	$x_1'x_2'$	$x_1'x_2$
x_3x_4		1		
x_3x_4'				
$x_3'x_4'$		1	1	
$x_3'x_4$	1	1	1	

Figure 6.56

The Quine–McCluskey Procedure

Remember that the key to reducing the canonical sum-of-products form for a truth function lies in recognizing terms of the sum that differ in only one factor. In the Karnaugh map, we see where such terms occur. A second method of reduction, the *Quine–McCluskey procedure*, organizes information from the canonical sum-of-products form into a table so that the search for terms differing by only one factor is easy to carry out. The procedure is a two-step process paralleling the use of the Karnaugh map. First we find groupings of terms (just as we looped together marked squares in the Karnaugh map); then we eliminate redundant groupings and make choices for terms that can belong to several groups.

6.36 Example   Let's illustrate the Quine–McCluskey procedure by using the truth function for Example 6.29. We did not write the actual truth function there, but the information is contained in the Karnaugh map. The truth function is shown in Figure 6.57. In Figure 6.58 the eight 4-tuples of 0s and 1s producing a function value of 1 are listed in a table, separated into four groupings according to the number of 1s. Note that terms of the canonical sum-of-products form differing by only one factor must be in adjacent groupings, which simplifies the search for such terms.

x_1	x_2	x_3	x_4	$f(x_1, x_2, x_3, x_4)$
1	1	1	1	0
1	1	1	0	0.
1	1	0	1	0
1	1	0	0	0
1	0	1	1	1
1	0	1	0	0
1	0	0	1	0
1	0	0	0	1
0	1	1	1	0
0	1	1	0	1
0	1	0	1	1
0	1	0	0	0
0	0	1	1	1
0	0	1	0	1
0	0	0	1	1
0	0	0	0	1

Figure 6.57

	x_1	x_2	x_3	x_4
(three 1s)	1	0	1	1
(two 1s)	0	1	1	0
	0	1	0	1
	0	0	1	1
(one 1)	1	0	0	0
	0	0	1	0
	0	0	0	1
(no 1s)	0	0	0	0

Figure 6.58

We compare the first term, 1011, with each of the three terms of the second group, 0110, 0101, and 0011, to locate terms differing by only one factor. Such a term is 0011. The combination 1011 and 0011 reduces to -011 when the changing variable x_1 is eliminated. (We write this reduced term with a dash in the x_1 position.) Next we start to form a second column in the table consisting of the reduced terms, again grouped according to the number of 1s (see Figure 6.59). We also mark the two terms 1011 and 0011 in the first column with a superscript 1; this 1 is a pointer that indicates the number of the reduced term in the second column that is formed from these two terms (numbering terms corresponds to putting loops in the Karnaugh map). Note that the reduced terms in the second column are numbered from top to bottom. We continue this process with all the terms. A numbered term may still be used in other combinations, just as a marked square in a Karnaugh map can be in more than one loop. When we are done, we obtain Figure 6.59.

x_1	x_2	x_3	x_4	
1	0	1	1	1
0	1	1	0	2
0	1	0	1	3
0	0	1	1	1,4,5
1	0	0	0	6
0	0	1	0	2,4,7
0	0	0	1	3,5,8
0	0	0	0	6,7,8

x_1	x_2	x_3	x_4
–	0	1	1
0	–	1	0
0	–	0	1
0	0	1	–
0	0	–	1
–	0	0	0
0	0	–	0
0	0	0	–

Figure 6.59

We then repeat this reduction process on the second column of Figure 6.59 to make a third column. Here not only the groupings but also the dashes help organize the search process, since terms differing by only one variable must have dashes in the same location. Again, numbers on terms that combine serve as pointers to the reduced terms in the third column. When the process cannot be continued, we reach the reduction table of Figure 6.60. The unnumbered terms are irreducible, so they represent the possible maximum-sized loops on a Karnaugh map.

x_1	x_2	x_3	x_4	
1	0	1	1	1
0	1	1	0	2
0	1	0	1	3
0	0	1	1	1,4,5
1	0	0	0	6
0	0	1	0	2,4,7
0	0	0	1	3,5,8
0	0	0	0	6,7,8

x_1	x_2	x_3	x_4	
–	0	1	1	
0	–	1	0	
0	–	0	1	
0	0	1	–	1
0	0	–	1	1
–	0	0	0	
0	0	–	0	1
0	0	0	–	1

x_1	x_2	x_3	x_4
0	0	–	–

Figure 6.60

For the second step of the process, we compare the original terms with the irreducible terms. A check in the comparison table in Figure 6.61 indicates that the original term in that column eventually led to the irreducible term in that row, which can be determined by following the pointers.

	1011	0110	0101	0011	1000	0010	0001	0000
−011	✓			✓				
0−10		✓				✓		
0−01			✓				✓	
−000					✓			✓
00−−				✓		✓	✓	✓

Figure 6.61

If a column in the comparison table has a check in only one row, the irreducible term for that row is the only one covering the original term, so it is an essential term and must appear in the final sum-of-products form. Thus, we see from Figure 6.61 that the terms −011, 0−10, 0−01, and −000 are essential and must be in the final expression. We also note that all columns with a check in row 5 also have checks in another row and so are covered by an essential, reduced term already in the expression. Thus, 00−− is redundant. As in Example 6.29, the minimal sum-of-products form is

$$x_2'x_3x_4 + x_1'x_3x_4' + x_1'x_3'x_4 + x_2'x_3'x_4' \qquad \square$$

In situations where there is more than one minimal sum-of-products form, the comparison table will have nonessential, nonredundant, reduced terms. A selection must be made from these reduced terms to cover all columns not covered by essential terms.

6.37 Example   We will use the Quine–McCluskey procedure on the problem presented in Example 6.32. The reduction table is given in Figure 6.62. The comparison table appears in Figure 6.63. We see from the comparison table that 011− and 10−0 are essential, reduced terms and that there are no redundant terms. The only original term not covered by essential terms

x_1	x_2	x_3	x_4		x_1	x_2	x_3	x_4
0	1	1	1	1	0	1	1	−
1	0	1	0	2,3	−	0	1	0
0	1	1	0	1,4	1	0	−	0
0	0	1	0	2,4	0	−	1	0
1	0	0	0	3				

Figure 6.62

	0111	1010	0110	0010	1000
011–	✓		✓		
–010		✓		✓	
10–0		✓			✓
0–10			✓	✓	

Figure 6.63

is 0010, column 4, and the choice of the reduced term for row 2 or for row 4 will cover it. Thus, the minimal sum-of-products form is

$$x_1'x_2x_3 + x_1x_2'x_4' + x_2'x_3x_4'$$

or

$$x_1'x_2x_3 + x_1x_2'x_4' + x_1'x_3x_4'$$ □

6.38 Practice Use the Quine–McCluskey procedure to find a minimal sum-of-products form for the truth function in Figure 6.64. □

x_1	x_2	x_3	$f(x_1, x_2, x_3)$
1	1	1	1
1	1	0	1
1	0	1	0
1	0	0	1
0	1	1	0
0	1	0	0
0	0	1	1
0	0	0	1

Figure 6.64

The Quine–McCluskey procedure applies to truth functions with any number of input variables, but for a large number of variables, the procedure is extremely tedious to do by hand. However, it is exactly the kind of systematic, mechanical process that lends itself to a computerized solution.

If the truth function f has few 0-values and a large number of 1-values, it may be simpler to implement the Quine–McCluskey procedure for the complement of the function, f', which will have 1-values where f has 0-values, and vice versa. Once a minimal sum-of-products expression is obtained for f', it can be complemented to obtain an expression for f, although the new expression will not be in sum-of-products form. (In fact, by DeMorgan's Laws, it will be equivalent to a product-of-sums form.) We can obtain the network for f from the sum-of-products network for f' by tacking an inverter on the end.

The whole object of minimizing a network is to simplify the internal configuration while preserving the external behavior. In Chapter 10 we will attempt the same sort of minimization on finite-state machine structures.

✔ Checklist

Techniques

Minimize the canonical sum-of-products form for a truth function by using a Karnaugh map.

Minimize the canonical sum-of-products form for a truth function by using the Quine–McCluskey procedure.

Main Ideas

Algorithms exist for reducing a canonical sum-of-products form to a minimized sum-of-products form.

Exercises Section 6.2

1. Write the minimal sum-of-products form for the Karnaugh maps of Figure 6.65.

2. Use a Karnaugh map to find the minimal sum-of-products form for the truth functions of Figure 6.66.

3. Use a Karnaugh map to find the minimal sum-of-products form for the following Boolean expressions:
   (a) $x_1'x_2'x_3x_4 + x_1x_2x_3'x_4 + x_1'x_2'x_3'x_4 + x_1x_2'x_3x_4' + x_1'x_2x_3x_4 + x_1'x_2x_3'x_4 + x_1'x_2'x_3x_4'$
   (b) $x_1'x_2'x_3'x_4' + x_1x_2x_3'x_4 + x_1'x_2'x_3'x_4 + x_1x_2x_3'x_4' + x_1'x_2x_3x_4 + x_1x_2'x_3'x_4'$

4. Use a Karnaugh map to find a minimal sum-of-products expression for the network of three variables shown in Figure 6.67. Sketch the new network.

5. Use a Karnaugh map to find a minimal sum-of-products form for the truth function in Figure 6.68. Don't-care conditions are shown by dashes.

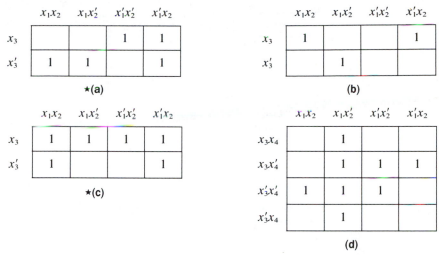

(a)

	x_1x_2	x_1x_2'	$x_1'x_2'$	$x_1'x_2$
x_3			1	1
x_3'	1	1		1

$\star$(a)

(b)

	x_1x_2	x_1x_2'	$x_1'x_2'$	$x_1'x_2$
x_3	1			1
x_3'		1		

(b)

(c)

	x_1x_2	x_1x_2'	$x_1'x_2'$	$x_1'x_2$
x_3	1	1	1	1
x_3'	1			1

$\star$(c)

(d)

	x_1x_2	x_1x_2'	$x_1'x_2'$	$x_1'x_2$
x_3x_4		1		
x_3x_4'		1	1	1
$x_3'x_4'$	1	1	1	
$x_3'x_4$		1		

(d)

(e)

	x_1x_2	x_1x_2'	$x_1'x_2'$	$x_1'x_2$
x_3x_4				1
x_3x_4'	1	1		
$x_3'x_4'$		1	1	
$x_3'x_4$			1	

(e)

Figure 6.65

x_1	x_2	x_3	$f(x_1, x_2, x_3)$
1	1	1	1
1	1	0	1
1	0	1	0
1	0	0	0
0	1	1	1
0	1	0	0
0	0	1	0
0	0	0	0

(a)

x_1	x_2	x_3	x_4	$f(x_1, x_2, x_3, x_4)$
1	1	1	1	1
1	1	1	0	1
1	1	0	1	1
1	1	0	0	1
1	0	1	1	0
1	0	1	0	1
1	0	0	1	0
1	0	0	0	1
0	1	1	1	1
0	1	1	0	1
0	1	0	1	1
0	1	0	0	1
0	0	1	1	0
0	0	1	0	0
0	0	0	1	0
0	0	0	0	0

(b)

Figure 6.66

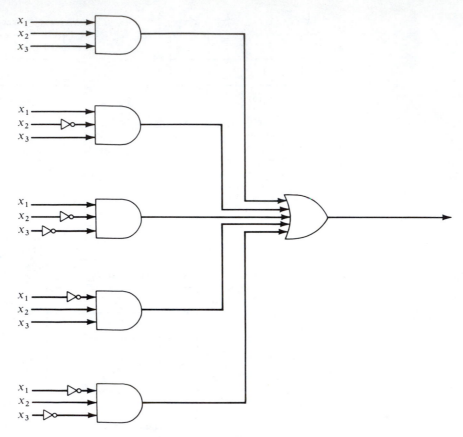

Figure 6.67

x_1	x_2	x_3	x_4	$f(x_1, x_2, x_3, x_4)$
1	1	1	1	0
1	1	1	0	1
1	1	0	1	0
1	1	0	0	–
1	0	1	1	0
1	0	1	0	–
1	0	0	1	0
1	0	0	0	0
0	1	1	1	0
0	1	1	0	1
0	1	0	1	0
0	1	0	0	1
0	0	1	1	1
0	0	1	0	0
0	0	0	1	–
0	0	0	0	0

Figure 6.68

⋆ 6. Use the Quine–McCluskey procedure to find a minimal sum-of-products form for the truth function illustrated by the map in Figure 6.65c.

╱ 7. Use the Quine–McCluskey procedure to find a minimal sum-of-products form for the network in Figure 6.69. Sketch the new network.

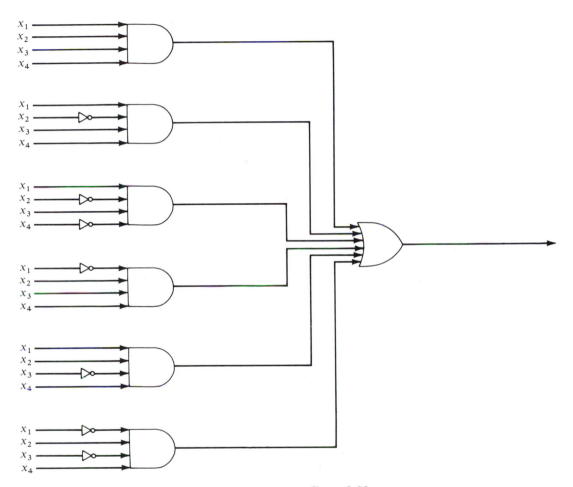

Figure 6.69

╱ 8. Use the Quine–McCluskey procedure to find the minimal sum-of-products form for the truth functions in Figure 6.70.

x_1	x_2	x_3	x_4	$f(x_1, x_2, x_3, x_4)$
1	1	1	1	0
1	1	1	0	1
1	1	0	1	0
1	1	0	0	0
1	0	1	1	0
1	0	1	0	1
1	0	0	1	1
1	0	0	0	1
0	1	1	1	0
0	1	1	0	0
0	1	0	1	0
0	1	0	0	1
0	0	1	1	1
0	0	1	0	1
0	0	0	1	0
0	0	0	0	1

(a)

x_1	x_2	x_3	x_4	$f(x_1, x_2, x_3, x_4)$
1	1	1	1	1
1	1	1	0	0
1	1	0	1	1
1	1	0	0	0
1	0	1	1	1
1	0	1	0	0
1	0	0	1	1
1	0	0	0	1
0	1	1	1	1
0	1	1	0	0
0	1	0	1	1
0	1	0	0	1
0	0	1	1	1
0	0	1	0	0
0	0	0	1	1
0	0	0	0	1

(b)

Figure 6.70

9. Use the Quine–McCluskey procedure to find the minimal sum-of-products form for the following Boolean expressions:

 ★ (a) $x_1 x_2' x_3 x_4' + x_1' x_2' x_3 x_4 + x_1' x_2 x_3 x_4 + x_1' x_2' x_3' x_4' + x_1' x_2 x_3 x_4' + x_1' x_2' x_3' x_4$

 (b) $x_1 x_2 x_3 x_4 + x_1 x_2' x_3 x_4 + x_1 x_2 x_3 x_4' + x_1 x_2' x_3 x_4' + x_1' x_2 x_3 x_4' + x_1 x_2 x_3' x_4' + x_1' x_2 x_3' x_4 + x_1' x_2 x_3' x_4$

 (c) $x_1 x_2 x_3 x_4 + x_1 x_2 x_3 x_4' + x_1' x_2 x_3 x_4' + x_1 x_2 x_3' x_4' + x_1' x_2' x_3 x_4' + x_1' x_2 x_3' x_4' + x_1 x_2' x_3' x_4 + x_1' x_2' x_3' x_4 + x_1 x_2 x_3' x_4$

 (d) $x_1' x_2 x_3' x_4 x_5' + x_1' x_2 x_3 x_4' x_5 + x_1 x_2 x_3 x_4 x_5 + x_1' x_2' x_3 x_4' x_5 + x_1 x_2' x_3 x_4 x_5 + x_1' x_2' x_3' x_4' x_5 + x_1 x_2' x_3 x_4' x_5 + x_1 x_2 x_3 x_4' x_5' + x_1 x_2 x_3' x_4 x_5 + x_1' x_2' x_3' x_4 x_5'$

10. Use the Quine–McCluskey procedure to find a minimal sum-of-products form for the truth function illustrated by the map in Figure 6.54.

Chapter 7

Algebraic Structures

From Chapter 5 we have a general notion of what a mathematical structure is. Here we consider some algebraic structures that simulate (among other things) various types of arithmetic. In Sections 7.1 and 7.2, we will look at numerous examples and also develop some basic theorems about these well-studied structures. The Simulation II ideas will be used in Section 7.3 to classify some instances of these structures. In Sections 7.4 and 7.5, we will further discuss the conditions under which one instance of an algebraic structure can simulate another.

To balance our emphasis on algebraic structures, we should inject a realistic note. When we notice that a certain concrete situation S is an instance of an algebraic structure, say a group, we may hope to reap all sorts of benefits from this observation. In particular, some theorem from group theory might reveal a property of S that would be difficult to discover had we not noticed that we had a group. Often the results are not that spectacular, and the observation serves merely to classify S or clarify some of its properties. Sometimes, however, we gain a great deal from the knowledge that S has a certain algebraic structure—such will be the case with the group codes of Chapter 8 and the general loop-free decomposition of finite-state machines in Section 10.4.

SECTION 7.1   SEMIGROUPS, MONOIDS, AND GROUPS—SIMULATION I

Definitions of Algebraic Structures

If we wanted to develop a structure as it developed historically, we could look at many different situations and try to abstract common properties, the process we called Simulation I in Chapter 5 and used there to define a Boolean algebra and a monoid. However, we really don't have time to pursue this approach. Instead, we will jump ahead and define those structures that have proved to be the most interesting and then give some examples.

Because algebraic structures are often arithmetic in nature, we are concerned with binary operations on sets and with the properties of these operations. As we have already mentioned, the associative property of a binary operation means that the grouping of elements does not matter; the commutative property of a binary operation means that the order of the elements does not matter. Identity elements also appeared in Chapter 5. Let's give the formal definitions of these terms.

7.1 Definition   Let S be a set and $\cdot$ denote a binary operation on S. Then $\cdot$ is **associative** if for all $x, y, z \in S$,

$$(x \cdot y) \cdot z = x \cdot (y \cdot z) \qquad\qquad \square$$

If $\cdot$ is associative on S, then we can write an expression such as $x \cdot y \cdot z$ without parentheses.

7.2 Definition   Let S be a set and $\cdot$ denote a binary operation on S. Then $\cdot$ is **commutative** if for all $x, y \in S$,

$$x \cdot y = y \cdot x \qquad\qquad \square$$

7.3 Definition   Let S be a set and $\cdot$ denote a binary operation on S. Then an element $i \in S$ is an **identity element** if for all $x \in S$,

$$x \cdot i = i \cdot x = x \qquad\qquad \square$$

A monoid is an algebraic structure; we recall the definition.

7.4 Definition $[S, \cdot]$ is a **monoid** if S is a nonempty set and $\cdot$ is a binary operation on S, and if the following are true:

(a) The operation $\cdot$ is associative.
(b) There is an identity element $i \in S$; that is,

$$x \cdot i = i \cdot x = x$$

for all $x \in S$. $\qquad\qquad \square$

Once again, we are using a multiplication dot here as a generic symbol representing a binary operation. In any specific case, the particular binary operation has to be defined. If the operation is addition, for example, the $+$ sign replaces the generic symbol. Even when the operation is multiplication and thus a multiplication dot is appropriate, we must be sure that we know what kind of multiplication is intended—multiplication of integers, multiplication of matrices, and so on. The notation x^2 stands for $x \cdot x$; thus in $[\mathbb{R}, +]$, where the operation is addition, x^2 becomes $x + x = 2x$. As an analogy with programming, we can think of the generic symbol as a formal parameter to be replaced by an actual parameter—the specific operation—when its value becomes known.

7.5 Example   From Chapter 5 we already have the following examples of monoids: $[M_2(\mathbb{Z}), \cdot]$, $[M_2^D(\mathbb{Z}), \cdot]$, $[\mathbb{R}, +]$, $[\mathbb{R}^+, \cdot]$, $[\mathbb{Z}, \cdot]$. If we review the properties of a Boolean algebra, we will see that $[B, +]$ is a monoid and $[B, \cdot]$ is a monoid for any Boolean algebra $[B, +, \cdot, ', 0, 1]$. Hence, for any sets, $[\mathscr{P}(S), \cup]$ and $[\mathscr{P}(S), \cap]$ are monoids. (Can you name an identity element for each of these monoids?)   □

As far as complexity is concerned, the monoid is in the middle of the three structures considered in this section. To get a *semigroup* we drop one of the requirements for a monoid, and to get a *group* we add another requirement.

7.6 Definition $[S, \cdot]$ is a **semigroup** if S is a nonempty set, $\cdot$ is a binary operation on S, and $\cdot$ is associative.   □

A semigroup $[S, \cdot]$ in which $\cdot$ is commutative is called (what else?) a **commutative semigroup.**

Naturally every structure that is a monoid is also a semigroup, so we already have some examples of semigroups. Here are a few more easy ones.

7.7 Practice   Verify that $[\mathbb{N}, +]$, $[\mathbb{N}, \cdot]$, $[\mathbb{Z}, +]$, $[\mathbb{R}, \cdot]$, $[\mathbb{R}^+, +]$, and $[M_2(\mathbb{Z}), +]$ are semigroups.   □

7.8 Practice   (a) Which of the semigroups of Practice 7.7 are commutative?

(b) Which of the semigroups of Practice 7.7 are monoids? Name the identities.   □

Now for the last structure.

7.9 Definition $[S, \cdot]$ is a **group** if S is a nonempty set and $\cdot$ is a binary operation on S, and if the following are true:

(a) The operation $\cdot$ is associative.

(b) There is an identity element $i \in S$; that is,

$$x \cdot i = i \cdot x = x$$

for all $x \in S$.

(c) Each $x \in S$ has an **inverse element** x^{-1} in S; that is,

$$x \cdot x^{-1} = x^{-1} \cdot x = i$$ □

We may refer to a group structure as "the group S" rather than "the group $[S, \cdot]$" when it is clear what the binary operation is.

7.10 Example   The monoid $[\mathbb{N}, +]$ fails to be a group because 2, for instance, has no inverse element. There is no $n \in \mathbb{N}$ such that $2 + n = n + 2 = 0$. The only element having an inverse is 0.   □

A group $[S, \cdot]$ in which $\cdot$ is commutative is called a **commutative,** or **abelian, group** (after the Norwegian mathematician Niels Abel).

Because the requirements that must be satisfied in going from semigroup to monoid to group keep getting stiffer, we expect some examples to drop out, but those remaining should have richer and more interesting personalities.

7.11 Practice   Which of the following monoids are groups? $[M_2(\mathbb{Z}), \cdot]$, $[M_2^D(\mathbb{Z}), \cdot]$, $[\mathbb{R}, +]$, $[\mathbb{R}^+, \cdot]$, $[\mathbb{Z}, \cdot]$, $[\mathscr{P}(S), \cup]$, $[\mathscr{P}(S), \cap]$, $[\mathbb{N}, \cdot]$, $[\mathbb{Z}, +]$, $[\mathbb{R}, \cdot]$, $[M_2(\mathbb{Z}), +]$ □

Examples of Algebraic Structures

Now we will look at some other examples of semigroups, monoids, and groups.

7.12 Example   An expression of the form

$$a_n x^n + a_{n-1} x^{n-1} + \cdots + a_0$$

where $a_i \in \mathbb{R}$, $i = 0, 1, \ldots, n$, and $n \in \mathbb{N}$ is a **polynomial in x with real-number coefficients** (or a **polynomial in x over** $\mathbb{R}$.) For each i, a_i is the **coefficient** of x^i. If i is the largest integer greater than 0 for which $a_i \neq 0$, the polynomial is of **degree** i; if no such i exists, the polynomial is of **zero degree.** Terms with zero coefficients are generally not written. Thus, $\pi x^4 - \frac{2}{3} x^2 + 5$ is a polynomial of degree 4, and the constant polynomial 6 is of zero degree. The set of all polynomials in x over $\mathbb{R}$ is denoted by $\mathbb{R}[x]$.

We define binary operations of $+$ and $\cdot$ in $\mathbb{R}[x]$ to be the familiar operations of polynomial addition and multiplication. For polynomials $f(x)$ and $g(x)$ members of $\mathbb{R}[x]$, the products $f(x) \cdot g(x)$ and $g(x) \cdot f(x)$ are equal because the coefficients are real numbers, and we can use all the properties of real numbers under multiplication and addition (properties such as commutativity and associativity). Similarly, for $f(x)$, $g(x)$, and $h(x)$ members of $\mathbb{R}[x]$, $(f(x) \cdot g(x)) \cdot h(x) = f(x) \cdot (g(x) \cdot h(x))$. The constant polynomial 1 is an identity because $1 \cdot f(x) = f(x) \cdot 1 = f(x)$ for every $f(x) \in \mathbb{R}[x]$. Thus, $[\mathbb{R}[x], \cdot]$ is a commutative monoid. It fails to be a group because only the nonzero constant polynomials have inverses. For example, there is no polynomial $g(x)$ such that $g(x) \cdot x = x \cdot g(x) = 1$, so the polynomial x has no inverse. However, $[\mathbb{R}[x], +]$ is a commutative group.   □

7.13 Practice   (a) For $f(x), g(x), h(x) \in \mathbb{R}[x]$, write the equations saying that $\mathbb{R}[x]$ under $+$ is commutative and associative.

(b) What is the identity element in $[\mathbb{R}[x], +]$?

(c) What is the inverse of $7x^4 - 2x^3 + 4$ in $[\mathbb{R}[x], +]$?   □

Polynomials play a special part in the history of group theory (the study of groups) because much research in group theory was prompted by the very practical problem of solving polynomial equations of the form $f(x) = 0, f(x) \in \mathbb{R}[x]$. The quadratic formula provides an algorithm for finding solutions for every $f(x)$ of degree 2, and the algorithm uses only the algebraic operations of addition, subtraction, multiplication, division, and taking roots. Other such algorithms exist for polynomials of degrees 3 and 4. One of the highlights of abstract algebra is the proof that no algorithm exists that works for every $f(x)$ of degree 5 and that uses only these operations. (Notice that this statement is much stronger than simply saying that no algorithm has yet been found; it says to stop looking for one.)

The next example introduces *modular arithmetic,* an important idea in computer science because of the finite nature of a computer. (There is more on this notion later in this chapter.)

7.14 Example   Let $\mathbb{Z}_5 = \{0, 1, 2, 3, 4\}$ and define *addition modulo 5,* denoted by $+_5$, on $\mathbb{Z}_5$ by $x +_5 y = r$, where r is the remainder when $x + y$ is divided by 5. For example, $1 +_5 2 = 3$ and $3 +_5 4 = 2$. *Multiplication modulo 5* is defined by $x \cdot_5 y = r$, where r is the remainder when $x \cdot y$ is divided by 5. Thus, $2 \cdot_5 3 = 1$ and $3 \cdot_5 4 = 2$. Then $[\mathbb{Z}_5, +_5]$ is a commutative group, and $[\mathbb{Z}_5, \cdot_5]$ is a commutative monoid.   □

7.15 Practice   (a) Complete the following tables defining $+_5$ and $\cdot_5$ on $\mathbb{Z}_5$:

$+_5$	0	1	2	3	4
0	0	1	2	3	4
1	1	2	3	4	0
2	2	3	4	0	1
3	3	4	0	1	2
4	4	0	1	2	3

$\cdot_5$	0	1	2	3	4
0	0	0	0	0	0
1	0	1	2	3	4
2	0	2	4	1	3
3	0	3	1	4	2
4	0	4	3	2	1

(b) What is an identity in $[\mathbb{Z}_5, +_5]$? In $[\mathbb{Z}_5, \cdot_5]$?   0   1

(c) What is an inverse of 2 in $[\mathbb{Z}_5, +_5]$?   3

(d) Which elements in $[\mathbb{Z}_5, \cdot_5]$ have inverses? 1, 2, 3, 4   □

Notice that when we use a table to define an operation on a finite set, it is easy to check for commutativity by looking for symmetry around the main diagonal. It is also easy to find an identity element because its row looks like the top of the table and its column looks like the side. And it is easy to locate an inverse of an element. Look along the row until you find a column where the identity appears; then check to see that changing the order of the elements still gives the identity. However, associativity (or the lack of it) is not immediately apparent from the table.

As we did on $\mathbb{Z}_5$, we can define operations of **addition modulo n** and **multiplication modulo n** on the set $\mathbb{Z}_n = \{0, 1, \ldots, n - 1\}$ where n is any positive integer. Again $[\mathbb{Z}_n, +_n]$ is a commutative group and $[\mathbb{Z}_n, \cdot_n]$ is a commutative monoid.

7.16 Practice   (a) Give the table for $\cdot_6$ on $\mathbb{Z}_6$.

(b) Which elements in $[\mathbb{Z}_6, \cdot_6]$ have inverses?   ☐

The next two examples give us algebraic structures where the elements are functions.

7.17 Example   Let A be a set and consider the set S of all functions f such that $f:A \to A$. The binary operation is function composition, denoted by $\circ$. Note that S is closed under $\circ$ and that function composition is associative (see Practice 7.18 below). Thus $[S, \circ]$ is a semigroup, called the **semigroup of transformations on A**. Actually $[S, \circ]$ is a monoid because the identity function i_A that takes each member of A to itself has the property that for any $f \in S$,

$$f \circ i_A = i_A \circ f = f$$ ☐

7.18 Practice   Prove that function composition on the set S defined above is associative.   ☐

7.19 Example   Again let A be a set and consider the set S_A of all bijections f such that $f:A \to A$ (permutations of A). Bijectiveness is preserved under function composition, function composition is associative, the identity function i_A is a permutation, and for any $f \in S_A$, the inverse function f^{-1} exists and is a permutation. Furthermore,

$$f \circ f^{-1} = f^{-1} \circ f = i_A$$

Thus, $[S_A, \circ]$ is a group, called the **group of permutations on A**.   ☐

If $A = \{1, 2, \ldots, n\}$ for some positive integer n, then S_A is called the **symmetric group of degree n** and denoted by S_n. Thus, S_3, for example, is the set of all permutations on $\{1, 2, 3\}$. There are six such permutations, which we will name as follows (using the cycle notation of Section 3.2):

$$\alpha_1 = i$$
$$\alpha_2 = (1, 2)$$
$$\alpha_3 = (1, 3)$$
$$\alpha_4 = (2, 3)$$
$$\alpha_5 = (1, 2, 3)$$
$$\alpha_6 = (1, 3, 2)$$

Then $\alpha_2 \circ \alpha_3 = (1, 2) \circ (1, 3) = (1, 3, 2) = \alpha_6$.

7.20 Practice   (a) Complete the group table for $[S_3, \circ]$.

$\circ$	α_1	α_2	α_3	α_4	α_5	α_6
α_1	α_1	α_2	α_3	α_4	α_5	α_6
α_2	α_2	α_1	α_6	α_5	α_4	α_3
α_3	α_3	α_5	α_1	α_6	α_2	α_4
α_4	α_4	α_6	α	α	α	α
α_5	α_5	α	α	α	α	α
α_6	α_6	α	α	α	α	α

(b) Is $[S_3, \circ]$ a commutative group? $\square$

$[S_3, \circ]$ is our first example of a noncommutative group ($[M_2(\mathbb{Z}), \cdot]$ was a noncommutative monoid).

The next example is very simple but particularly appropriate because it appears in several areas of computer science, including formal language theory and automata theory.

7.21 Example   Let A be a finite set; its elements are called **symbols** and A itself is called an **alphabet.** A **string,** or **word,** over A is a finite sequence of symbols from A. Thus, if $A = \{a, b\}$, then *abbaa, bbbbba,* and *a* are all strings over A. The number of symbols in a string is called its **length.** We also allow a 0-length string, a string with no symbols, which is denoted by λ. We let A^* denote the set of all strings over A. The binary operation $\cdot$ is **concatenation** (or catenation), meaning juxtaposition. Therefore, *abbaa* $\cdot$ *a* gives the string *abbaaa.* The operation of concatenation on A^* is associative, and the empty string λ is an identity because for any string $x \in A^*$,

$$x \cdot \lambda = \lambda \cdot x = x$$

Therefore, $[A^*, \cdot]$ is a monoid, called the **free monoid generated by** A.

The empty string λ should not be confused with the empty set $\varnothing$; even if A itself is $\varnothing$, then $A^* = \{\lambda\}$. If A is nonempty, then whatever the size of A, A^* is a denumerable (countably infinite) set. If A contains only one element, say $A = \{a\}$, then $\lambda, a, aa, aaa, \ldots$, is an enumeration of A^*. If A contains more than one element, then a lexicographical (alphabetical) ordering can be imposed on the elements of A. An enumeration of A is then obtained by counting the empty string first, then lexicographically ordering all strings of length 1 (there is a finite number of these), then lexicographically ordering all strings of length 2 (there is a finite number of these), and so forth.   □

7.22 Practice   For $A = \{a, b\}$

    (a) Is $[A^*, \cdot]$ a commutative monoid?
    (b) Is $[A^*, \cdot]$ a group?   □

Our final example simply gives us a procedure for constructing new groups from existing groups, which will be helpful in producing groups with certain properties.

7.23 Example   Let $[G, \circ]$ and $[H, *]$ be two groups. We can form a new group whose elements come from the set $G \times H$. For the binary operation $\cdot$ of the new group, we must define $(g_1, h_1) \cdot (g_2, h_2)$. We define $\cdot$ to be a componentwise operation, acting on the first components of the ordered pairs—those from G—with the binary operation available in G and acting on the second components of the ordered pairs with the binary operation available in H. Thus

$$(g_1, h_1) \cdot (g_2, h_2) = (g_1 \circ g_2, h_1 * h_2)$$

Then $[G \times H, \cdot]$ is a group, called the **direct product** of groups G and H.   □

7.24 Practice   (a) What is $(2, 3) \cdot (5, 6)$ in the direct product of $[\mathbb{R}^+, \cdot]$ and $[\mathbb{Z}, +]$?

    (b) What is $(2, 3) \cdot (5, 6)$ in the direct product of $[\mathbb{Z}_6, +_6]$ and $[\mathbb{Z}_8, +_8]$?

    (c) What is the inverse of $(1, 4)$ in the direct product of $[\mathbb{Z}_4, +_4]$ and $[\mathbb{Z}_6, +_6]$?   □

7.25 Practice   (a) Let 1_G be an identity in the group $[G, \circ]$ and 1_H be an identity in the group $[H, *]$. Show that $(1_G, 1_H)$ is an identity in the direct product $[G \times H, \cdot]$.

    (b) For (g, h) an element of the direct product $[G \times H, \cdot]$, what is $(g, h)^{-1}$?

    (c) If $[G, \circ]$ and $[H, *]$ are both commutative groups, is $[G \times H, \cdot]$ commutative?   □

Basic Results About Groups

We will now prove some basic theorems about groups. There are hundreds of theorems about groups and many books devoted exclusively to group theory, so we are barely scratching the surface here. The results we will prove follow almost immediately from the definitions involved.

By definition, a group $[G, \cdot]$ (or a monoid) has an identity element, and we have tried to be careful to refer to *an* identity element rather than *the* identity element. However, it is legal to say *the* identity because there is only one. To prove that the identity element is unique, suppose that i_1 and i_2 are both identity elements. Then

$$i_1 = i_1 \cdot i_2 = i_2$$

7.26 Practice   Justify the above equality signs.   □

Because $i_1 = i_2$, the identity element is unique. Thus, we have proved Theorem 7.27.

7.27 Theorem   In any group (or monoid) $[G, \cdot]$, the identity element i is unique.   □

Each element x in a group $[G, \cdot]$ has an inverse element, x^{-1}. Therefore, G contains many different inverse elements, but for each x, the inverse is unique.

7.28 Practice   Prove Theorem 7.29. (Hint: assume two inverses for x, namely y and z, and let i be the identity. Then $y = y \cdot i = y(xz) = $ etc.)   □

7.29 Theorem   For each x in a group $[G, \cdot]$, x^{-1} is unique.   □

If x and y belong to a group $[G, \cdot]$, then $x \cdot y$ belongs to G and must have an inverse element in G. Naturally, we expect that inverse to have some connection with x^{-1} and y^{-1}, which we know exist in G. We can show that $(x \cdot y)^{-1} = y^{-1} \cdot x^{-1}$; thus the inverse of a product is the product of the inverses in reverse order.

7.30 Theorem   For x and y members of a group $[G, \cdot]$, $(x \cdot y)^{-1} = y^{-1} \cdot x^{-1}$.
Proof: We must show that $y^{-1} \cdot x^{-1}$ has the two properties required of $(x \cdot y)^{-1}$.

$$(x \cdot y) \cdot (y^{-1} \cdot x^{-1}) = x \cdot (y \cdot y^{-1}) \cdot x^{-1}$$
$$= x \cdot i \cdot x^{-1}$$
$$= x \cdot x^{-1}$$
$$= i$$

Similarly, $(y^{-1} \cdot x^{-1}) \cdot (x \cdot y) = i$. Notice how associativity and the meaning of i and inverses all come into play in this proof.   □

7.31 Practice   Write 10 as $7 +_{12} 3$ and use Theorem 7.30 to find $(10)^{-1}$ in the group $[\mathbb{Z}_{12}, +_{12}]$.   □

Many familiar number systems such as $[\mathbb{Z}, +]$ and $[\mathbb{R}, +]$ are groups, and we make use of group properties when we do arithmetic or algebra in these systems. In $[\mathbb{Z}, +]$, for example, if we see the equation $x + 5 = y + 5$, we conclude that $x = y$. We are making use of the right cancellation law, which, we will soon see, holds in any group.

**7.32
Definition**   A set S with a binary operation $\cdot$ satisfies the **right cancellation law** if for $x, y, z \in S$, $x \cdot z = y \cdot z$ implies $x = y$. It satisfies the **left cancellation law** if $z \cdot x = z \cdot y$ implies $x = y$.   □

Now suppose that x, y, and z are members of a group $[G, \cdot]$ and that $x \cdot z = y \cdot z$. To conclude that $x = y$, we take advantage of z^{-1}. Thus,

$$x \cdot z = y \cdot z$$

implies

$$(x \cdot z) \cdot z^{-1} = (y \cdot z) \cdot z^{-1}$$
$$x \cdot (z \cdot z^{-1}) = y \cdot (z \cdot z^{-1})$$
$$x \cdot i = y \cdot i$$
$$x = y$$

Hence, G satisfies the right cancellation law.

7.33 Practice   Show that any group $[G, \cdot]$ satisfies the left cancellation law.   □

We have proved Theorem 7.34.

7.34 Theorem   Any group $[G, \cdot]$ satisfies the left and right cancellation laws.   □

7.35 Example   We know that $[\mathbb{Z}_6, \cdot_6]$ is not a group. Here the equation

$$4 \cdot_6 2 = 1 \cdot_6 2$$

holds, but of course $4 \neq 1$.   □

Again, working in $[\mathbb{Z}, +]$, we would solve the equation $x + 6 = 13$ by adding -6 to both sides, producing a unique answer of $x = 13 + (-6) = 7$. The property of being able to solve linear equations for unique solutions holds in all groups. Consider the equation $a \cdot x = b$ in the group

$[G, \cdot]$ where a and b belong to G and x is to be found. Then $x = a^{-1} \cdot b$ is an element of G satisfying the equation. Should x_1 and x_2 both be solutions to the equation $ax = b$, then $a \cdot x_1 = a \cdot x_2$ and, by left cancellation, $x_1 = x_2$. Similarly, the unique solution to $x \cdot a = b$ is $x = b \cdot a^{-1}$.

7.36 Theorem   Let a and b be any members of a group $[G, \cdot]$. Then the linear equations $a \cdot x = b$ and $x \cdot a = b$ have unique solutions in G.   □

7.37 Practice   Solve the equation $x +_8 3 = 1$ in $[\mathbb{Z}_8, +_8]$.   □

Theorem 7.36 tells us something about tables for finite groups. As we look along row a of the table, does element b appear twice? If so, then the table says that there are two distinct elements x_1 and x_2 of the group such that $a \cdot x_1 = b$ and $a \cdot x_2 = b$. But by Theorem 7.36, this double occurrence can't happen. Thus, a given element of a finite group appears at most once in a given row of the group table. However, to complete the row, each element must appear at least once. A similar result holds for columns. Therefore, in a group table, each element appears exactly once in each row and each column. This property alone, however, is not sufficient to insure that a table represents a group; the operation must also be associative (see Exercise 19 at the end of this section).

7.38 Practice   Assume that $\circ$ is an associative binary operation on $\{1, a, b, c, d\}$. Complete the table in Figure 7.1 to define a group with identity 1.   □

$\circ$	1	a	b	c	d
1	1	a	b	c	d
a	a	b	c	d	1
b	b	c	d	1	a
c	c	d	1	a	b
d	d	1	a	b	c

Figure 7.1

If $[S, \cdot]$ is a semigroup where S is finite with n elements, then n is said to be the **order of the semigroup,** denoted by $|S|$. If S is an infinite set, the semigroup is of infinite order.

7.39 Practice   (a) Name a commutative group of order 18.

(b) Name a noncommutative group of order 6.   □

More properties of groups appear in the Exercises at the end of this section.

✔ Checklist

Definitions

> associative binary operation (*p. 282*)
> commutative binary operation (*p. 282*)
> identity element (*p. 282*)
> monoid (*p. 282*)
> semigroup (*p. 283*)
> commutative semigroup (*p. 283*)
> group (*p. 283*)
> inverse element (*p. 283*)
> commutative (abelian) group (*p. 284*)
> polynomial in *x* with real-number coefficients
>   (polynomial in *x* over $\mathbb{R}$) (*p. 284*)
> coefficient (*p. 284*)
> degree of a polynomial (*p. 284*)
> polynomial of zero degree (*p. 284*)
> addition modulo *n* (*p. 286*)
> multiplication modulo *n* (*p. 286*)
> semigroup of transformations on a set *A* (*p. 286*)
> group of permutations on a set *A* (*p. 286*)
> symmetric group of degree *n* (*p. 286*)
> symbol (*p. 287*)
> alphabet (*p. 287*)
> string (word) (*p. 287*)
> length of a string (*p. 287*)
> concatenation of strings (*p. 287*)
> free monoid generated by a set *A* (*p. 287*)
> direct product of groups (*p. 288*)
> cancellation laws (*p. 290*)
> order of a semigroup (*p. 291*)

Techniques

Test whether a given set and operation have the properties necessary to form a semigroup, monoid, or group structure.

Main Ideas

There are many instances of semigroups, monoids, and groups.

In any group structure, the identity and inverse elements are unique, cancellation laws hold, and linear equations are solvable. These and other properties of groups follow from the definitions involved.

Exercises Section 7.1

★ 1. (a) A binary operation · is defined on the set $\{a, b, c, d\}$ by the table in Figure 7.2. Is · commutative? Is · associative?

(b) Let $S = \{p, q, r, s\}$. An associative operation · is partly defined on S by the table in Figure 7.3. Complete the table to preserve associativity. Is · commutative?

·	a	b	c	d
a	a	c	d	a
b	b	c	a	d
c	c	a	b	d
d	d	b	a	c

Figure 7.2

·	p	q	r	s
p	p	q	r	s
q	q	r	s	p
r	r	s	p	q
s	s	p	q	r

Figure 7.3

2. Define binary operations on the set $\mathbb{N}$ that are:
   (a) commutative but not associative
   (b) associative but not commutative
   (c) neither associative nor commutative
   (d) both associative and commutative

★ 3. Let $A = \{1, 2\}$.
   (a) Describe the elements and write the table for the semigroup of transformations on A.
   (b) Describe the elements and write the table for the group of permutations on A.

4. Determine if the following structures $[S, \cdot]$ are semigroups, monoids, groups, or none of these. Name the identity element in any monoid or group structure.
   ★(a) $S = \mathbb{N}; x \cdot y = \min(x, y)$
   ★(b) $S = \mathbb{R}; x \cdot y = (x + y)^2$
   ★(c) $S = \{a\sqrt{2} \mid a \in \mathbb{N}\}; \cdot = $ multiplication
   ★(d) $S = \{a + b\sqrt{2} \mid a, b \in \mathbb{Z}\}; \cdot = $ multiplication
   ★(e) $S = \{a + b\sqrt{2} \mid a, b \in \mathbb{Q}, a \text{ and } b \text{ not both } 0\}; \cdot = $ multiplication
   ★(f) $S = \{1, -1, i, -i\}; \cdot = $ multiplication (where $i^2 = -1$)
   ★(g) $S = \{1, 2, 4\}; \cdot = \cdot_6$
   (h) $S = \{1, 2, 3, 5, 6, 15, 30\}; x \cdot y = $ least common multiple of x and y
   (i) $S = \mathbb{N} \times \mathbb{N}; (x_1, y_1) \cdot (x_2, y_2) = (x_1, y_2)$
   (j) $S = \mathbb{N} \times \mathbb{N}; (x_1, y_1) \cdot (x_2, y_2) = (x_1 + x_2, y_1 y_2)$
   (k) $S = $ set of even integers; $\cdot = $ addition
   (l) $S = $ set of odd integers; $\cdot = $ addition
   (m) $S = $ set of all polynomials in $\mathbb{R}[x]$ of degree ≤ 3; $\cdot = $ polynomial addition

(n) S = set of all polynomials in $\mathbb{R}[x]$ of degree ≤ 3; $\cdot$ = polynomial multiplication

(o) $S = \left\{ \begin{bmatrix} 1 & z \\ 0 & 1 \end{bmatrix} \middle| z \in \mathbb{Z} \right\}$; $\cdot$ = matrix multiplication

(p) $S = \{1, 2, 3, 4\}$; $\cdot$ = $\cdot_5$

(q) $S = \mathbb{R} - \{-1\}$; $x \cdot y = x + y + xy$

(r) $S = \{f \mid f: \mathbb{N} \to \mathbb{N}\}$; $\cdot$ = function addition, that is, $(f + g)(x) = f(x) + g(x)$

5. Write a computer program that uses the definition of a group to determine whether a given 10×10 composition table of integers from 1 to 10 represents a group.

6. In any group $[G, \cdot]$, show that
    (a) $i^{-1} = i$
    (b) $(x^{-1})^{-1} = x$ for any $x \in G$

7. (a) Show that any group of order 2 is commutative by constructing a group table on the set $\{1, a\}$ with 1 as the identity.
    (b) Show that any group of order 3 is commutative by constructing a group table on the set $\{1, a, b\}$ with 1 as the identity. (You may assume associativity.)
    (c) Show that any group of order 4 is commutative by constructing a group table on the set $\{1, a, b, c\}$ with 1 as the identity. (You may assume associativity.) There will be four such tables, but three of them are isomorphic because the elements have simply been relabeled from one to the other. Find these three groups and indicate the relabeling. Thus, there are two essentially different groups of order 4, and both of these are commutative.

★ 8. Given an equilateral triangle, six permutations can be performed on the triangle that will leave its image in the plane unchanged. Three of these permutations are clockwise rotations in the plane of 120°, 240°, and 360° about the center of the triangle; these permutations are denoted R_1, R_2, and R_3, respectively. The triangle can also be flipped about any of the axes 1, 2, and 3 (see Figure 7.4); these permutations are denoted F_1, F_2, and F_3, respectively. Composition of permutations is a binary operation on the set D_3 of all six permutations. For example, $F_3 \circ R_2 = F_2$. The set D_3 under composition is a group, called the **group of symmetries of an equilateral triangle**. Complete the group table (Figure 7.5) for $[D_3, \circ]$. What is an identity element in $[D_3, \circ]$? What is an inverse element for F_1? For R_2?

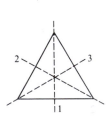

Figure 7.4

9. The set S_3, the symmetric group of degree 3, is isomorphic to D_3, the group of symmetries of an equilateral triangle (see Exercise 8). Find a bijection from the elements of S_3 to the elements of D_3 that preserves the operation. (Hint: R_1 of D_3 may be considered a permutation in S_3 sending 1 to 2, 2 to 3, and 3 to 1.)

$\circ$	R_1	R_2	R_3	F_1	F_2	F_3
R_1						
R_2						
R_3						
F_1						
F_2						
F_3	F_2					

Figure 7.5

10. Given a square, eight permutations can be performed on the square that will leave its image in the plane unchanged. Four of these are clockwise rotations in the plane of 90°, 180°, 270°, and 360° about the center of the square; these permutations are denoted R_1, R_2, R_3, and R_4, respectively. The square can also be flipped about any of the axes 1, 2, 3, or 4 (see Figure 7.6); these permutations are denoted F_1, F_2, F_3, and F_4, respectively. The composition of permutations is a binary operation on the set D_4 of all eight permutations. The set D_4 under composition is a group, the **group of symmetries of a square.** Write the group table for $[D_4, \circ]$.

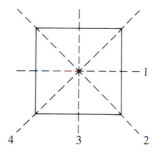

Figure 7.6

★ 11. Let $[S, \cdot]$ be a semigroup. An element $i_L \in S$ is a **left identity element** if for all $x \in S$, $i_L \cdot x = x$. An element $i_R \in S$ is a **right identity element** if for all $x \in S$, $x \cdot i_R = x$.
    (a) Show that if a semigroup $[S, \cdot]$ has both a left identity element and a right identity element, then $[S, \cdot]$ is a monoid.
    (b) Give an example of a finite semigroup with two left identities and no right identity.
    (c) Give an example of a finite semigroup with two right identities and no left identity.
    (d) Give an example of a semigroup with neither a right nor a left identity.

12. Let $[S, \cdot]$ be a monoid with identity i, and let $x \in S$. An element x_L^{-1} in S is a **left inverse** of x if $x_L^{-1} \cdot x = i$. An element x_R^{-1} in S is a **right inverse** of x if $x \cdot x_R^{-1} = i$.

  (a) Show that if every element in a monoid $[S, \cdot]$ has both a left inverse and a right inverse, then $[S, \cdot]$ is a group.

  (b) Let S be the set of all functions f such that $f:\mathbb{N} \to \mathbb{N}$. Then S under function composition is a monoid. Define a function $f \in S$ by $f(x) = 2x$, $x \in \mathbb{N}$. Then define a function $g \in S$ by

$$g(x) = \begin{cases} x/2 & \text{if } x \in \mathbb{N}, \ x \text{ even} \\ 1 & \text{if } x \in \mathbb{N}, \ x \text{ odd} \end{cases}$$

  Show that g is a left inverse for f. Also show that f has no right inverse.

  (c) Give an example of a monoid where at least one element has neither a right nor a left inverse.

★ 13. Let $[S, \cdot]$ be a semigroup. An element $0_R \in S$ is a **right zero** in the semigroup if for all $x \in S$, $x \cdot 0_R = 0_R$. An element $0_L \in S$ is a **left zero** if for all $x \in S$, $0_L \cdot x = 0_L$.

  (a) Show that if a semigroup $[S, \cdot]$ has both a left zero 0_L and a right zero 0_R, then $0_L = 0_R$; this element is called a **zero**.

  (b) Name any zero elements that exist in the semigroups of Exercise 4.

14. Let $[S, \cdot]$ be a monoid and let $x \in S$ be a right zero. If $|S| > 1$, show that x has no right inverse (see Exercises 12 and 13).

15. An element x of a semigroup $[S, \cdot]$ is **idempotent** if $x^2 = x$. Prove that a group has one and only one **idempotent** element.

16. For x a member of a semigroup $[S, \cdot]$, we can define x^n for any positive integer n by $x^1 = x$, $x^2 = x \cdot x$, and $x^n = x^{n-1} \cdot x$ for $n > 2$. Prove that in a finite group $[G, \cdot]$, for each $x \in G$ there is a positive integer k such that $x^k = i$.

★ 17. Let $[G, \cdot]$ be a group and let $x, y \in G$. Define a relation ρ on G by $x \, \rho \, y \leftrightarrow g \cdot x \cdot g^{-1} = y$ for some $g \in G$.

  (a) Show that ρ is an equivalence relation on G.

  (b) Prove that for each $x \in G$, $[x] = \{x\}$ if and only if G is commutative.

18. Let $[S, \cdot]$ be a semigroup having a left identity i_L (see Exercise 11) and the property that for every $x \in S$, x has a left inverse y such that $y \cdot x = i_L$. Prove that $[S, \cdot]$ is a group. (Hint: y also has a left inverse in S.)

19. Show that if $[S, \cdot]$ is a semigroup in which the linear equations $a \cdot x = b$ and $x \cdot a = b$ are solvable for any $a, b \in S$, then $[S, \cdot]$ is a group. (Hint: use Exercise 18.)

20. Prove that a finite semigroup satisfying the left and right cancellation laws is a group. (Hint: use Exercise 18.)

21. Show that a group $[G, \cdot]$ is commutative if and only if $(x \cdot y)^2 = x^2 \cdot y^2$ for each $x, y \in G$.

22. Show that a group $[G, \cdot]$ in which $x \cdot x = i$ for each $x \in G$ is commutative.

23. Prove that in the commutative monoid $[\mathbb{Z}_n, \cdot_n]$, every nonzero element has an inverse if and only if n is prime. (Hint: use the fact that if x and n are relatively prime, then there exist integers a and b with $ax + bn = 1$; then consider whether $a \in \mathbb{Z}_n$.)

SECTION 7.2   SUBSTRUCTURES

We know what structures are and we know what subsets are, so it should not be hard to guess what a substructure is. However, we will look at an example before we give any definitions. Addition is an associative binary operation on $\mathbb{R}^+$; thus $[\mathbb{R}^+, +]$ is a semigroup. Now let A be any nonempty subset of $\mathbb{R}^+$. For any x and y in A, x and y are also in $\mathbb{R}^+$, so that $x + y$ exists and is unique. The set A "inherits" a well-defined operation, $+$, from $[\mathbb{R}^+, +]$. Also, the associative property of $+$ is inherited because for $x, y, z \in A$, $x, y, z \in \mathbb{R}^+$, and the equation $(x + y) + z = x + (y + z)$ is true. Perhaps A under the inherited operation has all of the structure of $[\mathbb{R}^+, +]$ and is itself a semigroup. The only remaining property necessary for $[A, +]$ to be a semigroup is closure of A under $+$. This property is not inherited and depends upon the set A. For $E = \{2, 4, 6, \ldots\}$, closure holds and $[E, +]$ is a semigroup; for $O = \{1, 3, 5, \ldots\}$, closure does not hold and $[O, +]$ is not a semigroup.

Subsemigroups

7.40
Definition

Let $[S, \cdot]$ be a semigroup and $A \subseteq S$. Then $[A, \cdot]$ is a **subsemigroup** of $[S, \cdot]$ if $[A, \cdot]$ is itself a semigroup. ☐

For A to be a subsemigroup of S, A must be nonempty. Then, just as in the example above, the only test necessary to determine whether A is a subsemigroup is closure of A under the inherited operation.

7.41 Theorem

For $[S, \cdot]$ a semigroup and $A \subseteq S$, $A \neq \varnothing$, $[A, \cdot]$ is a subsemigroup of $[S, \cdot]$ if and only if A is closed under $\cdot$. ☐

7.42 Example   Let A be the nonempty subset of $M_2(\mathbb{Z})$ consisting of all matrices of the form

$$\begin{bmatrix} a & b \\ 0 & c \end{bmatrix}$$

where $a, b, c \in \mathbb{Z}$. Then $[A, +]$ is a subsemigroup of the semigroup $[M_2(\mathbb{Z}), +]$ because matrix addition of any two elements of A gives an element of A.   □

7.43 Practice   Which of the following are subsemigroups of the direct product of $[\mathbb{N}, \cdot]$ and $[\mathbb{N}, \cdot]$?

(a) $[A, \cdot]$ where $A = \{(x, y) \mid (x, y) \in \mathbb{N} \times \mathbb{N}$ and $y = 1\}$
(b) $[B, \cdot]$ where $B = \{(x, y) \mid (x, y) \in \mathbb{N} \times \mathbb{N}$ and $x \leq 2\}$
(c) $[C, \cdot]$ where $C = \{(x, y) \mid (x, y) \in \mathbb{N} \times \mathbb{N}$ and $x + y \leq 1\}$ □

7.44 Practice   If $[A, \cdot]$ is a subsemigroup of a commutative semigroup $[S, \cdot]$, will $[A, \cdot]$ be commutative?   □

7.45 Example   Let $I = \{f \mid f : \mathbb{N} \to \mathbb{N}$ and $f(x) \geq x\}$. Then $[I, \circ]$ is a subsemigroup of the semigroup of transformations on $\mathbb{N}$. To prove this, we must show closure of I under function composition. Let $f, g \in I$; then $(g \circ f)(x) = g(f(x)) \geq f(x) \geq x$, and $g \circ f \in I$.   □

We should mention some rather confusing terminology. The set of *all* functions on a set A into itself under function composition is called, as we said in the last section, *the semigroup of transformations on A*. Any subsemigroup of this (such as $[I, \circ]$ in the above example) is called **a transformation semigroup.** The distinction is that *a* transformation semigroup may not include all the functions on A into A, but *the* semigroup of transformations on A does. Transformation semigroups will arise in connection with finite-state machines in Chapter 10.

Submonoids

We would expect any subset of a monoid that is itself a monoid under the inherited operation to be called a submonoid. It is possible to live with this definition, but life is slightly less complicated if we require that the identity element of the submonoid agree with the identity element of the original monoid. Thus, we make the following definition.

7.46 Definition   Let $[S, \cdot]$ be a monoid with identity i, and $A \subseteq S$. Then $[A, \cdot]$ is a **submonoid** of $[S, \cdot]$ if $[A, \cdot]$ is itself a monoid with identity i.   □

To test whether $[A, \cdot]$ is a submonoid, we test for closure and whether $i \in A$ ($i \in A$ guarantees that $A \neq \varnothing$).

7.47 Example
(a) $[M_2^D(\mathbb{Z}), \cdot]$ is a commutative submonoid of the noncommutative monoid $[M_2(\mathbb{Z}), \cdot]$ (closure holds and

$$\begin{bmatrix} 1 & 0 \\ 0 & 1 \end{bmatrix}$$

belongs to $M_2^D(\mathbb{Z})$).

(b) The set of all polynomials in x over $\mathbb{R}$ with constant term equal to 1 forms a submonoid of the monoid $[\mathbb{R}[x], \cdot]$ (closure holds and the constant polynomial 1 belongs to the set). $\square$

7.48 Practice
Which of the sets on the left form submonoids of the monoids on the right?

(a) $\{0, 2, 4\}$; $[\mathbb{Z}_6, +_6]$
(b) $\{1, 2, 4\}$; $[\mathbb{Z}_6, \cdot_6]$
(c) $\{(0, x) | x \in \mathbb{N}\}$; direct product of $[\mathbb{N}, \cdot]$ and $[\mathbb{N}, \cdot]$ $\square$

Subgroups

Now for the last substructure. The definition of a subgroup, for reasons we shall see immediately, does not have to include a requirement about the identity.

7.49 Definition
Let $[G, \cdot]$ be a group and $A \subseteq G$. Then $[A, \cdot]$ is a **subgroup** of $[G, \cdot]$ if $[A, \cdot]$ is itself a group. $\square$

Suppose $[A, \cdot]$ is a subgroup of $[G, \cdot]$ and that the identity of $[A, \cdot]$ is i_A and the identity of $[G, \cdot]$ is i_G. We want to show that $i_A = i_G$. Notice that this equation does *not* follow from the uniqueness of a group identity because the element i_A, as far as we know, may not be an identity for all of G, and we cannot yet say that i_G is an element of A. However, $i_A = i_A \cdot i_A$ because i_A is the identity for $[A, \cdot]$, and $i_A = i_A \cdot i_G$ because i_G is the identity for $[G, \cdot]$. Because of the left cancellation law holding in the group $[G, \cdot]$, it follows that $i_A = i_G$. Hence, we did not need to specify in the definition that the subgroup have the same identity as the group. To test whether $[A, \cdot]$ is a subgroup of $[G, \cdot]$, we make sure $[A, \cdot]$ is a submonoid and then test for the one additional property necessary to make $[A, \cdot]$ a subgroup.

7.50 Practice
Complete Theorem 7.51. $\square$

7.51 Theorem
For $[G, \cdot]$, a group with identity i and $A \subseteq G$, $[A, \cdot]$ is a subgroup of $[G, \cdot]$ if

(a) _A is closed under ._
(b) _$i \in A$_
(c) _Every $x \in A$ has an inverse element in A_ $\square$

7.52 Example   (a) $[\mathbb{Z}, +]$ is a subgroup of the group $[\mathbb{R}, +]$.

(b) $[\{1, 4\}, \cdot_5]$ is a subgroup of the group $[\{1, 2, 3, 4\}, \cdot_5]$ (closure holds; $1 \in \{1, 4\}$, $1^{-1} = 1$, $4^{-1} = 4$).   □

7.53 Practice   (a) Show that $[\{0, 2, 4, 6\}, +_8]$ is a subgroup of the group $[\mathbb{Z}_8, +_8]$.

(b) Show that $[\{1, 2, 4\}, \cdot_7]$ is a subgroup of the group

$$[\{1, 2, 3, 4, 5, 6\}, \cdot_7].$$ □

If $[G, \cdot]$ is a group with identity i, then it is true that $[\{i\}, \cdot]$ and $[G, \cdot]$ are subgroups of $[G, \cdot]$. These somewhat trivial subgroups of $[G, \cdot]$ are called **improper subgroups.** Any other subgroups of $[G, \cdot]$ are **proper subgroups.**

7.54 Practice   Find all the proper subgroups of S_3, the symmetric group of degree 3. (You can find them by looking at the group table, see Practice 7.20.)   □

Another point of confusing terminology: the set of *all* bijections on a set A into itself under function composition is called *the group of permutations on A,* and any subgroup of this set (such as those in Practice 7.54) is called **a permutation group.** Again, the distinction is that *the* group of permutations on a set A includes all bijections on A into itself, but *a* permutation group may not. Permutation groups are of particular importance, not only because they were the first groups to be studied, but also because they are, if we consider isomorphic structures to be the same, the only groups. We will see this result of Arthur Cayley's in the next section.

There is an interesting subgroup we can always find in the symmetric group S_n for $n > 1$. We know that every member of S_n can be written as a composition of cycles, but it is also true that each cycle can be written as the composition of cycles of length 2, called **transpositions.** In S_7, for example, $(5, 1, 7, 2, 3, 6) = (5, 6) \circ (5, 3) \circ (5, 2) \circ (5, 7) \circ (5, 1)$, or $(1, 5) \circ (1, 6) \circ (1, 3) \circ (1, 2) \circ (2, 4) \circ (1, 7) \circ (4, 2)$. For any $n > 1$, the identity permutation i in S_n can be written as $i = (a, b) \circ (a, b)$ for any two elements a and b in the set $\{1, 2, \ldots, n\}$. This equation also shows that the inverse of the transposition (a, b) in S_n is (a, b). Now we borrow (without proof) one more fact: even though there are various ways to write a cycle as the composition of transpositions, for a given cycle the number of transpositions will always be even or it will always be odd. Consequently, we classify any permutation in S_n, $n > 1$, as **even** or **odd** according to the number of transpositions in any representation of that permutation. For example, in S_7, $(5, 1, 7, 2, 3, 6)$ is odd. If we denote by A_n the set of all even permutations in S_n, then A_n determines a subgroup of $[S_n, \circ]$. The composition of even permutations produces an even permutation, and $i \in A_n$. If $\alpha \in A_n$ and α as a product of transpositions is $\alpha = \alpha_1 \circ \alpha_2 \circ \cdots \circ \alpha_k$, then $\alpha^{-1} = \alpha_k^{-1} \circ \alpha_{k-1}^{-1} \circ \cdots \circ \alpha_1^{-1}$. Each inverse of a transposition is a transposition, so α^{-1} is also even.

The order of the group $[S_n, \circ]$ (the number of elements) is $n!$. What is the order of the subgroup $[A_n, \circ]$? We might expect half the permutations in S_n to be even and half to be odd. Indeed, this is the case. If we let O_n denote the set of odd permutations in S_n (which is not closed under function composition), then the mapping $f : A_n \to O_n$ defined by $f(\alpha) = \alpha \circ (1, 2)$ is a bijection.

7.55 Practice   Prove that $f : A_n \to O_n$ given by $f(\alpha) = \alpha \circ (1, 2)$ is one-to-one and onto.   □

Because there is a bijection from A_n onto O_n, each set has the same number of elements. But $A_n \cap O_n = \varnothing$ and $A_n \cup O_n = S_n$, so $|A_n| = |S_n|/2 = n!/2$.

7.56 Theorem   For $n \in \mathbb{N}$, $n > 1$, the set A_n of even permutations determines a subgroup, called the **alternating group**, of $[S_n, \circ]$ of order $n!/2$.   □

We encounter another interesting situation when we consider subgroups of the group $[\mathbb{Z}, +]$. For n any fixed element of $\mathbb{N}$, the set $n\mathbb{Z}$ is defined as the set of all integral multiples of n; $n\mathbb{Z} = \{nz \mid z \in \mathbb{Z}\}$. Thus, for example, $3\mathbb{Z} = \{0, \pm 3, \pm 6, \pm 9, \ldots\}$.

7.57 Practice   Show that for any $n \in \mathbb{N}$, $[n\mathbb{Z}, +]$ is a subgroup of $[\mathbb{Z}, +]$.   □

Not only is $[n\mathbb{Z}, +]$ a subgroup of $[\mathbb{Z}, +]$ for any fixed n, but sets of the form $n\mathbb{Z}$ are the only subgroups of $[\mathbb{Z}, +]$. To illustrate, let $[S, +]$ be any subgroup of $[\mathbb{Z}, +]$. If $S = \{0\}$, then $S = 0\mathbb{Z}$. If $S \neq \{0\}$, let m be a member of S, $m \neq 0$. Either m is positive or, if m is negative, $-m \in S$ and $-m$ is positive. The subgroup S, therefore, contains at least one positive integer. Let n be the smallest positive integer in S. We will now see that $S = n\mathbb{Z}$.

First, since 0 and $-n$ are members of S and S is closed under $+$, $n\mathbb{Z} \subseteq S$. To obtain inclusion in the other direction, let $s \in S$. Now we divide the integer s by the integer n to get an integer quotient q and an integer remainder r with $0 \leq r < n$. Thus, $s = nq + r$. Solving for r, $r = s + (-nq)$. But $nq \in S$, therefore $-nq \in S$, and $s \in S$, so by closure of S under $+$, $r \in S$. If r is positive, we have a contradiction of the definition of n as the smallest positive number in S. Therefore, $r = 0$ and $s = nq + r = nq$. We now have $S \subseteq n\mathbb{Z}$, and thus $S = n\mathbb{Z}$, which completes the proof of Theorem 7.58.

7.58 Theorem   Subgroups of the form $[n\mathbb{Z}, +]$ for $n \in \mathbb{N}$ are the only subgroups of $[\mathbb{Z}, +]$.   □

Our last example of a subgroup concerns the direct product of two groups $[G, \cdot]$ and $[H, \cdot]$ and shows why the direct product is useful for producing groups with certain properties. Let i_G denote the identity of G, and i_H the identity of H. Then $[G \times \{i_H\}, \cdot]$ is a subgroup of the group $[G \times H, \cdot]$ because if (g_1, i_H) and (g_2, i_H) are two members of $G \times \{i_H\}$, then $(g_1, i_H) \cdot (g_2, i_H) = (g_1 \cdot g_2, i_H \cdot i_H) = (g_1 \cdot g_2, i_H) \in G \times \{i_H\}$, and closure holds. The identity of $G \times H$, (i_G, i_H), belongs to $G \times \{i_H\}$, and for $(g, i_H) \in G \times \{i_H\}$, its inverse is (g^{-1}, i_H), which also belongs to $G \times \{i_H\}$. Similarly, $[\{i_G\} \times H, \cdot]$ is a subgroup of $[G \times H, \cdot]$.

There is an obvious bijection from G onto $G \times \{i_H\}$ given by $f(g) = (g, i_H)$. In fact, this bijection preserves the operation. Thus, for g_1 and g_2 members of G,

$$f(g_1 \cdot g_2) = (g_1 \cdot g_2, i_H) = (g_1, i_H) \cdot (g_2, i_H) = f(g_1) \cdot f(g_2)$$

Figure 7.7 illustrates this equation by a commutative diagram. It says that we can either operate and then map or map and then operate and still obtain the same result. Thus, G and $G \times \{i_H\}$ are isomorphic and essentially the same, except for relabeling. Any properties of $[G, \cdot]$ relating to how elements act under $\cdot$ will carry over to the mirror image $G \times \{i_H\}$. Therefore the direct product $[G \times H, \cdot]$ has a subgroup that acts like G. Let's make use of this fact to force the construction of an infinite noncommutative group, something we have not yet seen.

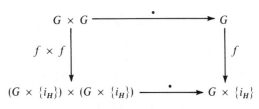

Figure 7.7

7.59 Example Take the noncommutative group $[S_3, \circ]$ ($\alpha_2 \circ \alpha_3 \neq \alpha_3 \circ \alpha_2$) and form its direct product with the group $[\mathbb{Z}, +]$. Because $\mathbb{Z}$ is infinite, $S_3 \times \mathbb{Z}$ is infinite. The subgroup $[S_3 \times \{0\}, \cdot]$ of $[S_3 \times \mathbb{Z}, \cdot]$ is structurally identical to $[S_3, \circ]$, so it will be noncommutative. In particular, $(\alpha_2, 0)$ and $(\alpha_3, 0)$ are members of $[S_3 \times \{0\}, \cdot]$, and thus of $[S_3 \times \mathbb{Z}, \cdot]$, and they do not commute. The group $[S_3 \times \mathbb{Z}, \cdot]$ is thus an infinite noncommutative group. $\square$

✔ Checklist

Definitions

subsemigroup (*p. 297*)
transformation semigroup (*p. 298*)
submonoid (*p. 298*)

subgroup (*p. 299*)
improper subgroup (*p. 300*)
proper subgroup (*p. 300*)
permutation group (*p. 300*)
transposition (*p. 300*)
even and odd permutations (*p. 300*)
alternating group (*p. 301*)

Techniques

Test whether a given subset of a semigroup, monoid, or group is a subsemigroup, submonoid, or subgroup.

Main Ideas

Subsets of a given algebraic structure may themselves be structures of the same type under the inherited operation.

The only subgroups of the group $[\mathbb{Z}, +]$ are of the form $[n\mathbb{Z}, +]$, where $n\mathbb{Z}$ is the set of all integral multiples of a fixed $n \in \mathbb{N}$.

Exercises Section 7.2

1. Let $A = \{p, q, r\}$. Show that the set of all strings over A with an even number of q's is a submonoid of the free monoid $[A^*, \cdot]$.

2. In each case, decide whether the structure on the left is a subgroup of the group on the right. If not, why not? (Note: here S^* denotes $S - \{0\}$.)

   ★ (a) $[\mathbb{Z}_5^*, \cdot_5]$; $[\mathbb{Z}_5, +_5]$

   ★ (b) $[P, +]$; $[\mathbb{R}[x], +]$ where P is the set of all polynomials in x over $\mathbb{R}$ of degree ≥ 3

   ★ (c) $[\mathbb{Z}^*, \cdot]$; $[\mathbb{Q}^*, \cdot]$

   (d) $[A, \circ]$; $[S, \circ]$ where S is the set of all bijections on $\mathbb{N}$ and A is the set of all bijections on $\mathbb{N}$ mapping 3 to 3

   (e) $[\mathbb{Z}, +]$; $[M_2(\mathbb{Z}), +]$

   (f) $[K, +]$; $[\mathbb{R}[x], +]$ where K is the set of all polynomials in x over $\mathbb{R}$ of degree $\leq k$ for some fixed k

   (g) $[\{0, 3, 6\}, +_8]$; $[\mathbb{Z}_8, +_8]$

★ 3. Find all the distinct subgroups of $[\mathbb{Z}_{12}, +_{12}]$.

4. Show that the sets $\{R_1, R_2, R_3, R_4\}$, $\{R_4, F_1\}$, $\{R_4, F_2\}$, $\{R_4, F_3\}$, and $\{R_4, F_4\}$ are subgroups of the group of symmetries of a square (see Exercise 10 of Section 5.1).

5. (a) Show that the subset

$$\alpha_1 = i$$
$$\alpha_2 = (1, 2) \circ (3, 4)$$
$$\alpha_3 = (1, 4) \circ (2, 3)$$
$$\alpha_4 = (1, 3) \circ (2, 4)$$

forms a subgroup of the symmetric group S_4.

(b) Show that the subset

$$\alpha_1 = i$$
$$\alpha_2 = (1, 2, 3, 4)$$
$$\alpha_3 = (1, 3) \circ (2, 4)$$
$$\alpha_4 = (1, 4, 3, 2)$$
$$\alpha_5 = (1, 2) \circ (3, 4)$$
$$\alpha_6 = (1, 4) \circ (2, 3)$$
$$\alpha_7 = (2, 4)$$
$$\alpha_8 = (1, 3)$$

forms a subgroup of the symmetric group S_4.

6. Let $A = \{0, 1\}$ and consider $[A^*, \cdot]$, the free monoid generated by A.
   (a) Let $R = \{x \in A^* | x$ has 1 as the rightmost symbol$\}$. Show that $[R, \cdot]$ is a subsemigroup of $[A^*, \cdot]$.
   (b) Describe the members of the smallest submonoid of $[A^*, \cdot]$ that contains 0 and 001.

★ 7. Find the elements of the alternating group A_4.

8. Let $A = R - \{0, 1\}$.
   (a) Show that $f(x) = 1 - x$ and $g(x) = 1/x$ are members of the group of permutations on A.
   (b) Find the elements of the smallest permutation group on A that contains f and g, and write the group table.

★ 9. (a) Let $[G, \cdot]$ be a group and let $[S, \cdot]$ and $[T, \cdot]$ be subgroups of $[G, \cdot]$. Show that $[S \cap T, \cdot]$ is a subgroup of $[G, \cdot]$.
   (b) Will $[S \cup T, \cdot]$ be a subgroup of $[G, \cdot]$? Prove or give a counterexample.

✓ ★ 10. In any semigroup $[S, \cdot]$, an element $a \in S$ with $a^2 = a$ is called **idempotent.** Prove that the set of idempotent elements in a commutative monoid forms a submonoid.

11. Let $[G, \cdot]$ be a commutative group with subgroups $[S, \cdot]$ and $[T, \cdot]$. Let $ST = \{s \cdot t | s \in S, t \in T\}$. Show that $[ST, \cdot]$ is a subgroup of $[G, \cdot]$.

★ 12. Let $[G, \cdot]$ be a commutative group with identity i. For a fixed positive integer k, let $B_k = \{x \mid x \in G, x^k = i\}$. Show that $[B_k, \cdot]$ is a subgroup of $[G, \cdot]$.

13. Let $[G, \cdot]$ be a commutative group with identity i. Let $A = \{x \in G \mid x^n = i$ for some positive integer $n\}$. Show that $[A, \cdot]$ is a subgroup of $[G, \cdot]$.

14. For any group $[G, \cdot]$, the **center** of the group is $A = \{x \in G \mid x \cdot g = g \cdot x$ for all $g \in G\}$.
    (a) Prove that $[A, \cdot]$ is a subgroup of $[G, \cdot]$.
    (b) Find the center of the group of symmetries of an equilateral triangle, $[D_3, \circ]$ (see Exercise 8 of Section 7.1).
    (c) Show that G is commutative if and only if $G = A$.
    (d) Let x and y be members of G with $x \cdot y^{-1} \in A$. Show that $x \cdot y = y \cdot x$.

★ 15. Let $[G, \cdot]$ be a group and $A = \{(x, x) \mid x \in G\}$. Show that $[A, \cdot]$ is a subgroup of $[G \times G, \cdot]$.

16. (a) Give an example of a subsemigroup of a group that is not a subgroup.
    (b) Let $[S, \cdot]$ be a subsemigroup of a finite group $[G, \cdot]$. Show that S is a subgroup.

17. (a) Let S_A denote the group of permutations on a set A, and let a be a fixed element of A. Show that the set H_a of all permutations in S_A leaving a fixed forms a subgroup of S_A.
    (b) If A has n elements, what is $|H_a|$?

18. (a) Let $[G, \cdot]$ be a group and $A \subseteq G$, $A \neq \varnothing$. Show that $[A, \cdot]$ is a subgroup of $[G, \cdot]$ if for each $x, y \in A$, $x \cdot y^{-1} \in A$. This subgroup test is sometimes more convenient to use than Theorem 7.51.
    (b) Use the test of part (a) to work Exercise 12 above.

19. (a) Let $[G, \cdot]$ be any group with identity i. For a fixed $a \in G$, a^0 denotes i and a^{-n} means $(a^n)^{-1}$. Let $A = \{a^z \mid z \in \mathbb{Z}\}$. Show that $[A, \cdot]$ is a subgroup of G.
    (b) The group $[G, \cdot]$ is a **cyclic group** if for some $a \in G$, $A = \{a^z \mid z \in \mathbb{Z}\}$ is the entire group G. In this case, a is a **generator** of $[G, \cdot]$. For example, 1 is a generator of the group $[\mathbb{Z}, +]$; remember that the operation is addition. Thus, $1^0 = 0$; $1^1 = 1$; $1^2 = 1 + 1 = 2$; $1^3 = 1 + 1 + 1 = 3, \ldots$; $1^{-1} = (1)^{-1} = -1$; $1^{-2} = (1^2)^{-1} = -2$; $1^{-3} = (1^3)^{-1} = -3, \ldots$. Every integer can be written as an integral "power" of 1, and $[\mathbb{Z}, +]$ is cyclic with generator 1. Show that the group $[\mathbb{Z}_7, +_7]$ is cyclic with generator 2.
    (c) Show that 5 is also a generator of the cyclic group $[\mathbb{Z}_7, +_7]$.

(d) Let $\mathbb{Z}_{11}^* = \mathbb{Z}_{11} - \{0\}$. Then $[\mathbb{Z}_{11}^*, \cdot_{11}]$ is a cyclic group; show that 2 is a generator.

(e) Show that 3 is a generator of the cyclic group $[\mathbb{Z}_4, +_4]$.

20. Let $[G, \cdot]$ be a cyclic group with generator a (see Exercise 19). Show that G is commutative.

21. Let $[G, \cdot]$ be a cyclic group (see Exercise 19). Show that every subgroup of G is cyclic.

SECTION 7.3   MORPHISMS—SIMULATION II

Homomorphisms

The ideas of homomorphism and isomorphism were introduced in Chapter 5. Now we want to apply these ideas to the algebraic structures of semigroups, monoids, and groups. Recall that if A and B are two instances of a structure, a homomorphism from A to B is a function $f: A \rightarrow B$ that *preserves the operations* of A in B. When we are working with particular structures, we express the property of preserving an operation by means of an equation. In effect, the equation always says, "operate and then map, or map and then operate," and it corresponds to a commutative diagram. Let's write the equation defining a homomorphism from one semigroup to another.

7.60 *Practice*   Complete Definition 7.61.   □

7.61   Let $[S, \cdot]$ and $[T, +]$ be semigroups. A function $f: S \rightarrow T$ is a **homo-**
Definition   **morphism** from $[S, \cdot]$ to $[T, +]$ if for all $x, y \in S$, $f(x \cdot y) = f(x) + f(y)$.   □

The commutative diagram corresponding to the equation $f(x \cdot y) = f(x) + f(y)$ appears in Figure 7.8. The left half of the equation, $f(x \cdot y)$, says "operate and then map" and traces the top and right paths of the diagram. The right half of the equation, $f(x) + f(y)$, says "map and then operate" and traces the left and bottom paths of the diagram. Notice that the operation on x and y (in S) is $\cdot$, but the operation on $f(x)$ and $f(y)$ (in

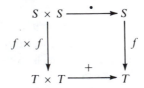

Figure 7.8

T) is +. We used two different symbols to emphasize this distinction. When working with homomorphisms, you must pay attention to the structure in which you are working and what the appropriate operation for that structure is.

7.62 Example   Let $\mathbb{N}^+$ denote the positive integers; then $[\mathbb{N}^+, +]$ is a semigroup. Let E be the set of positive even integers; then $[E, \cdot]$ is a semigroup. The function $f: \mathbb{N}^+ \to E$ defined by $f(x) = 2^x$ is a homomorphism. To prove this, note that for x and y in $\mathbb{N}^+$,

$$f(x + y) = 2^{x+y} = 2^x \cdot 2^y = f(x) \cdot f(y)$$

(Again, we must keep track of the appropriate operations. In comparing this example with the general definition, $[\mathbb{N}^+, +]$ corresponds to $[S, \cdot]$, so the generic $\cdot$ of the definition becomes + for this case. The semigroup $[E, \cdot]$ corresponds to $[T, +]$, so the generic + of the definition becomes $\cdot$.)   □

Because monoids and groups are also semigroups, Definition 7.61 applies equally well to monoids or groups.

7.63 Example   Consider the semigroup (actually a group) $[\mathbb{R}, +]$. The function $f: \mathbb{R} \to \mathbb{R}$ given by $f(x) = x^2$ is not a homomorphism. The required equation $f(x + y) = f(x) + f(y)$ becomes $(x + y)^2 = x^2 + y^2$, which is true only if $x = 0$ or $y = 0$.   □

7.64 Practice   In each case, decide whether the given function is a homomorphism from the semigroup on the left to the one on the right.

(a) $[\mathbb{N}, +]$; $[2\mathbb{N}, +]$ (where $2\mathbb{N} = \{2n \mid n \in \mathbb{N}\}$); $f(x) = 2x$
(b) $[\mathbb{N}, +]$, $[2\mathbb{N}, \cdot]$; $f(x) = 2x$
(c) $[M_2(\mathbb{Z}), +]$, $[\mathbb{Z}, +]$; $f\left(\begin{bmatrix} a & b \\ c & d \end{bmatrix}\right) = a$
(d) $[\mathbb{N}^2[x], +]$ (where $\mathbb{N}^2[x] = \{a_2x^2 + a_1x + a_0 \mid a_2, a_1, a_0 \in \mathbb{N}\}$), $[\mathbb{N}, +]$; $f(a_2x^2 + a_1x + a_0) = a_2$ □

Preserving Properties

Suppose that f is a homomorphism from a semigroup (or monoid or group) $[S, \cdot]$ to a semigroup (or monoid or group) $[T, +]$. The range of f, $f(S)$, is a subset of T. A homomorphism need not be an onto mapping, so we may have $f(S) \subset T$. The definition of a homomorphism requires that f preserve in its range the effect of the operation $\cdot$ in $[S, \cdot]$. Consequently, many of the properties of $[S, \cdot]$ are also preserved in the range, for example, the structure of $[S, \cdot]$. Thus, if $[S, \cdot]$ is a semigroup, then $f(S)$ is a semigroup under +. To prove this, we must verify closure (associativity is inherited from $[T, +])$. Let $f(x)$ and $f(y)$ belong to $f(S)$. We must prove $f(x) + f(y)$ belongs to $f(S)$, that is, that $f(x) + f(y)$ is $f(s)$ for some $s \in S$. However, $f(x) + f(y) = f(x \cdot y)$, and because x and y belong to S and S is closed under $\cdot$, $x \cdot y \in S$. This procedure verifies closure, and $[f(S), +]$ is a semigroup, a subsemigroup of $[T, +]$.

7.65 Practice   Let $[S, \cdot]$ and $[T, +]$ be monoids with identities i_S and i_T, respectively. Let f be a homomorphism from S to T. Show that $f(i_S)$ is an identity for $f(S)$ under $+$.   □

Now we know that if f is a homomorphism from a monoid $[S, \cdot]$ to a monoid $[T, +]$, the image $f(S)$ is a monoid under $+$. Note, however, that $[f(S), +]$ may not be a submonoid of $[T, +]$ because $f(i_S)$ may differ from i_T.

7.66 Practice   Let $[S, \cdot]$ and $[T, +]$ be groups with identities i_S and i_T, respectively. Let f be a homomorphism from S to T. For $f(x) \in f(S)$, show that

$$f(x^{-1}) + f(x) = f(x) + f(x^{-1}) = f(i_S)$$

(Note that $x \in S$, a group, so x^{-1} exists.)   □

When we have a homomorphism from a group $[S, \cdot]$ to a group $[T, +]$, we know that $[f(S), +]$ is a semigroup; from Practice 7.65 $f(i_S)$ is an identity for $f(S)$, and from Practice 7.66 every element in $f(S)$ has an inverse element (with respect to the identity $f(i_S)$) in $f(S)$. Thus, the range $f(S)$ is a subgroup of $[T, +]$, and it also follows that $f(i_S) = i_T$. We summarize this information in the following theorem.

7.67 Theorem   (a) Let f be a homomorphism from a semigroup $[S, \cdot]$ to a semigroup $[T, +]$. Then the range $f(S)$ is a semigroup under $+$.

(b) Let f be a homomorphism from a monoid $[S, \cdot]$ to a monoid $[T, +]$. Then the range $f(S)$ is a monoid under $+$.

(c) Let f be a homomorphism from a group $[S, \cdot]$ with identity i_S to a group $[T, +]$ with identity i_T. Then the range $f(S)$ is a group under $+$. Further, $f(i_S) = i_T$ and $f(x^{-1}) = -f(x)$ for all $x \in S$.   □

Theorem 7.67 says that a homomorphism preserves the structure of the domain in the range. Other properties of the domain structure relating entirely to the behavior of elements under the binary operation, called **algebraic properties,** are also preserved.

7.68 Practice   Let f be a homomorphism from a semigroup $[S, \cdot]$ to a semigroup $[T, +]$. Show that if $[S, \cdot]$ is commutative, then so is the semigroup $[f(S), +]$.   □

Nonalgebraic properties need not be preserved by a homomorphism. Thus, $f: \mathbb{Z} \to \{0\}$ given by $f(z) = 0$ for all $z \in \mathbb{Z}$ is a homomorphism from the infinite group $[\mathbb{Z}, +]$ to the 1-element group $[\{0\}, +]$. The order of a group is an example of a nonalgebraic property.

Homomorphic Images and Simulation

Because a homomorphism preserves algebraic properties, we can begin to think of one instance of a structure simulating another instance of that structure. A homomorphism from $[S, \cdot]$ to $[T, +]$ need not be onto or one-to-one. As we add these conditions one at a time to the homomorphism, we see what sorts of simulations are possible. In the following discussion, $[S, \cdot]$ and $[T, +]$ are two semigroups, two monoids, or two groups.

Suppose f is a homomorphism from S to T and that f is onto, that is, $f(S) = T$. Then T is called the **homomorphic image** of S under f. We know that the algebraic properties of S must also appear in T. From our discussion of Simulation II at the end of Chapter 5, we expect that S will simulate T and T will imperfectly simulate S. We will consider this situation more thoroughly in the next section; for now, we will simply look at another example.

7.69 Example   Let $f: \mathbb{Z} \to \mathbb{Z}_4$ be defined by $f(x) = x \cdot_4 1$. (Here $x \cdot_4 1$ denotes the *nonnegative* remainder when x is divided by 4.) Then we claim that f is a homomorphism from the group $[\mathbb{Z}, +]$ onto the group $[\mathbb{Z}_4, +_4]$. Clearly, f is onto. For $x, y \in \mathbb{Z}$, $f(x + y) = (x + y) \cdot_4 1 = x \cdot_4 1 +_4 y \cdot_4 1 = f(x) +_4 f(y)$; so f is a homomorphism.

To simulate $[\mathbb{Z}_4, +_4]$ by $[\mathbb{Z}, +]$, we use equivalence classes of $\mathbb{Z}$. We define a relation ρ on $\mathbb{Z}$ by $x \,\rho\, y \leftrightarrow f(x) = f(y) \leftrightarrow x \cdot_4 1 = y \cdot_4 1$. Then ρ is an equivalence relation on $\mathbb{Z}$ and partitions $\mathbb{Z}$ into the equivalence classes

$$[0] = \{\ldots, -12, -8, -4, 0, 4, 8, 12, \ldots\}$$
$$[1] = \{\ldots, -11, -7, -3, 1, 5, 9, 13, \ldots\}$$
$$[2] = \{\ldots, -10, -6, -2, 2, 6, 10, 14, \ldots\}$$
$$[3] = \{\ldots, -9, -5, -1, 3, 7, 11, 15, \ldots\}$$

(ρ is the relation of *congruence modulo 4*—see Example 3.31). There is an obvious one-to-one correspondence between the equivalence classes of $\mathbb{Z}$ and the elements of $\mathbb{Z}_4$.

Suppose now that n and m belong to $\mathbb{Z}_4$; we wish to compute $n +_4 m$ by simulation in $[\mathbb{Z}, +]$. We choose any integer z_n in the class of $\mathbb{Z}$ associated with n; we choose any integer z_m in the class of $\mathbb{Z}$ associated with m. (We choose preimages of n and m under f; remember f^{-1} does not exist.) We add $z_n + z_m$ in $[\mathbb{Z}, +]$; the sum belongs to a class of $\mathbb{Z}$, and we choose the member of $\mathbb{Z}_4$ corresponding to that class. This element of $\mathbb{Z}_4$ will be $n +_4 m$. For example, we compute $3 +_4 2$ in $[\mathbb{Z}_4, +_4]$ by choosing an element of $[3]$ and an element of $[2]$, say 11 and -2, respectively. We add these in $[\mathbb{Z}, +]$ ($11 + (-2) = 9$), and then note that $9 \in [1]$, so $3 +_4 2 = 1$.

The simulation of $[\mathbb{Z}, +]$ by $[\mathbb{Z}_4, +_4]$ is much less satisfactory and can give us answers only to within an equivalence class. Thus, if we try to add $z_n + z_m$ in $[\mathbb{Z}, +]$ by using $[\mathbb{Z}_4, +_4]$, we compute $f(z_n) +_4 f(z_m)$; this is some member p of $\mathbb{Z}_4$. But to get back to $\mathbb{Z}$, the best we can find is the class containing p. Thus, to add $13 + 22$, for example, we use $f(13) +_4 f(22)$, or $1 +_4 2 = 3$. However, in going back to $\mathbb{Z}$, $[3]$ contains an infinite number of values, only one of which is the correct answer of 35.   □

This example has implications for computer arithmetic. When we add integers on a computer, we would really like to do the arithmetic of $[\mathbb{Z}, +]$, but we cannot because $\mathbb{Z}$ is an infinite set and the capacity of the computer is finite (although large). Thus we need a finite arithmetic to simulate $[\mathbb{Z}, +]$. If we use $[\mathbb{Z}_n, +_n]$ for some (large) n, then, just as in the case of $\mathbb{Z}_4$ above, $[\mathbb{Z}_n, +_n]$ only imperfectly simulates $[\mathbb{Z}, +]$. An integer "answer" in the computer is actually only good to within modulo n. Naturally, we expect that the use of something finite to simulate something infinite will not work perfectly, but perhaps there is another finite arithmetic that can do a better job than $[\mathbb{Z}_n, +_n]$. We shall see in Section 7.5 that this is not the case, and that $[\mathbb{Z}_n, +_n]$ for some n is the only possible choice.

Isomorphisms

Now let's put a final condition on our homomorphism from S to T, and require it to be not only onto but also one-to-one.

7.70
Definition

Let $[S, \cdot]$ and $[T, +]$ be semigroups (or monoids or groups). A mapping $f : S \to T$ is an **isomorphism** from $[S, \cdot]$ to $[T, +]$ if:

(a) The function f is a bijection.
(b) For all $x, y \in S$, $f(x \cdot y) = f(x) + f(y)$.   □

7.71 Example

Let $M_2^0(\mathbb{Z})$ be the set of all 2×2 matrices of the form

$$\begin{bmatrix} 1 & z \\ 0 & 1 \end{bmatrix}$$

where $z \in \mathbb{Z}$. Then $M_2^0(\mathbb{Z})$ is a group under matrix multiplication. The function $f : M_2^0(\mathbb{Z}) \to \mathbb{Z}$ given by

$$f\left(\begin{bmatrix} 1 & z \\ 0 & 1 \end{bmatrix}\right) = z$$

is an isomorphism from $[M_2^0(\mathbb{Z}), \cdot]$ to $[\mathbb{Z}, +]$. Clearly, f is onto. To show that f is one-to-one, let

$$f\left(\begin{bmatrix} 1 & x \\ 0 & 1 \end{bmatrix}\right) = f\left(\begin{bmatrix} 1 & y \\ 0 & 1 \end{bmatrix}\right)$$

Then $x = y$ and

$$\begin{bmatrix} 1 & x \\ 0 & 1 \end{bmatrix} = \begin{bmatrix} 1 & y \\ 0 & 1 \end{bmatrix}$$

In addition, f is a homomorphism because for any

$$\begin{bmatrix} 1 & x \\ 0 & 1 \end{bmatrix} \quad \text{and} \quad \begin{bmatrix} 1 & y \\ 0 & 1 \end{bmatrix}$$

in $M_2^0(\mathbb{Z})$,

$$f\left(\begin{bmatrix} 1 & x \\ 0 & 1 \end{bmatrix} \cdot \begin{bmatrix} 1 & y \\ 0 & 1 \end{bmatrix}\right) = f\left(\begin{bmatrix} 1 & x+y \\ 0 & 1 \end{bmatrix}\right)$$

$$= x + y = f\left(\begin{bmatrix} 1 & x \\ 0 & 1 \end{bmatrix}\right) + f\left(\begin{bmatrix} 1 & y \\ 0 & 1 \end{bmatrix}\right)$$

$\square$

If f is an isomorphism from $[S, \cdot]$ to $[T, +]$, then f^{-1} exists and is a bijection. Further, f^{-1} is also a homomorphism, this time from T to S. To see this, let t_1 and t_2 belong to T and consider $f^{-1}(t_1 + t_2)$. Because $t_1, t_2 \in T$ and f is onto, $t_1 = f(s_1)$ and $t_2 = f(s_2)$ for some s_1 and s_2 in S. Thus,

$$\begin{aligned}
f^{-1}(t_1 + t_2) &= f^{-1}(f(s_1) + f(s_2)) \\
&= f^{-1}(f(s_1 \cdot s_2)) \\
&= (f^{-1} \circ f)(s_1 \cdot s_2) \\
&= s_1 \cdot s_2 \\
&= f^{-1}(t_1) \cdot f^{-1}(t_2)
\end{aligned}$$

Thus we can simply say that S and T are **isomorphic,** denoted by $S \simeq T$.

7.72 Practice   Let $5\mathbb{Z} = \{5z \,|\, z \in \mathbb{Z}\}$. Then $[5\mathbb{Z}, +]$ is a group. Show that $f : \mathbb{Z} \to 5\mathbb{Z}$ given by $f(x) = 5x$ is an isomorphism from $[\mathbb{Z}, +]$ to $[5\mathbb{Z}, +]$. $\square$

Checking whether a given function is an isomorphism from S to T, as in Practice 7.72, is not hard. Deciding whether S and T are isomorphic may be harder. To prove that they are isomorphic, we must produce a function. To prove that they are not isomorphic, we must show that no such function exists. Since we can't try all possible functions, we use ideas such as the following: there is no one-to-one correspondence between S and T, S is commutative but T is not, and so on.

The Mirror Picture

If S and T are two isomorphic algebraic structures, then we expect each to simulate the other. To illustrate, we know by Example 7.71 that $[M_2^0(\mathbb{Z}), \cdot]$ and $[\mathbb{Z}, +]$ are isomorphic groups, and we can calculate in either structure by hopping over to its mirror image, calculating there, and hopping back. To multiply

$$\begin{bmatrix} 1 & 7 \\ 0 & 1 \end{bmatrix} \quad \text{and} \quad \begin{bmatrix} 1 & -3 \\ 0 & 1 \end{bmatrix}$$

in $[M_2^0(\mathbb{Z}), \cdot]$, for example, we map to 7 and -3 in $[\mathbb{Z}, +]$, add there, getting 4, and map back to

$$\begin{bmatrix} 1 & 4 \\ 0 & 1 \end{bmatrix}$$

in $M_2^0(\mathbb{Z})$. The product of

$$\begin{bmatrix} 1 & 7 \\ 0 & 1 \end{bmatrix} \quad \text{and} \quad \begin{bmatrix} 1 & -3 \\ 0 & 1 \end{bmatrix}$$

in $[M_2^0(\mathbb{Z}), \cdot]$ is

$$\begin{bmatrix} 1 & 4 \\ 0 & 1 \end{bmatrix}$$

Conversely, to add $2 + 3$ in $[\mathbb{Z}, +]$, we go to

$$\begin{bmatrix} 1 & 2 \\ 0 & 1 \end{bmatrix} \quad \text{and} \quad \begin{bmatrix} 1 & 3 \\ 0 & 1 \end{bmatrix}$$

in $[M_2^0(\mathbb{Z}), \cdot]$. The result of multiplying

$$\begin{bmatrix} 1 & 2 \\ 0 & 1 \end{bmatrix} \quad \text{and} \quad \begin{bmatrix} 1 & 3 \\ 0 & 1 \end{bmatrix}$$

in $[M_2^0(\mathbb{Z}), \cdot]$ is

$$\begin{bmatrix} 1 & 5 \\ 0 & 1 \end{bmatrix}$$

which then maps to 5, and $2 + 3 = 5$ in $[\mathbb{Z}, +]$. Figure 7.9a shows how $[\mathbb{Z}, +]$ simulates $[M_2^0(\mathbb{Z}), \cdot]$, and Figure 7.9b shows how $[M_2^0(\mathbb{Z}), \cdot]$ simulates $[\mathbb{Z}, +]$.

The mirror pictures of Figure 7.9 show us the central fact about two

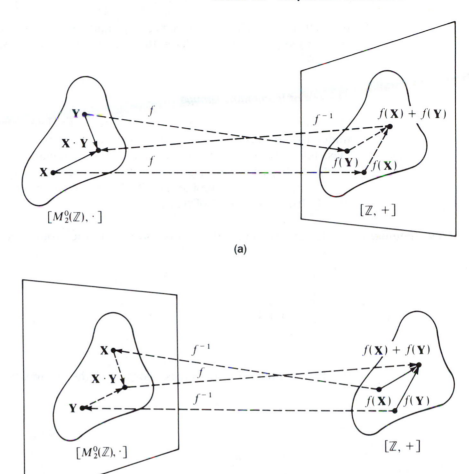

$$(a)$$

$$(b)$$

Figure 7.9

isomorphic structures, namely, that each is merely a relabeling of the other. There can be no basic algebraic differences between the behavior of two isomorphic structures, although the *names* of elements and operations may be different. If we ignore labels and concentrate on the underlying characteristics, then we make no distinction between isomorphic structures. Practice 7.73 shows that this classification by isomorphism is valid.

7.73 Practice   (a) Let $f: S \to T$ be an isomorphism from the semigroup $[S, \cdot]$ onto the semigroup $[T, +]$ and $g: T \to U$ be an isomorphism from $[T, +]$ onto the semigroup $[U, *]$. Show that $g \circ f$ is an isomorphism from S onto U.

(b) Let $\mathcal{S}$ be a collection of semigroups and define a binary relation ρ on $\mathcal{S}$ by $S \rho T \leftrightarrow S \simeq T$. Show that ρ is an equivalence relation on $\mathcal{S}$. □

Classes of Isomorphic Groups

We will finish this section by looking at some equivalence classes of groups under isomorphism. Often we pick out one member of an equivalence class and note that it is the typical member of that class, and that all other groups in the class look just like it (with different names).

A result concerning the nature of very small groups follows immediately from Exercise 7 of Section 7.1.

7.74 Theorem   (a) Every group of order 2 is isomorphic to the group whose group table is

·	1	a
1	1	a
a	a	1

(b) Every group of order 3 is isomorphic to the group whose group table is

·	1	a	b
1	1	a	b
a	a	b	1
b	b	1	a

(c) Every group of order 4 is isomorphic to one of the two groups whose group tables are

·	1	a	b	c
1	1	a	b	c
a	a	1	c	b
b	b	c	1	a
c	c	b	a	1

·	1	a	b	c
1	1	a	b	c
a	a	b	c	1
b	b	c	1	a
c	c	1	a	b

□

We can also prove that any group is essentially a permutation group. Suppose $[G, \cdot]$ is a group. We want to establish an isomorphism from G to a permutation group; each element g of G must be associated with a permutation α_g on some set. In fact, the set will be G itself; for any $x \in G$, we define $\alpha_g(x)$ to be $g \cdot x$. We must show that $\{\alpha_g | g \in G\}$ forms a permutation group, and that this permutation group is isomorphic to G. First we need to show that for any $g \in G$, α_g is indeed a permutation on G. From the definition $\alpha_g(x) = g \cdot x$, it is clear that $\alpha_g : G \to G$, but it must be shown that α_g is a bijection.

7.75 Practice Show that α_g as defined above is a permutation on G. $\square$

Now we consider $P = \{\alpha_g | g \in G\}$ and show that P is a group under function composition. P is nonempty because G is nonempty, and associativity always holds for function composition. We must show that P is closed and has an identity, and that each $\alpha_g \in P$ has an inverse in P. To show closure, let α_g and $\alpha_h \in P$. For any $x \in G$, $(\alpha_g \circ \alpha_h)(x) = \alpha_g(\alpha_h(x)) = \alpha_g(h \cdot x) = g \cdot (h \cdot x) = (g \cdot h) \cdot x$. Thus, $\alpha_g \circ \alpha_h = \alpha_{g \cdot h}$ and $\alpha_{g \cdot h} \in P$.

7.76 Practice (a) Let 1 denote the identity of G. Show that α_1 is an identity for P under function composition.

(b) For $\alpha_g \in P$, $\alpha_{g^{-1}} \in P$; show that $\alpha_{g^{-1}} = (\alpha_g)^{-1}$. $\square$

We know that $[P, \circ]$ is a permutation group, and it only remains to show that the function $f : G \to P$ given by $f(g) = \alpha_g$ is an isomorphism. Clearly, f is an onto function.

7.77 Practice Show that $f : G \to P$ defined above is

(a) one-to-one
(b) a homomorphism $\square$

We have now proved Theorem 7.78, first stated and proved by the English mathematician Cayley in the mid-1800s.

7.78 Theorem Every group is isomorphic to a permutation group. $\square$

✔ Checklist

Definitions

homomorphism (*p. 306*)
algebraic property (*p. 308*)
homomorphic image (*p. 309*)
isomorphism (*p. 310*)
isomorphic structures (*p. 311*)

Techniques

Test whether a given function from a structure S to a structure T is a homomorphism, an isomorphism, or neither.

Decide whether two structures are isomorphic.

Main Ideas

If f is a homomorphism from a structure S to a structure T, then the range $f(S)$ has the structure of S as well as the algebraic properties of S.

If S and T are isomorphic structures, they are identical except for relabeling, and each simulates the other.

The relation of isomorphism divides any collection of semigroups (or monoids or groups) into equivalence classes.

Except for isomorphisms, there is essentially only one group of order 2, one group of order 3, and two groups of order 4.

Every group is essentially a permutation group.

Exercises Section 7.3

1. In each case, decide whether the given function is a homomorphism from the group on the left to the one on the right. Are any of the homomorphisms also isomorphisms?

★ (a) $[\mathbb{Z}, +]$, $[\mathbb{Z}, +]$; $f(x) = 2$

★ (b) $[\mathbb{R}, +]$, $[\mathbb{R}, +]$; $f(x) = x + 1$

★ (c) $[\mathbb{Z} \times \mathbb{Z}, +]$, $[\mathbb{Z}, +]$; $f(x, y) = x + 2y$

★ (d) $[\mathbb{R}, +]$, $[\mathbb{R}, +]$; $f(x) = |x|$

★ (e) $[\mathbb{R}^*, \cdot]$, $[\mathbb{R}^*, \cdot]$ (where $\mathbb{R}^*$ denotes the set of nonzero real numbers); $f(x) = |x|$

(f) $[\mathbb{R}[x], +]$, $[\mathbb{R}, +]$; $f(a_n x^n + a_{n-1} x^{n-1} + \cdots + a_1 x + a_0) = a_n + a_{n-1} + \cdots + a_0$

(g) $[M_2^D(\mathbb{R}^+), \cdot]$, $[\mathbb{R} \times \mathbb{R}, +]$; $f\left(\begin{bmatrix} a & 0 \\ 0 & b \end{bmatrix}\right) = (a, b)$

(h) $[S_3, \circ]$, $[\mathbb{Z}_2, +_2]$; $f(\alpha) = \begin{cases} 1 & \text{if } \alpha \text{ is an even permutation} \\ 0 & \text{if } \alpha \text{ is an odd permutation} \end{cases}$

2. In each case, decide whether the given groups are isomorphic. If they are, produce an isomorphism function. If they are not, give a reason why they are not.

(a) $[\mathbb{Z}, +]$, $[12\mathbb{Z}, +]$ (where $12\mathbb{Z} = \{12z \,|\, z \in \mathbb{Z}\}$)

(b) $[\mathbb{Z}_5, +_5]$, $[5\mathbb{Z}, +]$

(c) $[5\mathbb{Z}, +]$, $[12\mathbb{Z}, +]$

(d) $[S_3, \circ]$, $[\mathbb{Z}_6, +_6]$

(e) $[\{a_1 x + a_0 | a_1, a_0 \in \mathbb{R}\}, +]$, $[\mathbb{C}, +]$

(f) $[\mathbb{Z}_6, +_6]$, $[S_6, \circ]$

(g) $[\mathbb{Z}_2, +_2]$, $[S_2, \circ]$

★ 3. Let b be a positive real number, $b \neq 1$. Let f be the function from the group $[\mathbb{R}^+, \cdot]$ to the group $[\mathbb{R}, +]$ given by $f(x) = \log_b x$ (recall that $y = \log_b x$ if and only if $b^y = x$).

  (a) Show that f is an isomorphism.

  (b) Let $b = 2$. Use $[\mathbb{R}, +]$ to simulate the computation $64 \cdot 512$ in $[\mathbb{R}^+, \cdot]$. (In the age B.C.—before calculators and computers— large numbers were multiplied by using tables of common logarithms ($b = 10$) to convert a multiplication problem to an addition problem.)

4. Let f be the function from the group $[\mathbb{Z}_{18}, +_{18}]$ to the group $[\mathbb{Z}_{12}, +_{12}]$ given by $f(x) = 4 \cdot_{12} x$. Then f can be shown to be a homomorphism.

  (a) The range $f(\mathbb{Z}_{18})$ of the function f is a subgroup of $[\mathbb{Z}_{12}, +_{12}]$. Find the elements of $f(\mathbb{Z}_{18})$.

  (b) Define an equivalence relation ρ on $\mathbb{Z}_{18}$ by $x \rho y \leftrightarrow f(x) = f(y)$. Describe the resulting equivalence classes.

  (c) Use $[\mathbb{Z}_{18}, +_{18}]$ (and the classes of part (b)) to simulate the computation $4 +_{12} 8$ in $[f(\mathbb{Z}_{18}), +_{12}]$.

  (d) Use $[f(\mathbb{Z}_{18}), +_{12}]$ to simulate the computation $14 +_{18} 8$ in $[\mathbb{Z}_{18}, +_{18}]$ to within an equivalence class.

5. Let $A = \{0, 1\}$, and let $[A^*, \cdot]$ denote the free monoid generated by A. Let f be a function from $[A^*, \cdot]$ to the monoid $[\mathbb{N}, +]$ given by $f(\alpha) = $ the number of 1s in the string α.

  (a) Show that f is a homomorphism.

  (b) Show that f is onto.

  (c) Show that f takes the identity of $[A^*, \cdot]$ to the identity of $[\mathbb{N}, +]$.

  (d) Define an equivalence relation ρ on A^* by $\alpha \rho \beta \leftrightarrow f(\alpha) = f(\beta)$. Describe the resulting equivalence classes.

  (e) Use $[A^*, \cdot]$ (and the classes of part (d)) to simulate the computation $3 + 4$ in $[\mathbb{N}, +]$.

  (f) Use $[\mathbb{N}, +]$ to simulate the computation $101 \cdot 1101$ in $[A^*, \cdot]$ to within an equivalence class.

6. Let S be the set of all functions from $\mathbb{R}$ to $\mathbb{R}$.

  (a) Let f be the function from the semigroup $[S, \circ]$ (the semigroup of transformations on $\mathbb{R}$) to the group $[\mathbb{R}, +]$ given by $f(g) = g(10)$. Show that f is not a homomorphism.

  (b) Define an operation $+$ on S by $(g + h)(x) = g(x) + h(x)$ for $x \in \mathbb{R}$. Show that $[S, +]$ is a group.

  (c) Let f be the function from the group $[S, +]$ to the group $[\mathbb{R}, +]$ given by $f(g) = g(10)$. Show that f is an onto homomorphism.

7. (a) Let $S = \{1, -1\}$. Show that $[S, \cdot]$ is a group where $\cdot$ denotes ordinary integer multiplication.
   (b) Let f be the function from the group $[S_n, \circ]$ to the group $[S, \cdot]$ given by

   $$f(\alpha) = \begin{cases} 1 & \text{if } \alpha \text{ is even} \\ -1 & \text{if } \alpha \text{ is odd} \end{cases}$$

   Prove that f is a homomorphism.

8. For any set S, $[\mathscr{P}(S), \cap]$ and $[\mathscr{P}(S), \cup]$ are monoids. Prove that they are isomorphic.

⋆ 9. Let $[G, \cdot]$ be a group with identity i and let $f: G \to G$ be given by $f(x) = i$. Prove that f is a homomorphism.

10. Let $[G, \cdot]$ be a commutative group with identity i.
    (a) Prove that for a fixed positive integer n, the function $f: G \to G$ given by $f(x) = x^n$ is a homomorphism.
    (b) Prove that the function $f: G \to G$ given by $f(x) = x^{-1}$ is an isomorphism.

11. A group $[G, \cdot]$ is called **divisible** if for any $g \in G$ and any positive integer n, the equation $x^n = g$ has at least one solution in G.
    (a) Show that $[\mathbb{Q}, +]$ is a divisible group. (Note that the equation $x^n = g$ becomes $nx = g$ when the group operation is addition.)
    (b) Let f be a homomorphism from a divisible group $[G, \cdot]$ to a group $[H, \cdot]$. Show that the range $f(G)$ is a divisible group.

⋆ 12. Let $[S, \cdot]$ and $[T, +]$ be semigroups, and suppose that $t \in T$ is an idempotent element, that is, $t + t = t$. Let f be a homomorphism from S onto T. Prove that the set $U = \{s \in S \mid f(s) = t\}$ is a sub-semigroup of S under $\cdot$.

⋆ 13. Let f be a homomorphism from a semigroup $[S, \cdot]$ onto a semigroup $[T, +]$.
    (a) Show that if x is an idempotent element in $[S, \cdot]$, then $f(x)$ is an idempotent element in $[T, +]$ (see Exercise 15, Section 7.1).
    (b) Show that if 0 is a zero in $[S, \cdot]$, then $f(0)$ is a zero in $[T, +]$ (see Exercise 13, Section 7.1).

14. (a) Let $[S, \cdot]$ be a semigroup. An isomorphism from S to S is called an **automorphism** on S. Let $\text{Aut}(S)$ be the set of all automorphisms on S, and show that $\text{Aut}(S)$ is a group under function composition.
    (b) For the group $[\mathbb{Z}_4, +_4]$, find the set of automorphisms and show its group table under $\circ$.

15. (a) Let $[G, +]$ be a commutative group, and let $\text{Hom}(G)$ be the set of all homomorphisms from G into G. (A homomorphism from a group G into itself is called an **endomorphism** of G.) Define an operation of addition on $\text{Hom}(G)$ as follows: for $f, g \in \text{Hom}(G)$,

$(f + g)(x) = f(x) + g(x)$ for each $x \in G$. Show that [Hom(G), +] is a commutative semigroup.

(b) Let 0_G be the identity element of the group [G, +], and define a function $i: G \to G$ by $i(x) = 0_G$ for all $x \in G$. Prove that $i \in$ Hom(G) and is an identity of [Hom(G), +].

(c) For $f \in$ Hom(G), define a function $-f: G \to G$ by $(-f)(x) = -f(x)$ for each $x \in G$. Show that $-f \in$ Hom(G) and is the inverse of f in [Hom(G), +]. This demonstration will conclude the proof that [Hom(G), +] is a commutative group.

16. Prove that a homomorphism f from a group G to a group H is an isomorphism if and only if there exists a homomorphism $g: H \to G$ such that $g \circ f = i_G$ and $f \circ g = i_H$, where i_G and i_H are the identity functions on G and H, respectively.

17. Let f be a homomorphism from a group G onto a group H. Show that f is an isomorphism if and only if the only element of G that is mapped to the identity of H is the identity of G.

★ 18. Let [G, ·] and [H, +] be groups.
(a) Show that the group [$G \times H$, ·] is isomorphic to the group [$H \times G$, ·].
(b) Prove that the function $f: G \times H \to G$ defined by $f(x, y) = x$ is a homomorphism.

★ 19. A bijection f is defined on a group [G, ·] by the equation $f(x) = x^{-1}$. Prove that if f is an isomorphism, then [G, ·] is a commutative group.

20. Prove that for any semigroup [S, ·] there is a homomorphism from S onto a transformation semigroup.

21. Let [G, ·] be a group and g a fixed element of G. Define $f: G \to G$ by $f(x) = g \cdot x \cdot g^{-1}$ for any $x \in G$. Prove that f is an isomorphism from G to G.

SECTION 7.4   HOMOMORPHISM THEOREMS

If we have some instance [A, ·] of an algebraic structure, we could certainly simulate it by constructing some [B, +] isomorphic to [A, ·]. (Remember the mirror picture.) But practically speaking, if we can "build" [B, +], we can probably "build" [A, ·] to begin with. Often we must settle for the imperfect simulation produced by a homomorphism from A onto B that is not an isomorphism. We have considered two examples of this situation, namely, the homomorphism example at the end of Section 5.2 and Example 7.69. (You might want to review these now.) In this section we will see why these examples work.

Defining the [S/f, *] Structure

Let's assume that $[S, \cdot]$ and $[T, +]$ are semigroups and that there is a homomorphism f from S to T. We will also assume that f is not a one-to-one function (everything we say would still be true, but trivial, if f is one-to-one). The range $f(S)$ is a subset of T, a semigroup under $+$, and the homomorphic image of S under f. We recall from our examples that the key to the relationship between S and its homomorphic image $f(S)$ is in the equivalence classes of the domain S. We define a binary relation ρ on S by $x \, \rho \, y \leftrightarrow f(x) = f(y)$.

7.79 Practice   Show that ρ as defined above is an equivalence relation on S.   □

Thus, ρ partitions S into equivalence classes. For $x \in S$, $[x]$ denotes the equivalence class to which x belongs. The image of x under f is some element in $f(S)$; the other members of $[x]$ are those members of S mapping under f to the same element. We denote the set of all distinct equivalence classes of S by S/f. A member of S/f will in general have more than one name.

7.80 Example   Let $f : \mathbb{N} \times \mathbb{N} \to \mathbb{N}$ be given by $f(x, y) = x$. Then f is a homomorphism from $[\mathbb{N} \times \mathbb{N}, +]$ (use componentwise addition) to $[\mathbb{N}, +]$ because $f((x_1, y_1) + (x_2, y_2)) = f(x_1 + x_2, y_1 + y_2) = x_1 + x_2 = f(x_1, y_1) + f(x_2, y_2)$. Here f is also an onto function, and $f(\mathbb{N} \times \mathbb{N}) = \mathbb{N}$. One equivalence class of $\mathbb{N} \times \mathbb{N}$ is $[(5, 7)]$; this class is also named $[(5, 2)]$, $[(5, 8001)]$, and so on, since f takes any of these ordered pairs to 5. In general, $[(x, y)] = \{(x, n) \mid n \in \mathbb{N}\}$.   □

7.81 Practice   Let $f : \mathbb{Z} \to \mathbb{Z}$ be given by $f(x) = x \cdot_3 1$. (Here $x \cdot_3 1$ denotes the *nonnegative* remainder when x is divided by 3.) Then f is a homomorphism from $[\mathbb{Z}, +]$ to $[\mathbb{Z}, +_3]$. Find the homomorphic image $f(\mathbb{Z})$, and show that $\mathbb{Z}/f = \{[16], [-12], [8]\}$.   □

Operating in S/f

We now want to create an algebraic structure using the set S/f. To do so, we must define a binary operation $*$ on the classes of S. Suppose that $[x]$ and $[y]$ are members of S/f. We want to define $[x] * [y]$ to be some unique member of S/f. A reasonable approach is to take x and y, representative members of the classes $[x]$ and $[y]$, respectively, form $x \cdot y$ in S, and then use the class $[x \cdot y]$ as the product $[x] * [y]$. Thus, $[x] * [y] = [x \cdot y]$. There is one problem with this definition of $*$, however, and it is a difficulty that arises whenever we operate on classes by choosing representatives of the classes. Since we want $*$ to be an operation on the *classes* $[x]$ and $[y]$, the resulting class $[x] * [y]$ must be independent of the particular representatives of $[x]$ and $[y]$ that we use for the computation. This property must hold in order for $*$ to be a well-defined operation. In our previous two homomorphism examples, we ignored this question, blithely saying "take any member of the class"; now we prove that the operation $*$ is well-defined.

Suppose, then, that we want to compute $[x] * [y]$ and that $[x] = [x']$ and $[y] = [y']$. We can therefore choose x' and y' as representatives of $[x]$ and $[y]$ and compute $[x' \cdot y']$. We want to show that $[x' \cdot y'] = [x \cdot y]$, which will be true if $f(x' \cdot y') = f(x \cdot y)$. Because f is a homomorphism, $f(x' \cdot y') = f(x') + f(y')$ and $f(x) + f(y) = f(x \cdot y)$. Further, because $[x] = [x']$ and $[y] = [y']$, it follows that $f(x) = f(x')$ and $f(y) = f(y')$. Thus,

$$f(x' \cdot y') = f(x') + f(y') = f(x) + f(y) = f(x \cdot y)$$

and $[x' \cdot y'] = [x \cdot y]$.

Not only do we have a binary operation on S/f, but $[S/f, *]$ is a semigroup.

7.82 Practice   Prove that $*$ is associative on S/f.   □

7.83 Example   Consider the semigroup $[(\mathbb{N} \times \mathbb{N})/f, *]$ where f is defined as in Example 7.80. Here, $[(3, 8)] * [(7, 14)] = [(3, 8) + (7, 14)] = [(10, 22)]$. However, it is also correct to write $[(3, 8)] * [(7, 14)] = [(10, 6)]$.

□

7.84 Practice   Which of the following computations in $[(\mathbb{N} \times \mathbb{N})/f, *]$ are correct?

(a) $[(0, 4)] * [(6, 2)] = [(0, 6)]$
(b) $[(4, 10)] * [(4, 11)] = [(8, 10)]$
(c) $[(15, 0)] * [(2, 12)] = [(17, 21)]$ □

Isomorphism from S/f to $f(S)$

Look again at Practice 7.81. Here we can name the elements of $\mathbb{Z}/f$ as $[0]$, $[1]$, and $[2]$. In the semigroup $[\mathbb{Z}/f, *]$, the computation $[1] * [2] = [3] = [0]$ is correct. This equation strongly reminds us of arithmetic in the homomorphic image $[\mathbb{Z}_3, +_3]$. The similarity is no accident. We next show that for the general case where f is a homomorphism from semigroup $[S, \cdot]$ to semigroup $[T, +]$, the semigroup $[S/f, *]$ is isomorphic to the homomorphic image $[f(S), +]$.

Since we are discussing four different semigroups here, perhaps Figure 7.10 can clarify the situation. The two blobs represent the semigroups S and T. The solid lines between the blobs denote the homomorphism f mapping S onto the semigroup $f(S)$ (the small ellipse on the

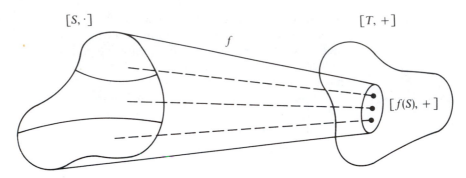

Figure 7.10

right). The dotted lines show the isomorphism we hope to establish between the classes in the semigroup S/f (the sections on the left) and the elements of $f(S)$.

To show that $[S/f, *]$ is isomorphic to $[f(S), +]$, we need a bijection $g: S/f \to f(S)$. It should be easy to guess g because f maps all the members of a given class to the same member of $f(S)$.

7.85 Practice   Let $[x] \in S/f$. What would you suggest as a definition for $g([x])$?   □

We define $g: S/f \to f(S)$ by $g([x]) = f(x)$. Notice that g is well-defined, because if $[x] = [y]$, then $f(x) = f(y)$ and $g([y]) = f(y) = f(x) = g([x])$. Now we will show that g is a bijection. To show that g is onto, let $t \in f(S)$ where $t = f(x)$ for some x in S. Then $[x] \in S/f$, and $g([x]) = f(x) = t$. To show that g is one-to-one, suppose that $g([x]) = g([y])$. Then $f(x) = f(y)$, and $[x] = [y]$.

Finally, we can show that g is a homomorphism by proving that $g([x] * [y]) = g([x]) + g([y])$.

7.86 Practice   Show that $g([x] * [y]) = g([x]) + g([y])$. (Hint: use the definition of $*$ and the fact that f is a homomorphism from S to T.)   □

We have now proved the following theorem.

7.87 Theorem   The Homomorphism Theorem for Semigroups
If f is a homomorphism from a semigroup S to a semigroup T, then the homomorphic image $f(S)$ is isomorphic to the semigroup S/f.   □

Shortly we will see similar results for monoids and groups. These results explain the imperfect simulations in our previous examples. If there is a homomorphism f from a structure $[A, \cdot]$ *onto* a structure $[B, +]$, then B is the homomorphic image $f(A)$ and is isomorphic to A/f. In this case, A (actually A/f) simulates B, but B can only simulate A/f, not A itself.

7.88 Example Consider once again the function $f:\mathbb{N} \times \mathbb{N} \to \mathbb{N}$ that is given by $f(x, y) = x$, a homomorphism from $[\mathbb{N} \times \mathbb{N}, +]$ onto $[\mathbb{N}, +]$. A typical member of $(\mathbb{N} \times \mathbb{N})/f$ is $[(x, y)] = \{(x, n)|n \in \mathbb{N}\}$. The isomorphism $g:(\mathbb{N} \times \mathbb{N})/f \to N$ associates $[(x, y)]$ with $f(x, y) = x$. To compute $2 + 5$ in $[\mathbb{N}, +]$, we find $g^{-1}(2) = [(2, 1)]$ and $g^{-1}(5) = [(5, 1)]$, compute $[(2, 1)] * [(5, 1)] = [(2, 1) + (5, 1)] = [(7, 2)]$, and then take $g([(7, 2)]) = f(7, 2) = 7$. However, to simulate the computation $(3, 4) + (5, 6)$ in $\mathbb{N} \times \mathbb{N}$ by using $[\mathbb{N}, +]$, we find $f(3, 4) + f(5, 6) = 3 + 5 = 8$, and we know only that the answer is some member of $\{(8, n)|n \in \mathbb{N}\}$. □

We want to extend Theorem 7.87 to monoids and groups. Thus, suppose that f is a homomorphism from a monoid $[S, \cdot]$ with identity i_S to a monoid $[T, +]$. Then the semigroup $[S/f, *]$ is also a monoid with identity $[i_S]$.

7.89 Practice Show that $[i_S]$ is an identity element for $[S/f, *]$; that is, show that for $[x] \in S/f$, $[x] * [i_S] = [i_S] * [x] = [x]$. □

We already know that the homomorphic image of a monoid is a monoid. Thus, Theorem 7.87 becomes the following.

7.90 Theorem The Homomorphism Theorem for Monoids
If f is a homomorphism from a monoid S to a monoid T, then the homomorphic image $f(S)$ is isomorphic to the monoid S/f. □

Finally, let f be a homomorphism from a group $[S, \cdot]$ with identity i_S to a group $[T, +]$. Then $[S/f, *]$ is a group. We must show that each $[x] \in S/f$ has an inverse with respect to the identity $[i_S]$.

7.91 Practice For $[x] \in S/f$, find an inverse element in S/f. □

Therefore, we arrive at Theorem 7.92.

7.92 Theorem The Homomorphism Theorem for Groups
If f is a homomorphism from a group S to a group T, then the homomorphic image $f(S)$ is isomorphic to the group S/f. □

7.93 Example Let $f:\mathbb{Z} \to \mathbb{Z}_8$ be defined by $f(x) = x \cdot_8 2$. Then f is a homomorphism from the group $[\mathbb{Z}, +]$ into the group $[\mathbb{Z}_8, +_8]$. The homomorphic image $f(\mathbb{Z})$ is $\{0, 2, 4, 6\}$. The distinct elements of $\mathbb{Z}/f$ are the classes $[0]$, $[1]$, $[2]$, and $[3]$ where

$$[0] = \{\ldots, -8, -4, 0, 4, 8, \ldots\}$$
$$[1] = \{\ldots, -7, -3, 1, 5, 9, \ldots\}$$
$$[2] = \{\ldots, -6, -2, 2, 6, 10, \ldots\}$$
$$[3] = \{\ldots, -5, -1, 3, 7, 11, \ldots\}$$

The isomorphism g from $\mathbb{Z}/f$ to $f(\mathbb{Z})$ is given by

$[0] \to 0$

$[1] \to 2$

$[2] \to 4$

$[3] \to 6$ □

7.94 Practice   For the above example, compute

(a) $g([2] * [3])$

(b) $g([2]) +_8 g([3])$ □

In Chapter 9 we will see a homomorphism theorem for the finite-state machine structure. Meanwhile, in the next section, we will expand on the situation for groups. When f is a homomorphism from a group $[S, \cdot]$ to a group $[T, +]$, we will see a new way to describe the members of S/f. We will also discover a sort of converse to the Homomorphism Theorem for Groups.

✔ Checklist

Techniques

Given a homomorphism f from a structure $[S, \cdot]$ to a structure $[T, +]$, find $f(S)$, find the distinct members of S/f, compute in $[S/f, *]$, find the isomorphism from S/f to $f(S)$, and use S/f to simulate $f(S)$.

Main Ideas

The Homomorphism Theorem for Semigroups, Monoids, and Groups: if f is a homomorphism from a semigroup (or monoid or group) S to a semigroup (or monoid or group) T, then the homomorphic image $f(S)$ is isomorphic to S/f.

Exercises Section 7.4

★ 1. Let $\mathbb{N}^2[x] = \{a_2x^2 + a_1x + a_0 \mid a_0, a_1, a_2 \in \mathbb{N}\}$ and let $+$ denote polynomial addition. Then $[\mathbb{N}^2[x], +]$ is a monoid.

(a) Prove that $f : \mathbb{N}^2[x] \to \mathbb{N}$ defined by $f(a_2x^2 + a_1x + a_0) = a_2$ is a homomorphism from $[\mathbb{N}^2[x], +]$ onto $[\mathbb{N}, +]$.

(b) Which of the following are correct computations in $[\mathbb{N}^2[x]/f, *]$?

(1) $[5x^2 + 3x + 2] * [3x^2 + 2x + 4] = [8x^2 + 3x + 2]$

(2) $[3x + 7] * [x + 1] = [3x^2 + 10x + 7]$

(3) $[x^2] * [2x] = [x^2 + 7]$

2. For each pair of groups S and T and the given homomorphism $f:S \to T$, find $f(S)$, find the distinct members of S/f, and indicate the isomorphism g from S/f to $f(S)$.

★ (a) $[\mathbb{Z}_{18}, +_{18}]$, $[\mathbb{Z}_{12}, +_{12}]$; $f(x) = 4 \cdot_{12} x$ (see Exercise 4 of Section 7.3)

(b) $[\mathbb{Z}, +]$, $[\mathbb{Z}_7, +_7]$; $f(x) = x \cdot_7 1$

(c) $[\mathbb{Z}_{12}, +_{12}]$, $[\mathbb{Z}_{18}, +_{18}]$; $f(x) = 6 \cdot_{18} x$

(d) $[M_2^0(\mathbb{Q}), \cdot]$ where

$$M_2^0(\mathbb{Q}) = \left\{ \begin{bmatrix} 1 & q \\ 0 & 1 \end{bmatrix} \middle| q \in \mathbb{Q} \right\}$$

and $\cdot$ is matrix multiplication, $[\mathbb{Q} \times \mathbb{Q}, +]$ where $+$ denotes componentwise addition;

$$f\left(\begin{bmatrix} 1 & q \\ 0 & 1 \end{bmatrix} \right) = (0, q)$$

3. Each problem below refers to the corresponding problem in Exercise 2.

★ (a) Compute $g([1] * [2])$ and $g([1]) +_{12} g([2])$

(b) Compute $g([3] * [3])$ and $g([3]) +_7 g([3])$

(c) Compute $g([2] * [2])$ and $g([2]) +_{18} g([2])$

(d) Compute

$$g\left(\begin{bmatrix} 1 & 5 \\ 0 & 1 \end{bmatrix} * \begin{bmatrix} 1 & -4 \\ 0 & 1 \end{bmatrix} \right)$$

and

$$g\left(\begin{bmatrix} 1 & 5 \\ 0 & 1 \end{bmatrix} \right) + g\left(\begin{bmatrix} 1 & -4 \\ 0 & 1 \end{bmatrix} \right)$$

4. Let $S = \{1, -1\}$ and $\cdot$ denote ordinary integer multiplication. The function $f:S_3 \to S$ given by

$$f(\alpha) = \begin{cases} 1 & \text{if } \alpha \text{ is even} \\ -1 & \text{if } \alpha \text{ is odd} \end{cases}$$

is a homomorphism from $[S_3, \circ]$ onto $[S, \cdot]$ (see Exercise 7, Section 7.3).

(a) Classify the elements of S_3 as even or odd permutations.

(b) Find the distinct members of S_3/f.

(c) Indicate the isomorphism g from S_3/f to S.

(d) Simulate the computation $(1)(-1)$ in $[S, \cdot]$ by using S_3/f.

5. Suppose you have a machine or black box that is a modulo 8 adder; that is, it performs arithmetic in $[\mathbb{Z}_8, +_8]$. Such a machine has 8 inputs and 8 outputs, denoted by 0, 1, 2, . . . , 7. You would like to use this machine to do arithmetic in $[\mathbb{Z}_4, +_4]$. The numbers 0, 1, 2, and 3 can be fed into the $\mathbb{Z}_8$ machine. Define a function from the outputs of the $\mathbb{Z}_8$ machine to $\{0, 1, 2, 3\}$ that enables this machine to simulate a modulo 4 adder.

6. Let f be a homomorphism from a group $[S, \cdot]$ with identity i_S to a group $[T, +]$ with identity i_T. Then $f(i_S) = i_T$. Show that $[i_S]$ is a subgroup of S.

SECTION 7.5   QUOTIENT GROUPS

In this section we will concentrate on further refinement of the Homomorphism Theorem for Groups (Theorem 7.92). Because you have now had lots of practice in keeping track of the various operations encountered in homomorphisms from one group to another, we will simply use $\cdot$ to denote an arbitrary group operation and sometimes write xy to denote $x \cdot y$. Of course, in any specific group the abstract $\cdot$ must be interpreted correctly.

Suppose then that $[G, \cdot]$ and $[H, \cdot]$ are groups and that f is a homomorphism from G to H. We know that G/f is a group structure, and we will first find a nice way to describe the equivalence classes making up this group. In this description we will use the equivalence class $[i_G]$, which is the identity of the group G/f.

The Kernel

7.95
Definition

For f a homomorphism from a group G with identity i_G to a group H with identity i_H, the set $[i_G] = \{x \in G \mid f(x) = i_H\}$ is called the **kernel** K of f. $\qquad\square$

7.96 Example

(a) Let $\mathbb{R}^* = \mathbb{R} - \{0\}$ and $\mathbb{R}^+$ be the set of all positive real numbers. Then the function f defined by $f(x) = |x|$ is a homomorphism from the group $[\mathbb{R}^*, \cdot]$ to the group $[\mathbb{R}^+, \cdot]$. The kernel K of f is $\{x \in \mathbb{R}^* \mid f(x) = 1\}$; therefore, $K = \{1, -1\}$.

(b) The function f defined by $f(x) = x \cdot_8 2$ is a homomorphism from $[\mathbb{Z}, +]$ to $[\mathbb{Z}_8, +_8]$. Here $K = \{x \in \mathbb{Z} \mid f(x) = x \cdot_8 2 = 0\} = \{\ldots, -8, -4, 0, 4, 8, \ldots\} = 4\mathbb{Z}$. $\qquad\square$

7.97 Practice

(a) The function f defined by $f(x) = x \cdot_3 1$ is a homomorphism from $[\mathbb{Z}, +]$ to $[\mathbb{Z}_3, +_3]$. Find the kernel K.

(b) The function f defined by $f(x) = x \cdot_8 4$ is a homomorphism from $[\mathbb{Z}_{12}, +_{12}]$ to $[\mathbb{Z}_8, +_8]$. Find the kernel K. $\qquad\square$

Because i_G is always an element of K, K is a nonempty subset of G. Furthermore, K is a subgroup of $[G, \cdot]$.

7.98 Practice Let $x, y \in K$ and show that $x \cdot y^{-1} \in K$, thus proving that K is a subgroup of $[G, \cdot]$ (see Exercise 18(a) of Section 7.2).   □

Although K is the only member of G/f that is a subgroup of G (because it is the only equivalence class containing i_G), all other members of G/f are related to K. Thus, let $[x]$ be a member of G/f and let $y \in [x]$. We can then write

$$f(y) = f(x)$$
$$f(y) = f(x) \cdot i_H$$
$$(f(x))^{-1} \cdot f(y) = i_H$$
$$f(x^{-1}) \cdot f(y) = i_H$$
$$f(x^{-1} \cdot y) = i_H$$

and $x^{-1} \cdot y \in K$. Let $x^{-1} \cdot y = k$ where $k \in K$. Then $y = xk$. If we let $xK = \{xk \mid k \in K\}$, we have shown that every member y of $[x]$ is a member of xK, or $[x] \subseteq xK$. However, we can also show that $xK \subseteq [x]$. To prove this, let $xk \in xK$. Then $f(xk) = f(x) \cdot f(k) = f(x) \cdot i_H$ (since $k \in K$) = $f(x)$, and $xk \in [x]$. Therefore, we know that $[x] = xK$. Any equivalence class in G/f consists of a representative of the class multiplied by all the various members of the kernel. Henceforth, we will use xK to denote the class $[x]$, and G/K to denote the collection G/f of classes.

7.99 Example (a) In the homomorphism of Example 7.96(a), $K = \{1, -1\}$. A typical member of $\mathbb{R}^*/K$ is $3K = \{3, -3\}$. As before, an equivalence class can have more than one name; here, $3K = -3K$.

(b) In the homomorphism of Example 7.96(b), $K = 4\mathbb{Z}$. The distinct members of $\mathbb{Z}/4\mathbb{Z}$ are (using additive notation this time)

$$0 + K = 0 + 4\mathbb{Z} = \{\ldots, -8, -4, 0, 4, 8, \ldots\}$$
$$1 + K = 1 + 4\mathbb{Z} = \{\ldots, -7, -3, 1, 5, 9, \ldots\}$$
$$2 + K = 2 + 4\mathbb{Z} = \{\ldots, -6, -2, 2, 6, 10, \ldots\}$$
$$3 + K = 3 + 4\mathbb{Z} = \{\ldots, -5, -1, 3, 7, 11, \ldots\}$$ □

7.100 Practice (a) For Practice 7.97(a), name the distinct members of $\mathbb{Z}/3\mathbb{Z}$ in terms of $K = 3\mathbb{Z}$.

(b) For Practice 7.97(b), name the distinct members of Z_{12}/K in terms of K.   □

We now know that for any $[x]$ in G/K, $[x] = xK = \{xk \mid k \in K\}$. A similar argument shows that $[x] = Kx = \{kx \mid k \in K\}$. Thus, for any $x \in G$, $xK = Kx$. (Note that this is set equality and not a claim that $xk = kx$ for $k \in K$.)

7.101
Practice

Show that $[x] = Kx$.   □

Multiplication Formula

The notation xK for members of the set G/K is useful for finding the product of two members of the *group* G/K. For $[x] = xK$ and $[y] = yK$, we know that multiplication in the group G/K is defined by $[x] \cdot [y] = [x \cdot y]$, but this equation leads to $xK \cdot yK = xyK$, which gives us an easy formula for multiplying in G/K. (By Practice 7.101, the formula $Kx \cdot Ky = Kxy$ is also valid.)

7.102
Example

(a) In $\mathbb{R}^*/K$, the product of $3K$ and $(-5)K$ is $3(-5)K = -15K = 15K$.

(b) In $\mathbb{Z}/4\mathbb{Z}$, the sum of $2 + 4\mathbb{Z}$ and $3 + 4\mathbb{Z}$ is $(2 + 3) + 4\mathbb{Z} = 5 + 4\mathbb{Z} = 1 + 4\mathbb{Z}$.   □

7.103
Practice

(a) Find the sum of $7 + 3\mathbb{Z}$ and $5 + 3\mathbb{Z}$ in $\mathbb{Z}/3\mathbb{Z}$.

(b) Find the additive inverse of $4 + 3\mathbb{Z}$ in $\mathbb{Z}/3\mathbb{Z}$.   □

Quotient Groups

In summary, we know that for f a homomorphism from group G to group H, the kernel K of f is a subgroup of G; for every $x \in G$, $xK = Kx$; and finally, $G/K = \{xK \mid x \in G\}$ is a group under the operation $xK \cdot yK = xyK$ and is isomorphic to the image group $f(G)$.

Now let's forget about homomorphisms (for the moment) and see how much of our construction for G/K still holds when, instead of K, we use an arbitrary subgroup of G. Thus we will ask, for S an arbitrary subgroup of G, the following questions:

1. Do the sets $xS = \{xs \mid s \in S\}$ for the various x's in G form a partition of G?
2. If so, will the multiplication formula $xS \cdot yS = xyS$ be a well-defined binary operation on these sets?
3. If so, will the sets form a group under this operation?

Only question 2 will cause us any difficulty. First, we need some terminology.

7.104
Definition

For S a subgroup of a group G, sets of the form $xS = \{xs \mid s \in S\}$ for $x \in G$ are called **left cosets** of S in G; sets of the form $Sx = \{sx \mid s \in S\}$ for $x \in G$ are called **right cosets** of S in G.   □

7.105
Example

The set $S = \{1, 5, 8, 12\}$ forms a subgroup of $[\mathbb{Z}_{13}^*, \cdot_{13}]$. The left coset $3S$ equals $\{3 \cdot_{13} 1, 3 \cdot_{13} 5, 3 \cdot_{13} 8, 3 \cdot_{13} 12\} = \{3, 2, 11, 10\}$.   □

7.106
Practice

(a) The set $S = \{0, 4, 8\}$ forms a subgroup of $[\mathbb{Z}_{12}, +_{12}]$. Find the members of the left coset $6 + S$ (note the additive notation).

(b) The set $S = \{i, (2, 3)\}$ forms a subgroup of the group $[S_3, \circ]$. Find the elements of the left cosets $(1, 2, 3)S$ and $(1, 3, 2)S$. Find the elements of the right cosets $S(1, 2, 3)$ and $S(1, 3, 2)$.   □

Coset Partitions

Question 1 above asks whether the collection of distinct left cosets of S in G partitions G. This partitioning will certainly occur if these cosets are the equivalence classes of some equivalence relation ρ on G. We define a relation ρ in such a way that if ρ is an equivalence relation, the left cosets will be its equivalence classes; then we show that ρ is in fact an equivalence relation.

Thus, for S a subgroup of G, we define a binary relation ρ on G by $x \rho y \leftrightarrow$ "x and y are members of the same left coset of S in G." Because $x = x \cdot i_G$, $x \in xS$; to say that y is a member of the same left coset is to say that $y = xs$ for some $s \in S$. Therefore, we can redefine ρ as $x \rho y \leftrightarrow y = xs$ for some $s \in S$.

7.107
Practice

Show that ρ as defined above is an equivalence relation on G.   □

We conclude that the distinct left cosets of S in G partition G. A similar argument shows that the right cosets partition G (but perhaps differently).

Normal Subgroups and Coset Multiplication

We propose treating the left cosets of S in G as objects and defining an operation $\cdot$ on the left cosets by $xS \cdot yS = xyS$. Question 2 asks whether such an operation is well-defined. Recall that every time we define an operation on classes by using representatives of the classes, we must be sure that the answer is independent of the representatives used.

7.108
Example

Consider the left cosets $(1, 2, 3)S$ and $(1, 3, 2)S$ where $S = \{i, (2, 3)\}$ in $[S_3, \circ]$. Suppose we define coset multiplication by $(1, 2, 3)S \cdot (1, 3, 2)S = (1, 2, 3)(1, 3, 2)S = S$. Then (see Practice 7.106b) $(1, 2) \in (1, 2, 3)S$ and $(1, 3) \in (1, 3, 2)S$, but $(1, 2)(1, 3) = (1, 3, 2) \notin S$. The natural rule for coset multiplication does not work!   □

Why isn't coset multiplication well-defined in Example 7.108? What property does the kernel K of some homomorphism have that the subgroup S of S_3 lacks? For K the kernel of a homomorphism, any left coset xK is equal to the right coset Kx. By Practice 7.106b, $(1, 2, 3)S \neq S(1, 2, 3)$

and $(1, 3, 2)S \neq S(1, 3, 2)$. We will see that what we need to make our coset multiplication well-defined is the equality of left and right cosets.

7.109
Definition

Let S be a subgroup of a group G. Then S is a **normal subgroup** of G if for all $x \in G$, $xS = Sx$. □

First we will show that normality is a sufficient condition for coset multiplication to be well-defined. Suppose, then, that S is a normal subgroup of a group $[G, \cdot]$ and that coset multiplication is to be defined by the rule $xS \cdot yS = xyS$. We must show that for $x_1 \in xS$ and $y_1 \in yS$, the product $x_1 y_1 \in xyS$. We know that $x_1 = xs_1$ and $y_1 = ys_2$ for some $s_1, s_2 \in S$. The product $x_1 y_1$ is $(xs_1)(ys_2) = x(s_1 y)s_2$. The element $s_1 y$ is a member of the right coset Sy, which by normality equals the left coset yS. Thus, $s_1 y \in yS$ and $s_1 y = ys_3$ for some $s_3 \in S$. Continuing our computation,

$$x_1 y_1 = (xs_1)(ys_2) = x(s_1 y)s_2 = x(ys_3)s_2 = xy(s_3 s_2) \in xyS$$

When S is normal, coset multiplication is well-defined.

Conversely, when coset multiplication $xS \cdot yS = xyS$ is well-defined, the subgroup S must be normal. To prove this, we assume the operation to be well-defined and then show that $xS = Sx$ for any $x \in G$. We will show set inclusion in each direction. First, let $sx \in Sx$. We need (temporarily) an arbitrary element $y \in G$ in order to consider the product $yS \cdot xS = yxS$. Because $ys \in yS$, $xs \in xS$, and coset multiplication is well-defined, we must have $(ys)(xs) \in yxS$. Thus, $ys \cdot xs = yxs_1$ for some $s_1 \in S$. Multiplying both sides of this equation by y^{-1} on the left and s^{-1} on the right, we get $sx = xs_1 s^{-1} \in xS$, which proves that $Sx \subseteq xS$.

Now we show the opposite inclusion. Let $xs \in xS$. Then $x^{-1} \in G$, and by the inclusion we have already proved, $Sx^{-1} \subseteq x^{-1}S$, so $sx^{-1} = x^{-1}s_1$ for some $s_1 \in S$. Multiplying both sides of this equation on the left and right by x, we get $xs = s_1 x \in Sx$, proving that $xS \subseteq Sx$. Therefore, $xS = Sx$ and S is normal. The answer to question 2 can be summarized as follows.

7.110
Theorem

Let S be a subgroup of a group G. Coset multiplication $xS \cdot yS = xyS$ is well-defined if and only if S is a normal subgroup. □

If S is normal, left and right cosets are the same, and we could also write the formula for coset multiplication in the form $Sx \cdot Sy = Sxy$.

The Group Structure

Given that we have a normal subgroup and that as a result coset multiplication is well-defined, the answer to question 3 is easy: the set of cosets under this operation forms a group.

7.111
Practice

Show that for S a normal subgroup of a group G, the collection of (left) cosets of S in G forms a group under coset multiplication.   □

7.112
Definition

For S a normal subgroup of a group G, the group of cosets of S in G under coset multiplication is called the **quotient group (factor group) of G modulo S,** denoted by G/S.   □

Of course, G/K is a quotient group where K is the kernel of a homomorphism. To find other normal subgroups, and therefore other examples of quotient groups, let's develop a test for normality. Suppose that S is a subgroup of G with the property that for any $x \in G$ and $s \in S$, $x^{-1} sx \in S$. This property insures that S is normal. To prove this, let $sx \in Sx$. Then by our hypothesis, $x^{-1}sx \in S$, so $x^{-1}sx = s_1$ where $s_1 \in S$, and $sx = xs_1 \in xS$. Thus, $Sx \subseteq xS$. Now let $xs \in xS$. Then by our hypothesis, $(x^{-1})^{-1}sx^{-1} = xsx^{-1} \in S$, so $xsx^{-1} = s_1$ where $s_1 \in S$, and $xs = s_1x \in Sx$. Thus, $xS \subseteq Sx$ and $xS = Sx$, so that S is normal.

7.113
Theorem

If S is a subgroup of G such that for any $x \in G$ and $s \in S$, $x^{-1}sx \in S$, then S is normal in G.   □

7.114
Example

(a) Any subgroup S of a commutative group G is normal because for $x \in G$, $s \in S$, $x^{-1}sx = sx^{-1}x = s \in S$.

(b) A_n is normal in S_n, $n > 1$. For $\alpha \in S_n$, if α is even then so is α^{-1}, and if α is odd, α^{-1} is too. For $\beta \in A_n$, β is even. Thus, $\alpha^{-1} \beta \alpha$ is even.   □

Fundamental Homomorphism Theorem

Recall that a quotient group G/S can be constructed without reference to a homomorphism function; all that is required is that S be a normal subgroup of G. However, we can show that for any quotient group G/S, there is a related homomorphism.

7.115
Practice

Let S be a normal subgroup of the group $[G, \cdot]$, and define a function $f : G \to G/S$ by $f(x) = xS$ (thus f maps each element of G to its coset). Show that f is a homomorphism from G onto G/S.   □

To tie things together, we will restate the Homomorphism Theorem for Groups (Theorem 7.92) in the terminology we have developed in this section. We will also include Practice 7.115, which is essentially the converse of this theorem, in our restatement.

7.116
Theorem

The Fundamental Homomorphism Theorem for Groups
If f is a homomorphism from a group $[G, \cdot]$ to a group $[H, \cdot]$, then the kernel K of f is a normal subgroup of G, and the homomorphic image $f(G)$ is isomorphic to the quotient group G/K. Conversely, any quotient group of G is a homomorphic image of G. ☐

Theorem 7.116 says that every homomorphic image of a group G is (to within an isomorphism) a quotient group of G, and every quotient group of G is a homomorphic image of G. Homomorphic images and quotient groups of a given group are practically the same things.

Computer Arithmetic

Now we can settle the issue about computer arithmetic raised after Example 7.69. We want to simulate, as best we can, $[\mathbb{Z}, +]$ by a finite structure, call it $[H, +]$. Because $[\mathbb{Z}, +]$ is a group, $[H, +]$ should be a group. To preserve the operation of $[\mathbb{Z}, +]$ in $[H, +]$, we need a homomorphism from $[\mathbb{Z}, +]$ onto $[H, +]$. Thus, H is a homomorphic image of $\mathbb{Z}$, and by the Fundamental Theorem (7.116), it is essentially a quotient group of $\mathbb{Z}$. What are the possible quotient groups of $\mathbb{Z}$? They must be groups of the form $\mathbb{Z}/S$ where S is a normal subgroup of $[\mathbb{Z}, +]$. Because $[\mathbb{Z}, +]$ is commutative, normality presents no problem, so we concentrate on possible subgroups of $[\mathbb{Z}, +]$. By Theorem 7.58, any subgroup S must be $n\mathbb{Z}$ for some n. Thus, H is isomorphic to $\mathbb{Z}/n\mathbb{Z}$. If $n = 0$, then $n\mathbb{Z} = \{0\}$, and the distinct members of $\mathbb{Z}/\{0\}$ are $0 + \{0\} = 0$, $1 + \{0\} = 1$, $-1 + \{0\} = -1, \ldots$. Hence, $\mathbb{Z}/\{0\}$ is infinite, although we require H to be finite. Therefore, $n > 0$, and the distinct members of $\mathbb{Z}/n\mathbb{Z}$ are $0 + n\mathbb{Z}, 1 + n\mathbb{Z}, \ldots, (n - 1) + n\mathbb{Z}$. The function $h : \mathbb{Z}/n\mathbb{Z} \to \mathbb{Z}_n$ given by $h(k + n\mathbb{Z}) = k$ is clearly an isomorphism from $[\mathbb{Z}/n\mathbb{Z}, +]$ to $[\mathbb{Z}_n, +_n]$. Therefore, $H \simeq \mathbb{Z}/n\mathbb{Z} \simeq \mathbb{Z}_n$, or $H \simeq \mathbb{Z}_n$. We conclude that to simulate $[\mathbb{Z}, +]$ by a finite structure, we must use $[\mathbb{Z}_n, +_n]$ for some n (or an isomorphic copy of $[\mathbb{Z}_n, +_n]$).

Conversely, $\mathbb{Z}/n\mathbb{Z}$ is a quotient group of $[\mathbb{Z}, +]$ for any positive integer n and, by the converse half of the Fundamental Theorem, is a homomorphic image of $[\mathbb{Z}, +]$. Therefore, for any positive integer n, $\mathbb{Z}/n\mathbb{Z}$ (which again is essentially $[\mathbb{Z}_n, +_n]$) can be used to (imperfectly) simulate $[\mathbb{Z}, +]$.

Lagrange's Theorem

The final result of this chapter is a counting theorem about finite groups. But first, suppose that G is any group, not necessarily finite, and that S is a subgroup of G. The set of left cosets of S in G forms a partition of G, and so does the set of right cosets of S in G. Left and right cosets coincide if and only if S is normal in G. Yet regardless of whether S is normal, we can find a one-to-one correspondence between the elements of any left coset xS and the elements of S itself. The obvious mapping to try is $xs \rightarrow s$.

7.117
Practice

Let S be a subgroup of a group G, and let $x \in G$. Show that the function $f:xS \rightarrow S$ given by $f(xs) = s$ is a bijection.   □

If S is finite, say $|S| = k$, then by Practice 7.117, every left coset of S in G has k elements. A similar argument shows that every right coset of S in G also has k elements. Now suppose that G itself is finite, say $|G| = n$. The left cosets of S in G partition G; the n elements of G are distributed among the left cosets; and each left coset has k elements. Thus, we must have an equation of the form $n = km$ where m is the number of left cosets of S in G, giving us Lagrange's Theorem, first proved about 1771.

7.118
Theorem

Lagrange's Theorem
The order of a subgroup of a finite group divides the order of the group.   □

Each right coset of S in G also has k elements, so the equation $n = km$ also holds for the partition of G into right cosets where there are m right cosets. Thus, the number of left cosets of S in G equals the number of right cosets of S in G. (Naturally, this statement is trivial if S is normal in G.)

7.119
Definition

Let S be a subgroup of a finite group G. The number of left cosets (right cosets) of S in G is called the **index** of S in G, denoted by $[G:S]$. $[G:S]$ divides $|G|$.   □

Note that if S is normal in G, then $[G:S] = |G/S|$.

Lagrange's Theorem helps us narrow down the possibilities for subgroups of a finite group. If $|G| = 12$, for example, we would not look for any subgroups of order 7 since 7 does not divide 12. Also, we would know that G cannot be used to simulate any group H of order 7. Such a simulation would involve a homomorphism from G onto H, and by the Fundamental Homomorphism Theorem for Groups, $G/K \simeq H$, so $|G/K| = 7$, and again 7 does not divide 12.

Once more, let $|G| = 12$. The fact that 6 divides 12 does not imply the existence of a subgroup of G of order 6. In fact, A_4 is a group of order $4!/2 = 12$, but it can be shown that A_4 has no subgroups of order 6. The converse to Lagrange's Theorem does not always hold. In certain cases the converse can be shown to be true—for example, in finite commutative groups (note that A_4 is not commutative).

7.120
Practice

(a) From Lagrange's Theorem, what are the possible orders of subgroups of the group $G = \mathbb{Z}_2 \times \mathbb{Z}_3 \times \mathbb{Z}_4$?

(b) For each number in the answer to (a), will G have a subgroup of this order? □

✔ Checklist

Definitions

kernel of a homomorphism (*p. 326*)
left and right cosets (*p. 328*)
normal subgroup (*p. 330*)
quotient group (factor group) (*p. 331*)
index of a subgroup (*p. 333*)

Techniques

Find the kernel of a homomorphism from a group G to a group H.

Given any normal subgroup S of G, find the distinct cosets of S in G and perform arithmetic in the quotient group G/S.

Main Ideas

If S is a normal subgroup of a group G, then the (left) cosets of S in G form a group G/S under the operation $xS \cdot yS = xyS$. Normality is necessary and sufficient for this operation to be well-defined.

The kernel K of a homomorphism f from a group G to a group H is a normal subgroup of G, and $G/K \simeq f(G)$.

Quotient groups of a group and homomorphic images of that group are essentially the same.

Computer simulation of $[\mathbb{Z}, +]$ must be done with $[\mathbb{Z}_n, +_n]$ for some n, but n can be any positive integer.

The order of a subgroup of a finite group divides the order of the group.

Exercises Section 7.5

1. Let $8\mathbb{Z} = \{8z \mid z \in \mathbb{Z}\}$. $8\mathbb{Z}$ is a normal subgroup of the commutative group $[\mathbb{Z}, +]$.
    (a) List the distinct cosets that are the elements of the quotient group $\mathbb{Z}/8\mathbb{Z}$ (name them and indicate their members).

(b) What is the sum $(3 + 8\mathbb{Z}) + (6 + 8\mathbb{Z})$ in $\mathbb{Z}/8\mathbb{Z}$?

2. The set $S = \{i, (1, 2) \circ (3, 4), (1, 4) \circ (2, 3), (1, 3) \circ (2, 4)\}$ is a subgroup of S_4.
   (a) Write all the elements of S_4.
   (b) Find all the left cosets of S in S_4.

★ 3. Let $S = \{0, 4, 8\}$. Then S is a normal subgroup of the commutative group $[\mathbb{Z}_{12}, +_{12}]$.
   (a) List the distinct cosets that are the elements of the quotient group $\mathbb{Z}_{12}/S$ (name them and indicate their members).
   (b) What is the sum $(2 + S) + (3 + S)$ in $\mathbb{Z}_{12}/S$?

4. Let $S = \{(0, 0), (1, 2), (2, 0), (0, 2), (1, 0), (2, 2)\}$. Then S is a normal subgroup of the commutative group $[\mathbb{Z}_3 \times \mathbb{Z}_4, +]$.
   (a) List the distinct cosets that are the elements of the quotient group $(\mathbb{Z}_3 \times \mathbb{Z}_4)/S$ (name them and indicate their members).
   (b) What is the sum of $(1, 1) + S$ and $(1, 1) + S$ in $(\mathbb{Z}_3 \times \mathbb{Z}_4)/S$?

★ 5. Let $S = \{1, 2, 4\}$. Then S is a normal subgroup of the commutative group $[\mathbb{Z}_7^*, \cdot_7]$ where $\mathbb{Z}_7^* = \mathbb{Z}_7 - \{0\}$.
   (a) List the distinct cosets that are the elements of the quotient group $\mathbb{Z}_7^*/S$.
   (b) What is the product of $3S$ and $4S$ in $\mathbb{Z}_7^*/S$?

6. In each case below, two groups G and H are listed and a homomorphism $f : G \to H$ is given. Describe the kernel K and describe the quotient group G/K.
   (a) $[\mathbb{Z}, +], [\mathbb{Z}_{24}, +_{24}]; f(x) = 3 \cdot_{24} x$
   (b) $[\mathbb{Z}_{12}, +_{12}], [\mathbb{Z}_8, +_8]; f(x) = 2 \cdot_8 x$
   (c) $[\mathbb{R}^* \times \mathbb{R}^*, \cdot], [\mathbb{R}^*, \cdot]$ where $\mathbb{R}^* = \mathbb{R} - \{0\}; f(x, y) = xy$
   (d) $[\mathbb{R}[x], +], [\mathbb{R}, +]; f(a_n x^n + a_{n-1} x^{n-1} + \cdots + a_0) = a_0$
   (e) $[M_2(\mathbb{Q}), +], [\mathbb{Q}, +]; f\left(\begin{bmatrix} a & b \\ c & d \end{bmatrix}\right) = a - d$

★ 7. The function f defined by $f(x) = x \cdot_k 1$ is a homomorphism from $[\mathbb{Z}, +]$ to $[\mathbb{Z}_k, +_k]$. Describe the kernel K.

8. A function $f : \mathbb{Z} \times \mathbb{Z} \to \mathbb{Z}$ is defined by $f(x, y) = x + y$.
   (a) Show that f is a homomorphism from $[\mathbb{Z} \times \mathbb{Z}, +]$ onto $[\mathbb{Z}, +]$.
   (b) Describe the kernel K of f.
   (c) Which of the following elements belong to the coset $(2, 3) + K$?
       $(4, 1), (3, 4), (1, 4), (5, 0)$
   (d) Write the function that makes $(\mathbb{Z} \times \mathbb{Z})/K \simeq \mathbb{Z}$.

★ 9. A function $f : \mathbb{Q} \times \mathbb{Q} \times \mathbb{Q} \to \mathbb{Q} \times \mathbb{Q}$ is defined by $f(x, y, z) = (y, y)$.
   (a) Find $f(\mathbb{Q} \times \mathbb{Q} \times \mathbb{Q})$ and show that f is a homomorphism from $[\mathbb{Q} \times \mathbb{Q} \times \mathbb{Q}, +]$ onto $[f(\mathbb{Q} \times \mathbb{Q} \times \mathbb{Q}), +]$.

(b) Describe the kernel K of f.

(c) Which of the following elements belong to the coset $(3, 4, 5) + K$? $(2, 3, 4)$, $(4, 4, 4)$, $(4, 5, 6)$, $(5, 4, 3)$

(d) Is the computation $((-7, 2, 6) + K) + ((5, 1, 9) + K) = (10, 3, 2) + K$ in $(\mathbb{Q} \times \mathbb{Q} \times \mathbb{Q})/K$ correct?

(e) Write the function that makes $(\mathbb{Q} \times \mathbb{Q} \times \mathbb{Q})/K \simeq f(\mathbb{Q} \times \mathbb{Q} \times \mathbb{Q})$.

10. A function $f : \mathbb{Q}^* \to \mathbb{Q}^+$ is defined by $f(x) = x^2$.

(a) Find $f(\mathbb{Q}^*)$ and show that f is a homomorphism from $[\mathbb{Q}^*, \cdot]$ onto $[f(\mathbb{Q}^*), \cdot]$.

(b) Describe the kernel K of f.

(c) List all the elements of the product $(3K)(7K)$ in $\mathbb{Q}^*/K$.

(d) Write the function that makes $\mathbb{Q}^*/K \simeq f(\mathbb{Q}^*)$.

11. For two groups $[G, \cdot]$ and $[H, \cdot]$, define $f : G \times H \to G$ by $f(x, y) = x$.

(a) Prove that f is an onto homomorphism.

(b) Describe the kernel K of f.

(c) Write the function that makes $(G \times H)/K \simeq G$.

12. Let F denote the set of all functions $f : \mathbb{R} \to \mathbb{R}$. For $f, g \in F$, define addition of functions as follows: for $x \in \mathbb{R}$, $(f + g)(x) = f(x) + g(x)$.

(a) Prove that $[F, +]$ is a group.

(b) Let $a \in \mathbb{R}$. Define a function $\alpha : F \to \mathbb{R}$ by $\alpha(f) = f(a)$. Prove that α is a homomorphism from $[F, +]$ onto $[\mathbb{R}, +]$.

(c) Describe the kernel K of α.

(d) Let $g + K$ denote a member of F/K. For any member h of $g + K$, what is the value of $h(a)$?

(e) Write the function that makes $F/K \simeq \mathbb{R}$.

★ 13. Let f be a homomorphism from a group G onto a group H. Show that f is an isomorphism if and only if the kernel of f is $\{i_G\}$.

★ 14. For any group $[G, \cdot]$, the **center** of the group is $A = \{x \in G \,|\, x \cdot g = g \cdot x \text{ for all } g \in G\}$. $[A, \cdot]$ is a subgroup of $[G, \cdot]$ (see Exercise 14 of Section 7.2). Show that A is a normal subgroup.

15. For $n > 1$, find the order of the quotient group S_n/A_n.

16. Let n and k be fixed positive integers with k a divisor of n. Show that $[\mathbb{Z}_n, +_n]$ has a quotient group of order k.

17. (a) Prove that a quotient group of a commutative group is commutative.

(b) A group $[G, \cdot]$ is **divisible** if for any $g \in G$ and any positive integer n, the equation $x^n = g$ has at least one solution in G. Prove that a quotient group of a divisible group is divisible.

★ 18. Prove the converse of Theorem 7.113: if S is a normal subgroup of a group G, then for any $x \in G$ and $s \in S$, $x^{-1}sx \in S$.

19. Let $G = \{(a, b) \mid a, b \in \mathbb{R}, a \neq 0\}$.
    (a) Prove that $[G, \cdot]$ is a group where $(a, b) \cdot (c, d) = (ac, ad + b)$.
    (b) Let $S = \{(a, 0) \mid a \in \mathbb{R}, a \neq 0\}$, $T = \{(1, b) \mid b \in \mathbb{R}\}$. Show that S and T are subgroups of G.
    (c) Prove that S is not a normal subgroup of G but that T is. (Hint: use Exercise 18).

20. (a) Let S and T be normal subgroups of a group G. Show that $S \cap T$ is a normal subgroup of G. (Hint: use Exercise 18).
    (b) Let S_1 and S_2 be normal subgroups of the groups G_1 and G_2, respectively. Show that $S_1 \times S_2$ is a normal subgroup of the group $G_1 \times G_2$.

21. (a) Let H be a subgroup of a group G. Show that $g^{-1} H g$ is a subgroup of G for any $g \in G$.
    (b) Let H be a finite subgroup of a group G and suppose that H is the only subgroup of G with order $|H|$. Show that H is normal in G.

22. Let f be a homomorphism from a group G to a group H, and let S be a normal subgroup of G. Show that $f(S)$ is a normal subgroup of $f(G)$.

★ 23. Let S and T be normal subgroups of a group G with $T \subseteq S$. Show that T is normal in S and that S/T is a normal subgroup of G/T.

24. Let S and T be normal subgroups of a group G with $T \subseteq S$.
    (a) Show that if $xT = yT$ in G/T, then $xS = yS$ in G/S.
    (b) Let $f : G/T \to G/S$ be defined by $f(xT) = xS$. By part (a), f is a well-defined function. Show that f is a homomorphism.
    (c) Prove that the quotient group S/T is the kernel of f.
    (d) Prove that $(G/T)/(S/T) \simeq G/S$. (Note the suggestion of cancellation here.)

25. Let G be a group and for each $g \in G$, define a function $f_g : G \to G$ by $f_g(x) = gxg^{-1}$.
    (a) Prove that each f_g is an isomorphism from G onto G.
    (b) Let $F = \{f_g \mid g \in G\}$. Show that F is a group under function composition.
    (c) Let A be the center of the group G, $A = \{g \in G \mid xg = gx$ for all $x \in G\}$. Prove that $G/A \simeq F$.

26. Let S be a subgroup of a finite group $[G, \cdot]$ with $[G:S] = 2$. Show that S is a normal subgroup of G.

27. Let S be a subgroup of a finite group $[G, \cdot]$ with $[G:S] = 2$. Prove that $x^2 \in S$ for any $x \in G$. (Hint: use Exercise 26, Theorem 7.74, and the fact that the composition of homomorphisms is a homomorphism.)

★ 28. Let S and T be subgroups of a finite group $[G, \cdot]$ with $T \subseteq S$. Show that $[G:T] = [G:S] \cdot [S:T]$.

29. We know that for any finite group G and subgroup S, the number of left cosets of S in G equals the number of right cosets of S in G. Thus, if L is the set of all left cosets of S in G and R is the set of all right cosets of S in G, then there is a bijection from L onto R. Show that such a bijection exists even if G is infinite.

Chapter 8

Coding Theory

One area of computer science that relies heavily on algebraic structures, such as the group structure defined in Chapter 7, is information coding. Errors can be introduced into data transmission or data storage through a variety of means such as hardware failure, interference, or the general random glitches to which sensitive, complex electronic equipment is subject. One can protect against such errors by coding the information before it is sent or stored and then decoding it when it is received or retrieved. The method of coding and decoding should in some way maximize the probability of correcting, or at least detecting, any errors. This chapter introduces a kind of code called a group code, where the encoded words form a group under an appropriate binary operation. These codes barely scratch the surface of the work that has been done in coding theory. More sophisticated coding methods involve algebraic structures more complex than groups.

SECTION 8.1   ENCODING

At the mention of coding information, many people think of double agents scribbling mysterious symbols on a piece of paper that is passed in a folded newspaper to someone sitting on a park bench (at least that's how it's always done in the movies). Coding and decoding information for the purpose of insuring its secrecy is known as **cryptology** and involves much interesting mathematics. Exercises 1 through 3 at the end of this section discuss some coding and decoding techniques suitable for cryptology. However, in many other types of codes secrecy is not the main object. What is your ZIP code? What is your area code? What is your Social Security number? The postal service uses ZIP codes to represent geographic localities in numerical form. The telephone company uses area codes to represent clusters of switching networks in numerical form. And many organizations, both government and private, use your Social Security number to represent *you* in numerical form!

Coding, in general, is simply a translation of information from one form to another, more convenient form. Some other familiar examples are the Universal Product Code—the series of black vertical lines containing product information found on many grocery items and other goods—and the code a computer uses (ASCII, for example, or EBCDIC) to convert alphabetic characters into binary form for computer representation.

General Ideas About Information Coding

In this chapter we will be interested in codes intended to protect the information being transmitted. Here we do not mean protection in the cryptologic sense, but protection against corruption in the information itself. In the communication model shown in Figure 8.1, the channel medium could be air (to transmit radio waves, voices, or satellite signals), wires (to transmit electronic signals), or magnetic tape (to transmit sequences of binary digits). Because any channel is subject to disturbances that can corrupt the information being transmitted, what is received may not be what was sent. The disturbances can be due to hardware failures, random interference from other sources (**noise**), or deterioration (since transmission can take place over time in the sense of data storage as well as over space).

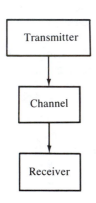

Figure 8.1

Let's play the children's game of telephone, which involves voice transmission. Suppose the transmitted message is "the black hat" but the received message (in the case of voice transmission, we might say the perceived message) is "the black cat." Because the received message makes sense and *could* have been the transmitted message, there is no way to detect that an error has occurred. An alternative is to encode the message to be transmitted by repeating it. We will call an encoded version of a message a **code word.** Thus, the code word for "the black hat" would be "the black hat the black hat," and the code word for "the black cat" would be "the black cat the black cat." A received message of "the black hat the black cat" would alert the receiver that an error has occurred in transmission. However, the received message is equally close to either of the two code words, so there would be no way to be certain what the correct code word was. The receiver would have to request a retransmission. Notice that two errors could still go undetected.

Now let's make our code words three copies of the message: "the black hat the black hat the black hat" and "the black cat the black cat the black cat." A received message of "the black hat the black cat the black hat" would signal that either one or two errors have occurred. If we assume it likelier that one error has occurred, we decode the message to the closest code word, "the black hat the black hat the black hat." This process, called **maximum-likelihood decoding,** gives us the correct code word for a received message when no more than one error has occurred. Since we can detect up to two errors and correct the effects of one error, we have designed a **double-error detecting, single-error correcting code.**

Out of this silly example we get three useful ideas: maximum-likelihood decoding, redundancy in coding, and distance of received messages from code words. These ideas are related, and we will look at each again. From this example, we can see that encoding messages with a certain amount of redundancy increases the capability of detecting and perhaps correcting errors. On the other hand, the redundancy also "wastes" transmission time and makes the code words longer than the original messages, thus introducing more opportunities for errors to take place. In any given application, the benefits and drawbacks of using a particular coding scheme must be weighed.

Binary Codes

We will assume that any information we want to transmit can be represented as a sequence of binary digits, that is, 0s and 1s. We will also assume that errors occurring during transmission are independent—a given disturbance affects only one binary digit and not the surrounding digits. Our final assumption is that it is just as likely for a 0 to be corrupted to a 1 as it is for a 1 to be corrupted to a 0 (although in actuality 1s get corrupted more frequently than 0s), and that the probability of either event happening is relatively small, say .01. For a sequence of three binary digits, the probability of correct transmission of all three digits is $(.99)(.99)(.99) = (.99)^3$. The probability of a single digit being in error (which can occur in any of three places) is $(.01)(.99)(.99) + (.99)(.01)(.99) + (.99)(.99)(.01) = 3(.01)(.99)^2$. Similarly, the probability of two digits being in error is $3(.01)^2(.99)$, and the probability of errors in all three digits is $(.01)^3$. Maximum-likelihood decoding simply assumes the most probable situation—that the smallest possible number of errors has occurred. A single-error correcting code, for example, does not always decode correctly, but it will do so in the likeliest cases, namely, in any case where only one error has occurred (or no errors have occurred).

Parity Checks

A simple redundancy in coding binary words involves adding a digit, called a **parity bit digit,** to the end of a word. Thus, code words for m-tuple messages become $(m + 1)$-tuples. In an **even parity check,** the last digit is chosen so as to make the total number of 1s in the $(m + 1)$-tuple code word an even number; in an **odd parity check,** the last digit is chosen so as to make the total number of 1s odd. For example, in an even parity check code, the code word for 1011 is 10111. Any single error in transmission can be detected because it produces an odd number of 1s in the received word. Because it is impossible to tell which digit is in error, however, this code has no correcting capabilities. Also, any even number of errors in transmission cannot be detected, and any odd number of errors is indistinguishable from a single error. Thus, this code is single-error detecting. Such codes are used when storing data on auxiliary memory media for computers, such as magnetic tapes and disks.

The codes we will consider in this chapter are generalizations of the even-parity-check code in which the m-tuple message becomes the first m components of an n-tuple code word, and the additional $n - m$ binary digits are all special sorts of parity checks. This technique is often used to detect and correct errors in the main memory of a computer where the computer word length is m bits. For a machine advertising an m-bit word length, the hardware itself incorporates the check digits and the actual word length stored is n. Before we consider the general case, we need to define the distance between code words. (Richard W. Hamming pioneered the study of error-detecting and error-correcting codes in 1950.)

Distance

8.1 Definition   Let X and Y be binary n-tuples. The **Hamming distance** between X and Y, $H(X, Y)$, is the number of components in which X and Y differ.   □

8.2 Practice   For $X = 01011$ and $Y = 11001$, what is $H(X, Y)$? (Note that we'll often leave out parentheses and commas when describing our n-tuples.)   □

8.3 Practice   Show that Hamming distance is a *metric* on the set of binary n-tuples; that is, show that for all binary n-tuples X, Y, and Z:

    (a) $H(X, Y) \geq 0$
    (b) $H(X, Y) = 0$ if and only if $X = Y$
    (c) $H(X, Y) = H(Y, X)$
    (d) $H(X, Z) \leq H(X, Y) + H(Y, Z)$ □

Now we will assume that we have a code, that is, a collection of code words, all of which are binary n-tuples.

8.4 Definition   The **minimum distance** of a code is the minimum Hamming distance between all possible pairs of distinct code words.   □

Each error that occurs in the transmission of a code word adds one unit to the Hamming distance between that code word and the received word. According to our maximum-likelihood decoding rule, we will decode a received word as the closest code word in terms of Hamming distance. Thus, it is not surprising that the error-detecting and error-correcting capabilities of our code are directly related to distance. Let's consider this further.

Suppose we picture the code words as specific binary n-tuples distinguished from the set S of all binary n-tuples—as in Figure 8.2. Suppose also that the minimum distance of the code is at least $d + 1$. Then any time a code word is corrupted by d or fewer errors, it will be changed to an n-tuple that is not another code word, and the occurrence of errors can be detected. Conversely, if any combination of d or fewer errors can be detected, code words must be at least $d + 1$ apart. For example, if the only two code words are 00000 and 11111, the minimum distance is 5. If 00000 is corrupted by 4 errors to, let's say, 10111, this will be detected because the received word does not match either code word.

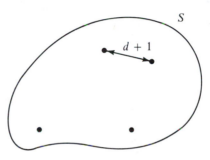

Figure 8.2

8.5 Theorem A code is d-error detecting (can detect all combinations of d or fewer errors) if and only if its minimum distance is at least $d + 1$. ☐

Now suppose that the minimum distance of the code is at least $2d + 1$. Then any time a code word X is corrupted by d or fewer errors, the received word X' will be such that $H(X, X') \le d$, but for any other code word Y, $H(X', Y) \ge d + 1$; X' will be correctly decoded as X. Conversely, to correct any received word with d or fewer errors, the minimum distance of the code must be at least $2d + 1$ so that neighborhoods of radius d around code words do not intersect (see Figure 8.3). Again, suppose the only two code words are 00000 and 11111, with minimum distance 5. If 00000 is corrupted by 2 errors to 11000, it will be correctly decoded to 00000, the closest code word, but if 00000 is corrupted by 3 errors to 11001, it will be incorrectly decoded to 11111, the closest code word.

8.6 Theorem A code is d-error correcting (can correct all combinations of d or fewer errors) if and only if its minimum distance is at least $2d + 1$. ☐

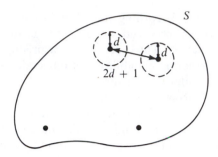

Figure 8.3

8.7 Example Suppose a code has a minimum distance of 6. Then it can detect any combination of ≤ 5 errors and correct any combination of ≤ 2 errors. If code words X and Y are such that $H(X, Y) = 6$, then if there are words X' and X'' produced by 5 errors on X and 4 errors on X, respectively, they will be incorrectly decoded as Y; if there is a word X''' produced by 3 errors on X, it can be arbitrarily decoded correctly as X or incorrectly as Y (see Figure 8.4). This code is double-error correcting, 5-error detecting. $\square$

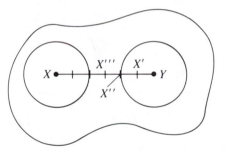

Figure 8.4

Theorems 8.5 and 8.6 show that, as we would expect, it requires a stronger condition to correct errors than merely to detect them. One might wonder what good merely detecting an error does if the error cannot be corrected. The answer is that it prompts the receiver to request a retransmission from the sender. Such retransmissions are possible in data transmission, say in a computer network, but not in data storage, where the error may be detected long after the "sender" is no longer able to retransmit.

Group Codes

Suppose once again that our code words are binary n-tuples. The set of all binary n-tuples can be expressed as $\mathbb{Z}_2^n$, and this set forms a group under componentwise addition modulo 2; we'll denote this group by $[\mathbb{Z}_2^n, +_2]$.

8.8 Practice   In the group $[\mathbb{Z}_2^5, +_2]$, calculate:

(a) $(01101) +_2 (11011)$
(b) $-(10110)$ □

The set of code words is a subset of $\mathbb{Z}_2^n$, and we want these code words to be sufficiently widely scattered in $\mathbb{Z}_2^n$ so that the minimum distance of the code is large enough to allow some error correction. The minimum distance of the code, as we shall see, is easy to compute if the code words form a subgroup of $[\mathbb{Z}_2^n, +_2]$. In this case we have a **group code**. The n-tuple of all 0s is the group identity, denoted by 0_n.

8.9 Definition   The **weight** of a code word X in $\mathbb{Z}_2^n$, $W(X)$, is the number of 1s it contains.   □

8.10 Theorem   The minimum distance of a group code is the minimum weight of all the nonzero code words.

Proof: Let d be the minimum distance of a group code; then there are distinct code words X and Y with $H(X, Y) = d$. Because we have a group code, closure holds and $X +_2 Y = Z$ is a code word. The code word $Z \neq 0_n$ because X and Y are distinct; in fact, Z will have 1s in exactly those components where X and Y differ, so $W(Z) = H(X, Y) = d$. Thus, the minimum weight of the code is $\leq d$. If the minimum weight is $< d$, let W be a nonzero code word with $W(W)$ the minimum weight. Then $H(W, 0_n) = W(W) < d$ (remember 0_n is a code word), which contradicts the fact that d is the minimum distance of the code. Therefore, the minimum distance equals the minimum weight.   □

8.11 Example   The set $\{00000, 01111, 10101, 11010\}$ is a group code in $\mathbb{Z}_2^5$. The minimum distance of the code is 3.   □

8.12 Practice   Verify that closure holds for the code words in Example 8.11.   □

Generating Group Codes

How can we produce subgroups of $\mathbb{Z}_2^n$ to use as code words, and how can we control the minimum distance of the code? Again, some algebraic ideas come to our rescue. Let **H** be any $n \times r$ binary matrix with $r < n$. If $X \in \mathbb{Z}_2^n$, we can treat X as a $1 \times n$ matrix and then perform the matrix multiplication $X \cdot \mathbf{H}$, where all multiplications and additions are done modulo 2. The result of this multiplication is a $1 \times r$ matrix that we can treat as a binary r-tuple.

8.13 Example   Let $r = 3$, $n = 5$, and

$$\mathbf{H} = \begin{bmatrix} 1 & 1 & 0 \\ 0 & 1 & 1 \\ 1 & 0 & 1 \\ 1 & 0 & 0 \\ 1 & 1 & 0 \end{bmatrix}$$

Then

$$[11010] \begin{bmatrix} 1 & 1 & 0 \\ 0 & 1 & 1 \\ 1 & 0 & 1 \\ 1 & 0 & 0 \\ 1 & 1 & 0 \end{bmatrix} = [001]$$

$\square$

We can thus think of multiplication by $\mathbf{H}$ as a mapping from the group $[\mathbb{Z}_2^n, +_2]$ to the group $[\mathbb{Z}_2^r, +_2]$. Furthermore, this mapping is a group homomorphism (see Exercise 6, this section.) Now let's consider exactly those members X of $\mathbb{Z}_2^n$ such that $X \cdot \mathbf{H} = 0_r$, the zero of the group $[\mathbb{Z}_2^r, +_2]$. This set is the kernel of the group homomorphism and is therefore a subgroup of $[\mathbb{Z}_2^n, +_2]$. We take this set to be the set of code words. Then we can easily determine the minimum weight (minimum distance) of the code simply by looking at $\mathbf{H}$. If $\mathbf{H}$ has d distinct rows that add to 0_r in $[\mathbb{Z}_2^r, +_2]$, say $i_1, i_2, \ldots, i_d$, we can choose an X in $\mathbb{Z}_2^n$ having 1s exactly in the $i_1, i_2, \ldots, i_d$ components. Then $X \cdot \mathbf{H} = 0_r$, so that X is a code word and $W(X) = d$. On the other hand, if X is a code word with $W(X) = d$ and X has 1s exactly in components $i_1, i_2, \ldots, i_d$, then the equation $X \cdot \mathbf{H} = 0_r$ forces rows $i_1, i_2, \ldots, i_d$ of $\mathbf{H}$ to sum to 0_r. Therefore, the minimum weight of the code equals the minimum number of distinct rows of $\mathbf{H}$ that add to 0_r. In particular, to produce a single-error correcting code, we must have distance at least 3; thus we would have to choose an $\mathbf{H}$ with no row consisting of all 0s and no two rows alike (these would add to 0_r).

8.14 Example   The code words of Example 8.11 were generated using the matrix $\mathbf{H}$ where

$$\mathbf{H} = \begin{bmatrix} 1 & 0 & 1 \\ 1 & 1 & 1 \\ 1 & 0 & 0 \\ 0 & 1 & 0 \\ 0 & 0 & 1 \end{bmatrix}$$

$\mathbf{H}$ has no row of all 0s and no two rows alike, but rows 1, 2, and 4 add to $(0, 0, 0)$. Again, we see that the minimum distance of this code is 3.

$\square$

8.15 Practice For each code word X of Example 8.11, verify that $X \cdot \mathbf{H} = 0_3$ where $\mathbf{H}$ is given above. □

In computing the product $X \cdot \mathbf{H}$, we multiply elements of X by corresponding elements of the columns of $\mathbf{H}$ and then sum. For each column of $\mathbf{H}$, the pattern of 1s in the column determines which components of X contribute to the sum. If the sum is to be 0 (as is true when $X \cdot \mathbf{H} = 0_r$), then those selected components of X must sum to 0 and therefore must consist of an even number of 1s. Thus, for a code word X, each column of $\mathbf{H}$ performs an even parity check on selected components of X. $\mathbf{H}$ is called a **parity check matrix.**

Canonical Parity-Check Matrix

For an $n \times r$ parity check matrix $\mathbf{H}$, we have not yet said anything about the *size* of the code that $\mathbf{H}$ generates (how many code words we can have). We can best address this question by assuming that our parity check matrix has the **canonical form**

$$\mathbf{H} = \begin{bmatrix} \mathbf{B} \\ \text{- - -} \\ \mathbf{I}_r \end{bmatrix}$$

where $\mathbf{I}_r$ is the $r \times r$ identity matrix and $\mathbf{B}$ is an arbitrary $(n - r) \times r$ binary matrix. The matrix of Example 8.14 is in canonical form. The $\mathbf{I}_r$ portion of $\mathbf{H}$ has the effect that each column of $\mathbf{H}$ selects a distinct component from among the last r components of X when the multiplication $X \cdot \mathbf{H}$ is performed. Each of the last r components of X therefore controls the even parity check for one of the r multiplications that are done. For X to be a code word, the first $n - r$ components can be arbitrary, but the final r components are then determined. The maximum number of code words is the maximum number of ways to select binary $(n - r)$-tuples, or 2^{n-r}. Let $m = n - r$. We can code all members of $\mathbb{Z}_2^m$ as code words in $\mathbb{Z}_2^n$ by leaving the first m components alone and then choosing the last r components so that the even parity check works for each column of $\mathbf{H}$. This procedure helps us encode members of $\mathbb{Z}_2^m$ in $\mathbb{Z}_2^n$; such a code is called an (n, m) code. The first m components of a code word are the **information digits,** and the last r components are the **check digits.**

8.16 Example The matrix $\mathbf{H}$ of Example 8.14 is

$$\mathbf{H} = \begin{bmatrix} 1 & 0 & 1 \\ 1 & 1 & 1 \\ 1 & 0 & 0 \\ 0 & 1 & 0 \\ 0 & 0 & 1 \end{bmatrix}$$

This matrix is in canonical form, where $n = 5$, $r = 3$, and $m = n - r = 2$. Matrix $\mathbf{H}$ can thus generate $2^2 = 4$ code words. The 4 members of $\mathbb{Z}_2^2$ are 00, 01, 10, and 11. Each can be coded as a member of $\mathbb{Z}_2^5$ by keeping the first two digits and adding the appropriate check digits. To code 10, for instance, we have

$$[10C_1C_2C_3] \begin{bmatrix} 1 & 0 & 1 \\ 1 & 1 & 1 \\ 1 & 0 & 0 \\ 0 & 1 & 0 \\ 0 & 0 & 1 \end{bmatrix} = [000]$$

Thus $C_1 = 1$, $C_2 = 0$, and $C_3 = 1$. We encode 10 as 10101.   □

8.17 Practice   Use the encoding procedure and the matrix $\mathbf{H}$ above to code 00, 01, and 11 in $\mathbb{Z}_2^5$. Compare the results with Example 8.11.   □

Hamming Codes

For a given $n \times r$ canonical parity-check matrix $\mathbf{H}$, we now know how to encode all of $\mathbb{Z}_2^m = \mathbb{Z}_2^{n-r}$ as a subgroup of $\mathbb{Z}_2^n$ and how to determine from $\mathbf{H}$ the minimum distance of the resulting code. Now let's turn the problem around. Suppose we want to encode $\mathbb{Z}_2^m$ as, say, a single-error correcting code. How big will the code words have to be (what is n), or equivalently, how many check digits must be added (what is r)? Once we know the dimensions of a parity check matrix $\mathbf{H}$, how can we find a canonical $\mathbf{H}$ that generates the code?

If the code is to be single-error correcting, no two rows of $\mathbf{H}$ can be alike. Since each row has r elements, there can be no more than 2^r rows; however, we cannot have the zero row, so the number of possible rows is $\leq (2^r - 1)$. Thus, $n \leq 2^r - 1$ and $m = n - r \leq 2^r - r - 1$. This inequality says that to code the set of all binary m-tuples in a single-error correcting code requires r check digits where r is such that $m \leq 2^r - r - 1$. Then $\mathbf{H}$ can be taken to be any matrix of the form

$$\begin{bmatrix} \mathbf{B} \\ \cdots \\ \mathbf{I}_r \end{bmatrix}$$

where $\mathbf{B}$ is an $m \times r$ binary matrix with at least two 1s in each row (and no rows alike). If r and m are such that $m = 2^r - r - 1$, then the code is said to be a **perfect code**. In this case, the rows of the matrix $\mathbf{B}$ consist of all the $2^r - r - 1$ r-tuples with at least two 1s, and the code generated by $\mathbf{H}$ is called a **Hamming code**. We have shown the following result.

8.18 Theorem   For any number $n = 2^r - 1$ there exists a perfect, single-error correcting Hamming code of length $2^r - 1$ and size $2^{2^r - r - 1}$.   □

8.19 Example   The code of Example 8.16 is a (5, 2) code; $n = 5$, $m = 2$, and $r = 3$. The code is single-error correcting and satisfies the relation $m \leq 2^r - r - 1$, but it is not a perfect code. The perfect code that has 3 check digits is a (7, 4) code. A matrix generating such a code is

$$\mathbf{H} = \begin{bmatrix} 1 & 1 & 0 \\ 0 & 1 & 1 \\ 1 & 1 & 1 \\ 1 & 0 & 1 \\ 1 & 0 & 0 \\ 0 & 1 & 0 \\ 0 & 0 & 1 \end{bmatrix}$$

□

8.20 Practice   The **H** given in Example 8.19 can encode $\mathbb{Z}_2^4$ in $\mathbb{Z}_2^7$. Give the list of 16 code words generated by the encoding procedure.   □

An Encoding Circuit

We can actually design a logic network to perform the encoding process for a group code, given an $n \times r$ canonical parity-check matrix **H**. We will need one new type of logic element that we did not discuss in Chapter 6. The AND gates, OR gates, and inverters of Chapter 6 all produced output instantaneously. Now we want to consider a sequence of input signals that arrive at specific times, designated t_1, t_2, and so forth. Our new element, called a **delay element** or **shift element,** receives an input signal at time t and outputs that same signal at time $t + 1$. A string of m such elements, called a **shift register,** if fed a sequence $x_1, x_2, \ldots, x_m$ of binary values (left to right), will be filled with these values after m units of time and will then output x_1 at time t_{m+1}, x_2 at time t_{m+2}, and so on, until all m input values have been shifted right and output in order (see Figure 8.5).

We also need a logic device to perform addition modulo 2. The output for addition modulo 2 of two inputs is given by the sum digit s of the half-adder of Section 6.1. From Figure 6.15a, we can see that the circuit of Figure 8.6 computes $x_1 +_2 x_2$. Because addition modulo 2 is associative, this circuit can be combined with similar ones to give a circuit for addition modulo 2 of any number of inputs. The output of such a circuit is 1 if and only if there is an odd number of 1s in the input. We will denote such a device by $\boxed{+_2}$.

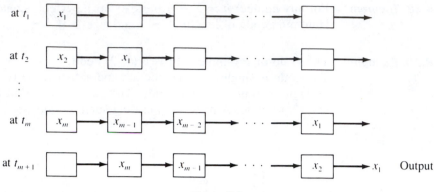

Figure 8.5

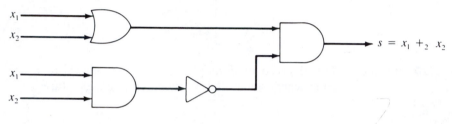

Figure 8.6

For

$$\mathbf{H} = \begin{bmatrix} \mathbf{B} \\ \cdots \\ \mathbf{I}_r \end{bmatrix}$$

the encoding is performed by adding r check digits to a binary m-tuple $x_1 x_2 \cdots x_m$ to produce a binary n-tuple $X = x_1 x_2 \cdots x_n$ with $X \cdot \mathbf{H} = 0_r$. The components of column j of $\mathbf{H}$ determine the jth check digit in X. The jth check digit should be 1, to produce even parity, if and only if the product

$$[x_1 x_2 \cdots x_m] \cdot \begin{bmatrix} b_{1j} \\ b_{2j} \\ . \\ . \\ . \\ b_{mj} \end{bmatrix}$$

produces odd parity. We therefore want to do addition modulo 2 on those components of $x_1 x_2 \cdots x_m$ that correspond to 1s in $b_{1j} b_{2j} \cdots b_{mj}$. The jth check digit is 1 if and only if the result of this computation is 1.

The encoding network works as follows. We load the word to be encoded, $x_1x_2 \cdots x_m$, into a shift register (left to right) and at the same time output those digits one at a time as the information digits of the code word. By the time x_m has been output, the shift register is full. We then simultaneously compute the check digits by r $\boxed{+_2}$ devices, one for each of the r columns of **B**. These are dumped into another shift register, which then cranks them out one at a time.

8.21 Example   For the code of Example 8.16, the canonical parity-check matrix **H** is

$$
\mathbf{H} = \begin{bmatrix} 1 & 0 & 1 \\ 1 & 1 & 1 \\ 1 & 0 & 0 \\ 0 & 1 & 0 \\ 0 & 0 & 1 \end{bmatrix}
$$

Here $m = 2$ and $r = 3$. The encoding circuit is shown in Figure 8.7.

Note that the $\boxed{+_2}$ device for the first check digit, x_3, adds both x_1 and x_2 modulo 2 because both elements of column 1 of **B** are 1. The $\boxed{+_2}$ device for x_4 only "adds" x_2 modulo 2 because $b_{12} = 0$. The $\boxed{+_2}$ device for the third check digit again adds both x_1 and x_2.

To code 10, we load the shift register for information digits with $x_1 = 1$ and $x_2 = 0$. Then the value for x_3 is $x_1 +_2 x_2 = 1 +_2 0 = 1$, the value for x_4 is $x_2 = 0$, and the value for x_5 is $x_1 +_2 x_2 = 1$. The code word is thus 10101.   □

8.22 Practice   Simulate the action of the circuit of Figure 8.7 as it encodes 00, 01, and 11 in $\mathbb{Z}_2^5$. Compare the results with Practice 8.17.   □

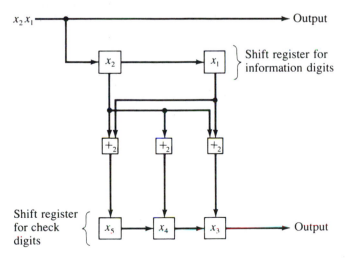

Figure 8.7

In this section we have primarily been concerned with defining codes and developing an encoding procedure. The next section addresses the corresponding decoding procedure.

✔ Checklist

Definitions

cryptology (*p. 339*)
code word (*p. 340*)
maximum-likelihood decoding (*p. 341*)
double-error detecting code (*p. 341*)
single-error correcting code (*p. 341*)
parity bit digit (*p. 341*)
even parity check (*p. 341*)
odd parity check (*p. 342*)
Hamming distance (*p. 342*)
minimum distance of a code (*p. 342*)
group code (*p. 345*)
weight of a code word (*p. 345*)
parity check matrix (*p. 347*)
canonical parity-check matrix (*p. 347*)
information digits (*p. 347*)
check digits (*p. 347*)
perfect code (*p. 348*)
Hamming code (*p. 348*)
delay element (*p. 349*)
shift element (*p. 349*)
shift register (*p. 349*)

Techniques

Given an $n \times r$ canonical parity-check matrix, find m such that $\mathbb{Z}_2^m$ can be encoded and produce the code words for $\mathbb{Z}_2^m$.

Given m such that $\mathbb{Z}_2^m$ is to be encoded as a single-error correcting code, find a canonical parity-check matrix to generate the code.

Design an encoding circuit for a group code, given a canonical parity-check matrix.

Main Ideas

The error-detecting and error-correcting capabilities of a binary code are functions of the minimum distance of the code.

In a group code, the minimum distance is the minimum weight of the nonzero code words.

A parity check matrix **H** can be used to generate a group code, in which case the minimum distance of the code can be determined from **H**.

With an $n \times r$ canonical parity-check matrix, it is easy to encode $\mathbb{Z}_2^m$ in $\mathbb{Z}_2^n$, where $m = n - r$.

For any number $n = 2^r - 1$, there exists a perfect single-error correcting Hamming code of length n and size $2^{2^r - r - 1}$.

Logic networks exist to encode group codes from the canonical parity-check matrix.

Exercises Section 8.1

★ 1. A simple code for transmitting secret messages is a shift cypher, or a Caesar cypher, so-called because Julius Caesar used it to transmit military information. An integer s is chosen as the amount of shift, $1 \le s < 26$, and each letter in the message is encoded as the letter that is s units farther along in the alphabet, with the last s letters of the alphabet shifted in a cycle to the first s letters. For example, if $s = 4$, then A is encoded as E, B is encoded as F, W is encoded as A, and so on, as shown in Figure 8.8. Decoding a message requires knowledge of s. Thus, for $s = 4$, the message KIVSRMQS is easily decoded as GERONIMO.

A	B	C	D	E	F	G	H	I	J	K	L	M	N	O	P	Q	R	S	T	U	V	W	X	Y	Z
↓	↓	↓	↓	↓	↓	↓	↓	↓	↓	↓	↓	↓	↓	↓	↓	↓	↓	↓	↓	↓	↓	↓	↓	↓	↓
E	F	G	H	I	J	K	L	M	N	O	P	Q	R	S	T	U	V	W	X	Y	Z	A	B	C	D

Figure 8.8

(a) The centurian who was supposed to inform you of s was killed en route, but you have received the message

    MXXSMGXUEPUHUPQP

in a Caesar cypher. Find the value of s and decode the message.

(b) Deciphering codes where letters of the alphabet are coded by a one-to-one function to letters of the alphabet can sometimes be made easier by making a frequency count of the letters in the received text. In general, E is the most frequently used letter of the English alphabet, T is the next most frequently used, and so on. Thus, a long passage of code in which K is the most frequent letter suggests trying to decode K as E. Frequency counts of combinations of letters can also be used. If the following paragraph were coded with such a function, what would make it difficult to decode?

> Unusual paragraphs of this sort go far toward driving a spy crazy. If a paragraph such as this should turn up in your mailbox, how would you go about solving it to find out what your contact was actually saying? Or is it only proof that your mailbox is too dusty?

(c) Decode the received word HAL under a Caesar cypher where $s = 25$.

2. Let us assume that we have a one-to-one function f mapping letters of the alphabet onto $\mathbb{Z}_{26}$, the integers from 0 to 25. We will use the simplest such mapping where $f(A) = 1$, $f(B) = 2, \ldots, f(Y) = 25$, $f(Z) = 0$. (The code we are about to describe can of course be made harder to decipher by a trickier function.) We "homomorphically" extend f to strings of letters by defining $f(x_1 x_2 \cdots x_n) = f(x_1)$, $f(x_2), \ldots, f(x_n)$. The message to be sent is broken down into blocks of k units in length; f is applied to each block X to generate a member X' of $\mathbb{Z}_{26}^k$. The key to the encoding is a $k \times k$ matrix $\mathbf{M}$ with entries in $\mathbb{Z}_{26}$ and having a multiplicative inverse $\mathbf{M}^{-1}$. The block X is coded as $f^{-1}(X'\mathbf{M})$ where all arithmetic is done in $\mathbb{Z}_{26}$. The decoding process consists of applying the function f, multiplying by $\mathbf{M}^{-1}$, and applying f^{-1}. Thus,

$$f^{-1}([f(f^{-1}(X'\mathbf{M}))]\mathbf{M}^{-1}) = f^{-1}([X'\mathbf{M}]\mathbf{M}^{-1}) = f^{-1}(X') = X$$

For example, if

$$\mathbf{M} = \begin{bmatrix} 3 & 5 \\ 2 & 3 \end{bmatrix}$$

then

$$\mathbf{M}^{-1} = \begin{bmatrix} 23 & 5 \\ 2 & 23 \end{bmatrix}$$

If X is the block "IF," then $X' = (9, 6)$, and the code word is

$$f^{-1}\left((9, 6) \begin{bmatrix} 3 & 5 \\ 2 & 3 \end{bmatrix} \right) = f^{-1}(13, 11) = MK$$

The decoding process (which requires knowledge of both $\mathbf{M}$ and f) translates MK back to $(13, 11)$ and then applies $\mathbf{M}^{-1}$:

$$(13, 11) \begin{bmatrix} 23 & 5 \\ 2 & 23 \end{bmatrix} = (9, 6)$$

Finally, $f^{-1}(9, 6) = IF$.

(a) Given a matrix encoding where the code matrix $\mathbf{M}$ is

$$\begin{bmatrix} 3 & 2 \\ 7 & 5 \end{bmatrix}$$

(**M** has previously been delivered to you by a trusted agent), find $\mathbf{M}^{-1}$ and decode the message MXOSHI.

(b) Errors introduced in the transmission of code words with matrix encoding are not independent. If the code word MXOSHI from part (a) is corrupted to MXOTHI, what word does the decoding process produce?

3. The disadvantage of standard cryptologic codes is that for efficient decoding a key to the decoding process (such as the shift factor s of Exercise 1 and the matrix $\mathbf{M}$ of Exercise 2) must be passed to the receiver separately. A coding process has been developed (see R. L. Rivest, A. Shamir, and L. Adleman, "A Method for Obtaining Digital Signatures and Public-Key Cryptosystems," *Communications of the ACM* 21, no. 2 (February 1978): 120–126) that allows any participant P in a communications network to make public to the network his encoding scheme. Doing so, however, does not disclose the key to the decoding process. Any member S of the network wishing to send a secret communication to P may do so by encoding it with P's known coding scheme, and only P will be able to decode it. Thus secure communication of a decoding key is not necessary. (The method has the further advantage that S can encode the message to P in such a way as to include S's "signature," proof that this particular message did indeed come from S.)

The method works as follows. Two large prime numbers p and q are chosen at random, and $p \cdot q = n$ is computed. Letting $m = (p - 1)(q - 1)$, a large random number y is chosen such that the greatest common divisor of y and m is 1. This step guarantees the existence of an integer x, $0 < x < m$, such that $x \cdot_m y = 1$. Efficient computer algorithms exist for producing p, q, y, and x. The message to be coded is translated into a string of integers in $\mathbb{Z}_{26}$, and the resulting string treated as a single number T, $0 \le T \le n - 1$, or as a sequence of such numbers. The coding process consists of computing $T^x \cdot_n 1$.

The coding scheme is made public by announcing n and x. To decode $T^x \cdot_n 1$, compute $(T^x \cdot_n 1)^y \cdot_n 1$. It can be shown that this computation produces T as a result. The key to the decoding, y, is not made public. Both the coding and decoding processes can be efficiently done by computer. To crack the code, however, requires finding the prime factors p and q of n, which at present cannot be done efficiently. If n is, say, a 200-digit number, it is estimated that finding its prime factors using a high-speed computer and the best current algorithms could require three billion years! Thus, such a code is currently unbreakable in practice, if not in theory. (However, researchers at Sandia Laboratories have factored a 69-digit number in less than 33 hours on a supercomputer; see *Time*, February 13, 1984, p. 47.)

As an example, we'll use this coding and decoding scheme with trivially small numbers. Let $p = 3$ and $q = 5$. Then $n = 15$ and $m = 8$. We choose $y = 11$.

(a) Compute x.

(b) Represent the letter C as a 3 and find the code for 3.

(c) Decode your answer to part (b) to retrieve the 3.

4. A simple code for transmitting numeric data is to add a single check digit d obtained as follows: if a string of digits $x_1 x_2 \cdots x_n$ is to be sent, $0 \le x_i \le 9$, then the check digit d consists of the units digit of the sum

$$x_1 + 2x_2 + x_3 + 2x_4 + \cdots + x_n^*$$

$$\text{where } x_n^* = \begin{cases} x_n & \text{if } n \text{ is odd} \\ 2x_n & \text{if } n \text{ is even} \end{cases}$$

The code word is then $x_1 x_2 \cdots x_n d$. This code can detect many errors where digits are transposed (a common error when data are manually copied).

★ (a) Determine whether an error has occurred if the received word is 21347.

★ (b) What will the code word be for the digits 15247? What will the received word be if the digits 4 and 7 are transposed? What will the received word be if the digits 5 and 2 are transposed, as well as the digits 4 and 7? Which of these errors can be detected?

(c) Write a computer program using this scheme to detect errors in received strings of up to 20 digits.

5. Let's define a function $A(n, d)$ giving the size (number of code words) of the largest code of length n (code words are binary n-tuples) with minimum distance d. (Here we are considering arbitrary codes, not necessarily group codes.) There is no known equation to express this function, but some of its values can be computed. Find

(a) $A(n, n)$

(b) $A(n, 1)$

(c) $A(n, 2)$, $n \ge 2$

6. Let $\mathbf{H}$ be an $n \times r$ binary matrix mapping $\mathbb{Z}_2^n \to \mathbb{Z}_2^r$ by the operation $X \cdot \mathbf{H}$ for $X \in \mathbb{Z}_2^n$, where we treat n-tuples and r-tuples as $1 \times n$ and $1 \times r$ matrices, respectively. Show that this operation is a homomorphism from the group $[\mathbb{Z}_2^n, +_2]$ to the group $[\mathbb{Z}_2^r, +_2]$ by showing that for $X, Y \in \mathbb{Z}_2^n$, $(X +_2 Y)\mathbf{H} = X \cdot \mathbf{H} +_2 Y \cdot \mathbf{H}$.

7. Consider the canonical parity-check matrix

$$\mathbf{H} = \begin{bmatrix} 1 & 1 & 1 \\ 0 & 1 & 1 \\ 1 & 0 & 1 \\ 1 & 0 & 0 \\ 0 & 1 & 0 \\ 0 & 0 & 1 \end{bmatrix}$$

(a) Show that the code generated by **H** is single-error correcting.

(b) Write the set of binary *m*-tuples **H** encodes, and write the code word for each.

8. Let

$$H = \begin{bmatrix} 1 & 0 & 1 \\ 1 & 1 & 0 \\ 0 & 1 & 1 \\ 1 & 0 & 0 \\ 0 & 1 & 0 \\ 0 & 0 & 1 \end{bmatrix}$$

be a canonical parity-check matrix.

(a) Show that the code generated by **H** is single-error correcting.

(b) Write the set of binary *m*-tuples that **H** encodes, and write the code word for each.

★ 9. Consider the canonical parity-check matrix

$$H = \begin{bmatrix} 1 & 1 & 0 & 1 \\ 0 & 1 & 1 & 1 \\ 0 & 1 & 0 & 1 \\ 1 & 0 & 0 & 1 \\ 1 & 1 & 0 & 0 \\ 1 & 0 & 0 & 0 \\ 0 & 1 & 0 & 0 \\ 0 & 0 & 1 & 0 \\ 0 & 0 & 0 & 1 \end{bmatrix}$$

(a) Show that the code generated by **H** is single-error correcting.

(b) Write the set of binary *m*-tuples that **H** encodes, and write the code word for each.

(c) Is this a perfect code?

10. Write a computer program that, given a canonical $n \times r$ parity-check matrix **H** for a single-error correcting code with $r \le 4$ and $n \le 2^r - 1$, will write the set of binary *m*-tuples that **H** encodes and the code word for each.

★ 11. Which of the following are perfect codes? Which are single-error correcting?

(a) (5, 3)

(b) (12, 7)

(c) (15, 11)

★ 12. For each computer word length given, what is the minimum number of check digits required to code the set of computer words as a single-error correcting code?

(a) 32 (for IBM machines)

(b) 36 (for DEC machines)

(c) 60 (for CDC machines)

13. Give an example of a canonical parity-check matrix that will generate a single-error correcting code for the set of words in $\mathbb{Z}_2^6$.

14. Give a canonical parity-check matrix for a single-error correcting (15, 11) code. Is this a perfect code?

★ 15. Design an encoding circuit for the code of Exercise 7.

16. Design an encoding circuit for the code of Exercise 8.

17. Design an encoding circuit for the (7, 4) Hamming code of Example 8.19. Simulate the action of the circuit to encode 0010 and 0101.

18. Let **H** be an $n \times r$ matrix for a perfect, single-error correcting Hamming code, $r \geq 3$. Let **H**′ be the $n \times (r + 1)$ matrix obtained by attaching a column of 1s to **H**. Show that **H**′ generates a code of minimum distance 4.

19. Many Hamming codes are a special case of codes called **cyclic codes.** Here we represent a binary n-tuple by a polynomial of degree $n - 1$ with coefficients of 0 or 1. Thus

$$1101 \leftrightarrow 1 \cdot x^3 + 1 \cdot x^2 + 0 \cdot x + 1 = x^3 + x^2 + 1$$

and

$$10010 \leftrightarrow x^4 + x$$

In a cyclic code, the code words are represented by polynomials that are polynomial multiples of a **generating polynomial** $g(x)$. In an (n, m) cyclic Hamming code, the generating polynomial $g(x)$ has degree $r = n - m$. We encode a binary m-tuple $p = p_1 p_2 \cdots p_m$ by first changing it to a binary n-tuple p^* with r 0s added on the right (low order) end; $p^* = p_1 p_2 \cdots p_m 00 \cdots 0$. Let $p^*(x)$ be the polynomial representation of p^*. We then divide $p^*(x)$ by the generating polynomial $g(x)$, following the rules of modulo 2 arithmetic on the coefficients. (Remember that in $[\mathbb{Z}_2, +_2]$, $-1 = 1$, so subtraction is the same as addition.) This produces a quotient polynomial $q(x)$ and a remainder polynomial $r(x)$ with

$$p^*(x) = q(x)g(x) + r(x) \qquad \text{degree of } r(x) < r$$

Let

$$t(x) = p^*(x) - r(x) = q(x)g(x)$$

The binary n-tuple t represented by $t(x)$ is then the code word for p. Note that $t(x)$ is a polynomial multiple of $g(x)$.

For the Hamming code of Example 8.19, where $n = 7$, $m = 4$, and $r = 3$, the generating polynomial is $g(x) = x^3 + x^2 + 1$. To encode 1100, for example, the following procedure is carried out:

1. Change 1100 to 1100000.
2. Form the polynomial representation $p^*(x)$ of 1100000, $x^6 + x^5$.
3. Divide $x^6 + x^5$ by $x^3 + x^2 + 1$:

$$
\begin{array}{r}
x^3 \qquad\qquad +1 \\
x^3 + x^2 + 1 \;\overline{\big)\; x^6 + x^5 \qquad\qquad} \\
x^6 + x^5 \qquad\quad +x^3 \\
\overline{\qquad\qquad\qquad x^3} \\
x^3 + x^2 + 1 \\
\overline{\qquad\qquad\qquad x^2 + 1}
\end{array}
$$

Thus $q(x) = x^3 + 1$ and $r(x) = x^2 + 1$.

4. Form the polynomial

$$
\begin{aligned}
t(x) &= p^*(x) - r(x) \\
&= x^6 + x^5 - (x^2 + 1) \\
&= x^6 + x^5 + x^2 + 1
\end{aligned}
$$

5. The code word is the binary 7-tuple associated with $x^6 + x^5 + x^2 + 1$, 1100101.

Use the above procedure to encode $\mathbb{Z}_2^4$ in $\mathbb{Z}_2^7$. Compare your answers with Practice 8.20.

SECTION 8.2   DECODING

In this section we put ourselves at the receiving end of the communications channel and attempt to decode received words. As before, we assume that our code words are binary n-tuples. Regardless of the encoding procedure used, a direct decoding process involves comparing the received n-tuple X with all the code words. Then we decode X as the closest code word in terms of Hamming distance. There may not be a unique closest code word, and even if there is, our process may cause us to decode incorrectly, depending on the number of errors incurred in transmission and the minimum distance of the code (recall Theorem 8.6). If no errors have occurred, then X is itself a code word and will be correctly decoded to itself. The difficulty with this decoding process is that if the code is of size 2^m, we have to do 2^m comparisons in order to find the code word closest to the received word X; we must have an array of all 2^m code words against which to compare. For $m = 32$, for example (see Exercise 12 of the previous section), 2^{32} is an extremely large number, and this approach is not practical.

Decoding a Group Code

We can improve upon the direct decoding process. Suppose that our code is a group code generated by an $n \times r$ canonical parity-check matrix $\mathbf{H}$. Then we can devise a decoding procedure in which only 2^r elements must ultimately be stored. In a single-error correcting code, if $m = 32$, then r can be 6, and $2^6 = 64$, which is a reasonable number of pieces of information to keep available. Let's see how the decoding procedure works.

We recall that the set $\mathscr{C}$ of code words is the kernel of the homomorphism that $\mathbf{H}$ induces from $[\mathbb{Z}_2^n, +_2]$ to $[\mathbb{Z}_2^r, +_2]$. Thus $\mathscr{C}$ is a normal subgroup of $[\mathbb{Z}_2^n, +_2]$, and $\mathbb{Z}_2^n$ can be partitioned into cosets of the form $X + \mathscr{C}$, $X \in \mathbb{Z}_2^n$. Set $\mathscr{C}$ has 2^m elements, and the number of cosets (by Lagrange's Theorem) is $2^n/2^m = 2^{n-m} = 2^r$. The cosets provide the key to the decoding. Suppose a word $X \in \mathbb{Z}_2^n$ is received. Then X belongs to the coset $X + \mathscr{C}$. Each element E_i of this coset is an n-tuple of the form $X + C_i$ where $C_i \in \mathscr{C}$. Because E_i and C_i are in $\mathbb{Z}_2^n$, they have the property that $-E_i = E_i$ and $-C_i = C_i$. The equation

$$E_i = X + C_i$$

can be written as

$$X = C_i + E_i \tag{1}$$

or

$$X + E_i = C_i \tag{2}$$

From (1) we see that 1s in E_i occur in exactly those components where X and C_i differ. Thus, the weight of E_i equals the distance between X and C_i. The code word C_i closest to X is the one for which the corresponding E_i has minimum weight. To decode X, we look for the element in the coset of X having minimum weight and, according to (2), add that element to X. The result is the code word to which we decode X.

The coset element having minimum weight is called the **coset leader,** and it may not be unique. If two n-tuples of minimum weight occur in the same coset, one is arbitrarily chosen as the coset leader, which simply means that no word in this particular coset is sufficiently close to a code word to allow accurate decoding. Remember that in a d-error correcting code, the decoding procedure can be applied to any received n-tuple, but it will properly correct only those with $\leq d$ errors.

To summarize this decoding procedure, when X is received, we must find the coset to which X belongs and then add that coset's leader to X.

8.23 Example   Consider the code of Example 8.11. Here $n = 5$ and $\mathscr{C} = \{00000, 01111, 10101, 11010\}$. Suppose the 5-tuple $X = 11011$ is received. By inspection, we see that the closest code word is 11010, and we would decode X as 11010. Let's use the decoding procedure. X belongs to the coset $X + \mathscr{C}$. The members of this coset are

$$11011 + 00000 = 11011$$
$$11011 + 01111 = 10100$$
$$11011 + 10101 = 01110$$
$$11011 + 11010 = 00001$$

The coset leader is 00001. Adding this to X, we get

$$11011 + 00001 = 11010$$

and we decode X to 11010. ☐

Syndrome

Given a received X, if we generate all the members of the coset $X + \mathscr{C}$ by adding the code words to X, as in Example 8.23, we still must have the 2^m code words on hand, so we have not gained anything. Suppose, however, that we have somehow arrived at a list of the 2^r coset leaders. We can then identify the coset leader corresponding to X by recalling some work from Chapter 7. There we learned that if f is a homomorphism from a group S to a group T with kernel K, the cosets $s + K$, where $s \in S$, are equivalence classes of S where elements are equivalent if and only if they map to the same place under f. Translating this result to our group code, the parity check matrix $\mathbf{H}$ provides a homomorphism from $\mathbb{Z}_2^n$ to $\mathbb{Z}_2^r$ with kernel $\mathscr{C}$, and elements of $\mathbb{Z}_2^n$ are in the same coset if and only if they map to the same place under this homomorphism. Thus, X and Y are in the same coset of $\mathscr{C}$ in $\mathbb{Z}_2^n$ if and only if $X \cdot \mathbf{H} = Y \cdot \mathbf{H}$.

8.24 Definition   In a binary group code generated by the $n \times r$ parity check matrix $\mathbf{H}$, for any $X \in \mathbb{Z}_2^n$, the r-tuple $X \cdot \mathbf{H}$ is the **syndrome** of X. ☐

8.25 Theorem   Let $\mathbf{H}$ be an $n \times r$ parity check matrix generating a group code $\mathscr{C}$. Then for $X, Y \in \mathbb{Z}_2^n$, X and Y are in the same coset of $\mathscr{C}$ in $\mathbb{Z}_2^n$ if and only if X and Y have the same syndrome. ☐

Theorem 8.25 can also be proved directly (see Exercise 3 in this section).

8.26 Practice   In Example 8.23 we found the four members of one coset. The parity check matrix that generated the code for this example is

$$\mathbf{H} = \begin{bmatrix} 1 & 0 & 1 \\ 1 & 1 & 1 \\ 1 & 0 & 0 \\ 0 & 1 & 0 \\ 0 & 0 & 1 \end{bmatrix}$$

Compute the syndrome for each member of the coset. ☐

8.27 Example   Again considering Example 8.23, suppose we know that 00000, 00001, and 00010 are some of the coset leaders, and we receive the word $X = 01101$. To decode X, we compute its syndrome

$$X \cdot \mathbf{H} = [01101] \begin{bmatrix} 1 & 0 & 1 \\ 1 & 1 & 1 \\ 1 & 0 & 0 \\ 0 & 1 & 0 \\ 0 & 0 & 1 \end{bmatrix} = [010]$$

and the syndrome of each coset leader

$$[00000] \cdot \mathbf{H} = [000]$$
$$[00001] \cdot \mathbf{H} = [001]$$
$$[00010] \cdot \mathbf{H} = [010]$$

We conclude that X and the coset leader 00010 belong to the same coset, and we decode X as $01101 + 00010 = 01111$.   □

Finding Coset Leaders

To decode we now need to have available only the 2^r coset leaders (plus the $n \times r$ encoding matrix). But how can the coset leaders be determined? There's no good answer to this general problem. In theory, of course, we can simply examine each member of $\mathbb{Z}_2^n$ in turn, compute its syndrome, and for each distinct syndrome, save the new member of $\mathbb{Z}_2^n$ if its weight is less than that of the previous member of $\mathbb{Z}_2^n$ having that syndrome. However, this process involves searching through 2^n n-tuples, which can be an unacceptably large set. In certain cases, we can narrow down to a reasonable size the set of n-tuples qualifying as possible coset leaders.

For example, suppose our code is a perfect, single-error correcting code. (A more general case is considered in Exercise 4 at the end of this section.) Then $n = 2^r - 1$, and the rows of $\mathbf{H}$ are binary representations of the integers $1, 2, \ldots, 2^r - 1$. The code word 0_n is the coset leader corresponding to the syndrome 0_r. Any other syndrome is a binary r-tuple representing a digit d, $1 \le d \le 2^r - 1$. Let row i_d be the row of $\mathbf{H}$ representing d; then the n-tuple with 1 in component i_d and 0s elsewhere is the coset leader for this syndrome.

On the other hand, each n-tuple with a weight of 1 has a unique syndrome. Thus the set of coset leaders for the $2^r = n + 1$ cosets consists precisely of 0_n (the coset leader for the coset consisting of code words) plus the n n-tuples with a weight of 1. In a perfect, single-error correcting code, there are no arbitrary choices of coset leaders between two possibilities both of minimum weight. Every received word is within one unit of a unique code word.

Applied to Figure 8.3, this result means that if the circles of the figure have radius 1, they should cover the whole space. In fact, there are 2^m circles, each with $n + 1$ elements, and $2^m(n + 1) = 2^m \cdot 2^r = 2^{m+r} = 2^n$, which is the size of the whole space. The circle partition of the set $\mathbb{Z}_2^n$ is "orthogonal" to the coset partition in the sense that the intersection of any block from one partition with any block from the other contains just one element of $\mathbb{Z}_2^n$.

8.28 Example   The perfect (7, 4) code of Example 8.19 has the coset leaders and corresponding syndromes of Figure 8.9. A received word of 1101101 is decoded by computing its syndrome,

$$[1101101] \cdot \mathbf{H} = [101]$$

Coset leaders	Syndromes
0000000	000
0000001	001
0000010	010
0000100	100
0001000	101
0010000	111
0100000	011
1000000	110

Figure 8.9

The received word is then decoded to

$$1101101 + 0001000 = 1100101$$

□

8.29 Practice   In the perfect (7, 4) code, how would the received word 1000100 be decoded?

□

8.30 Example   The (5, 2) code of Example 8.14 is not a perfect code (nor does it satisfy the requirements of Exercise 4). Since n is small, we can find its coset leaders by brute force. Figure 8.10 shows the eight cosets of $\mathbb{Z}_2^5$, together with the coset leader and syndrome for each. Note that in two cosets, there was an arbitrary choice of coset leader between two candidates.

Coset leaders				Syndromes
00000	01111	10101	11010	000
00001	01110	10100	11011	001
00010	01101	11000	10111	010
00011	01100	11001	10110	011
00100	11110	01011	10001	100
10000	00101	01010	11111	101
00110	01001	10011	11100	110
01000	00111	10010	11101	111

Figure 8.10

A received word of 10101 will be decoded as 10101 (this is a code word). A received word of 11000 will be decoded as 11010. A received word of 10011 will be decoded as 10101, or because its coset leader has weight 2, it can be flagged to indicate that at least two errors have occurred and that decoding cannot be done with certainty.   □

8.31 Example   The matrix of Example 8.14 allows encoding of $\mathbb{Z}_2^2$ in $\mathbb{Z}_2^5$. Using Exercise 4 at the end of this section, we choose a different canonical parity-check matrix,

$$\mathbf{H}^* = \begin{bmatrix} 1 & 0 & 1 \\ 0 & 1 & 1 \\ 1 & 0 & 0 \\ 0 & 1 & 0 \\ 0 & 0 & 1 \end{bmatrix}$$

The set of code words is now {00000, 01011, 10101, 11110}. The coset leaders (except for the last one) and syndromes are given in Figure 8.11. We do not have to search through all of $\mathbb{Z}_2^5$ to find the coset leaders. For the coset with syndrome 110, for example, it is clear from the form of $\mathbf{H}^*$ that we can add rows 3 and 4, so we put 1s in these components and 0s elsewhere. It is also clear from the form of $\mathbf{H}^*$ that no 5-tuple with weight 1 can produce this syndrome, so 00110 is the coset leader (11000 will also work). (Because our original $\mathbf{H}$ for this problem was quite small, this sort of procedure could also have been used in Example 8.30. The advantage to organizing $\mathbf{H}$ as suggested in Exercise 4 is that coset leaders can be found more systematically, which is what we are looking for in the case of large n.)   □

Coset leaders	Syndromes
00000	000
00001	001
00010	010
01000	011
00100	100
10000	101
00110	110
—	111

Figure 8.11

8.32 Practice   In Example 8.31, what is the coset leader for the syndrome 111?   □

Even easier decoding (at the expense of a slight complication in encoding) occurs if we use a noncanonical parity-check matrix as discussed in Exercise 5 of this section.

Given a parity check matrix and a table of coset leaders and syndromes such as that in Figure 8.9, a decoding circuit can be constructed to take a received word, compute its syndrome, and add the appropriate coset leader to the received word, giving the decoded word as output.

✔ Checklist

Definitions

coset leader (*p. 360*)
syndrome (*p. 361*)

Techniques

Given an $n \times r$ parity check matrix for a group code, classify the elements of $\mathbb{Z}_2^n$ into cosets with respect to the set of code words, identify coset leaders, and decode received words.

Main Ideas

For a group code $\mathscr{C}$ generated by an $n \times r$ parity check matrix **H**, each word X in $\mathbb{Z}_2^n$ is decoded by using its syndrome to locate the coset of $\mathscr{C}$ in $\mathbb{Z}_2^n$ to which it belongs and then adding the coset leader to X.

The form of **H** may make identification of the coset leaders easier.

Exercises Section 8.2

★ 1. In the perfect (7, 4) code of Example 8.28, decode the following received words (use Figure 8.9):

0011011

1110101

0100011

2. In the (5, 2) code of Example 8.30, decode the following received words (use Figure 8.10):

11011

01011

11001

3. Let **H** be an $n \times r$ parity check matrix generating a group code $\mathscr{C}$. Show that X and Y are in the same coset of $\mathscr{C}$ in $\mathbb{Z}_2^n$ if and only if $X \cdot \mathbf{H} = Y \cdot \mathbf{H}$.

4. Let $\mathbf{H}$ be an $n \times r$ parity check matrix for a group code where $2^{r-1} \leq n < 2^r$ and the rows of $\mathbf{H}$ are binary representations of the integers $1, 2, \ldots, n$. Prove that no coset leader has a weight greater than 2.

★ 5. Let $\mathbf{H}$ be an $n \times r$ parity check matrix where the ith row is the binary representation of the integer i, $1 \leq i \leq n$. Then $\mathbf{H}$ generates a single-error correcting code. (Because $\mathbf{H}$ is not in canonical form, the encoding process is not quite the same.) Prove that any received word X in which a single error has occurred can be correctly decoded by knowing only the syndrome of X.

6. Consider the (6, 3) code of Exercise 7, Section 8.1.
   (a) Partition $\mathbb{Z}_2^6$ into cosets, indicating coset leaders and syndromes as in Figure 8.10.
   (b) How would each of the following words be decoded? Flag any for which you know that more than one error has occurred.

   > 010011
   >
   > 110110
   >
   > 011010
   >
   > 110010

★ 7. Consider the (6, 3) code of Exercise 8, Section 8.1.
   (a) Write a table of coset leaders and syndromes, such as in Figure 8.11.
   (b) Decode the following received words. Flag any for which you know that more than one error has occurred.

   > 100110
   >
   > 010011
   >
   > 110111
   >
   > 111111

8. Consider the (6, 3) code of Exercise 7, Section 8.1. The $\mathbf{H}$ given there encodes all of $\mathbb{Z}_2^3$ in $\mathbb{Z}_2^6$.
   (a) Using Exercise 5 above, write a new matrix $\mathbf{H}^*$ encoding $\mathbb{Z}_2^3$ and allowing for easy decoding.
   (b) Write the new code word for each member of $\mathbb{Z}_2^3$.
   (c) If the word $X = 011101$ is received, decode it without generating a list of coset leaders.

9. Consider the (9, 5) code of Exercise 9, Section 8.1. Try to make up a table of coset leaders and syndromes such as in Figure 8.11 without generating all the members of each coset.

★ 10. (a) Give a canonical parity-check matrix $\mathbf{H}$ for a perfect single-error correcting (15, 11) code (see Exercise 14 of Section 8.1).
   (b) Decode the received word

   > 011000010111001

11. (a) Write a canonical parity-check matrix for a single-error correcting (8, 4) code.
    (b) Write the table of coset leaders and syndromes.
    (c) Decode the following received words. Flag any for which you know that more than one error has occurred.

    01110011

    10010000

    00111011

12. Given an $n \times r$ canonical parity-check matrix for a single-error correcting (n, m) code with $r \le 3$ and $n \le 2^r - 1$, write a computer program to decode received binary n-tuples.

13. In a cyclic code (see Exercise 19, Section 8.1), the syndrome of a received word can be found by representing the word in polynomial form, dividing by the generating polynomial $g(x)$, which has degree r, and looking at the remainder $r(x)$, degree $r(x) < r$. The corresponding r-tuple is the syndrome of the received word. The word can be decoded by using a table of coset leaders and syndromes. Thus in the perfect (7, 4) code of Example 8.28, with generating polynomial $x^3 + x^2 + 1$, the syndrome for the received word of 1101101 is computed as follows:

$$
\begin{array}{r}
x^3 \\
x^3 + x^2 + 1 \overline{)\; x^6 + x^5 + x^3 + x^2 + 1} \\
\underline{x^6 + x^5 + x^3 } \\
x^2 + 1
\end{array}
$$

Then $r(x) = x^2 + 1$, and the corresponding r-tuple, 101, is the syndrome.

Using the above procedure, find the syndromes of

1000100

0011011

1110101

0100011

Chapter 9

Finite-State Machines

The algebraic structures of Chapter 7 were basically attempts to model, or simulate, arithmetic. In this chapter we will consider an attempt to simulate much more general computation. In fact, a finite-state machine, the structure introduced in the first section of this chapter, simulates computational devices such as modern digital computers. We will also learn in Section 9.2 how one finite-state machine can simulate another. Then in Section 9.3 we will see how adequately this structure models computation in the most general sense by characterizing its capabilities as a *recognizer*.

SECTION 9.1   MACHINES—SIMULATION I

Roe Bott stops at the vending machine for a cup of coffee before heading for his computer science class. Figure 9.1 shows what appears on the front of the vending machine. Roe inserts his money (this turns the machine on) and presses the button for coffee and then the button for sugar. Little does he realize that he is operating a *sequential network,* or a *finite-*

Coffee ○	Coin ⊟
Tea ○	
Black ○	
Cream ○	
Sugar ○	
Cream and Sugar ○	

Figure 9.1

state machine. Roe's insertion of the right amount of money and pressing buttons constitute input into the machine. This input is contributed at discrete (distinct or separate) moments of time, which we will denote by t_0, t_1, t_2, and so on. The machine has certain responses to these inputs. The response to a given input may occur at any time after that input and before the next input, but to synchronize the operation, we will look at the machine only at the fixed times, or clock pulses, t_0, t_1, t_2, and so on. Thus, the responses to an input at time t_i will appear at time t_{i+1}.

The machine has two types of responses to inputs. One type of response is the visible output of the machine, which at some moments of time is a drink of some type and at other moments of time is nothing. A less visible type of response is the state of the machine. Thus, after Roe inserts his money, the machine goes from an off state to a waiting state. After Roe presses the button for coffee, the machine goes from a waiting state to a coffee state. The output that occurs with all of these states is nothing. After Roe presses the sugar button, the machine goes to a coffee–sugar state and there is an output of coffee with sugar. Finally, the machine turns off and becomes ready to function for another customer. Figure 9.2 describes the action of the machine for Roe's input.

As we noted, the machine's reactions to an input appear at the next clock pulse. The output of coffee with sugar is a direct result of the machine being in the coffee–sugar state. However, reaching the coffee–sugar state at t_3 is a function of both the input at t_2 *and* the state at t_2. In turn, the state at t_2 is a function of both the input and state at t_1. Essentially, the coffee state allows the machine to remember the coffee input and lets that input have an effect on the machine's behavior for longer than just one clock pulse.

Let's try to abstract some of the important features of the vending machine operation.

1. Operations of the machine are *synchronized,* or so we may assume, by discrete clock pulses.
2. The machine proceeds in a *deterministic* fashion; that is, its actions in response to a given sequence of inputs are completely predictable. Randomness, probability, or magic play no part here.
3. The machine responds to *inputs*.

Time	t_0	t_1	t_2	t_3	t_4
Input	coin	coffee button	sugar button	—	—
State	off	waiting	coffee	coffee–sugar	off
Output	nothing	nothing	nothing	coffee with sugar	nothing

Figure 9.2

4. There is a *finite number of states* that the machine can attain. At any given moment, the machine is in exactly one of these states. Which state it will be in next is a function of both the present state and the present input. The present state, however, depends upon the previous state and input, and so forth back to the initial operation. Thus, the state of the machine at any moment serves as a form of memory of past inputs.

5. The machine is capable of *output*. The nature of the output is a function of the present state of the machine, meaning that it also depends upon past inputs.

Definition

We have noted the essential aspects of the vending machine, considering them in the abstract. By doing this, we are building a mathematical model describing the vending machine. (Recall the discussion of Simulation I in Section 5.1.)

Our model is called a finite-state machine (the formal definition appears below). A model is useful only if it describes a wide variety of situations, and indeed, this one does. Even the modern digital computer is a finite-state machine. Its operations are synchronized by very rapid clock pulses; it operates in a deterministic fashion and can respond to inputs. A computer is composed of a large number of bistable (on–off) elements. If there are n such elements, there are altogether 2^n on–off configurations that the computer can assume. These configurations are the states of the computer, and this number is finite although very large. The present state of the computer (its present memory configuration) reflects its history of past inputs. Finally, the output at any moment depends upon the present state of the machine.

9.1 Definition $M = [S, I, O, f_S, f_O]$ is a **finite-state machine** if S is a finite set of states, I is a finite set of input symbols (the **input alphabet**), O is a finite set of output symbols (the **output alphabet**), and f_S and f_O are functions where $f_S: S \times I \to S$ and $f_O: S \to O$. The machine is always initialized to begin in a fixed starting state s_0.   □

The function f_S is the next-state function. It maps a (state, input) pair to a state. Thus, the state at clock pulse t_{i+1}, state (t_{i+1}), is obtained by applying the next-state function to the state at time t_i and the input at time t_i:

state $(t_{i+1}) = f_S(\text{state } (t_i), \text{input } (t_i))$

The function f_O is the output function. When f_O is applied to a state at time t_i, we get the output at time t_i:

output $(t_i) = f_O(\text{state } (t_i))$

Notice that the effect of applying function f_O is available instantly, but the effect of applying function f_S is not available until the next clock pulse.

Examples of Finite-State Machines

To describe a particular finite-state machine, we have to define the three sets and two functions involved.

9.2 Example   A finite-state machine M is described as follows: $S = \{s_0, s_1, s_2\}$, $I = \{0, 1\}$, $O = \{0, 1\}$. Because the two functions f_S and f_O act on finite domains, they can be defined by a **state table,** as in Figure 9.3. The machine M begins in state s_0, which has an output of 0. If the first input symbol is a 0, the next state of the machine is then s_1, which has an output of 1.

	Next state		
	Present input		
Present state	0	1	Output
s_0	s_1	s_0	0
s_1	s_2	s_1	1
s_2	s_2	s_0	1

Figure 9.3

If the next input symbol is a 1, the machine stays in state s_1 with an output of 1. By continuing this analysis, we see that an input sequence consisting of the characters 01101 (read left to right) would produce the following effect:

Time	t_0	t_1	t_2	t_3	t_4	t_5
Input	0	1	1	0	1	—
State	s_0	s_1	s_1	s_1	s_2	s_0
Output	0	1	1	1	1	0

The initial 0 of the output string is spurious—it merely reflects the starting state, not the result of any input.

In a similar way, the input sequence 1010 produces an output of 00111.

Another way to define the functions f_S and f_O (in fact all of M) is by a directed graph called a **state graph.** Each state of M with its corresponding output is the label of a node of the graph. The next-state function is given by directed arcs of the graph, each arc showing the input symbol(s) that produces that particular state change. The state graph for M appears in Figure 9.4. ☐

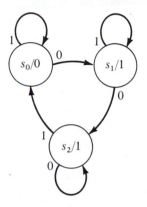

Figure 9.4

9.3 Practice For the machine M of Example 9.2, what output sequence is produced by the input sequence 11001? ☐

9.4 Practice A machine M is given by the state graph of Figure 9.5. Give the state table for M. ☐

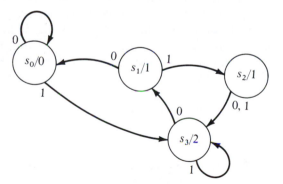

Figure 9.5

9.5 Practice A machine M is described by the state table of Figure 9.6.

(a) Draw the state graph for M.
(b) What output corresponds to an input sequence of 2110? ☐

Present state	Next state			Output
	Present input			
	0	1	2	
s_0	s_0	s_1	s_1	0
s_1	s_1	s_0	s_0	1

Figure 9.6

The machine of Example 9.2 is not particularly interesting. If finite-state machines model real-world computers, they should be able to do something. Let's try to build a finite-state machine that will add two binary numbers. The input will consist of a sequence of pairs of binary digits, each of the form 00, 01, 10, or 11. These represent the digits of the two numbers to be added, read from right to left. Thus, the least significant digits of the numbers to be added are given first, and the output gives the least significant digits of the answer first. Recall the basic facts of binary addition:

$$\begin{array}{cccc} 0 & 0 & 1 & 1 \\ \underline{0} & \underline{1} & \underline{0} & \underline{1} \\ 0 & 1 & 1 & 10 \end{array}$$ (here a carry to the next column takes place)

A moment's thought shows us that we can encounter four cases in adding the digits in any given column: (1) the output should be 0 with no carry; (2) the output should be 0 but there needs to be a carry to the next column; (3) the output should be 1 with no carry; and (4) the output should be 1 with a carry to the next column. We will let these cases be represented by the states s_0, s_1, s_2, and s_3, respectively, of the machine; s_0, as always, is the starting state. We have already indicated the output for each state, but we need to determine the next state, based on the present state and the input. For example, suppose we are in state s_1 and the input is 11. The output for the present state is 0, but there is a carry, so in the next column we are adding $1 + 1 + 1$, which results in an output of 1 and a carry. The next state is s_3.

9.6 Practice   In the binary adder under construction:

(a) What is the next state if the present state is s_2 and the input is 11?

(b) What is the next state if the present state is s_3 and the input is 10?   □

After considering all possible cases, we have the complete state graph of Figure 9.7. The operation of this machine in adding the two numbers 011 and 101 (low-order digits first) can be traced as follows:

Time	t_0	t_1	t_2	t_3	t_4
Input	11	10	01	00	—
State	s_0	s_1	s_1	s_1	s_2
Output	0	0	0	0	1

The output is 1000 when we ignore the initial 0, which does not reflect the action of any input. Converting this arithmetic to decimal form, we have computed $3 + 5 = 8$. Note the symmetry of this machine with respect to the inputs of 10 and 01, reflecting that binary addition is commutative.

9.7 Practice   Compute the sum of 01110110 and 01010101 by using the binary adder machine of Figure 9.7.   □

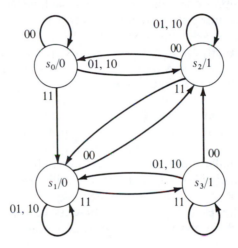

Figure 9.7

We have already noted that a given input signal may affect the behavior of a finite-state machine for longer than just one clock pulse. Because of the limited memory of past inputs that can be incorporated into the states of a machine, we can use these machines as *recognizers*. A machine can be built to recognize, say by producing an output of 1, when the input it has received matches a certain description. (This operation is essentially the one carried on by the lexical analyzer, or scanner, in a compiler. Input strings are broken down into recognizable substrings, and the substrings are then treated as units in the next stage of the compilation process. The scanner should be able to recognize, for example, a sequence of input symbols constituting a key word or a reserved word in the programming language.) We will discuss the capabilities of finite-state machines as recognizers more fully in Section 9.3. Here we will simply construct some examples.

9.8 Example The machine described in Figure 9.8 is a parity check machine. When the input received through time t_i contains an even number of 1s, then the output at time t_{i+1} is 1; otherwise, the output is 0. □

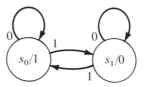

Figure 9.8

9.9 Example Suppose we want to design a machine having an output of 1 exactly when the input string received to that point ends in 101. As a special case, an input sequence consisting of just 101 could be handled by progressing directly from state s_0 to states s_1, s_2, and s_3 with outputs of 0 except for s_3, which has an output of 1. This much of the design results in Figure 9.9a. From Figure 9.9a, we want to be in state s_2 whenever the input has been such that one more 1 should take us to s_3 (with an output of 1); thus we should be in s_2 whenever the two most recent input symbols were 10, regardless of what came before. In particular, a string of 1010 should put us in s_2; hence, the next-state function for s_3 with an input of 0 is s_2. Similarly, we can use s_1 to "remember" that the most recent input symbol received was 1, and that a 01 will take us to s_3. In particular, 1011 should put us in s_1; hence, the next-state function for s_3 with an input of 1 is s_1. The rest of the next-state function can be determined the same way; Figure 9.9b shows the complete state graph.

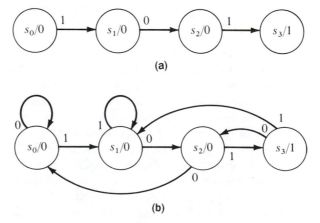

Figure 9.9

Notice that the machine is in state s_2 at the end of an input of 0110 and at the end of an input of 011010—in fact, at the end of any input ending in 10; yet s_2 cannot distinguish between these inputs. *Each state of M represents a class of indistinguishable input histories, s_3 being the state representing all inputs ending in 101.* □

9.10 Practice   Draw the state graph for a machine producing an output of 1 exactly when the input string received to that point ends in 00. □

Models of Network Protocols

A typical computer network has a number of stations, or nodes, that communicate with one another over various electronic connections or paths. The paths, unfortunately, are subject to physical disturbances that may damage a message or even cause it to be lost altogether. We will assume that a message is a sequence of binary digits. Then we can append extra digits to the message when it is sent to help the receiver detect transmission errors (see Chapter 8). Suppose there is a long stream of messages from node A to node B; perhaps a data file is being sent, for example. The goal of the network is to ensure that B passes on exactly what A sent. Since B must not pass on a message in which an error is detected, A must retransmit any such message. And since there cannot be any gaps in the data file either, a message must also be retransmitted if it is entirely lost on the first try. On the other hand, since we cannot allow parts of the data file to be duplicated, if A for some reason retransmits a message that B has already passed on, B must be prevented from passing on the duplicate message. Clearly there is a need for A and B to exchange signals and for a set of rules to govern how such signals should be sent and what they should mean. Such a set of rules for communication is called a **protocol.**

Finite-state machines can be used to represent communication protocols. For example, suppose we assume the following protocol for the situation where A is sending messages to B. When B receives a message and detects no errors, it sends an acknowledgment back to A. When A receives the acknowledgment, it sends the next message to B. However, if B detects an error in the message it receives or if the message is lost and never received by B, node B does not send an acknowledgment. Node A, if it does not receive an acknowledgment from B within a certain time, retransmits the same message. This requires that A set a timer when it sends each message and that A retransmit the message when it "times out." So far the arrangement is pretty simple. However, suppose that B receives the message with no errors and sends back its acknowledgment, but the acknowledgment gets lost in transmission. Then A times out and retransmits the previous message, which arrives correctly at B. We don't want B to pass on the duplicate message. To solve this problem, A attaches an extra identifying bit (sequence number) of 0 or 1 to each message. Thus B can recognize the duplicate of a message it has just received and not pass it on.

The states of the finite-state machine represent the situation of the network at any given moment. This includes what condition sender A is in, what condition receiver B is in, and the condition of the channel between A and B. For example, in this protocol A might have sent a message with sequence number 0; B might have received that message with no problem and would now be expecting a message with sequence number 1; and the channel might be carrying B's acknowledgment back to A. We would represent this state by the ordered triple $(0, 1, K)$: A has sent 0, B is expecting 1, and the channel is carrying the acknowledgment.

Figure 9.10 shows the possible states of a finite-state machine for this protocol. The states are ordered triples where the first component represents the message A has sent, 0 or 1, and the second component represents the message B expects next, 0 or 1. The third component represents the state of the channel. The four possibilities here are:

0 (the channel is carrying a message with identifying bit 0)
1 (the channel is carrying a message with identifying bit 1)
K (the channel is carrying an acknowledgment from B to A)
L (the channel has just lost whatever it was carrying)

Note that not all possible ordered triples will be states. For example, the combination $(0, 0, K)$ cannot occur because if A has sent message 0 and B is still expecting message 0, then the message has not yet arrived correctly at B and the channel cannot be carrying an acknowledgment back to A.

$(0, 0, 0)$	$(0, 0, L)$	$(0, 1, L)$
$(0, 1, K)$	$(1, 1, L)$	$(1, 0, L)$
$(1, 1, 1)$	$(1, 0, 1)$	
$(1, 0, K)$	$(0, 1, 0)$	

Figure 9.10

9.11 Example   The state $(1, 0, 1)$ represents the situation where A has sent message 1, which is therefore in the channel, but B is expecting message 0. This can only occur when B has already received a correct message 1 and passed it on, but the acknowledgment was lost and A retransmitted. When B receives this retransmission, it will not pass it on because it does not have the identifying number that B wants.   □

9.12 Practice   Describe the situation represented by the state $(1, 1, L)$.   □

Input to this finite-state machine consists of events that move the machine from a state to a next state. Examples of such events are: *B* receives a correct message, *A* times out and retransmits, *A* gets an acknowledgment and sends the next message, the channel loses a message, and so on.

Figure 9.11 shows the complete finite-state machine for this protocol. There is no output given for the various states, and the next-state function is described by the arcs of the graph. The initial state is (0, 0, 0), where *A* has sent message 0, *B* is expecting message 0, and message 0 is moving along the channel. The cycle from (0, 0, 0) to (0, 1, K) to (1, 1, 1) to (1, 0, K) and back to (0, 0, 0) represents the successful transmission of a message with identifying bit 0 and then a message with identifying bit 1, after which the next message, with bit 0, is sent. The "side trips" at each of these four nodes represent unusual events where a message or its acknowledgment is lost or damaged.

9.13 Example   The sequence of events described by the cycle from (0, 1, K) to (0, 1, L) to (0, 1, 0) and back to (0, 1, K) is the following: (0, 1, K) means that *A*

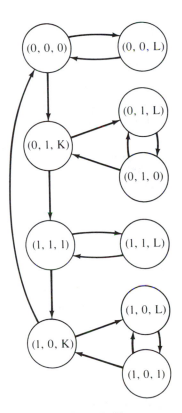

Figure 9.11

has sent message 0, B has received it and is now expecting to receive message 1 next; B's acknowledgment for message 0 is on the channel. However, the channel loses the acknowledgment, so the machine moves to state (0, 1, L). Node A times out and retransmits message 0, which is again on the channel, represented by state (0, 1, 0). The retransmission arrives correctly, and B puts another acknowledgment on the line, state (0, 1, K). Unfortunately, this loop could be repeated any number of times, although such a situation is not likely. $\square$

Representing a communication protocol by a finite-state machine helps to detect the possibility of **deadlock** in the protocol. In a deadlock, the system gets into a configuration where no matter what either node does, B cannot pass on messages successfully. This is represented in the state graph of the finite-state machine by a section of the graph that can be reached from the initial state but from which there is no exit and in which no messages can be passed on. The possibility of deadlock is, of course, an undesirable attribute of a protocol.

✔ Checklist

Definitions

finite-state machine (*p. 371*)
input alphabet (*p. 371*)
output alphabet (*p. 371*)
state table (*p. 372*)
state graph (*p. 372*)
protocol (*p. 377*)
deadlock (*p. 380*)

Techniques

Compute the output string for a given finite-state machine and a given input string.

Draw a state graph from a state table and vice versa.

Construct a finite-state machine to act as a recognizer for a certain type of input.

Main Ideas

Finite-state machines have a synchronous, deterministic mode of operation and limited memory capabilities available through the finite number of states.

Exercises Section 9.1

★ 1. For each input sequence and machine given, compute the corresponding output sequence (starting state is always s_0).

(a) 011011010

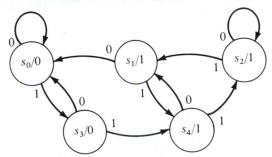

Figure 9.12

(b) *abccaab*

| | Next state | | | |
| | Present input | | | |
Present state	*a*	*b*	*c*	Output
s_0	s_2	s_0	s_3	*a*
s_1	s_0	s_2	s_3	*b*
s_2	s_2	s_0	s_1	*a*
s_3	s_1	s_2	s_0	*c*

Figure 9.13

(c) 0100110

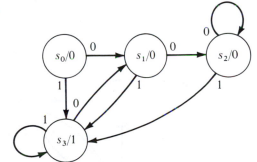

Figure 9.14

2. (a) For the machine described in Figure 9.12, find all input sequences
yielding an output sequence of 0011110.

    (b) For the machine described in Figure 9.13, find all input sequences yielding an output sequence of *abaaca*.

    (c) For the machine described in Figure 9.14, what will be the output for an input sequence $a_1a_2a_3a_4a_5$ where $a_i \in \{0, 1\}$, $1 \le i \le 5$?

In Exercises 3 through 6, write the state table for the machine, and compute the output sequence for the given input sequence.

3. 00110

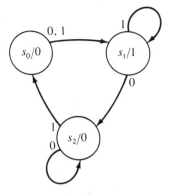

Figure 9.15

4. 1101100

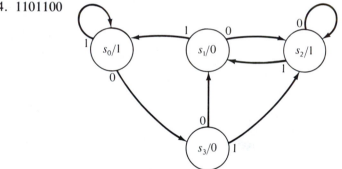

Figure 9.16

★ 5. 01011

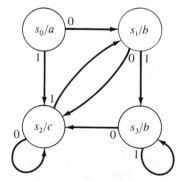

Figure 9.17

6. *acbabc*

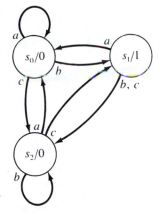

Figure 9.18

In Exercises 7 through 10, draw the state graph for the machine, and compute the output sequence for the given input sequence.

 7. 10001

| | Next state | | |
| | Present input | | |
Present state	0	1	Output
s_0	s_0	s_2	1
s_1	s_1	s_0	0
s_2	s_0	s_1	0

Figure 9.19

8. 0011

| | Next state | | |
| | Present input | | |
Present state	0	1	Output
s_0	s_2	s_3	0
s_1	s_0	s_1	1
s_2	s_1	s_3	0
s_3	s_1	s_2	1

Figure 9.20

★ 9. *acbbca*

Present state	Next state			Output
	Present input			
	a	*b*	*c*	
s_0	s_1	s_1	s_1	0
s_1	s_2	s_2	s_1	0
s_2	s_0	s_2	s_1	1

Figure 9.21

10. 21021

Present state	Next state			Output
	Present input			
	0	1	2	
s_0	s_3	s_1	s_2	1
s_1	s_3	s_0	s_1	2
s_2	s_2	s_1	s_1	0
s_3	s_1	s_4	s_0	0
s_4	s_1	s_4	s_2	2

Figure 9.22

11. Construct a finite-state machine that will compute $x + 1$ where x is the input given in binary form, least significant digit first.

12. (a) Construct a finite-state machine that will compute the 2's complement of p where p is a binary number input with the least significant digit first. (See Exercise 10, Section 6.1.)

    (b) Use the machine of part (a) to find the 2's complement of 1100 and of 1011.

★ 13. (a) Construct a *delay machine* having input and output alphabet $\{0, 1\}$ that, for any input sequence $a_1a_2a_3 \cdots$, produces an output sequence of $00a_1a_2a_3 \cdots$.

    (b) Explain (intuitively) why a finite-state machine cannot be built that, for any input sequence $a_1a_2a_3 \cdots$, produces the output sequence $0a_10a_20a_3 \cdots$.

★ 14. You have an account at First National Usury Trust and a card to operate their bank machine. Once you have entered your card, the bank machine will allow you to process a transaction only if you enter your correct code number, which is 417. Draw a state graph for a finite-state machine designed to recognize this code number. The output alphabet should have three symbols: "bingo" (correct code), "wait" (correct code so far), and "dead" (incorrect code). The input alphabet is $\{0, 1, 2, \ldots, 9\}$. To simplify notation, you may designate an arc by I-{3}, for example, meaning that the machine will take this path for an input symbol that is any digit except 3. (At FNUT, you only get one chance to enter the code correctly.)

15. Construct finite-state machines that act as recognizers for the input described by producing an output of 1 exactly when the input received to that point matches the description. (The input and output alphabet in each case is {0, 1}.)

    (a) The set of all strings where the number of 0s is a multiple of 3
    (b) The set of all strings containing at least four 1s
    (c) The set of all strings containing exactly one 1
    (d) The set of all strings beginning with 000
    (e) The set of all strings where the second input is 0 and the fourth input is 1
    (f) The set of all strings consisting entirely of any number (including none) of 01 pairs or consisting entirely of two 1s followed by any number (including none) of 0s
    (g) The set of all strings ending in 110
    (h) The set of all strings containing 00

16. A paragraph of English text is to be scanned and the number of words beginning with "con" counted. Design a finite-state machine that will output a 1 each time such a word is encountered. The output alphabet is {0, 1}. The input alphabet is the 26 letters of the English alphabet, a finite number of punctuation symbols (period, comma, etc.), and a special character β for blank. To simplify your description, you may use I-{m}, for example, to denote any input symbol not equal to m.

17. (a) In a certain computer language (BASIC, for example), any decimal number N must be presented in one of the following forms:

    $$sd^* \qquad sd^*.d^* \qquad d^* \qquad d^*.d^* \qquad (1)$$

    where s denotes the sign (i.e., $s \in \{+, -\}$), d is a digit (i.e., $d \in \{0, 1, 2, \ldots, 9\}$), and d^* denotes a string of digits where the string may be of any length, including length zero (the empty string). Thus, the following would be examples of valid decimal numbers:

    $$+2.74 \qquad -.58 \qquad 129 \qquad +$$

    Design a finite-state machine that recognizes valid decimal numbers by producing an output of 1. The input symbols are $+$, $-$, ., and the 10 digits. To simplify notation, you may use d to denote any digit input symbol.

    (b) Modify the machine of part (a) to recognize any sequence of decimal numbers (as defined in part (a)) separated by commas. For example, such a machine would recognize

    $$+2.74, -.58, 129, +$$

The input alphabet should be the same as for the machine of part (a) with the addition of the symbol c for comma.

(c) In Pascal, a decimal number must be presented in a form similar to that for BASIC except that any decimal point that appears must have at least one digit before it and after it. Write an expression similar to expression (1) in part (a) to describe the valid form for a decimal number. How would you modify the machine of part (a) to recognize such a number?

★ 18. Let M be a finite-state machine with n states. The input alphabet is $\{0\}$. Show that for any input sequence that is long enough, the output of M must eventually be periodic. What is the maximum number of inputs before periodic output begins? What is the maximum length of a period?

19. Write a computer program that, given the state table description of a finite-state machine with no more than 50 states and no more than 5 input symbols, will write the output string for any given input string.

In Exercises 20 through 26, describe the event that causes the transition from the first state to the second state in Figure 9.11.

20. $(0, 0, 0)$ to $(0, 1, K)$ ★ 21. $(0, 1, K)$ to $(1, 1, 1)$

22. $(1, 1, 1)$ to $(1, 1, L)$ 23. $(1, 0, K)$ to $(1, 0, L)$

24. $(1, 0, L)$ to $(1, 0, 1)$ 25. $(1, 0, 1)$ to $(1, 0, K)$

26. $(1, 0, 1)$ to $(1, 0, L)$

SECTION 9.2   MORPHISMS—SIMULATION II

After we have defined a structure to simulate diverse phenomena (Simulation I), we can consider what it means for one instance of a structure to simulate another instance of that structure (Simulation II). (Perhaps we can build a simpler machine to do a given task.) As we saw in the algebraic structures of Chapter 7, if A and B are two instances of a structure, then simulation is directly tied to homomorphism. For finite-state machines we can reasonably define notions of both homomorphism and simulation, but this time simulation will be a weaker idea than homomorphism. Unlike our previous structures, we can view the behavior of a finite-state machine both internally and externally. External behavior considers the machine as a box where only input and output are of interest; internal behavior concerns not only input and output but also state changes of the machine. Since the definition of a finite-state machine certainly takes into account internal behavior, we will take this view first. For simplicity, we will assume that all the machines under consideration have the same input alphabet and the same output alphabet, usually $I = O = \{0, 1\}$.

Homomorphism

9.14
Definition

Let $M = [S, I, O, f_S, f_O]$ and $M' = [S', I, O, f'_S, f'_O]$ be two finite-state machines. A **homomorphism** from M to M' is a function $g:S \to S'$ such that for any $i \in I$ and $s \in S$,

$$g(f_S(s, i)) = f'_S(g(s), i) \tag{1}$$

and

$$f_O(s) = f'_O(g(s)) \tag{2}$$

□

For a homomorphism from M to M', we thus associate with each state of M a corresponding state in M'. Equation (2) says that a state and its corresponding state have the same output. Equation (1) says that for any input symbol, a state and its corresponding state will proceed under the appropriate next-state functions to corresponding states. We can illustrate equation (1) by the commutative diagram of Figure 9.23 (where $g:S \times I \to S' \times I$ means that (s, i) maps to $(g(s), i)$). If we think of applying the next-state function as an operation, then this diagram says that in a machine homomorphism, just as in any other homomorphism, you can operate and then map, or you can map and then operate—the result will be the same.

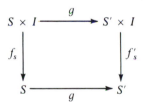

Figure 9.23

9.15 Example

Figure 9.24 shows two machines, M and M', defined by state graphs. The function $g: \{0, 1, 2\} \to \{a, b\}$ given by

$$g: \quad 0 \to a$$
$$1 \to b$$
$$2 \to b$$

is a homomorphism from M onto M'. Clearly, a state in M and its corresponding state in M' have the same output, and it only requires six cases to test that for any input symbol, a state in M and its corresponding state in M' move to corresponding states. For example, Figure 9.25 gives an instance of the general commutative diagram starting with state 1, input 1 in M. □

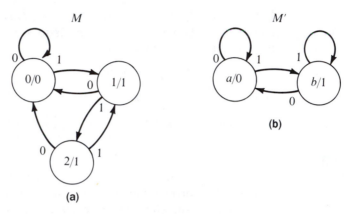

Figure 9.24

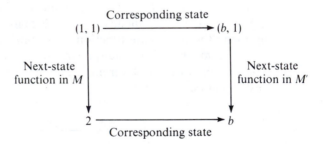

Figure 9.25

9.16 Practice   For the homomorphism in Example 9.15, draw the instance of the commutative diagram that begins with state 2, input 0 in M. □

In any finite-state machine M, we can consider $f_O(s)$ to represent the output beginning at state s and processing the empty input string λ, which is a special case of the following situation: If we begin at state s and process a nonempty input string α, that is, trace the operation of M through the various states determined by f_S, there will be a corresponding output string. We will also use f_O to designate this more general output function; thus $f_O : S \times I^* \to O^*$ where I^* denotes the set of all strings over I and O^* denotes the set of all strings over O.

9.17 Practice Let $M = [S, I, O, f_S, f_O]$ and $M' = [S', I, O, f'_S, f'_O]$ be finite-state machines with a homomorphism g from M to M'. The definition of a homomorphism says that for any state s in S, $f_O(s, \lambda) = f'_O(g(s), \lambda)$. Show that for any $s \in S$ and $\alpha \in I^*$,

$$f_O(s, \alpha) = f'_O(g(s), \alpha)$$

In other words, show that a state in M and its corresponding state in M' will produce the same output strings for the same input strings. $\square$

A Homomorphism Theorem

Suppose that g is a homomorphism from machine $M = [S, I, O, f_S, f_O]$ *onto* machine $M' = [S', I, O, f'_S, f'_O]$. We define a binary relation ρ on S by

$$s_m \, \rho \, s_n \leftrightarrow g(s_m) = g(s_n)$$

States in M are therefore related when they have the same corresponding states in M'. Clearly, ρ is an equivalence relation on S and will partition S into distinct equivalence classes. We want to treat the classes themselves as states of a new machine, which we will denote by M/g. The new machine will have input alphabet I and output alphabet O, but we must be able to define an output function and a next-state function. Any state of M/g is an equivalence class $[s]$ of states of M. All members of $[s]$ correspond to the same state in M' under the homomorphism g. From the definition of a homomorphism, a state and its corresponding state have the same output. Thus, all states in $[s]$ have the same output associated with them, and we will define this common output symbol to be the output associated with state $[s]$ of M/g.

To define the next-state function for M/g, consider state $[s_m]$ and an input symbol i. In machine M, the next state for s_m under i is given by $f_S(s_m, i)$. We can define the next state for $[s_m]$ under i to be the class $[f_S(s_m, i)]$. We are thus operating on a class by operating on a representative of the class, and we must be sure that such an operation is well-defined—that the result is independent of whatever representative of the class is chosen. Thus, suppose that $s_n \in [s_m]$. We then have

$$g(s_m) = g(s_n)$$

Also, by the definition of homomorphism,

$$g(f_S(s_m, i)) = f'_S(g(s_m), i)$$

and

$$g(f_S(s_n, i)) = f'_S(g(s_n), i)$$

Putting these facts together, we can write

$$g(f_S(s_m, i)) = f'_S(g(s_m), i) = f'_S(g(s_n), i) = g(f_S(s_n, i))$$

This equation says that the next states in M for s_m and s_n under i map under g to the same state in M', so they belong to the same equivalence class in M. Therefore, we can define the next state of a class under i to be the class of the next state of a representative, and the choice of representative does not matter.

9.18 Example   In Example 9.15, g is an onto homomorphism from M to M'. The states of M/g are [0] and [1] where $[0] = \{0\}$ and $[1] = \{1, 2\}$. The state table for M/g appears in Figure 9.26.   □

| | Next state | | |
| | Present input | | |
Present state	0	1	Output
[0]	[0]	[1]	0
[1]	[0]	[1]	1

Figure 9.26

If the situation for machines continues to parallel the situation for algebraic structures in Section 7.4, we would expect that for g a homomorphism from M onto M', the homomorphic image M' is isomorphic to M/g. This is indeed the case, but we first make sure we have the definition well in mind.

9.19 Practice   Complete the definition. An **isomorphism** from machine $M = [S, I, O, f_S, f_O]$ to machine $M' = [S', I, O, f'_S, f'_O]$ is a function $g: S \rightarrow S'$ such that g is a homomorphism and a _____ .   □

If g is an isomorphism from M to M', then the number of states in S equals the number of states in S'. The inverse function g^{-1} exists. Furthermore, because g^{-1} is an isomorphism from M' to M (see the next Practice problem), we can speak of M and M' as being isomorphic.

9.20 Practice   If g is an isomorphism from M to M', show that g^{-1} is an isomorphism from M' to M.   □

Isomorphic finite-state machines are structurally identical. Their states may have different names but their state graphs, if the nodes are relabeled, look exactly the same.

For g a homomorphism from M onto M', we want to define an isomorphism h from M/g to M'. Let $[s]$ be any state in M/g. Define h by $h([s]) = g(s)$. Any representative of $[s]$ will have the same image under g as s has, so h is well-defined. We must show that h is a bijection and a homomorphism. To see that h is onto, let $s' \in S'$. Then $s' = g(s)$ for some $s \in S$, $[s]$ is a state in M/g, and $h[s] = s'$. To see that h is one-to-one, suppose $h([s_m]) = h([s_n])$. Then $g(s_m) = g(s_n)$, so s_m and s_n are in the same class and $[s_m] = [s_n]$. The output of $[s]$ in machine M/g is defined as the output of $g(s)$ in machine M'; thus, a state in M/g and its corresponding state (under h) in M' have the same output. Finally, the next state of $[s]$ in machine M/g under input i is defined as $[f_S(s, i)]$, and the next state of $h([s]) = g(s)$ under input i is $f'_S(g(s), i)$. Because $h[f_S(s, i)] = g(f_S(s, i)) = f'_S(g(s), i)$, corresponding states (under h) proceed to corresponding states. The function h is therefore an isomorphism from M/g to M'. All of these facts can be summarized in the following theorem.

9.21 Theorem   The Homomorphism Theorem for Machines
Let g be a homomorphism from a finite-state machine M onto a finite-state machine M'. Then the homomorphic image M' is isomorphic to the machine M/g.   □

9.22 Practice   Refer to Examples 9.15 and 9.18.

(a) Draw the state graph for M/g.
(b) Define the isomorphism h from M/g to M'.   □

Quotient Machines

There is a sort of converse to Theorem 9.21. Let $M = [S, I, O, f_S, f_O]$ be a finite-state machine. Whenever S can be partitioned so that a new machine can be defined using the blocks of the partition as states, the new machine is called a **quotient machine** of M. For a quotient machine to exist, all states in a given partition block must have the same output, and for each input symbol, all states in a given partition block must proceed under the next-state function f_S to states that are all in the same partition block. The output function and next-state function for the quotient machine are then defined in the obvious way. Given that a quotient machine exists, it is always possible to define a machine homomorphism from M onto the quotient machine, namely the function g that maps a state s of M to the state $[s]$ in the quotient machine. The function g is clearly onto. Also, a state in M and its corresponding state (under g) in the quotient machine have the same output and proceed, under any input symbol, to corresponding states. Hence, g is a homomorphism. When we combine this result with that stated in Theorem 9.21, we get the following theorem.

9.23 Theorem   The Fundamental Homomorphism Theorem for Machines
Every homomorphic image of a finite-state machine M is essentially (to within an isomorphism) a quotient machine of M, and every quotient machine of M is a homomorphic image of M.   □

Compare Theorem 9.23 with the Fundamental Homomorphism Theorem for Groups (Theorem 7.116).

9.24 Example   Figure 9.27 shows the state table for a machine M. We can construct a quotient machine by partitioning S into blocks $A = \{s_0, s_2\}$, $B = \{s_1\}$, and $C = \{s_3, s_4\}$. Note that all members of a given block have the same output, and for a given input symbol, all proceed under the next-state function to states in the same block.   ☐

Present state	Next state — Present input 0	1	Output
s_0	s_2	s_0	0
s_1	s_3	s_0	1
s_2	s_0	s_0	0
s_3	s_3	s_1	1
s_4	s_4	s_1	1

Figure 9.27

9.25 Practice   Write the state table for the quotient machine of Example 9.24. Define the homomorphism g from M onto the quotient machine.   ☐

Simulation

We now take the external view of the behavior of a finite-state machine where only input strings and their corresponding output strings are noted. Our definition of simulation pertains to reflecting the external behavior of one machine in that of another.

9.26 Definition   Let $M = [S, I, O, f_S, f_O]$ and $M' = [S', I, O, f'_S, f'_O]$ be two finite-state machines. M **simulates** M' if there is a function $g\colon S' \to S$ such that for any string $\alpha \in I^*$ and $s' \in S'$,

$$f'_O(s', \alpha) = f_O(g(s'), \alpha)$$ ☐

Thus, M simulates M' if every state in M' has a corresponding state in M producing the same output string for any input string. In particular, if $\alpha = \lambda$, then $f'_O(s', \lambda) = f_O(g(s'), \lambda)$; thus a state and its corresponding state have the same output symbol.

9.27 Practice   For machines A and B described in Figure 9.28, show that A simulates B but that B does not simulate A.   ☐

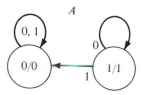

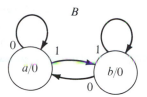

Figure 9.28

Simulation and Homomorphism

What is the connection between machine homomorphism and machine simulation? By Practice 9.17, we know that if there is a homomorphism from M to M', then the homomorphism function is also a simulation function, and M' simulates M. If there is a homomorphism from M onto M', then every state s' in M' has at least one preimage in S; if a function is defined to map s' to one of its preimages, then this function is a simulation function, and M simulates M'.

9.28 Example Consider the two machines of Example 9.15. The function

$$g: \quad 0 \to a$$
$$1 \to b$$
$$2 \to b$$

guarantees that M' simulates M. Thus, for example, the output string produced by starting in state 2 of M and processing any given input string α is the same as that produced by starting in state b of M' and processing α. The function

$$h: \quad a \to 0$$
$$b \to 1$$

guarantees that M simulates M'.

Here M is the more complicated machine. Because M' is a homomorphic image of M, we know that M' is isomorphic to a quotient machine of M. This means that, in some sense, a picture of M' is embedded in M, so it is not surprising that the more complex structure can simulate the simpler structure. What is surprising is that the simpler structure, M', also simulates the more complex structure, M. Note, however, that M' simulates only the external behavior of M. There is no homomorphism from M' to M—M' cannot mimic the state behavior of M. □

9.29 Definition Machines M and M' are **equivalent** if each simulates the other. □

It is easy to show (Exercise 6, this section) that the relation of equivalence is an equivalence relation on any collection of finite-state machines.

We have established that whenever there is a homomorphism from M onto M', then M and M' are equivalent. We have already mentioned, however, that simulation is a weaker notion than homomorphism. Although a homomorphism produces simulation, simulation can be present without a homomorphism. In the following example two equivalent machines are given, each with the same number of states, but there is not a homomorphism from either one to the other.

9.30 Example Two machines A and B are described by the state graphs in Figure 9.29. For any state in either A or B, the output upon processing any input string

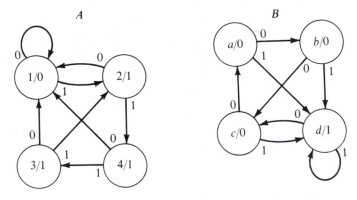

Figure 9.29

α is, aside from the initial output symbol, α itself. Both A and B are "copy machines." Thus, A simulates B by the mapping

$$a \rightarrow 1$$
$$b \rightarrow 1$$
$$c \rightarrow 1$$
$$d \rightarrow 2$$

and B simulates A by the mapping

$$1 \rightarrow a$$
$$2 \rightarrow d$$
$$3 \rightarrow d$$
$$4 \rightarrow d$$

A and B are therefore equivalent machines.

If a homomorphism g existed from A to B, it would have to map 1 to a, b, or c and map 2, 3, and 4 to d. Suppose $g(1) = a$. Then consider the next states for both 1 and a under input 0. The next state for 1 under

0 is 1; the next state for a under 0 is b. But $g(1) \neq b$. Because corresponding states do not proceed to corresponding states, g is not a machine homomorphism. Similar difficulties occur if 1 maps to b or to c. Since these are all the possibilities, there is no homomorphism from A to B.

The proof that there is no homomorphism from B to A works the same way. $\square$

Although one machine can simulate another without a homomorphism, if we want to show that M simulates M', we will generally establish the existence of a homomorphism from M onto M'. Simulation is also established if we show the existence of a homomorphism from some submachine of M onto M'; not all the states of M need be used.

✔ Checklist

Definitions

> machine homomorphism (*p. 387*)
> machine isomorphism (*p. 390*)
> quotient machine (*p. 391*)
> machine simulation (*p. 392*)
> equivalent machines (*p. 393*)

Techniques

> Verify that a given function is a machine homomorphism.

Main Ideas

> Homomorphism from one machine to another (Simulation II—internal behavior).

> The partition of the states of a machine into blocks that can sometimes serve as states of a quotient machine.

> A homomorphic image of M is isomorphic to a quotient machine of M, and a quotient machine of M is a homomorphic image of M.

> Simulation of one machine by another (Simulation II—external behavior).

> Although machines can simulate each other when no homomorphism exists, a homomorphism from one machine onto another means that each machine simulates the other.

Exercises Section 9.2

1. Figure 9.30 shows two finite-state machines M and M'. The function g given by

$$g: \quad 1 \to a$$
$$2 \to b$$
$$3 \to c$$
$$4 \to c$$

is a homomorphism from M to M'. Verify that this is true by testing all possible cases.

M

| | Next state | | |
| | Present input | | |
Present state	0	1	Output
1	3	4	0
2	3	1	0
3	2	3	1
4	2	3	1

M'

| | Next state | | |
| | Present input | | Output |
Present state	0	1	
a	c	c	0
b	c	a	0
c	b	c	1

Figure 9.30

★ 2. For each of the finite-state machine pairs M and M' of Figure 9.31, define a homomorphism g from M to M'.

★ 3. For any onto homomorphisms in Exercise 2, draw the state graph of M/g and define the isomorphism from M/g to M'.

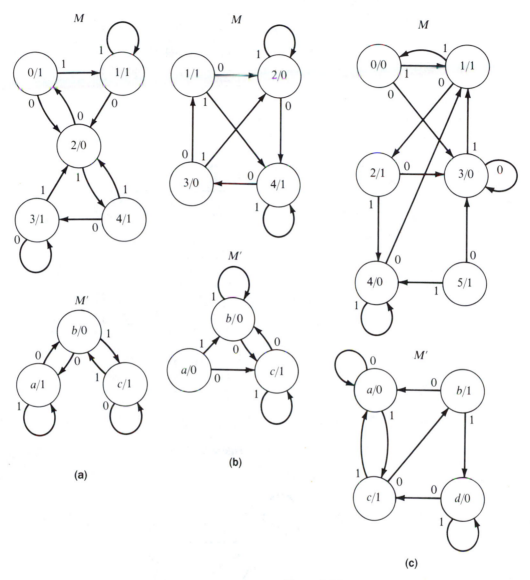

Figure 9.31

4. Partition the state set of each finite-state machine in Figure 9.32 so as to define a quotient machine. Write the state table for each quotient machine. Define the homomorphism from the original machine onto the quotient machine.

5. Partition the state set of the finite-state machine in Figure 9.33 so as to define a quotient machine. Draw the state graph of the quotient

	Next state		Output
	Present input		
Present state	0	1	
s_0	s_1	s_0	1
s_1	s_4	s_3	1
s_2	s_0	s_5	0
s_3	s_4	s_1	1
s_4	s_5	s_5	0
s_5	s_3	s_0	1

(a)

	Next state			Output
	Present input			
Present state	a	b	c	
s_0	s_2	s_1	s_4	0
s_1	s_2	s_0	s_4	1
s_2	s_3	s_4	s_2	0
s_3	s_3	s_1	s_3	0
s_4	s_3	s_0	s_1	1

(b)

Figure 9.32

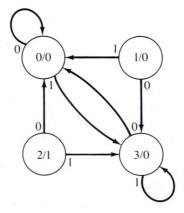

Figure 9.33

machine. Define the homomorphism from the original machine onto the quotient machine.

6. Prove that the binary relation of equivalence on any collection of finite-state machines is an equivalence relation. Describe the corresponding equivalence classes.

7. Construct two finite-state machines M and M', different from any examples given in this section, so that no homomorphism exists from M to M' but M' simulates M.

SECTION 9.3   MACHINES AS RECOGNIZERS

We have already seen some examples of finite-state machines as recognizers that signal by giving an output of 1 whenever an input string belonging to a particular set of possible input strings has been received. The machine of Example 9.8, for instance, recognizes the set of all strings consisting of an even number of 1s. Now we want to see exactly what sets finite-state machines can recognize. Remember that recognition is possible because machine states can have a limited memory of past inputs. Even though the machine is finite, a particular input signal can affect the behavior of a machine "forever." However, not every input signal can do so, and some classes of inputs require remembering so much information that no machine can detect them.

To avoid writing down outputs, we will designate those states of a finite-state machine with an output of 1 as **final states** and denote them in the state graph with a double circle. Then we can give the following definition of recognition.

9.31
Definition
A finite-state machine M with input alphabet I **recognizes** a subset S of I^* if M, beginning in state s_0 and processing an input string α, ends in a final state if and only if $\alpha \in S$.   □

9.32 Practice   Describe the sets recognized by the machines in Figure 9.34.   □

Regular Sets

We want a compact, symbolic way to describe sets such as those appearing in the answer to Practice 9.32. We will describe such sets by using *regular expressions;* each regular expression describes a particular set. First, we will define what regular expressions are; then we will see how a regular

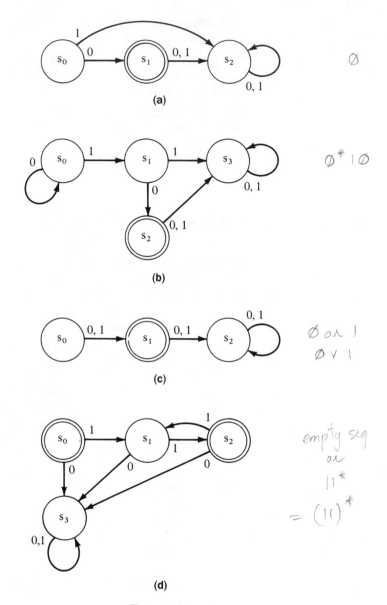

Figure 9.34

expression describes a set. We assume here that I is some finite set of symbols; later I will be the input alphabet for a finite-state machine.

9.33
Definition

Regular expressions over I are:

    (a) the symbol $\varnothing$ and the symbol λ
    (b) the symbol i for any $i \in I$

(c) the expressions (AB), $(A \vee B)$, and $(A)^*$ if A and B are regular expressions

(This definition of a regular expression over I is still another example of a recursive definition.)

Furthermore, any set represented by a regular expression according to the conventions described below is a **regular set:**

(a) $\varnothing$ represents the empty set.
(b) λ represents the set $\{\lambda\}$ containing the empty string.
(c) i represents the set $\{i\}$.
(d) For regular expressions A and B:

(AB) represents the set of all elements of the form $\alpha\beta$ where α belongs to the set represented by A and β belongs to the set represented by B.

$(A \vee B)$ represents the union of A's set and B's set.

$(A)^*$ represents the set of all concatenations of members of A's set. □

We note that λ, the empty string, is a member of the set represented by A^*. In writing regular expressions, we can eliminate parentheses when no ambiguity results.

In our discussion, we will be a little sloppy and say things like "the regular set $0^* \vee 10$" instead of "the set represented by the regular expression $0^* \vee 10$."

9.34 Example   Here are some regular expressions and a description of the set each one represents.

(a) $1^*0(01)^*$ (a′) Any number (including none) of 1s, followed by a single 0, followed by any number (including none) of 01 pairs.

(b) $0 \vee 1^*$ (b′) A single 0 or any number (including none) of 1s.

(c) $(0 \vee 1)^*$ (c′) Any string of 0s or 1s, including λ.

(d) $11((10)^*11)^*(00^*)$ (d′) A nonempty string of pairs of 1s interspersed with any number (including none) of 10 pairs, followed by at least one 0. □

9.35 Practice   Which strings belong to the set described by the regular expression?

(a) 10100010; $(0^*10)^*$
(b) 011100; $(0 \vee (11)^*)^*$
(c) 000111100; $((011 \vee 11)^*(00)^*)^*$ □

9.36 *Practice*   Write regular expressions for the sets recognized by the machines of
Practice 9.32.   □

A regular set may be described by more than one regular expression.
For example, the set of all strings of 0s and 1s, which we already know
from Example 9.34(c) to be described by $(0 \lor 1)^*$, is also described by
the regular expression $((0 \lor 1^*)^* \lor (01)^*)^*$. We might, therefore, write the
equation

$$(0 \lor 1)^* = ((0 \lor 1^*)^* \lor (01)^*)^*$$

Although we may be quite willing to accept this particular equation, how
can we decide in general whether two regular expressions are equal, that
is, that they represent the same set? An efficient algorithm that will make
this decision for any two regular expressions has not been found. Indeed,
we will see in Section 11.3 why it seems likely that no such procedure
ever will be found. (Exercise 12 at the end of Section 10.1 outlines a
nonefficient procedure.)

Kleene's Theorem

We have introduced regular sets because, as we will see, these are exactly
the sets finite-state machines are capable of recognizing. Thus, any set
recognized by some finite-state machine is regular, and conversely, any
regular set can be recognized by some finite-state machine. This result
was first proved by the American mathematician Stephen Kleene in 1956.

A Machine Recognizes a Regular Set

First, we will show that any set recognized by a finite-state machine is
regular.

We have represented finite-state machines by directed graphs. Temporarily, we will enlarge the set of machines to include structures whose
graphs may not have a full complement of arrows; thus some states under
a given input symbol may have no next state defined. If we call such
structures Machines (with a capital "M"), then a (finite-state) machine
is a special case of a Machine. Although we are ultimately interested in
the set of strings taking a given machine from its starting state to any final
state, we first consider the set of strings taking a Machine from any one
state to another, not necessarily different, state. By using induction on
the size of the Machine, we will prove that such a set is regular.

For the basis step, assume that we have a Machine M with only one
state, s_0. Let $K = \{i_1, i_2, \ldots, i_k\}$ be the set of input symbols for which
the next-state function on s_0 is defined. We want to find a regular expression for the set of all strings taking M from s_0 to s_0. Since there is nowhere

else to go, any input string from K^* does this. Thus, the regular expression is $(i_1 \vee i_2 \vee \cdots \vee i_k)^*$. Note that the set includes λ, which certainly takes M from s_0 to s_0.

Now we assume that in any k-state Machine, the set of strings taking the Machine from any state s_m to any state s_n is regular. Finally, we let M be a Machine with $k + 1$ states, and we let s_m and s_n be states in M. We consider the two cases $s_m = s_n$ and $s_m \neq s_n$.

For the case $s_m = s_n$, we first consider nonempty strings taking M from s_m back to s_m for the first time. Such strings will be of two types:

1. a single input symbol $i \in I$
2. a string of the form $i_p \alpha i_q$ where $i_p, i_q \in I$, i_p moves M from s_m to a different state s_{m1}, α is a string moving M from s_{m1} to some, not necessarily different, state s_{1m} but keeping it away from s_m, and i_q takes M from s_{1m} back to s_m

Let A be the set of all input strings taking M from s_{m1} to s_{1m} without going through s_m. If we disconnect s_m, the rest of the Machine is a k-state Machine, and A is regular by the inductive hypothesis. For a fixed i_p and i_q, $i_p A i_q$ is thus a regular set. The set B of all strings of type 2 above is the union of a finite number of such sets (taking the union over the various i_p's and i_q's); hence B is regular. And the set C of all strings described by types 1 and 2 is the union of B with a finite number of single input symbols; C is also regular. Now C^* denotes the set of concatenations of members of C and describes the set of all input strings taking M from s_m to s_m; C^* is regular.

Now we need to handle the second case where $s_m \neq s_n$. Again, we first consider the set E of all strings moving M from s_m to s_n for the first time. Any such string is of the form αi where α takes M from s_m to some $s_{1n} \neq s_n$ but keeps it away from s_n, and i takes M from s_{1n} to s_n. Let D be the set of all input strings taking M from s_m to s_{1n} without going through s_n. If we disconnect s_n, the rest of the Machine is a k-state Machine, and D is regular by the inductive hypothesis. For a fixed i, Di is thus a regular set. The set E consists of the union of a finite number of such sets (taking the union over the various i's); E is also regular. Now let F denote the set of all strings taking M from s_n to s_n; we know F is regular by the previous case. The regular set EF is the set of all input strings taking M from s_m to s_n.

We now have shown that the set of input strings taking a Machine M from any one state to any one state is regular. The set of strings taking a (finite-state) machine M from s_0 to any final state is the union of a finite number of such sets, so it is regular. If M has no final states, the empty set $\varnothing$ is the only set recognized, and $\varnothing$ is also regular.

We have now proved the first half of Kleene's Theorem.

9.37 Theorem Any set recognized by a finite-state machine is regular. □

Theorem 9.37 says that, given a finite-state machine M, there exists a regular expression describing the set of strings that M recognizes. The proof of Theorem 9.37, however, does not tell us how to find such an expression easily.

Nondeterministic Finite-State Machines

The other half of Kleene's Theorem says that for any regular set, there is a finite-state machine that recognizes it. To prove this, we will introduce a new kind of machine, called a **nondeterministic finite-state machine**—defined as an ordinary finite-state machine except that for each state–input pair, the next state need not be uniquely determined and there is, in fact, a set of possible next states; this set could even be the empty set.

9.38 Example   Figure 9.35 shows the state table and the state graph for a nondeterministic machine M.   □

Present state	Next state Present input 0	Next state Present input 1	Output
s_0	s_0, s_1	s_1, s_2	0
s_1	s_1	s_1	0
s_2	s_1, s_2	s_0	1

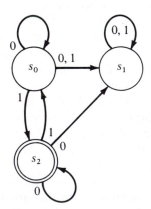

Figure 9.35

As a nondeterministic machine acts upon an input string α, the first input symbol processed leads M from the starting state to a set of possible next states. Each of these states, upon processing the second symbol, has a set of possible next states; the union of these sets is the set of possible states for M after processing two symbols of α. If we continue this procedure, we can find the set of possible states for M after processing α. If any of the states in this set is a final state of M, then we say that M **recognizes** α. The set of strings so recognized is the set recognized by M.

9.39 Example   The nondeterminsitic machine in Figure 9.35 recognizes the regular set $0*1(0 \vee 10*1)*$. For any string α in this set, M has a possible sequence of moves that would put M in a final state at the end of processing α. $\square$

A nondeterministic machine M does not operate by choosing at each clock pulse some next state out of a set of possible next states. Rather, it operates like a parallel processor, keeping track at all times of all its possible configurations. In fact, we can simulate M's behavior by running in parallel a bunch of deterministic machines, each of which traces out a different possible sequence of moves for M. We can also simulate M's behavior by constructing a single big deterministic machine with enough states to represent all of M's possible configurations.

9.40 Theorem   For any nondeterministic machine M recognizing a set S, there is a deterministic machine M' also recognizing S.

Proof: The states of M' are sets of states of M. If s_0 is the starting state of M, then the set $\{s_0\}$ is the starting state for M'. For each state $\{s_{i1}, s_{i2}, \ldots, s_{in}\}$ of M' and each input symbol i, we find the next state of M' by taking the union of the set of next states for s_{i1} under i in M, s_{i2} under i in M, and so on. A state of M' is labeled a final state if and only if it contains a final state of M. $\square$

9.41 Example   Figure 9.36 shows the state table for the deterministic counterpart of the nondeterministic M of Example 9.38. For example, the next state of $\{s_1, s_2\}$ under 1 is $\{s_0, s_1\}$ because the set of next states in M for s_1 under 1 is $\{s_1\}$ and the set of next states in M for s_2 under 1 is $\{s_0\}$. From the

Present state	Next state		Output
	Present input		
	0	1	
$A = \{s_0\}$	$\{s_0, s_1\}$	$\{s_1, s_2\}$	0
$B = \{s_1\}$	$\{s_1\}$	$\{s_1\}$	0
$C = \{s_2\}$	$\{s_1, s_2\}$	$\{s_0\}$	1
$D = \{s_0, s_1\}$	$\{s_0, s_1\}$	$\{s_1, s_2\}$	0
$E = \{s_1, s_2\}$	$\{s_1, s_2\}$	$\{s_0, s_1\}$	1

Figure 9.36

state graph for M, Figure 9.37, we see that M indeed recognizes the set $0*1(0 \lor 10*1)*$, and we also see that states B and C are unreachable from the starting state A and thus could be eliminated.   □

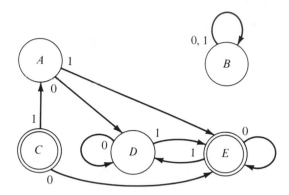

Figure 9.37

In Example 9.41, the number of states in the deterministic machine is close to the number of states in the original nondeterministic machine. This situation is unusual; if M has n states, M' could have as many as $2^n - 1$ states.

9.42 Practice   Find a deterministic machine M' recognizing the same set as the non-deterministic machine M whose state table is given in Figure 9.38.   □

	Next state		
	Present input		
Present state	0	1	Output
s_0	s_0	s_1	*0*
s_1	s_1	s_0, s_1	*1*

Figure 9.38

A Regular Set Has a Machine Recognizer

Theorem 9.40 says that we gain no recognition capabilities by considering nondeterministic machines. Therefore, the proof of Kleene's Theorem will be complete if we can show that for any regular set, there is a non-deterministic finite-state machine that recognizes it. We will prove that such a machine exists by showing how to construct it. Because the definition of a regular expression is inductive, we must construct our machine inductively.

We let I be the set of symbols, and consider the various types of regular expressions.

1. $\emptyset$ and λ. A trivial machine with a single, nonfinal state, as in Figure 9.39a, recognizes $\emptyset$. Figure 9.39b shows a deterministic machine that recognizes λ (a deterministic machine is a special case of a nondeterministic machine).

2. $i \in I$. Figure 9.39c shows a deterministic machine that recognizes i.

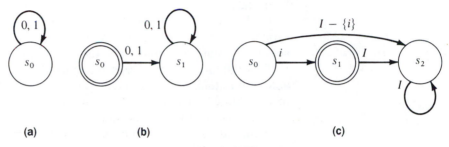

(a)   (b)   (c)

Figure 9.39

We now assume the inductive hypothesis that for regular expressions A and B there are nondeterministic recognizers M_A and M_B. To avoid mixups, we'll also assume that the states in M_A and the states in M_B have different names.

3. AB. The basic idea here is to connect the two machines M_A and M_B in series to create a machine M_{AB} that recognizes AB. The set of states for M_{AB} is the union of the sets of states of M_A and M_B. The starting state for M_{AB} is the starting state of M_A, and the final states of M_{AB} are the final states of M_B. Whenever a state–input pair in M_A could take M_A to a final state, we want to allow the possibility of jumping instead to the starting state of M_B, so that we can begin to process strings $\beta \in B$ in M_B. Hence, we modify the state table for M_A so that whenever the set of next states contains a final state in M_A, we add the starting state of M_B to that set. Then for any $\alpha\beta \in AB$, there is a sequence of moves taking M_{AB} from its starting state through the actions of M_A on α to the point of recognition, then transferring to perform the actions of M_B on β until β is recognized by M_B; hence, $\alpha\beta$ is recognized by M_{AB}.

4. $A \vee B$. The basic idea here is to connect the two machines M_A and M_B in parallel to create a machine $M_{A \vee B}$ that recognizes $A \vee B$. The states of $M_{A \vee B}$ are the states of M_A plus the states of M_B plus one additional state $\bar{s}$, designated as the starting state for $M_{A \vee B}$. The final states of $M_{A \vee B}$ are the final states of M_A plus the final states of M_B. When we process the first symbol i

of a string γ, we want to allow the possibility of simulating either M_A's actions in processing i beginning in its starting state s_A or M_B's actions in processing i beginning in its starting state s_B. We define the set of next states for $\bar{s}$ under i to be the union of the set of next states of s_A under i and the set of next states of s_B under i. Thus, $M_{A \vee B}$ processes γ by simulating either M_A or M_B and recognizes γ if it is recognized by either M_A or M_B.

5. A^*. M_{A^*} uses the set of states of M_A plus an additional starting state $\bar{s}$, which must be a final state in order to recognize λ. The final states of M_A are also final states of M_{A^*}. If i is the first symbol of a string γ, then M_{A^*} should simulate M_A's actions in processing i beginning in M_A's starting state s_A. Thus, we let the set of next states of $\bar{s}$ under i be the set of next states of s_A under i. If an initial segment of γ is recognized by A, we want to be able to reinitialize at once. Hence, we modify M_A so that the set of next states for any final state and input j contains the set of next states of $\bar{s}$ under j. This modification allows the first character after the initial segment to be processed just as M_A would do it starting in s_A. Thus M_{A^*} recognizes A^*.

This procedure will have to be modified slightly to take care of troublesome cases involving λ. To construct a machine for 1^*0^*, for example, we would want to leave the starting state for the machine of 1^* as a final state, even though only the final states of 0^* should remain final according to our directions for the expression AB. Similarly, a machine for 1^*0 would call for a transfer on 0 from the starting state of the machine for 1^* to the final state of the machine for 0.

9.43 Example   Let's build the nondeterministic recognizer for the regular set $1 \vee 00^*$. First, we build the machine for 0^*. We begin with the machine for 0, Figure 9.40a, and modify it as described in the directions for A^* to get Figure 9.40b, the machine for 0^*. Figure 9.40c shows the machine for 00^*, and finally, Figure 9.40d is a nondeterministic machine recognizing $1 \vee 00^*$.   □

The procedure just described should be viewed as a canonical procedure; that is, it is completely general and always works. However, for any particular case, we can probably come up with a much simpler machine. Thus, suppose we want a (deterministic) finite-state machine to recognize the regular set $1 \vee 00^*$. The canonical procedure is to produce the nondeterministic machine of Figure 9.40d and then use the procedure described in the proof of Theorem 9.40 to find the corresponding deterministic machine, which will be a horrendous contraption. However, it is not hard to think up the machine of Figure 9.41, a four-state machine recognizing $1 \vee 00^*$.

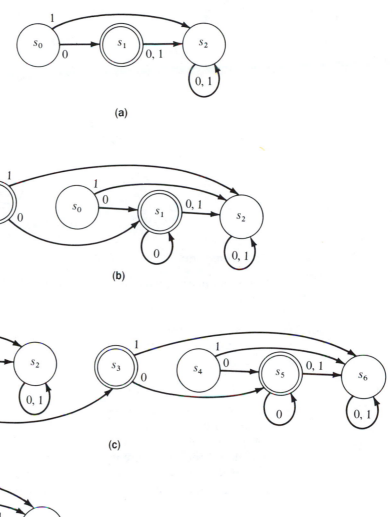

(a)

(b)

(c)

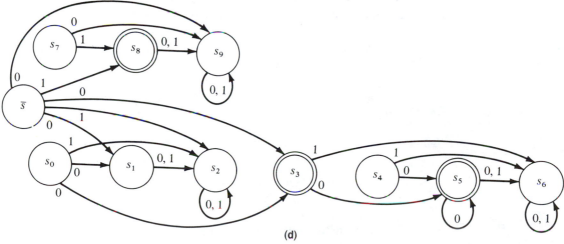

(d)

Figure 9.40

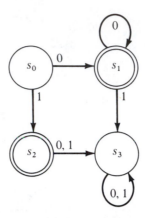

Figure 9.41

Final Comments

Let's summarize what we know about regular sets and finite-state machines.

9.44 Theorem Kleene's Theorem
A set is regular if and only if it is recognized by some finite-state machine.

□

Theorem 9.44 outlines the limitations as well as the capabilities of finite-state machines, because there are certainly many sets that are not regular. For example, $S = \{0^n 1^n \mid n \geq 0\}$ is not regular where a^n stands for a string of n copies of a. (Notice that $0*1*$ does not do the job.) By Kleene's Theorem, there is no finite-state machine capable of recognizing S. Yet S seems like such a reasonable set, and surely we humans could count a string of 0s followed by 1s and see whether we had the same number of 0s as 1s. This lapse suggests some deficiency in our use of a finite-state machine as a model of a computational device. We will investigate this further in Chapter 11.

✔ Checklist

Definitions

final state (*p. 399*)
recognition (by deterministic machine) (*p. 399*)
regular expression (*p. 400*)
regular set (*p. 401*)
nondeterministic finite-state machine (*p. 404*)
recognition (by nondeterministic machine) (*p. 405*)

Techniques

Find a regular expression given the description of a regular set.

Decide whether a given string belongs to a given regular set.

Given a nondeterministic finite-state machine, construct a deterministic machine that recognizes the same set of input strings.

Given a regular set, construct the canonical nondeterministic machine that recognizes it.

Main Ideas

The class of nondeterministic finite-state machines cannot recognize any more sets than the class of finite-state machines.

The class of sets recognizable by finite-state machines is the class of all regular sets; hence, there are limitations to the recognition capabilities of finite-state machines.

Exercises Section 9.3

1. Give a regular expression for the set recognized by each finite-state machine in Figure 9.42.

2. Give a regular expression for the set recognized by each finite-state machine in Figure 9.43.

3. Give a regular expression for the set recognized by each nondeterministic finite-state machine in Figure 9.44.

4. Give a regular expression for each set described below.
   (a) The set of all strings of 0s and 1s beginning with 0 and ending with 1
   (b) The set of all strings of 0s and 1s having an odd number of 0s
   ★(c) {101, 1001, 10001, 100001, . . .}
   (d) The set of all strings of 0s and 1s containing at least one 0
   (e) The set of all strings of a's and b's where each a is followed by two b's
   (f) The set of all strings of 0s and 1s containing exactly two 0s

★ 5. Does the given string belong to the given regular set?
   (a) 01110111; (1*01)*(11 ∨ 0*)
   (b) 11100111; ((1*0)* ∨ 0*11)*
   (c) 011100101; 01*10*(11*0)*
   (d) 1000011; (10* ∨ 11)*(0*1)*

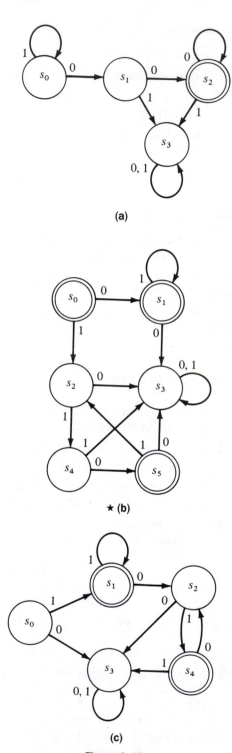

(a)

★ (b)

(c)

Figure 9.42

| Present state | Next state | | Output |
| | Present input | | |
	0	1	
s_0	s_3	s_1	0
s_1	s_1	s_2	0
s_2	s_3	s_3	1
s_3	s_3	s_3	0

(a)

| Present state | Next state | | Output |
| | Present input | | |
	0	1	
s_0	s_3	s_1	1
s_1	s_1	s_2	1
s_2	s_2	s_2	0
s_3	s_0	s_2	0

(b)

| Present state | Next state | | Output |
| | Present input | | |
	0	1	
s_0	s_2	s_1	1
s_1	s_3	s_1	1
s_2	s_3	s_4	0
s_3	s_3	s_3	0
s_4	s_5	s_3	0
s_5	s_2	s_3	1

(c)

Figure 9.43

6. (a) Write a regular expression for the set of all alphanumeric strings beginning with a letter. (The set of legal identifiers in FORTRAN is a subset of this set.)
   (b) Write a regular expression for the set of all arithmetic expressions indicating the addition or subtraction of two positive integers.

7. In the proof of Theorem 9.37, where is the idea of a Machine (as opposed to a machine) needed?

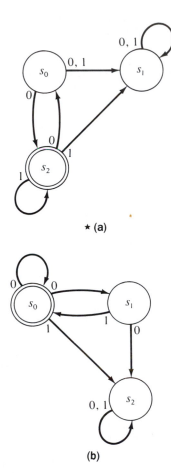

★ (a)

(b)

Figure 9.44

8. Give a regular expression for the set described and construct, by intuition, a nondeterministic finite-state machine to recognize that set. (The input alphabet is {0, 1}.)
   (a) {01, 11}
   (b) The set of all strings ending in 00 or 11
   (c) The set of all strings containing 111

9. For each machine in Figure 9.45, use the proof of Theorem 9.40 to find a deterministic machine recognizing the same set as the nondeterministic machine. Write a regular expression for the set.

10. For each machine in Figure 9.46, use the proof of Theorem 9.40 to find the state table for a deterministic machine recognizing the same set as the nondeterministic machine. Write a regular expression for the set.

Present state	Next state		Output
	Present input		
	0	1	
s_0	s_0, s_1	s_1	1
s_1	s_0	s_0, s_1	0

★ (a)

Present state	Next state		Output
	Present input		
	0	1	
s_0	s_1, s_2	s_0	1
s_1	s_0, s_1	s_2	0
s_2	s_2	s_2	0

(b)

Present state	Next state		Output
	Present input		
	0	1	
s_0	s_2	s_1, s_3	0
s_1	s_2	s_1	1
s_2	s_2	s_2	0
s_3	s_4	s_2	0
s_4	s_2	s_5	0
s_5	s_2	s_3	1

(c)

Figure 9.45

11. (a) Prove that if A is a regular set, then the set A^R consisting of the reverse of all strings in A is also regular.
    (b) For any string α, let α^R be the reverse string. Do you think the set $\{\alpha\alpha^R \mid \alpha \in I^*\}$ is regular?

12. Let A be the regular set recognized by the finite-state machine M of Figure 9.42a. Then A^R is a regular set (see Exercise 11 above).
    (a) Create a nondeterministic finite-state machine by using the final state of M as the start state, the start state of M as the only final

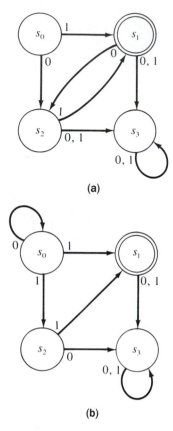

(a)

(b)

Figure 9.46

state, and reversing all the arrows. Does this machine recognize A^R?

(b) Construct by intuition a deterministic machine to recognize A^R.

13. (a) Construct the canonical nondeterministic recognizer for the regular set $0 \vee 1$.

(b) Use the method of Theorem 9.40 to construct the deterministic equivalent of the nondeterministic machine of part (a). This new machine is the canonical recognizer for $0 \vee 1$.

(c) By intuition, construct a deterministic machine to recognize $0 \vee 1$.

⋆ 14. (a) Construct the canonical nondeterministic recognizer for the regular set 1^*.

(b) Use the method of Theorem 9.40 to construct the deterministic equivalent of the nondeterministic machine of part (a). This new machine is the canonical recognizer for 1^*.

    (c) By intuition, construct a deterministic machine to recognize 1*.

15. (a) Construct the canonical nondeterministic recognizer for the regular set 1*0.
    (b) Contemplate (but do not actually construct) the deterministic equivalent of the nondeterministic machine of part (a) constructed according to the method of Theorem 9.40.
    (c) By intuition, construct a deterministic machine to recognize 1*0.

16. (a) Construct the canonical nondeterministic recognizer for the regular set $(0 \vee 01)^*$.
    (b) Contemplate (but do not actually construct) the deterministic equivalent of the nondeterministic machine of part (a) constructed according to the method of Theorem 9.40.
    (c) By intuition, construct a deterministic machine to recognize $(0 \vee 01)^*$.

17. Prove that if A is a regular set whose symbols come from the alphabet I, then $I^* - A$ is a regular set.

Chapter 10

Machine Design
and Construction

Chapter 9 introduced finite-state machines and gave some indications of how they can be useful. In this chapter we deal with practical considerations in the actual physical implementation of a finite-state machine. A method for possibly reducing the number of states needed in the machine is developed in Section 10.1. In Section 10.2, we will see how any particular finite-state machine can be physically constructed from the simple logic gates of Chapter 6 plus one additional element. Finally, Sections 10.3 and 10.4 explore possibilities for building a particular finite-state machine from a storehouse of stock machines.

SECTION 10.1   MACHINE MINIMIZATION

If we want to construct a finite-state machine M, the number of internal states is a factor in the cost of construction. If we can find a machine having the same external behavior as M but fewer states, we will have a machine that can do the job we want done but is more economical to build.

Unreachable States

First, let's observe that we can remove any **unreachable states** of M.

10.1 Example   Let M be given by the state table of Figure 10.1. Although the state table contains the same information as the state graph (Figure 10.2), the graph shows us at a glance that state s_2 can never be reached from the starting state s_0. If we simply remove state s_2, we have the state graph of Figure

	Next state		
	Present input		
Present state	0	1	Output
s_0	s_1	s_3	0
s_1	s_3	s_0	0
s_2	s_1	s_3	1
s_3	s_0	s_1	1

Figure 10.1

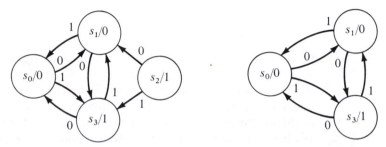

Figure 10.2   **Figure 10.3**

10.3 for a machine M' with one less state that behaves exactly like M; that is, it gives the same output as M for any input string. Notice that M and M' are not equivalent. Although M clearly simulates M', M' does not simulate M because state s_2 of M has no corresponding state in M'. But because s_2 can never be reached under any possible input string, it does not matter that there is no corresponding state for it; machine equivalence is a stronger idea than we need here.   □

10.2 Practice   What state(s) are unreachable from s_0 in the machine of Figure 10.4? Try to get your answer directly from the state table.   □

	Next state		
	Present input		
Present state	0	1	Output
s_0	s_1	s_4	0
s_1	s_4	s_1	1
s_2	s_2	s_2	1
s_3	s_3	s_1	0
s_4	s_0	s_0	1

Figure 10.4

Because the state graph of a finite-state machine is a directed graph, it has an associated adjacency matrix. Algorithm *ShortestPath* (Section 4.3) can be used with this matrix to detect unreachable states by setting the weight of each arc to 1.

Minimization Procedure

We will assume from now on that all unreachable states have been removed from M. We continue to look for a reduced machine M' with the same external behavior as M. This means that if we process any input string α through M from the starting state s_0 and if we process α through M' from the starting state s'_0, both machines produce identical output strings.

Because all states in M are now reachable, every state in M must have a corresponding state in M' that will produce the same output string for all input strings. To see this, suppose there is some state s_i of M with no corresponding state in M'. Then we could argue as follows: Find an input string β taking M from s_0 to s_i. Such a string exists because s_i is reachable from s_0. Apply β to M' starting at s'_0, and let M' be in state s'_i at the end of processing β. Because s'_i is not a corresponding state for s_i, there is some input string γ such that $f_O(s_i, \gamma) \neq f'_O(s'_i, \gamma)$. The concatenation string $\beta\gamma$ is an input string producing different output when fed into machine M than when fed into machine M'. Thus M' would not have the same external behavior as M, which contradicts the nature of M'. It follows that M' must simulate M and conversely. In short, M and M' must be equivalent.

We know that machine equivalence is an equivalence relation on finite-state machines. Therefore, we want to find a minimal representative of $[M]$, one with the fewest states. The procedure for constructing this minimal machine involves equivalences of the states of M.

10.3 Definition

Two states s_i and s_j of M are **equivalent** if for any $\alpha \in I^*$, $f_O(s_i, \alpha) = f_O(s_j, \alpha)$. $\qquad\square$

Equivalent states of a machine produce identical output strings for any input string. Notice that we are now comparing states within a single machine and not between two machines as we were just a moment ago when talking about simulation.

10.4 Practice

Prove that state equivalence is an equivalence relation on the states of a machine. $\qquad\square$

A Quotient Machine

For the time being, we will postpone the problem of how to identify equivalent states in a given machine M. Let's simply assume that we have somehow found which states are equivalent and have partitioned the states of M into the corresponding equivalence classes. These classes can serve as states for a quotient machine of M, provided (1) all states in the same class have the same output; and (2) for each input symbol, all states in the same class proceed under the next-state function to states that are all in the same class.

10.5 Practice   Show that conditions (1) and (2) are satisfied when M is partitioned into classes of equivalent states. $\square$

We will use M^* to denote the quotient machine of M that we have found. By the Fundamental Homomorphism Theorem for Machines (Theorem 9.23), M^* is a homomorphic image of M. Therefore, M^* and M are equivalent machines. M^* belongs to $[M]$, and the number of states of M^* is less than or equal to the number of states of M. We still need to show that there is no member of $[M]$ with fewer states than M^*.

Let $M_1 \in [M]$. Then M_1 and M^* are equivalent. In particular, M_1 simulates M^*, and there is a mapping g from the states of M^* to the states of M_1 such that a state in M^* and the corresponding state in M_1 produce the same output string for any given input string. The mapping g must be one-to-one. To see this, suppose s_i^* and s_j^* are two states of M^* with $g(s_i^*) = g(s_j^*)$. Because s_i^* and s_j^* have the same corresponding state in M_1, it also follows that s_i^* and s_j^* are themselves equivalent states in M^*. By the construction of M^*, $s_i^* = s_j^*$. Therefore g is a one-to-one map from the states of M^* to the states of M_1, and M_1 has at least as many states as M^*.

Equivalent States

The minimization problem for M thus boils down to finding the equivalent states of M. Perhaps we should note first that the obvious approach of directly trying to satisfy the definition of equivalent states will not work. Given two states s_i and s_j of M, we cannot actually compare the outputs corresponding to each possible input string. Fortunately, the problem is not as infinite as it first sounds; we only need to identify k-equivalent states.

10.6 Definition   Two states s_i and s_j of M are **k-equivalent** if for any $\alpha \in I^*$ where α has $\leq k$ symbols, $f_O(s_i, \alpha) = f_O(s_j, \alpha)$. $\square$

Clearly, k-equivalence is an equivalence relation on the states of M. It is possible to test two states of M for k-equivalence directly since we can actually produce the finite number of input strings having $\leq k$ symbols. But it turns out that we don't have to do this. We can begin by finding 0-equivalent states. These are states producing the same output for 0-length input strings, that is, states having the same associated output symbol. Thus, we can identify the 0-equivalence classes directly from the description of M.

10.7 Example   Let M be defined by the state table of Figure 10.5. The 0-equivalence classes of the states of M are

$$\{0, 2, 5\} \quad \text{and} \quad \{1, 3, 4, 6\} \qquad \qquad \square$$

Present state	Next state Present input 0	1	Output
0	2	3	0
1	3	2	1
2	0	4	0
3	1	5	1
4	6	5	1
5	2	0	0
6	4	0	1

Figure 10.5

Our procedure to find k-equivalent states is a recursive one; we know how to find 0-equivalent states, and we will show how to find k-equivalent states once we have identified states that are $(k - 1)$-equivalent. Suppose, then, that we already know which states are $(k - 1)$-equivalent. If states s_i and s_j are k-equivalent, they must produce the same output strings for any input string of length $\leq k$, in particular, for any string of length $\leq (k - 1)$. Thus, s_i and s_j must at least be $(k - 1)$-equivalent. But also they must produce the same output strings for any k-length input string.

An arbitrary k-length input string consists of a single arbitrary input symbol followed by an arbitrary $(k - 1)$-length input string. Applying such a k-length string to states s_i and s_j (which themselves have the same output symbol), the single input symbol moves s_i and s_j to next states s_i' and s_j'; then s_i' and s_j' must produce identical output strings for the remaining, arbitrary $(k - 1)$-length string, which will surely happen if s_i' and s_j' are $(k - 1)$-equivalent. Therefore, *to find k-equivalent states, look for $(k - 1)$-equivalent states whose next states under any input symbol are $(k - 1)$-equivalent.*

10.8 Example   Consider again the machine M of Example 10.7. We know the 0-equivalent states. To find 1-equivalent states, we look for 0-equivalent states with 0-equivalent next states. For example, the states 3 and 4 are 0-equivalent; under the input symbol 0, they proceed to states 1 and 6, respectively, which are 0-equivalent states, and under the input symbol 1 they both proceed to 5, which of course is 0-equivalent to itself. Therefore, states 3 and 4 are 1-equivalent. But states 0 and 5, themselves 0-equivalent, proceed under the input symbol 1 to states 3 and 0, respectively, which are not 0-equivalent states. So states 0 and 5 are not 1-equivalent; the input string 1 will produce an output string of 01 from state 0 and 00 from state 5. The 1-equivalence classes for M are

$$\{0, 2\}, \{5\}, \{1, 3, 4, 6\}$$

To find 2-equivalent states, we look for 1-equivalent states with 1-equivalent next states. Thus, states 1 and 3, although 1-equivalent, proceed under input 1 to states 2 and 5, respectively, which are not 1-equivalent states. Therefore, states 1 and 3 are not 2-equivalent. The 2-equivalence classes for M are

$$\{0, 2\}, \{5\}, \{1, 6\}, \{3, 4\}$$

The 3-equivalence classes for M are the same as the 2-equivalence classes.

□

10.9 Definition   Given two partitions π_1 and π_2 of a set S, π_1 is a **refinement** of π_2 if each block of π_1 is a subset of a block of π_2.   □

In Example 10.8, each successive partition of the states M into equivalence classes is a refinement of the previous partition. This refinement will always happen; k-equivalent states must also be $(k - 1)$-equivalent, so the blocks of the $(k - 1)$-partition can only be further subdivided. However, the subdivision process cannot continue indefinitely (at worst it can only go on until each partition block contains only one state); there will eventually be a point where $(k - 1)$-equivalent states and k-equivalent states coincide. (In Example 10.8, 2-equivalent and 3-equivalent states coincide.) Once this happens, all next states for members of a partition block under any input symbol fall within a partition block. Thus, k-equivalent states are also $(k + 1)$-equivalent and $(k + 2)$-equivalent, and so on. Indeed, these states are equivalent.

Summary

The total procedure (called the Moore reduction algorithm) for finding equivalent states is to start with 0-equivalent states, then 1-equivalent states, and so on, until the partition no longer subdivides. A pseudocode description of this algorithm follows.

Algorithm MooreReduction

find 0-equivalent states
repeat
  while untested equivalence classes remain do
    begin
      select untested equivalence class
      while untested state pairs in current class remain do
        begin
          select untested state pair in current class
          unset exit flag
          while untried input symbols remain and exit flag not set do
            begin
              select untried input symbol
              for both states in current pair, find next state under current
                input symbol
              if the next states are not equivalent, set exit flag
            end while
          if exit flag set, note that current states will not stay equivalent
        end while
      form new equivalence classes
    end while
until set of new equivalence classes = set of old equivalence classes

You may have forgotten by now, but we wanted to use the partition into equivalent states to form a quotient machine of M that would be a minimal machine M^* equivalent to M.

**10.10
Example**

For the machine M of Examples 10.7 and 10.8, the quotient machine M^* will have states

$$A = \{0, 2\}$$
$$B = \{5\}$$
$$C = \{1, 6\}$$
$$D = \{3, 4\}$$

The state table for M^* (Figure 10.6) is obtained from that for M. Machine M^* (starting state A) is equivalent to M; it will reproduce M's output for any input string, but it has four states instead of seven. □

**10.11
Example**

We will minimize M where M is given by the state table of Figure 10.7. The 0-equivalence classes of M are

$$\{0, 2, 4\}, \{1, 3\}$$

Next state			
	Present input		
Present state	0	1	Output
A	A	D	0
B	A	A	0
C	D	A	1
D	C	B	1

Figure 10.6

Next state			
	Present input		
Present state	0	1	Output
0	3	1	1
1	4	1	0
2	3	0	1
3	2	3	0
4	1	0	1

Figure 10.7

The 1-equivalence classes of M are

$\{0\}, \{2, 4\}, \{1, 3\}$

No refinement is possible. Let

$A = \{0\}$

$B = \{2, 4\}$

$C = \{1, 3\}$

The reduced machine is shown in Figure 10.8.   □

Next state			
	Present input		
Present state	0	1	Output
A	C	C	1
B	C	A	1
C	B	C	0

Figure 10.8

10.12
Practice

Minimize the machines whose state tables are shown in Figures 10.9 and 10.10.   □

	Next state		
	Present input		
Present state	0	1	Output
0	2	1	1
1	2	0	1
2	4	3	0
3	2	3	1
4	0	1	0

Figure 10.9

	Next state		
	Present input		
Present state	0	1	Output
0	1	3	1
1	2	0	0
2	1	3	0
3	2	1	0

Figure 10.10

We will assume from here on that all finite-state machines have been minimized.

✔ **Checklist**

Definitions

unreachable state (*p. 419*)
equivalent states (*p. 421*)
k-equivalent states (*p. 422*)
partition refinement (*p. 424*)

Techniques

Minimize finite-state machines.

Main Ideas

Unreachable states can be removed from a machine.

Once unreachable states are removed, a machine with the same external behavior as M must be equivalent to M. A machine equivalent to M with the least possible number of states is the quotient machine of M formed from classes of equivalent states.

Exercises Section 10.1

★ 1. Identify any unreachable states of each M in Figure 10.11.

Present state	Next state Present input 0	1	Output
s_0	s_2	s_0	0
s_1	s_2	s_1	1
s_2	s_2	s_0	1

(a)

Present state	Next state Present input a	b	c	Output
s_0	s_1	s_0	s_3	0
s_1	s_1	s_3	s_0	1
s_2	s_3	s_2	s_1	0
s_3	s_1	s_1	s_0	0

(b)

Figure 10.11

2. Use Algorithm *ShortestPath* to find any unreachable states of machine M in Figure 10.12. Show the final d-values and s-values of Algorithm *ShortestPath*.

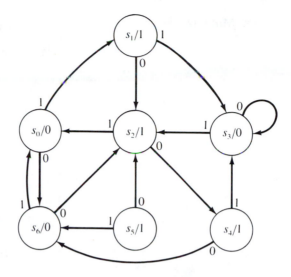

Figure 10.12

★ 3. Minimize M.

	Next state		
	Present input		
Present state	0	1	Output
0	3	6	1
1	4	2	0
2	4	1	0
3	2	0	1
4	5	0	1
5	3	5	0
6	4	2	1

Figure 10.13

4. Minimize M.

	Next state		
	Present input		
Present state	0	1	Output
0	5	3	1
1	5	2	0
2	1	3	0
3	2	4	1
4	2	0	1
5	1	4	0

Figure 10.14

5. Minimize M.

Present state	Next state Present input 0	1	Output
0	1	2	0
1	2	3	1
2	3	4	0
3	2	1	1
4	5	4	1
5	6	7	0
6	5	6	1
7	8	1	0
8	7	3	0

Figure 10.15

6. Minimize M.

Present state	Next state Present input 0	1	Output
0	7	1	1
1	0	3	1
2	5	1	0
3	7	6	1
4	5	6	0
5	2	3	0
6	3	0	1
7	4	0	0

Figure 10.16

7. Minimize M.

Present state	Next state Present input 0	1	Output
0	1	3	0
1	2	4	1
2	5	4	0
3	1	2	2
4	2	1	1
5	4	0	2

Figure 10.17

★ 8. Minimize M.

Present state	Next state Present input a b c	Output
0	1 4 0	1
1	4 2 3	0
2	3 4 2	1
3	4 0 1	0
4	1 0 2	0

Figure 10.18

9. Minimize M.

Present state	Next state Present input 0 1	Output
0	1 3	0
1	2 0	0
2	0 3	0
3	2 1	0

Figure 10.19

10. Minimize M.

Present state	Next state Present input 0 1	Output
0	1 3	1
1	2 0	0
2	4 3	1
3	0 1	1
4	2 4	0

Figure 10.20

11. Let M be a finite-state machine and M^* be the minimized machine for M. Let M^{**} be the minimization of any machine equivalent to M. Prove that M^* and M^{**} are isomorphic, thus showing that the reduced machine for $[M]$ is essentially unique.

★ 12. Given any two regular expressions A and B, the following long, drawn-out procedure can be used to determine whether A and B represent the same set. We first use the procedure of Section 9.3 (building the nondeterministic machine and then converting it to a deterministic machine) to build a finite-state machine M_A recognizing the regular set A and a finite-state machine M_B recognizing the regular set B. We remove unreachable states from these machines. Then $A = B$ if and only if M_A and M_B recognize the same set. But M_A and M_B recognize the same set if, for any string α, processing α through M_A from its starting state and processing α through M_B from its starting state produces the same output strings, meaning that M_A and M_B are equivalent machines (see the discussion preceding Definition 10.3).

The next part of the procedure is to minimize M_A and M_B; denote the minimized machines by M_A^* and M_B^*. By Exercise 11, $A = B$ if and only if M_A^* and M_B^* are isomorphic. Because there are only a finite number of possible bijections, we can test whether each bijection is an isomorphism by considering the behavior of each pair of corresponding states under each input symbol.

Try to decide intuitively which of the following equalities are true. If you have trouble, you can always resort to the above procedure!

(a) $0*0 = 00*$

(b) $0*1* = (0 \vee 1)*(01)*$

(c) $1*0 = 0 \vee 1*10$

(d) $1*0* = 1*10* \vee 1*00*$

SECTION 10.2   BUILDING MACHINES

Sequential Networks

We know that the output of a finite-state machine is a function of its present state, and that the present state of the machine is a function of past inputs. Thus, the states of a machine have certain memory capabilities. On the other hand, in the logic networks of Chapter 6, which are combinations of AND gates, OR gates, and inverters, the output is virtually instantaneous and a function only of the present input. Yet we will see that we can build any finite-state machine by using a logic network—provided we introduce one additional element. The new element must provide the memory missing from our previous logic networks.

Delay Elements

The element we need is a **delay element** (the simplest of a class of elements known as *flip-flops*). It has a single input that takes on values of 0 or 1, and it will output at time $t + 1$ the input signal it received at time t. Figure 10.21 represents the delay element *at time t + 1*. The delay element has a memory that captures input for the duration of one clock pulse. (We used delay elements in the encoding circuits of Chapter 8. Figure 10.21 shows signals propagating from right to left, instead of left to right, because that is how delay elements will be used in this chapter.)

The networks of Chapter 6, which had no delay elements, are also called **combinational networks**; when one or more delay elements are introduced, the network is known as a **sequential network**. Input sequences can be run through sequential networks (hence the name) provided that the clock pulse synchronizes input signals as well as delay elements.

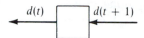

$d(t)$ ⟵ ☐ ⟵ $d(t + 1)$

Figure 10.21

10.13
Example

In Figure 10.22 the effect of the delay element is to feed the output from the terminal AND gate of the network back into the initial OR gate at the next clock pulse. Figure 10.23 shows the behavior of this network for a certain sequence of input values. The initial output of the delay element is assumed to be 0.   ☐

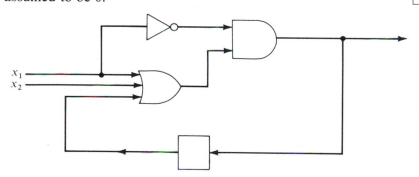

Figure 10.22

Time	0	1	2	3	4
x_1	1	0	0	1	0
x_2	1	1	0	0	0
Output	0	1	1	0	0

Figure 10.23

There are some general restrictions on the arrangements allowed in sequential networks just as there are in combinational networks. As in combinational networks, input or output lines cannot be tied together except by passing through an element, and both types of lines can be split to serve as input to more than one element. Unlike combinational networks, however, loops (where the output of an element becomes part of its input) are allowed, provided that at least one delay element is incorporated into the loop. The delay element prevents the confusion that results when an element tries to act upon its current output.

Networks for Finite-State Machines

We can build any finite-state machine by using a sequential network. A cluster of delay elements in the network functions as the states of the machine, retaining some memory of past inputs. The general structure of such a network is shown in Figure 10.24. It consists of two parts: (1) a combinational network (no delay elements) and (2) some loops containing all the delay elements.

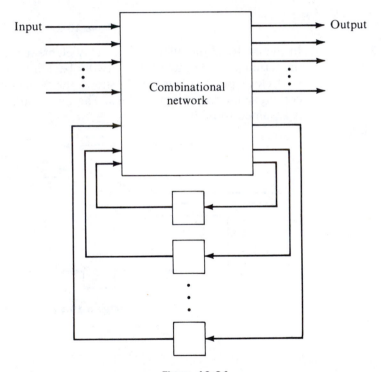

Input → Combinational network → Output

Figure 10.24

Now we need some specifics about the construction. All of our logic elements are binary, and input and output values can only be 0 or 1. If the finite-state machine has an input alphabet other than $\{0, 1\}$, we must encode the input symbols in binary form by using n-tuples of binary numbers to represent the input symbols (making sure that n is big enough to encode all the symbols). The set of binary n-tuples has 2^n members; thus, if we have k input symbols to encode, we pick n just big enough so that $2^n \geq k$, or $n \geq \log_2 k$. Because $\log_2 k$ may not be an integer, we want n to be the smallest integer greater than or equal to $\log_2 k$. We represent this integer by $\lceil \log_2 k \rceil$ ($\lceil x \rceil$ denotes the least integer greater than or equal to x; $\lceil x \rceil$ is called the **ceiling function** of x). If the output alphabet of our machine is not $\{0, 1\}$, we do a similar encoding for output.

10.14
Example

Suppose we want to build a sequential network for a finite-state machine with six input symbols. We compute $\lceil \log_2 6 \rceil = 3$. Our sequential network will have three input wires. Notice that there are eight combinations of 0s and 1s that can appear on these three wires. We will assign the six input symbols to six of these combinations, and two combinations will remain unused. □

If the finite-state machine we are building has q states, we compute $\lceil \log_2 q \rceil = p$. Our network will then have p delay elements; different combinations of 0s and 1s on the inputs (or outputs) of the delay elements will represent the different states.

The essence of the construction is to translate the state table of the finite-state machine into truth functions for the outputs and delay elements and then to construct the logic network for each truth function, as we did in Chapter 6. Let's look at an example.

10.15
Example

We will build a sequential network for the finite-state machine of Example 9.2. The input and output alphabet for this machine is $\{0, 1\}$, and the state table appears in Figure 10.25.

Present state	Next state		Output
	Present input		
	0	1	
s_0	s_1	s_0	0
s_1	s_2	s_1	1
s_2	s_2	s_0	1

Figure 10.25

Because there are only two input symbols and two output symbols, we need only one input line x (taking on values of 0 or 1) and one output line y for the network. There are three states, so we need $\lceil \log_2 3 \rceil = 2$ delay elements. We arbitrarily associate states with configurations of the inputs or outputs of the delay elements as in Figure 10.26.

Now we use the information contained in the state table to write three truth functions. One truth function describes the behavior of the output $y(t)$; it is a function of the two variables $d_1(t)$ and $d_2(t)$, representing the present state. The other two truth functions describe the behavior of

	d_1	d_2
s_0	0	0
s_1	0	1
s_2	1	0

Figure 10.26

$d_1(t + 1)$ and $d_2(t + 1)$, representing the next state; these are functions of $x(t)$, $d_1(t)$, and $d_2(t)$, the present input and the present state. Figure 10.27 shows these truth functions.

$x(t)$	$d_1(t)$	$d_2(t)$	$y(t)$	$d_1(t + 1)$	$d_2(t + 1)$
0	0	0	0	0	1
1	0	0	0	0	0
0	0	1	1	1	0
1	0	1	1	0	1
0	1	0	1	1	0
1	1	0	1	0	0

Figure 10.27

In constructing the third line of Figure 10.27, for example, $x(t) = 0$, $d_1(t) = 0$, and $d_2(t) = 1$, meaning that the present input is 0 and the present state is s_1. The output associated with state s_1 is 1, so $y(t) = 1$. The next state associated with input 0 and present state s_1 is s_2, so $d_1(t + 1) = 1$ and $d_2(t + 1) = 0$. Notice that there are some don't-care conditions for these functions because the configuration $d_1(t) = 1$ and $d_2(t) = 1$ does not occur.

The canonical sum-of-products form for each of these truth functions is

$$y(t) = d_1'd_2 + d_1d_2' \qquad (y \text{ is not a function of } x)$$
$$d_1(t + 1) = x'd_1'd_2 + x'd_1d_2'$$
$$d_2(t + 1) = x'd_1'd_2' + xd_1'd_2$$

The logic networks for these expressions go into the large box in Figure 10.24. Thus Figure 10.28 is a wiring diagram for our finite-state machine. ☐

10.16
Practice

One possible behavior pattern for the finite-state machine of Example 9.2 is shown in Figure 10.29. Follow the wiring diagram of Figure 10.28 to see that it reproduces this behavior. Note that d_1 and d_2 are initially both 0 to correspond to the start state s_0. ☐

10.17
Practice

Construct a sequential network for the parity check machine of Example 9.8. ☐

Minimization

Now we have a procedure for building any finite-state machine with a sequential network using the proper arrangement of four simple types of elements: AND gates, OR gates, inverters, and delay elements. There

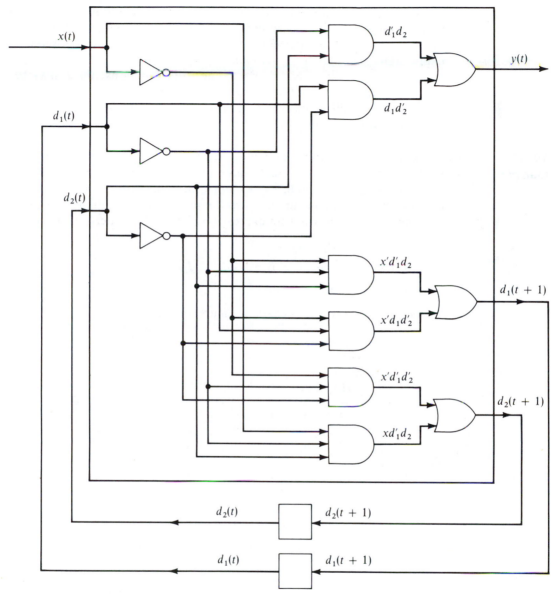

Figure 10.28

Time	t_0	t_1	t_2	t_3	t_4	t_5
Input	0	1	1	0	1	—
State	s_0	s_1	s_1	s_1	s_2	s_0
Output	0	1	1	1	1	0

Figure 10.29

are several factors involved in minimizing these sequential networks. If the machine itself is not already minimized, we can use Algorithm *MooreReduction* to minimize it. We may be able to minimize the Boolean expressions, and thus the corresponding logic networks, as in Chapter 6. If there are don't-care conditions to the truth functions, these may be assigned values of either 0 or 1 to aid in minimizing the Boolean expression. Finally, a different encoding of machine states as binary n-tuples may minimize the network.

10.18
Example

Consider again the finite-state machine analyzed in Example 10.15. The truth functions of Figure 10.27 have don't-care conditions. If we use Karnaugh maps (Section 6.2) to examine these functions, we see that we can minimize by assigning values to the don't-care conditions as shown in Figure 10.30. The Boolean expressions for the logic network become

$$y(t) = d_1 + d_2$$
$$d_1(t + 1) = x'd_1 + x'd_2 = x'(d_1 + d_2)$$
$$d_2(t + 1) = xd_2 + x'd_1'd_2'$$

The simplified network appears in Figure 10.31.   □

$x(t)$	$d_1(t)$	$d_2(t)$	$y(t)$	$d_1(t + 1)$	$d_2(t + 1)$
0	0	0	0	0	1
1	0	0	0	0	0
0	0	1	1	1	0
1	0	1	1	0	1
0	1	0	1	1	0
1	1	0	1	0	0
0	1	1	1	1	0
1	1	1	1	0	1

Figure 10.30

✔ Checklist

Definitions

delay element (*p. 432*)
combinational network (*p. 433*)
sequential network (*p. 433*)
ceiling function (*p. 435*)

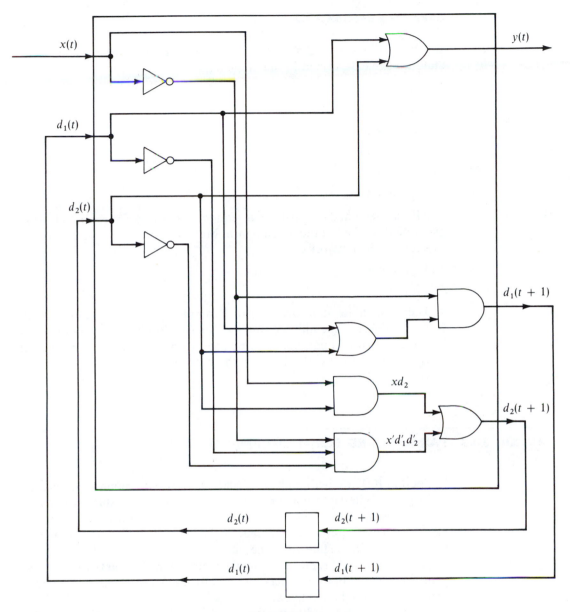

Figure 10.31

Techniques

Construct sequential networks for finite-state machines.

Main Ideas

Only four simple types of logic elements are needed to construct a network for any finite-state machine.

Exercises Section 10.2

For Exercises 1 through 6, construct a sequential network for each finite-state machine described in the given figure.

★ 1. Figure 9.20

2. Figure 9.43b

3. Figure 9.16

★ 4. Figure 9.9b

5. Figure 9.5

6. Figure 9.7

For Exercises 7 and 8, construct a sequential network for each finite-state machine described in the given figure. Make use of don't-care conditions to simplify the networks.

★ 7. Figure 9.6

8. Figure 9.12

9. Using the finite-state machine of Example 10.15 and Example 10.18, find a different encoding of states as binary n-tuples that, together with the don't-care conditions, can be used to produce a simpler network than that of Figure 10.31. Draw the network.

SECTION 10.3 PARALLEL AND SERIAL DECOMPOSITIONS

In Section 10.1, we learned how to find an equivalent minimized machine for a given finite-state machine M. Although this procedure usually reduces the number of states required, it still leaves us with a single, perhaps very complex finite-state machine. In Section 10.2 we built sequential networks for finite-state machines using four basic logic elements. The sequential network replaces a single machine with a network of very simple elements, but there may be lots of these elements and their interconnections may be complex.

We now want to discuss the middle ground between these two extremes—how to replace a single finite-state machine with a network of a few simpler finite-state machines. The ability to do such a replacement might allow us to stock up on some standard small machines and use them in various combinations to build new machines. (Think of writing a computer program calling for various standard procedures or subroutines.) This area of investigation is called **structure theory** or **decomposition theory** of machines, and it will occupy us for the rest of this chapter.

Suppose that M is our original machine. A decomposition for M will be a network of smaller machines. The network viewed as a whole, however, is itself a machine M'. We want M' to simulate M. In the first two types of decomposition we will consider, something stronger will happen: a submachine of M' will be isomorphic to M.

Our first two decomposition types borrow ideas from basic electronics. Given two black boxes, we can think of connecting them in parallel (Figure 10.32a) or series (Figure 10.32b). If the black boxes are finite-state machines, we want to view the resulting network as a finite-state machine.

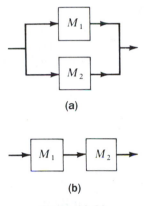

(a)

(b)

Figure 10.32

Parallel Decompositions

Let M_1 and M_2 be two finite-state machines connected in parallel as in Figure 10.32a. We want to represent the total network as a finite-state machine M_p. Clearly, the present state of M_p should depend equally upon the present states of M_1 and M_2. This goal suggests using ordered pairs of states from the two machines for the state set of M_p; in fact, we define the entire machine M_p in terms of ordered pairs of elements from M_1 and M_2, with the next-state function and output function operating componentwise.

10.19
Definition

Given two finite-state machines $M_1 = [S_1, I_1, O_1, f_{S_1}, f_{O_1}]$ and $M_2 = [S_2, I_2, O_2, f_{S_2}, f_{O_2}]$, the **parallel connection** of M_1 and M_2 is the machine $M_p = [S_1 \times S_2, I_1 \times I_2, O_1 \times O_2, f_S, f_O]$ where $f_S((s_1, s_2), (i_1, i_2)) = (f_{S_1}(s_1, i_1), f_{S_2}(s_2, i_2))$, $f_O(s_1, s_2) = (f_{O_1}(s_1), f_{O_2}(s_2))$. □

10.20
Practice

The state tables for machines M_1 and M_2 are given in Figure 10.33. Complete the state table (Figure 10.34) for the parallel connection M_p of M_1 and M_2. □

M_1	Input		Output
	0	1	
A	A	A	1
B	B	A	0

M_2	Input		Output
	0	1	
a	b	a	1
b	a	b	1

Figure 10.33

M_p	Input				Output
	$(0, 0)$	$(0, 1)$	$(1, 0)$	$(1, 1)$	
(A, a)	(A, b)	(A, a)	(A, b)	(A, a)	$(1, 1)$
(A, b)	(A, a)	—	(A, a)	—	$(1, 1)$
(B, a)	(B, b)	(B, a)	—	(A, a)	$(0, 1)$
(B, b)	—	—	(A, a)	(A, b)	—

Figure 10.34

We want to begin with an arbitrary machine M and find two machines M_1 and M_2 whose parallel connection M_p will simulate M—indeed, we want M_p to have a submachine isomorphic to M. The machine M_p is called a **parallel decomposition** of M. When will a parallel decomposition of M exist, and if it does, how can we find M_1 and M_2?

If a parallel decomposition of M exists, then the original machine M will probably be more complex than either M_1 or M_2, and in some sense it must carry on the activities of both M_1 and M_2. This suggests that M_1 and M_2 will be closely related to quotient machines of M. It turns out that we can temporarily ignore outputs, choosing them later so that they work. We define a partition of a machine in which all states in a block, under a given input symbol, proceed to states that are all in the same partition block as having the **substitution property** or as being an **S.P. partition.** It is clear how to define the next-state function for a machine whose states are the blocks of an S.P. partition.

For π_1 and π_2, two S.P. partitions of M, we can define (except for outputs) machines M_1 and M_2 using the blocks of π_1 and π_2, respectively, as states. Any state of M belongs to a block of π_1 and a block of π_2. Applying the next-state function for M_1 results in a block of π_1 that contains the next state of M. Applying the next-state function for M_2 results in a block of π_2 that also contains the next state of M. For M_p to simulate M, these two new blocks of π_1 and π_2 must uniquely identify the next state of M. We therefore impose the additional property that the intersection of any block from π_1 with any block from π_2 contains at most one element of M. Partitions with such a property are called **orthogonal.**

10.21
Example

M is a machine given by the state table of Figure 10.35.

M	Input 0	1	Output 0
0	3	2	0
1	5	2	0
2	4	1	0
3	1	4	1
4	0	3	0
5	2	3	0

Figure 10.35

The partitions

$$\pi_1 = \{\overline{0, 1, 2}; \overline{3, 4, 5}\}$$

and

$$\pi_2 = \{\overline{0, 5}; \overline{1, 4}; \overline{2, 3}\}$$

are orthogonal S.P. partitions where each overbar identifies a partition block. Let's use π_1 and π_2 to construct a parallel decomposition of M. (Note that M is already a reduced machine.) The machines M_1 corresponding to π_1 and M_2 corresponding to π_2 have next-state functions

M_1	0	1
$A = \{0, 1, 2\}$	B	A
$B = \{3, 4, 5\}$	A	B

M_2	0	1
$a = \{0, 5\}$	c	c
$b = \{1, 4\}$	a	c
$c = \{2, 3\}$	b	b

Because M receives input of only 0 and 1, in constructing the parallel connection M_p of M_1 and M_2 to simulate M, we need only consider inputs of $(0, 0)$ and $(1, 1)$, which we will write as 0 and 1, respectively. Then the next-state function for M_p is

M_p	0	1
(A, a)	(B, c)	(A, c)
(A, b)	(B, a)	(A, c)
(A, c)	(B, b)	(A, b)
(B, a)	(A, c)	(B, c)
(B, b)	(A, a)	(B, c)
(B, c)	(A, b)	(B, b)

Now we want to define an isomorphism g from a submachine of M_p to M. Because π_1 and π_2 are orthogonal, each state of M_p corresponds to a single state of M, suggesting that g be defined by

g: $(A, a) \rightarrow 0$
    $(A, b) \rightarrow 1$
    $(A, c) \rightarrow 2$
    $(B, a) \rightarrow 5$
    $(B, b) \rightarrow 4$
    $(B, c) \rightarrow 3$

The construction of M_p should make clear that g serves as an isomorphism from M_p onto M at least as far as preserving the next-state function. We will do a sample calculation, in which f_S denotes the next-state function of M and f_p denotes the next-state function of M_p. Then $g(f_p((B, a), 0)) = g(A, c) = 2$ and $f_S(g(B, a), 0) = f_S(5, 0) = 2$. The commutative diagram appears in Figure 10.36.

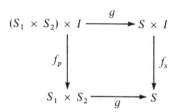

Figure 10.36

Now it is time to take care of the outputs. The output for M_p will be ordered pairs of the output symbols for M_1 and M_2. For M_p to simulate M, we must extend our definition of homomorphism to allow a mapping g' from the output symbols of M_p to those of M. If each state of M_1 and each state of M_2 has a distinct output symbol, then each ordered pair in the output of M_p is associated with at most one state of M, and therefore one output symbol of M. This defines our mapping. In some cases, however, we can limit the outputs of M_1 and M_2 to $\{0, 1\}$ and choose the output symbol for each state so that we can define a mapping. For this example, we can define the function g' by

g': $(0, 0) \rightarrow 0$
    $(0, 1) \rightarrow 0$
    $(1, 0) \rightarrow 0$
    $(1, 1) \rightarrow 1$

and assign the outputs for M_1 and M_2 so that $(1, 1)$ in M_p occurs exactly at (B, c), the state of M_p corresponding to 3, which is the only state of M with output 1. (Other output functions g' and other output assignments can be used.) The complete tables for M_1, M_2, and M_p appear in Figure 10.37. The parallel connection M_p of M_1 and M_2 is isomorphic to M by means of the function g (and g'). The state table for M_p is a relabeling of the original table for M.   □

M_1	Input 0	Input 1	Output
A	B	A	0
B	A	B	1

M_2	Input 0	Input 1	Output
a	c	c	0
b	a	c	0
c	b	b	1

M_p	Input 0	Input 1	Output
(A, a)	(B, c)	(A, c)	$(0, 0)$
(A, b)	(B, a)	(A, c)	$(0, 0)$
(A, c)	(B, b)	(A, b)	$(0, 1)$
(B, a)	(A, c)	(B, c)	$(1, 0)$
(B, b)	(A, a)	(B, c)	$(1, 0)$
(B, c)	(A, b)	(B, b)	$(1, 1)$

Figure 10.37

10.22
Practice

For Example 10.21, compute:

(a) $g(f_p((A, c), 1))$
(b) $f_s(g(A, c), 1)$ □

The following theorem on the parallel decomposition of machines merely formalizes the discussion preceding Example 10.21 and the construction of Example 10.21 itself; we will omit the proof.

10.23
Theorem

A finite-state machine M has a parallel decomposition if and only if there are two orthogonal S.P. partitions on M.   □

Notice that the partition π_1 consisting of a single block—all the states of M—is trivially S.P., and so is the partition π_2 into single states of M. Furthermore, these are orthogonal partitions, but they result in a parallel decomposition where M_2 is essentially M and M_1 does nothing. For a nontrivial decomposition, we must find nontrivial, orthogonal S.P. partitions.

10.24
Example

The state table for a machine M is given in Figure 10.38. The state tables for the component machines in a parallel decomposition of M appear in Figure 10.39. The isomorphism g (and the output function g') for these machines is given by

$$g:\ (A, a) \to 1 \qquad g':\ (0, 0) \to 0$$
$$(A, b) \to 3 \qquad\qquad (0, 1) \to 1$$
$$(B, a) \to 4 \qquad\qquad (1, 0) \to 1$$
$$(B, b) \to 0 \qquad\qquad (1, 1) \to 0$$
$$(B, c) \to 2$$

M	Input 0	1	Output
0	0	1	0
1	2	0	0
2	4	3	0
3	0	4	1
4	2	3	1

Figure 10.38

M_1	Input 0	1	Output
$A = \{1, 3\}$	B	B	0
$B = \{0, 2, 4\}$	B	A	1

M_2	Input 0	1	Output
$a = \{1, 4\}$	c	b	0
$b = \{0, 3\}$	b	a	1
$c = \{2\}$	a	b	1

Figure 10.39

(Again, other output assignments and other output functions g' can be used.) Here, g is an isomorphism from a submachine of the decomposition onto M since state (A, c) is never used. The total number of states in the decomposition of M is larger than the number of states in M itself, but the component machines are smaller. ☐

**10.25
Practice**

For a parallel decomposition of the machine M of Figure 10.40:

M	Input 0	1	Output
0	1	3	0
1	0	2	1
2	3	1	0
3	2	0	1

Figure 10.40

(a) Write the state tables of component machines.
(b) Give the isomorphism g (and the output function g'). ☐

Serial Decompositions

Now let's consider two machines connected in series, as in Figure 10.32b. Actually, we'll modify the wiring for a series connection. Figure 10.32b suggests that an input signal enters only M_1. Then it takes one clock pulse before the output of M_1 (via its next state) reflects the impact of this input signal, and two clock pulses before the output of M_2 reflects it. When more than two machines are connected serially, this delay is propagated. Consequently, if we want to connect serial and parallel networks, we must be concerned with timing. To avoid this problem, we'll direct the input signal to both M_1 and M_2 and compute next-state functions simultaneously. We will also read both outputs simultaneously. Our wiring diagram is thus modified to that of Figure 10.41a.

All this talk about simultaneous operations may sound as if we have gone back to the parallel case; however, the crucial difference is that the output of M_1 serves as input to M_2 (see the segment enclosed by dashes). In fact, we will allow M_2 to have the maximum information from M_1 by requiring that the output of M_1 be the present state of M_1. For simplicity, we will put the same requirement on M_2, giving us Figure 10.41b.

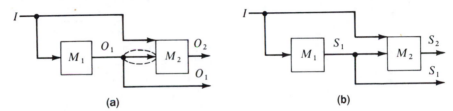

(a)   (b)

Figure 10.41

10.26
Definition

Let $M_1 = [S_1, I_1, O_1, f_{S_1}, f_{O_1}]$ and $M_2 = [S_2, I_2, O_2, f_{S_2}, f_{O_2}]$ be two finite-state machines with

$$O_1 = S_1, \quad f_{O_1}(s_1) = s_1$$
$$O_2 = S_2, \quad f_{O_2}(s_2) = s_2$$

and

$$I_2 = S_1 \times I_1$$

The **serial connection** of M_1 and M_2 is the machine

$$M_s = [S_1 \times S_2, I_1, O, f_s, f_O]$$

where

$$O = S_1 \times S_2$$
$$f_O(s_1, s_2) = (f_{O_1}(s_1), f_{O_2}(s_2)) = (s_1, s_2)$$
$$f_S((s_1, s_2), i) = (f_{S_1}(s_1, i), f_{S_2}(s_2, (s_1, i)))$$

□

10.27
Practice

The state tables for machines M_1 and M_2 are given in Figure 10.42. Complete the state table (Figure 10.43) for the serial connection of M_1 and M_2.   □

	Input		
M_1	0	1	Output
A	A	A	A
B	B	A	B

	Input				
M_2	$(A, 0)$	$(A, 1)$	$(B, 0)$	$(B, 1)$	Output
a	b	a	a	b	a
b	a	b	a	b	b

Figure 10.42

	Input		
M_s	0	1	Output
(A, a)	(A, b)	(A, a)	(A, a)
(A, b)	—	(A, b)	(A, b)
(B, a)	—	—	(B, a)
(B, b)	(B, a)	(A, b)	(B, b)

Figure 10.43

Given a machine M, we want to find two machines whose serial connection M_s will have a submachine isomorphic to M. The machine M_s is a **serial decomposition** of M. If M has a serial decomposition, then each state s of M is associated with a unique (s_1, s_2) state of M_s, and under an input symbol i, if $(s, i) \to r$ in M, where r is associated with (r_1, r_2) in M_s, then we must have $(s_1, i) = r_1$. Except for outputs, this association identifies a machine homomorphism from M onto M_1, making M_1 (except for outputs) a quotient machine of M. Thus M_1 is the machine of an S.P. partition π of M. Conversely, if M has an S.P. partition π, and we make the "head machine" in a serial decomposition of M the machine for π, then we can build a "tail machine" as follows.

Find a partition π_2 of M such that π_1 and π_2 are orthogonal. (This is easy; make one block of π_2 consist of the first element in each block of π_1, the second block of π_2 consist of the second element in each block of π_1, etc. We are not requiring that π_2 be S.P.) Any state of M belongs to a block of π_1 and a block of π_2. Let M_2 have as states the blocks of π_2. Because π_1 and π_2 are orthogonal and M_2 knows (from its present state) the block of π_2 and (from the output of M_1) the block of π_1, M_2 knows the exact, present state s of M. Consulting M for the next state of s under the present input tells us the next-state function for M_2.

10.28
Example

A machine M is given by the state table of Figure 10.44. An S.P. partition of M is $\pi_1 = \{\overline{0, 3}; \overline{1, 6}; \overline{2, 4, 5}\}$. Thus, we define M_1 by Figure 10.45. Now we let $\pi_2 = \{\overline{0, 1, 2}; \overline{3, 6, 4}; \overline{5}\}$. The state table for M_2 will have the format of Figure 10.46.

M	Input 0	1	Output
0	1	6	1
1	3	3	1
2	3	5	1
3	6	6	1
4	0	2	0
5	3	4	0
6	0	0	1

Figure 10.44

M_1	Input 0	1	Output
$A = \{0, 3\}$	B	B	A
$B = \{1, 6\}$	A	A	B
$C = \{2, 4, 5\}$	A	C	C

Figure 10.45

M_2	$(A, 0)$	$(A, 1)$	Input $(B, 0)$	$(B, 1)$	$(C, 0)$	$(C, 1)$	Output
$a = \{0, 1, 2\}$							a
$b = \{3, 6, 4\}$							b
$c = \{5\}$							c

Figure 10.46

To complete the state table, consider the next state for a under input $(A, 0)$. The a–A pair uniquely selects state 0 of M, which under input 0 goes to state $1 \in \{0, 1, 2\}$. Thus, the next state for M_2 should be a. Other entries are computed in the same way; the completed table for M_2 appears in Figure 10.47. The dashes indicate don't-care conditions, which are combinations of states of M_1 and M_2 that do not represent states of M. These conditions could be specified in any way that might economize the construction of M_2. Without actually writing the machine M_s, we can see that the isomorphism function g (and output function g') given by

M_2	Input (A, 0)	(A, 1)	(B, 0)	(B, 1)	(C, 0)	(C, 1)	Output
a	a	b	b	b	b	c	a
b	b	b	a	a	a	a	b
c	–	–	–	–	b	b	c

Figure 10.47

g:		g':	
$(A, a) \rightarrow 0$		$(A, a) \rightarrow 1$	
$(A, b) \rightarrow 3$		$(A, b) \rightarrow 1$	
$(B, a) \rightarrow 1$		$(B, a) \rightarrow 1$	
$(B, b) \rightarrow 6$		$(B, b) \rightarrow 1$	
$(C, a) \rightarrow 2$		$(C, a) \rightarrow 1$	
$(C, b) \rightarrow 4$		$(C, b) \rightarrow 0$	
$(C, c) \rightarrow 5$		$(C, c) \rightarrow 0$	

makes a submachine of M_s isomorphic to M.   □

10.29
Practice

In machine M of Example 10.28, state 4 under input 1 goes to state 2. What is the equivalent computation in M_s?   □

The discussion preceding Example 10.28 was essentially a proof of the following theorem.

10.30
Theorem

A finite-state machine M has a serial decomposition if and only if there is an S.P. partition on M.   □

Again, a nontrivial decomposition would require that both M_1 and M_2 have fewer states than M itself, in which case the S.P. partition is nontrivial.

10.31
Practice

Write the state tables of component machines and give the isomorphism g and the output function g' for a serial decomposition of the machine M of Figure 10.48.   □

M	Input 0	1	Output
0	3	0	1
1	1	0	1
2	3	2	0
3	3	0	0

Figure 10.48

A parallel decomposition of a machine seems a less complex arrangement than a serial decomposition of the same machine. Real computers are being built that operate in a parallel mode. For example, making all computations in a matrix multiplication in parallel and fitting the answers into the proper array is faster than the serial approach of computing one entry, then the next, and so on. Research is being done on the types of computational tasks that best lend themselves to parallel processing.

✔ Checklist

Definitions

structure theory (decomposition theory) (*p. 440*)
parallel connection (*p. 441*)
parallel decomposition (*p. 442*)
substitution property (*p. 442*)
S.P. partition (*p. 442*)
orthogonal partitions (*p. 442*)
serial connection (*p. 447*)
serial decomposition (*p. 448*)

Techniques

Find a nontrivial parallel decomposition of a given machine M if one exists.

Find a nontrivial serial decomposition of a given machine M if one exists.

Main Ideas

A finite-state machine M has a nontrivial parallel decomposition if and only if there are two nontrivial orthogonal S.P. partitions on M.

A finite-state machine M has a nontrivial serial decomposition if and only if there is a nontrivial S.P. partition on M.

Exercises Section 10.3

★ 1. (a) Given the machine M of Figure 10.49, write the state tables for component machines (with output alphabets {0, 1}) in a parallel decomposition of M; also give the isomorphism g and the output function g'.

	Input		
M	0	1	Output
0	4	3	0
1	2	1	1
2	1	3	1
3	2	0	1
4	0	1	0

Figure 10.49

(b) The following computations are done in M; do the corresponding computations in the parallel decomposition of M.

$$f_s(4, 0) = 0$$
$$f_s(0, 1) = 3$$
$$f_s(3, 0) = 2$$

2. (a) Given the machine M of Figure 10.50, write the state tables for component machines (with output alphabets $\{0, 1\}$) in a parallel decomposition of M; also give the isomorphism g and the output function g'.

	Input		
M	0	1	Output
0	2	1	1
1	3	1	0
2	4	6	0
3	5	6	1
4	5	5	0
5	4	5	0
6	3	5	0

Figure 10.50

(b) The following computations are done in M; do the corresponding computations in the parallel decomposition of M.

$$f_s(4, 0) = 5$$
$$f_s(2, 1) = 6$$
$$f_s(6, 0) = 3$$

3. (a) Given the machine M of Figure 10.51, write the state tables for component machines (with output alphabets $\{0, 1\}$) in a parallel decomposition of M; also give the isomorphism g and the output function g'.

M	Input 0	1	Output
0	4	1	1
1	3	1	0
2	3	1	0
3	2	4	1
4	0	4	0

Figure 10.51

(b) The following computations are done in M; do the corresponding computations in the parallel decomposition of M.

$f_S(0, 0) = 4$

$f_S(2, 0) = 3$

$f_S(3, 1) = 4$

4. (a) Given the machine M of Figure 10.52, write the state tables for component machines in a parallel decomposition of M; also give the isomorphism g and the output function g'. Can you use $\{0, 1\}$ as the output alphabet for each component machine?

M	Input 0	1	Output
0	1	2	0
1	1	5	1
2	5	3	1
3	4	0	0
4	4	1	0
5	5	4	0

Figure 10.52

(b) The following computations are done in M; do the corresponding computations in the parallel decomposition of M.

$f_S(3, 0) = 4$

$f_S(1, 1) = 5$

$f_S(0, 1) = 2$

★ 5. (a) Given the machine M of Figure 10.53, write the state tables for component machines in a serial decomposition of M; also give the isomorphism g and the output function g'.

	Input		
M	0	1	Output
0	1	2	0
1	0	4	0
2	5	3	1
3	4	2	0
4	3	1	1
5	3	1	0

Figure 10.53

(b) The following computations are done in M; do the corresponding computations in the serial decomposition of M.

$f_S(2, 0) = 5$

$f_S(4, 1) = 1$

$f_S(3, 1) = 2$

6. (a) Given the machine M of Figure 10.54, write the state tables for component machines in a serial decomposition of M; also give the isomorphism g and the output function g'.

	Input		
M	0	1	Output
0	4	2	1
1	2	3	1
2	0	4	0
3	1	2	1
4	0	3	0

Figure 10.54

(b) The following computations are done in M; do the corresponding computations in the serial decomposition of M.

$f_S(0, 0) = 4$

$f_S(2, 0) = 0$

$f_S(4, 1) = 3$

7. (a) Given the machine M of Figure 10.55, write the state tables for component machines in a serial decomposition of M; also give the isomorphism g and the output function g'.

M	Input 0	1	Output
0	5	4	1
1	6	5	0
2	3	6	0
3	1	0	1
4	5	3	0
5	2	1	1
6	4	2	0

Figure 10.55

(b) The following computations are done in M; do the corresponding computations in the serial decomposition of M.

$$f_S(2, 0) = 3$$

$$f_S(3, 1) = 0$$

$$f_S(2, 1) = 6$$

SECTION 10.4   CASCADE DECOMPOSITIONS

The simulations of a machine M by either parallel or serial decompositions (Section 10.3) have two features. First, there is an isomorphism from a submachine of the decomposition onto M. Second, the decompositions are *loop-free:* no machine output becomes the input of an earlier machine. (The individual component machines may have a loop structure; indeed, if we recall how we built machines from logic elements in Section 10.2, loops were essential.) We want to relax the isomorphism property but preserve the loop-free construction. A more general loop-free construction, of which the parallel and serial forms are special cases, appears in Figure 10.56. It is called a **cascade connection** of machines M_1 and M_2.

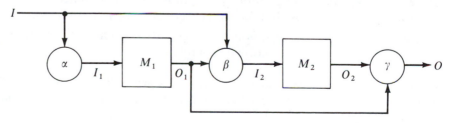

Figure 10.56

Let $M_1 = [S_1, I_1, O_1, f_{S_1}, f_{O_1}]$ and $M_2 = [S_2, I_2, O_2, f_{S_2}, f_{O_2}]$. The network as a whole, M_c, has input I, output O, and states $S_1 \times S_2$. In Figure 10.56, α, β, and γ are transformations on, respectively, I (to convert input to I_1 for use by M_1), $I \times O_1$ (to convert information to I_2 for use by M_2), and $O_1 \times O_2$ (to convert outputs to O). The output function f_O for M_c is

$$f_O(s_1, s_2) = \gamma(f_{O_1}(s_1), f_{O_2}(s_2))$$

Thus, M_c lets M_1 and M_2 each compute its output for its present state and runs the results through γ. The next-state function f_S for M_c is

$$f_S((s_1, s_2), i) = (f_{S_1}(s_1, \alpha(i)), f_{S_2}(s_2, \beta(i, f_{O_1}(s_1))))$$

Thus, M_c lets M_1 and M_2 each compute its next state based on its present state and appropriately transformed input. The cascade construction can be extended to more than two machines.

We now want to know when an arbitrary machine M can be simulated by a cascade of machines and how these machines can be found. Before we explore this question, we need a few more algebraic ideas.

Algebraic Structures and Finite-State Machines

Presently, we will tie together algebraic structures, namely semigroups and monoids, and finite-state machines. First, however, we will make a quick list (without proofs) of other algebraic results that we will need.

10.32 Definition

A semigroup B **divides** a semigroup A if B is a homomorphic image of a subsemigroup of A. □

10.33 Theorem

If a group G divides a finite semigroup A, then G is a homomorphic image of a subgroup of A. □

10.34 Definition

A group $[G, \cdot]$ is **simple** if it has no proper normal subgroups. □

10.35 Example

(a) Any group of prime order is simple, since by Lagrange's Theorem (7.118) it has no proper subgroups of any kind.

(b) $[\mathbb{Z}_6, +_6]$ is not a single group because $\{0, 2, 4\}$ and $\{0, 3\}$ are both proper normal subgroups. □

10.36 Definition

A normal subgroup $[S, \cdot]$ of a group $[G, \cdot]$ is **maximal** if $S \neq G$ and S is not a proper subset of any other proper normal subgroup of G. □

10.37 Example

Both $\{0, 2, 4\}$ and $\{0, 3\}$ are maximal normal subgroups in $[\mathbb{Z}_6, +_6]$. Also, $\{0\}$ is a maximal normal subgroup of $\{0, 2, 4\}$ and of $\{0, 3\}$. □

Let's assume that a group $[G, \cdot]$ with identity i contains a maximal normal subgroup G_1, that G_1 in turn contains a maximal normal subgroup G_2, G_2 contains a maximal normal subgroup G_3, and so on. Also assume that we can only perform this process a finite number of times. Thus,

$$\{i\} = G_k \subset G_{k-1} \subset \cdots \subset G_2 \subset G_1 \subset G_0 = G$$

with each G_i a maximal normal subgroup of G_{i-1}, $1 \le i \le k$. This arrangement is called a **composition series** of G. Associated with it is a set of quotient groups G_{i-1}/G_i, $1 \le i \le k$, called the **factor set** of the composition.

10.38
Example

Two composition series for $[\mathbb{Z}_6, +_6]$ are

$$\{0\} \subset \{0, 2, 4\} \subset \mathbb{Z}_6$$

and

$$\{0\} \subset \{0, 3\} \subset \mathbb{Z}_6$$

The factor set for the first series contains $\mathbb{Z}_6/\{0, 2, 4\}$, which has two elements and is isomorphic to $[\mathbb{Z}_2, +_2]$, and $\{0, 2, 4\}/\{0\}$, which has three elements and is isomorphic to $[\mathbb{Z}_3, +_3]$. The factor set for the second series contains $\mathbb{Z}_6/\{0, 3\} \simeq \mathbb{Z}_3$ and $\{0, 3\}/\{0\} \simeq \mathbb{Z}_2$. □

In Example 10.38, each factor set for $\mathbb{Z}_6$ contains the same number of elements. Furthermore, to within isomorphism, each factor set contains the same elements. That this is typical is given by the Jordan–Hölder Theorem.

10.39
Theorem

Jordan–Hölder Theorem
Any two factor sets of a group G have the same number of elements, and the members of any factor set are isomorphic to the members of any other factor set. □

Every finite group G must have a composition series. The Jordan–Hölder Theorem allows us to identify a set of essentially unique groups associated with G; these groups are called **composition factors** of G.

10.40
Practice

(a) A composition series for $[\mathbb{Z}_{12}, +_{12}]$ is

$$\{0\} \subset \{0, 6\} \subset \{0, 3, 6, 9\} \subset \mathbb{Z}_{12}$$

Find the members of the factor set.

(b) Two other composition series for $\mathbb{Z}_{12}$ are given below. Show that the members of the factor sets are isomorphic to the members of the factor set of part (a).

$$\{0\} \subset \{0, 4, 8\} \subset \{0, 2, 4, 6, 8, 10\} \subset \mathbb{Z}_{12}$$
$$\{0\} \subset \{0, 6\} \subset \{0, 2, 4, 6, 8, 10\} \subset \mathbb{Z}_{12} \qquad \square$$

Semigroup of a Machine

Given a finite-state machine M, each input symbol, through applying the next-state function, has the effect of a transformation on the set S of states of M. The composition of these transformations represents the effects of strings of input symbols. (Here we will compose from left to right since input strings are read from left to right.) Thus, the set of all transformations on S accomplished by input strings is closed under composition (and associative) and is therefore a semigroup, called the **semigroup of** M, denoted by S_M. If we consider the identity transformation to be the result of applying the empty string of input symbols, then every such semigroup is actually a monoid, but for some reason it is still called the semigroup of the machine. The mapping that takes a string of input symbols to its corresponding transformation of states is a homomorphism from I^*, the free monoid generated by I (Example 7.21), onto this monoid.

10.41
Example

M	Input 0	1
0	2	1
1	0	1
2	2	0

Figure 10.57

The next-state function for a machine M is given in Figure 10.57. The transformations on $S = \{0, 1, 2\}$ resulting from all input strings of length 0 to 3 are given in Figure 10.58. Input strings of 4 or more characters do not produce any new transformations. The set S_M has eight elements; if we label the eight columns of Figure 10.58 by i, 0, 1, 2, and so on, left to right, then we have the semigroup table for S_M as given in Figure 10.59. Recall that the binary operation is composition. For example, $3 \circ 4$ in S_M is the transformation produced by an input string 0110, but since 011 has the same effect as 11, we can write 0110 as 110, which accomplishes transformation 6. Thus, $3 \circ 4 = 6$. $\qquad \square$

	i	0	1	2	3	4	5	6
				100			111	
		010	101	000			011	110
S	λ	0	1	00	01	10	11	001
0	0	2	1	2	0	0	1	0
1	1	0	1	2	1	0	1	0
2	2	2	0	2	0	2	1	0

Figure 10.58

○	i	0	1	2	3	4	5	6
i	i	0	1	2	3	4	5	6
0	0	2	3	2	6	0	5	6
1	1	4	5	2	1	6	5	6
2	2	2	6	2	6	2	5	6
3	3	0	5	2	3	6	5	6
4	4	2	1	2	6	4	5	6
5	5	6	5	2	5	6	5	6
6	6	2	5	2	6	6	5	6

Figure 10.59

10.42
Practice

(a) Find the state transformations in the semigroup of M where the next-state function for M is given by Figure 10.60.

	Input	
M	0	1
0	0	1
1	0	0

Figure 10.60

(b) Show the semigroup table for S_M.   □

Machine of a Semigroup

As a converse to the semigroup of a machine, we can view any finite semigroup S as a finite-state machine $M(S)$ whose inputs and states are the elements of S and where the semigroup table for S defines the next-state function for $M(S)$. The machine $M(S)$ is the **machine of the semigroup** S.

We can now start with a machine M, form its semigroup S_M, and then consider that semigroup as a machine $M(S_M)$. There is a relationship between the two machines; we can find a homomorphism from $M(S_M)$ onto M. Every state q in $M(S_M)$ stands for a class $[q]$ of input strings in M. Each member of $[q]$ takes the starting state s_0 of M to the same state of M, denoted by $s_0[q]$. Now we define a mapping g from the states of $M(S_M)$ to the states of M by $g(q) = s_0[q]$. To show that g preserves the next-state function, let $q \in M(S_M)$ and let $x \in I$, the input alphabet for M (here we consider only a subset of the input alphabet of $M(S_M)$). If we operate and then map, we compute the next-state in $M(S_M)$, $q \circ x$, and then $g(q \circ x) = s_0[q \circ x]$. If we map and then operate, we take $g(q) = s_0[q]$ and then apply the next-state function in M. But the next state of $s_0[q]$ under x is $s_0[q \circ x]$. The mapping g is onto because every state in M is reachable from s_0 (s_0 is reachable from s_0 by λ). We have never bothered to define the output for $M(S_M)$, but once g is determined, output for $M(S_M)$ can be defined so that corresponding states have corresponding output. Thus, we have a homomorphism from $M(S_M)$ onto M, and $M(S_M)$ simulates M.

10.43
Example

For the machine of Example 10.41, the homomorphism g from $M(S_M)$ onto M is given by

$$g: \quad i \rightarrow 0$$
$$0 \rightarrow 2$$
$$1 \rightarrow 1$$
$$2 \rightarrow 2$$
$$3 \rightarrow 0$$
$$4 \rightarrow 0$$
$$5 \rightarrow 1$$
$$6 \rightarrow 0$$

In $M(S_M)$, the next state of 3 under 1 is 5, and $g(5) = 1$; $g(3) = 0$ and in M the next state of 0 under 1 is 1. ☐

10.44
Practice

Find a homomorphism from $M(S_M)$ onto M for the machine of Practice 10.42. ☐

Before we get to some results on cascade decompositions, we need to define a simple finite-state machine that will be a stock item in such networks.

10.45
Definition

A **flip-flop** is a two-state, three-input machine defined by the state table of Figure 10.61. ☐

F	i	Input 0	1
0	0	0	1
1	1	0	1

Figure 10.61

Theorems on Cascade Decompositions

The most general result on cascade decompositions was given in 1965 by K. B. Krohn and J. L. Rhodes; we will state it but omit the proof.

10.46 Theorem

Let M be a finite-state machine with semigroup S_M. Then M can be simulated by a cascade connection of flip-flops and machines of simple groups dividing S_M. □

Note that for any machine M, S_M will be a finite semigroup. Groups dividing S_M are, by Theorem 10.33, homomorphic images of subgroups of S_M. And by the Fundamental Homomorphism Theorem for Groups, such groups are isomorphic to quotient groups of subgroups of S_M. Thus, as component machines, we are essentially looking for (are you ready?) simple quotient groups of subgroups of the semigroup S_M! If S_M is itself a group (that is, if inputs to M produce permutations of the states), we may already know a lot about its structure and be able to find the needed substructures without much difficulty. But, in fact, we have a simpler cascade decomposition when the semigroup of M is a group.

10.47 Theorem

Let M be a finite-state machine whose semigroup is a group G. Then M can be simulated by a cascade connection of the machines of the composition factors of G. □

Notice that the Jordan–Hölder Theorem guarantees that the composition factors of G are unique to within isomorphism.

Now for the proof of Theorem 10.47. Let M be a machine with semigroup G, a group. Since we already know that $M(G)$ simulates M, we can concentrate on a simulation of $M(G)$. Suppose that a composition series for G has the form

$$\{i\} = G_k \subset G_{k-1} \subset \cdots \subset G_2 \subset G_1 \subset G_0 = G$$

What we will actually prove is that, given any group G^* and a normal subgroup S of G^*, we can simulate $M(G^*)$ by a cascade of $M(S)$ and $M(G^*/S)$. This result will allow us to simulate $M(G_0)$ by a cascade of $M(G_1)$ and $M(G_0/G_1)$, and then to simulate $M(G_1)$ by a cascade of $M(G_2)$ and $M(G_1/G_2)$, and so on. Putting these decompositions together, we will simulate $M(G_0)$ by a cascade of $M(G_0/G_1)$, $M(G_1/G_2)$, . . . , $M(G_k/\{i\}) = M(G_k)$, and $M(\{i\})$. However, $M(\{i\})$ is a trivial machine and is not needed in the network.

Thus we let G (instead of G^*) be any group with S a normal subgroup of G. Our claim is that the cascade decomposition shown in Figure 10.62 can simulate $M(G)$ for the right choices of α, β, and γ. The set I will be the input alphabet for $M(G)$, which consists of the elements of G. We will choose γ so that O consists of the states of $M(G)$, that is, the elements

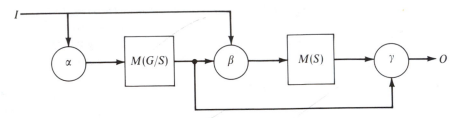

Figure 10.62

of G; we can easily assign outputs to these states by using the output function for $M(G)$. Then we must associate with each state of $M(G)$ (each $g \in G$) a state of the cascade machine so that corresponding states will produce the same output strings when given the same input strings. We assume that the output for $M(G/S)$ equals the present state, and similarly for $M(S)$.

For each coset $[g]$ in G/S, we pick a fixed representative member g'. Then for any $g \in G$, $g \in [g] = [g'] = Sg'$, and $g = sg'$ for some $s \in S$. We associate with g the state $([g], s)$ of the cascade machine. Our choice for α and γ is relatively clear; α will map $g \in G$ to $[g]$, and γ will map $([g], s)$ to sg' where g' is the fixed representative of $[g]$. Now let's figure out a definition for β that will make things work.

Suppose that $M(G)$ is presently in state g^* and receives an input of g. The next state for $M(G)$ is $g^* \cdot g$. The present state of the cascade machine must be the corresponding state for g^*, namely $([g^*], s^*)$ where $g^* = s^*g^{*'}$. The input to the cascade machine is g. The action of $M(G/S)$ is clear; its present state is $[g^*]$, and its input is $[g]$, so its next state is $[g^*][g] = [g^*g]$. The present state of $M(S)$ is s^*, and its input is $\beta(g, [g^*])$, a yet-to-be-determined member of S. The next state of $M(S)$ is $s^* \cdot \beta(g, [g^*])$. Thus, the next state of the cascade is $([g^*g], s^* \cdot \beta(g, [g^*]))$. γ will map this to

$$s^* \cdot \beta(g, [g^*])(g^*g)'$$

But if we want this to equal the next state of $M(G)$, g^*g, then

$$g^*g = s^* \cdot \beta(g, [g^*])(g^*g)'$$

or (since $g^* = s^*g^{*'}$)

$$s^*g^{*'}g = s^* \cdot \beta(g, [g^*])(g^*g)'$$

or

$$\beta(g, [g^*]) = (g^{*'}g)((g^*g)')^{-1}$$

This argument gives us our definition of β. All we must do now is show that $\beta(g, [g^*])$ is indeed a member of S, and the proof of Theorem 10.47 will be complete.

10.48
Practice

Show that $\beta(g, [g^*]) = (g^{*\prime}g)(g^*g)^{\prime})^{-1}$ is a member of S. (Hint: first show that $g^{*\prime}g$ and $(g^*g)^\prime$ are in the same coset of S in G.)   □

10.49
Example

We know that

$$\{0\} \subset \{0, 2, 4\} \subset \mathbb{Z}_6$$

is a composition series for $[\mathbb{Z}_6, +_6]$. We can use $M(\mathbb{Z}_6/\{0, 2, 4\})$ and $M(\{0, 2, 4\})$ to simulate $M(\mathbb{Z}_6)$. The elements of $Z_6/\{0, 2, 4\}$ are $[0] = \{0, 2, 4\}$ and $[1] = \{1, 3, 5\}$, and we will take 0 and 1 as the fixed coset representatives. If, for example, $M(\mathbb{Z}_6)$ is in state 5 receiving an input of 3, its next state is $5 +_6 3 = 2$. The present state of the cascade machine is $([5], 4)$ because $5 = 4 +_6 1$. The next state of the cascade machine is

$$([5 +_6 3], 4 +_6 \beta(3, [5]))$$

or

$$([2], 4 +_6 (1 +_6 3) +_6 (-(5 +_6 3)^\prime)) = ([2], 4 +_6 (4) +_6 (-0))$$
$$= ([2], 2)$$

and

$$\gamma([2], 2) = 2$$

because

$$2 = 2 +_6 0$$ □

✔ **Checklist**

Definitions

cascade connection (*p. 455*)
division of a semigroup (*p. 456*)
simple group (*p. 456*)
maximal subgroup (*p. 456*)
composition series of a group (*p. 457*)
factor set (*p. 457*)
composition factors of a group (*p. 457*)
semigroup of a machine (*p. 458*)
machine of a semigroup (*p. 459*)
flip-flop (*p. 460*)

Techniques

Given a finite-state machine, construct the table for its semigroup (monoid).

Find a homomorphism from $M(S_M)$ onto M for a given machine M.

For a group G with normal subgroup S, simulate a computation in $M(G)$ by using a cascade decomposition of $M(G)$.

Main Ideas

For any finite group G, there is a set of essentially unique groups obtained by forming the quotient groups of successive subgroups in a composition series of G.

For a given machine M, there is a homomorphism from $M(S_M)$ onto M; thus $M(S_M)$ simulates M.

If M is a finite-state machine whose semigroup is a group G, then M can be simulated by a cascade connection of machines of composition factors of G.

Exercises Section 10.4

1. If
$$\{i\} = G_k \subset G_{k-1} \subset \cdots G_3 \subset G_2 \subset G_1 \subset G_0 = G$$
    is a composition series of a finite group G and if $|G_{i-1}/G_i| = m_i$, $1 \le i \le k$, show that $|G| = m_1 \cdot m_2 \cdots \cdots m_k$.

2. Find two composition series for $[\mathbb{Z}_{15}, +_{15}]$, and show that the members of the respective factor sets are isomorphic.

★ 3. Find three composition series for $[\mathbb{Z}_{60}, +_{60}]$, and show that the members of the respective factor sets are isomorphic.

4. Show that the group $[\mathbb{Z}, +]$ does not have a composition series. (Hint: assume that $\{0\} = G_k \subset G_{k-1} \subset \cdots \subset G_1 \subset \mathbb{Z}$ is a composition series for $\mathbb{Z}$. Use Theorem 7.58 to show that $\{0\}$ is not a maximal subgroup of G_{k-1}.)

5. In each case below, the next-state function for a machine M is given. Find the state transformations in the semigroup S_M, and show the semigroup table for S_M. Is S_M a group?

★ (a)

M	Input 0	1
0	1	2
1	2	1
2	2	2

(b)

M	Input 0	1
0	2	1
1	1	0
2	0	1

(c)

M	Input 0	1
0	1	2
1	2	1
2	0	0

6. Find a homomorphism for $M(S_M)$ onto M for each machine of Exercise 5.

7. Let M be any finite-state machine with k states, and suppose that each input symbol performs a permutation on the states of M. Show that S_M is a group.

	Input	
M	0	1
0	0	1
1	1	0

Figure 10.63

8. (a) The next-state function for a machine M is given in Figure 10.63. Show that the semigroup S_M is isomorphic to $[\mathbb{Z}_2,\ +_2]$.

   (b) For the machine M of part (a), show that the semigroup of the parallel connection of M and M is isomorphic to $[\mathbb{Z}_2 \times \mathbb{Z}_2,\ +]$.

   (c) Let S_{M_1} and S_{M_2} be the semigroups of two machines M_1 and M_2. Show that the semigroup of the parallel connection of M_1 and M_2 is isomorphic to the semigroup $S_{M_1} \times S_{M_2}$.

★ 9. In $M(\mathbb{Z}_6)$, if the present state is 3 and the input is 4, the next state is $3 +_6 4 = 1$. Simulate this computation in the cascade connection of $M(\mathbb{Z}_6/\{0, 2, 4\})$ and $M(\{0, 2, 4\})$ (see Example 10.49).

10. Use the composition series

$$\{0\} \subset \{0, 3\} \subset \mathbb{Z}_6$$

to simulate the following computations in $M(\mathbb{Z}_6)$ by means of the cascade connection of $M(\mathbb{Z}_6/\{0, 3\})$ and $M(\{0, 3\})$ (see Example 10.49).

   ★ (a) $f_S(2, 5) = 2 +_6 5 = 1$

   (b) $f_S(4, 1) = 4 +_6 1 = 5$

11. Use the composition series

$$\{0\} \subset \{0, 5, 10\} \subset \mathbb{Z}_{15}$$

to simulate the following computations in $M(\mathbb{Z}_{15})$ by means of the cascade connection of $M(\mathbb{Z}_{15}/\{0, 5, 10\})$ and $M(\{0, 5, 10\})$ (see Example 10.49).

   (a) $f_S(6, 7) = 13$

   (b) $f_S(8, 11) = 4$

Chapter 11

Computability

In the first section of this chapter we consider an adequate model, or simulator, of an algorithm or effective procedure, namely the Turing machine. In Section 11.2 we discuss a universal simulator for all such machines (the theoretical inspiration for the stored-program computer). We also find that there are some problems for which no algorithmic solution can ever be found. In Section 11.3, for those problems for which solution algorithms do exist, we discuss how to measure the relative efficiency of these algorithms.

SECTION 11.1 TURING MACHINES—SIMULATION I

Algorithms

A Case Study

Kleene's Theorem (Theorem 9.44) tells us that finite-state machines can recognize only regular sets. Because $S = \{0^n 1^n \mid n \geq 0\}$ is not regular, no finite-state machine can recognize it. We probably consider ourselves to be finite-state machines and imagine that our brains, being composed of a large number of cells, can take on only a finite, although immensely large, number of configurations, or states. We feel sure, however, that if someone presented us with an arbitrarily long string of 0s followed by an arbitrarily long string of 1s, we could detect whether the number of 0s and 1s was the same. Let's think of some techniques we might use.

For small strings of 0s and 1s, we could just look at the strings and decide. Thus, we can tell without great effort that $000111 \in S$ and that $00011 \notin S$. However, for the string 00000000000000011111111111111111, we must devise another procedure, and we would probably resort to counting. We would count the number of 0s received, and when we got to the first 1, we would write the number of 0s down (or remember it) for future reference; then we would begin counting 1s. (This process is what we did mentally for smaller strings.) However, we have now made use of some extra memory because when we finished counting 1s, we would have to retrieve the number representing the total number of 0s to make a comparison. But such information retrieval is what the finite-state machine cannot do; its only capacity for remembering input is to have a given input symbol send it to a particular state. Suppose we attempt to build a finite-state machine to recognize S. We could count the number of 0s seen by having each new 0 move us to a new state of the machine. However, since the number of states of any given machine is a finite number, this plan fails if the number of 0s read in is larger than this finite number. Thus our machine clearly could not process $0^n 1^n$ for all n. In fact, if we try to solve this problem on a real computer, we encounter the same difficulty. If we set a counter as we read in 0s, we might get an overflow because our counter can go only so high. To process $0^n 1^n$ for arbitrarily large n requires an unlimited auxiliary memory for storing the value of our counter, which in practice cannot exist.

Another way we humans might attack the problem of recognizing S is to wait until the entire string was presented. Then we would go to one end of the string and cross out a 0, go to the other end and cross out a 1, go back and forth to cross out another 0–1 pair, and continue this operation until we run out of 0s or 1s. The string belongs to S if and only if we run out of both at the same time. Although this approach sounds rather different from the first one, it still requires remembering each of the inputs, since we must go back and read them once the string is complete. The finite-state machine, of course, cannot reread input.

We have come up with two computational procedures to decide, given a string of 0s and 1s, whether that string belongs to S. Both procedures require some form of additional memory unavailable in a finite-state machine. Evidently, the finite-state machine is not a model of the most general form of computational procedure. Before we consider a better model, we'll try to elaborate on what we mean by a computational procedure.

Intuitive Description

We will use the terms *algorithm*, *effective procedure*, and *computational procedure* interchangeably. In fact, we have already mentioned algorithms a number of times, presenting algorithms that are methods for solving various problems (finding the shortest path between two points in a graph, for instance). Yet we have not formally defined the term *algorithm*—and we won't do so now, either! Instead, we will appeal to a commonly held, intuitive understanding of an algorithm as being a "recipe" for carrying out a task. We will further assume that any input to which an algorithm is to be applied has been encoded in numeric form, usually nonnegative integers, just as input for an actual digital computer program is encoded and then stored in binary form.

An algorithm is characterized by certain properties. It consists of a list of precise instructions in some language, say English; each instruction must be finite, and the list itself must be finite. Each instruction must be one that can be mechanically carried out. If the algorithm is a method for solving a particular problem, then when applied to appropriate input, the algorithm must stop (halt) and produce the correct answer if an answer exists. If an answer does not exist, the algorithm may halt and declare that an answer does not exist, or it may go on indefinitely while searching for an answer. (This latter possibility, which might be caused by an infinite loop, is one that we want to avoid when implementing actual computer programs, but it is convenient when discussing algorithms theoretically to allow for the possibility of this type of behavior for problems that have no solution.)

The Turing Machine

To simulate more general computational procedures than the finite-state machine can handle, we use a Turing machine, proposed by the British mathematician Alan M. Turing in 1936. A Turing machine is essentially a finite-state machine with the added ability to reread its input and also to erase and write over its input; it also has unlimited auxiliary memory. Thus, the Turing machine overcomes the deficiencies we noted in finite-state machines.

A Turing machine consists of a finite-state machine and a tape divided into cells, each cell containing at most one symbol from an allowable finite alphabet. At any one instant, only a finite number of cells on the tape are nonblank. We use the special symbol b to denote a blank cell. The finite-state unit, through its read–write head, reads one cell of the tape at any given moment (see Figure 11.1). By the next clock pulse, depending upon the present state of the unit and the symbol read, the unit either does nothing (halts) or completes three actions. These actions are: (1) printing a symbol from the alphabet on the cell read (it might be the same symbol that's already there), (2) going to the next state (it might

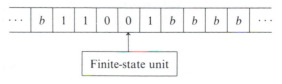

Figure 11.1

be the same state as before), and (3) moving the read–write head one cell left or right.

We can describe the actions of any particular Turing machine by a set of quintuples of the form (s, i, i', s', d) where s and i indicate the present state and the tape symbol being read, i' denotes the symbol printed, s' denotes the new state, and d denotes the direction in which the read–write head moves (R for right, L for left). Thus, a machine in the configuration illustrated by Figure 11.2a, if acting according to the instructions contained in the quintuple $(2, 1, 0, 1, R)$, would move to the configuration illustrated in Figure 11.2b.

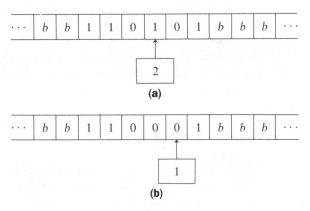

(a)

(b)

Figure 11.2

11.1
Definition

Let S be a finite set of states and I a finite set of tape symbols (the **tape alphabet**) including a special symbol b. A **Turing machine** is a set of quintuples of the form (s, i, i', s', d) where $s, s' \in S$; $i, i' \in I$; and $d \in \{R, L\}$ and where no two quintuples begin with the same s and i symbols. □

The restriction that no two quintuples begin with the same s and i symbols ensures that the action of the Turing machine is deterministic and completely specified by its present state and symbol read. If a Turing machine gets into a configuration for which its present state and symbol read are not the first two symbols of any quintuple, the machine halts.

Just as in the case of ordinary finite-state machines, we specify a fixed starting state, denoted by 0, in which the machine begins any computation. We also assume an initial configuration for the read–write head, namely, a position over the farthest left nonblank symbol on the tape. (If the tape is initially all blank, the read–write head can be positioned anywhere to start.)

11.2 Example

A Turing machine is defined by the set of quintuples

$$(0, 0, 1, 0, R)$$
$$(0, 1, 0, 0, R)$$

(0, *b*, 1, 1, *L*)

(1, 0, 0, 1, *R*)

(1, 1, 0, 1, *R*)

The action of this Turing machine when processing a particular initial tape is shown by the sequence of configurations in Figure 11.3. Since there are no quintuples defining the action to be taken when in state 1 reading *b*, the machine halts with final tape

$\cdots$	*b*	1	0	0	0	0	*b*	$\cdots$	

□

The tape serves as a memory medium for a Turing machine, and, in general, the machine can reread cells of the tape. Since it can also write on the tape, the nonblank portion of the tape can be as long as desired, although there are still only a finite number of nonblank cells at any time. Hence the machine has available an unbounded, though finite, amount of storage. Because Turing machines overcome the limitations of finite-state machines, Turing machines should have considerably higher capabilities. In fact, a finite-state machine is a very special case of a Turing machine, one that always prints the old symbol on the cell read, always moves to the right, and always halts on the symbol *b*.

11.3 Practice   Given the Turing machine

(0, 0, 0, 1, *R*)

(0, 1, 0, 0, *R*)

(0, *b*, *b*, 0, *R*)

(1, 0, 1, 0, *R*)

(1, 1, 1, 0, *L*)

(a) What is the final tape, given the initial tape

$\cdots$	*b*	1	0	*b*	$\cdots$

(Since it is tedious to draw all the little squares, you don't need to do so; just write down the contents of the final tape.)

(b) Describe the behavior of the machine when started on the tape

$\cdots$	*b*	0	1	*b*	$\cdots$

(c) Describe the behavior of the machine when started on the tape

$\cdots$	*b*	0	0	*b*	$\cdots$

□

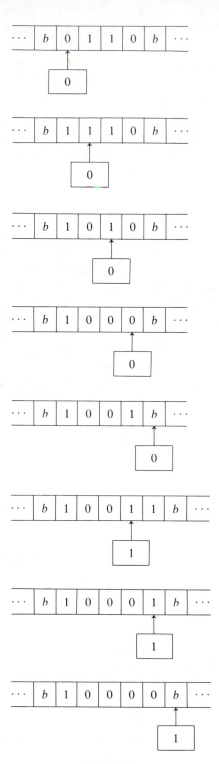

Figure 11.3

Practice 11.3(b) and 11.3(c) illustrate two ways in which a Turing machine can fail to halt: by endlessly cycling or by moving forever along the tape.

Turing Machines as Set Recognizers

Although the Turing machine computations we have seen so far are not particularly meaningful, we will use Turing machines to do two kinds of jobs. First, we'll use the Turing machine as a recognizer, much as we considered finite-state machines as recognizers in Chapter 9. We can even give a very similar definition, provided we first define a final state for a Turing machine. A **final state** in a Turing machine is one that is not the first symbol in any quintuple. Thus, upon entering a final state, whatever the symbol read, the Turing machine halts.

11.4
Definition

A Turing machine T with tape alphabet I **recognizes** (**accepts**) a subset S of I^* if T, beginning in standard initial configuration on a tape containing a string α of tape symbols, halts in a final state if and only $\alpha \in S$.   □

Note that Definition 11.4 leaves open two possible behaviors for T when applied to a string α of tape symbols not in S. T may halt in a nonfinal state, or T may fail to halt at all.

We can now build a Turing machine to recognize our old friend $S = \{0^n 1^n \mid n \geq 0\}$. The machine is based on our second approach to this recognition problem, sweeping back and forth across the input and crossing out 0–1 pairs.

11.5 Example

We want to build a Turing machine that will recognize $S = \{0^n 1^n \mid n \geq 0\}$. We will use one additional special symbol, call it X. Thus the tape alphabet $I = \{0, 1, b, X\}$. State 6 is the only final state. The quintuples making up T are given below, together with a description of their function.

$(0, b, b, 6, R)$	Recognizes the empty tape, which is in S.
$(0, 0, X, 1, R)$	Erases the leftmost 0 and begins to move right.
$(1, 0, 0, 1, R)$	
$(1, 1, 1, 1, R)$	Moves right in state 1 until it reaches the end
$(1, X, X, 2, L)$	of the string; then moves left in state 2.
$(1, b, b, 2, L)$	
$(2, 1, X, 3, L)$	Erases the rightmost 1 and begins to move left.
$(3, 1, 1, 3, L)$	Moves left over 1s.
$(3, 0, 0, 4, L)$	Goes to state 4 if more 0s are left.
$(3, X, X, 5, R)$	Goes to state 5 if no more 0s in string.
$(4, 0, 0, 4, L)$	Moves left over 0s.

$(4, X, X, 0, R)$ Finds left end of string and begins sweep again.

$(5, X, X, 6, R)$ No more 1s in string; machine accepts.

The columns in Figure 11.4 show key configurations in the machine's behavior on the tape

$\cdots$	b	0	0	0	1	1	1	b	$\cdots$

which, of course, it should accept.   □

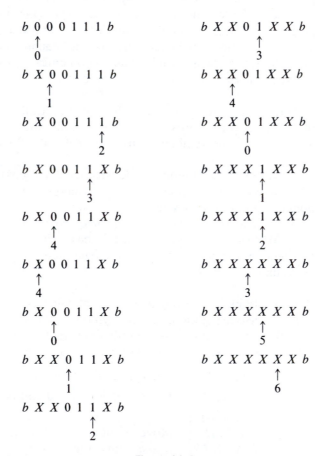

Figure 11.4

11.6 Practice   For the Turing machine of Example 11.5, describe the final configuration upon processing input tapes

(a) $\cdots$ | b | 0 | 0 | 1 | 1 | 1 | b | $\cdots$

(b) $\cdots$ | b | 0 | 0 | 0 | 1 | 1 | b | $\cdots$

(c) $\cdots$ | b | 0 | 0 | 0 | 0 | 1 | 1 | b | $\cdots$ $\square$

11.7 Practice   Design a Turing machine to recognize the set of all strings of 0s and 1s ending in 00. (This set can be described by the regular expression $(0 \vee 1)*00$, so you should be able to use a Turing machine that changes no tape symbols and always moves to the right.) $\square$

11.8 Practice   Modify the Turing machine of Example 11.5 to recognize $\{0^n 1^{2n} \mid n \geq 0\}$. $\square$

The set of quintuples defining a given Turing machine can always be altered so that whenever the machine is about to halt in a nonfinal state, it goes instead to some new state that causes it to move forever right. Therefore, we could have defined recognition of a set S by a Turing machine T by saying that T halts when it is begun on a tape containing a string α if $\alpha \in S$, and T fails to halt when begun on a tape containing a string α if $\alpha \notin S$. We can, in other words, force "halt in a nonfinal state" to become "do not halt." We cannot, however, do the opposite, because we may not be able to recognize every "do not halt" situation; we will say more about our ability to recognize halting situations in Section 11.2.

If a set S is recognized by a Turing machine T, there is an effective procedure for generating a list of members of S, as follows. Remember that S is a subset of I^*. The set I^* is denumerable (see Example 7.21), and its members can be enumerated by a list $\alpha_1, \alpha_2, \alpha_3, \ldots$. Each α_i is a candidate for membership in S. We can make lots of copies of T, say T_1, T_2, T_3, and so on, and start them up on tapes containing strings $\alpha_1, \alpha_2, \alpha_3$, and so forth. Periodically, we go back and check the progress of each T_i. Whenever a T_i has halted in a final state, we know that $\alpha_i \in S$. For those T_i's that are still computing, we in general cannot conclude whether $\alpha_i \in S$ because the computation may or may not halt at some future time.

Suppose, however, that the set S is recognized by a Turing machine T that halts on all input strings (as in Example 11.5). We then have an effective procedure for deciding membership in S. We pick any string $\alpha \in I^*$ and run T on α. The computation eventually halts. If T halts in a final state, $\alpha \in S$; otherwise, $\alpha \notin S$.

This distinction between *generating* members of a set and *deciding* membership in a set is important. In Section 12.2 we will see that sets exist which can be recognized by Turing machines, so there is an effective procedure for generating the members of these sets, but for which there is no Turing machine to decide membership; that is, there is no recognition machine that halts on all input.

Turing Machines as Function Computers

The second job for which we will use Turing machines is to compute functions. Given a particular Turing machine T and a string α of tape symbols, we begin T in standard initial configuration on a tape containing α. If T eventually halts with a string β on the tape, we may consider β as the value of a function evaluated at α. Using function notation, $T(\alpha) = \beta$. The domain of the function T consists of all strings α for which T eventually halts. We can also think of T as computing **number-theoretic functions**, functions from a subset of $\mathbb{N}^k$ into $\mathbb{N}$ for any $k \geq 1$. We will think of a string of 1s of length $n + 1$ as the unary representation of the nonnegative integer n; we'll denote this encoding of n by $\bar{n}$. Then a tape containing the string $\bar{n}_1 * \bar{n}_2 * \cdots * \bar{n}_k$ can be thought of as the representation of the k-tuple $(n_1, n_2, \ldots, n_k)$ of nonnegative integers. If T begun in the standard initial configuration on such a tape eventually halts with a final tape that is the representation $\bar{m}$ of a nonnegative integer m, then T has acted as a k-variable function T^k, where $T^k(n_1, n_2, \ldots, n_k) = m$. If T begun in standard initial configuration on such a tape either fails to halt or halts with the final tape not a representation of a nonnegative integer, then the function T^k is undefined at $(n_1, n_2, \ldots, n_k)$.

There is thus an infinite sequence $T^1, T^2, \ldots, T^k, \ldots$ of number-theoretic functions computed by T associated with each Turing machine T. For each k, the function T^k is a **partial function** on $\mathbb{N}^k$, meaning that its domain may be a proper subset of $\mathbb{N}^k$. A special case of a partial function on $\mathbb{N}^k$ is a **total function** on $\mathbb{N}^k$, where the function is defined for all k-tuples of nonnegative integers.

11.9 Example   Let a Turing machine T be given by the quintuples

$$(0, 1, 1, 0, R)$$
$$(0, b, 1, 1, R)$$

If T is begun in standard initial configuration on the tape

| $\cdots$ | b | 1 | 1 | 1 | b | $\cdots$ |

then T will halt with final configuration

$$b \ 1 \ 1 \ 1 \ 1 \ b$$
$$\uparrow$$
$$1$$

Therefore, T defines a 1-variable function T^1 that maps $\bar{2}$ to $\bar{3}$. In general, T maps $\bar{n}$ to $\overline{n + 1}$, so $T^1(n) = n + 1$, a total function of one variable.

$\square$

In Example 11.9, we began with a Turing machine and observed a particular function it computed, but we can also begin with a number-theoretic function and try to find a Turing machine to compute it.

**11.10
Definition**

A **Turing-computable function** is a number-theoretic function computed by some Turing machine. □

A Turing-computable function f can in fact be computed by an infinite number of Turing machines. Once a machine T is found to compute f, we can always include extraneous quintuples in T, producing other machines that also compute f.

**11.11
Example**

We want to find a Turing machine that computes the function f defined as follows:

$$f(n_1, n_2) = \begin{cases} n_2 - 1 & \text{if } n_2 \neq 0 \\ \text{undefined} & \text{if } n_2 = 0 \end{cases}$$

Thus f is a partial function of two variables. Let's consider the Turing machine given by the following quintuples. The actions performed by various sets of quintuples are described.

$\left.\begin{array}{l}(0, 1, 1, 0, R) \\ (0, *, *, 1, R)\end{array}\right\}$ Passes right over $\bar{n}_1$ to $\bar{n}_2$.

$(1, 1, 1, 2, R)$ Counts first 1 in $\bar{n}_2$.

$(2, b, b, 3, R)$ $n_2 = 0$; halts.

$\left.\begin{array}{l}(2, 1, 1, 4, R) \\ (4, 1, 1, 4, R) \\ (4, b, b, 5, L)\end{array}\right\}$ Finds the right end of $\bar{n}_2$.

$(5, 1, b, 6, L)$ Erases last 1 in $\bar{n}_2$.

$\left.\begin{array}{l}(6, 1, 1, 6, L) \\ (6, *, b, 7, L)\end{array}\right\}$ Passes left to $\bar{n}_1$, erasing $*$.

$(7, 1, b, 7, L)$ Erases $\bar{n}_1$.

$(7, b, b, 8, L)$ $\bar{n}_1$ erased; halts with $\overline{n_2 - 1}$ on tape.

If T is begun on the tape

$\cdots$	b	1	1	$*$	1	1	1	1	b	$\cdots$

then T will halt with final configuration

$b\ b\ b\ b\ b\ 1\ 1\ 1\ b$
$\uparrow$
8

This configuration agrees with the requirement that $f(1, 3) = 2$. If T is begun on the tape

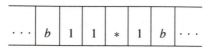

then T will halt with final configuration

$$b \; 1 \; 1 \; * \; 1 \; b \; b$$
$$\uparrow$$
$$3$$

Because the final tape is not $\overline{m}$ for any nonnegative integer m, the function computed by T is undefined at $(1, 0)$—just as we want. It is easy to see that this Turing machine computes f and that f is therefore a Turing-computable function. □

11.12
Practice

Design a Turing machine to compute the function

$$f(n) = \begin{cases} n - 2 & \text{if } n \geq 2 \\ 1 & \text{if } n < 2 \end{cases}$$ □

The domain of a Turing-computable function f is a set that can be generated by building many copies of a machine T to compute f, feeding some input into them, and periodically checking to see which have halted on a final tape representing a nonnegative integer. If T is still running on a given input, we in general cannot decide whether T will eventually halt, so we cannot say whether the input is in the domain of the function. If T halts on all input, we can run T on a given input and then simply look at the final tape, decide whether it represents a nonnegative integer, and so determine if the input belongs to the domain of f. As in the case of set recognition, there are Turing-computable functions (so we can *generate* the domain) not computable by Turing machines that halt on all input (so we cannot *decide* what belongs to the domain).

Other Formulations

There are other definitions of the Turing machine. We could, for instance, allow the machine's tapes to have several tracks, so that the machine could read more than one symbol at a time (see Figure 11.5a), or we could allow it to have multiple tapes and finite-state units that operate independently (see Figure 11.5b). We could also require that the tape be a singly infinite tape; that is, instead of extending indefinitely in both directions, the tape extends indefinitely in only one direction and the other end is fixed (see Figure 11.6).

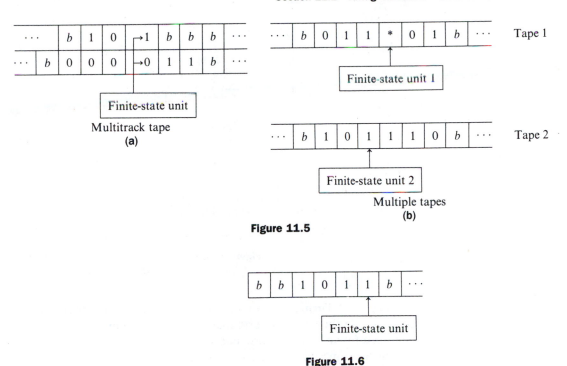

Figure 11.5

Figure 11.6

We can even require that the Turing machine be allowed only two tape symbols, the blank symbol plus one additional symbol. All of these definitions of a Turing machine are equivalent to each other and to the original definition, because any job done by a Turing machine of one kind can be done by a machine of any other kind (with perhaps some encoding and decoding of tape symbols to and from a new alphabet). A Turing machine computation is a pretty versatile idea; we'll see just how versatile shortly.

The Church–Turing Thesis

Is the Turing machine a better model of an effective procedure than the finite-state machine? Although our concept of effective procedure is an intuitive one, we are quite likely to agree that any Turing-computable function f is a function whose values can be found by an effective procedure or algorithm. In fact, if f is computed by the Turing machine T, then the set of quintuples of T is itself the algorithm; as a finite list of finite instructions that can be carried out mechanically, it satisfies the various conditions common to anyone's notion of an algorithm. Therefore, we are probably willing to accept the proposal illustrated by Figure 11.7. The figure shows "computable by effective procedure" as a "cloudy,"

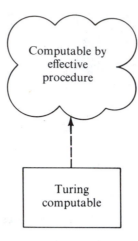

Figure 11.7

intuitive idea and "Turing computable" as a mathematically precise, well-defined idea, and the arrow asserts that any Turing-computable function is computable by an effective procedure.

Given the simplicity of the definition of a Turing machine, it is a little startling to contemplate Figure 11.8, which asserts that any function computable by any means that we might consider to be an effective procedure is also Turing computable. Combining Figures 11.7 and 11.8, we get the Church–Turing Thesis (Figure 11.9), named after Turing and another famous mathematician, Alonzo Church.

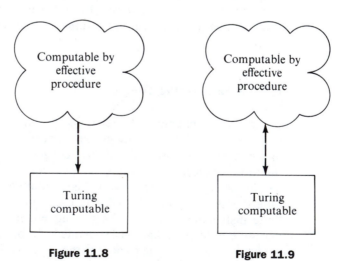

Figure 11.8   **Figure 11.9**

11.13
Church–Turing
Thesis

A number-theoretic function is computable by an effective procedure if and only if it is Turing computable.   □

Because the Church–Turing Thesis equates an intuitive idea with a mathematical idea, it can never be formally proved and must remain a thesis, not a theorem. What, then, is its justification?

One piece of evidence is that whenever a procedure generally agreed to be an effective procedure has been proposed to compute a function, someone has been able to design a Turing machine to compute that function. (Of course, there is always the nagging thought that someday this might not happen.)

Another piece of evidence is that other mathematicians, several of them at about the same time that Turing developed the Turing machine, proposed other models of an effective procedure. On the surface, each proposed model seems quite unrelated to any of the others. However, because all of these models are formally defined, just as Turing computability is, it is possible to determine on a formal, mathematical basis whether any of them are equivalent. All of these models, as well as Turing computability, have been proven equivalent; that is, they all define the same class of functions, which suggests that Turing computability embodies everyone's concept of an effective procedure. Figure 11.10 illustrates what has been done; here solid lines represent mathematical proofs, and dashed lines correspond to the Church–Turing Thesis. The dates indicate when the various models were proposed.

The Church–Turing Thesis is now widely accepted as a working tool by researchers dealing with effective computability. If, in a research paper, a procedure is set forth for computing a function and the procedure intuitively seems to be effective, then the Church–Turing Thesis is invoked, and the function is declared to be Turing computable (or one of

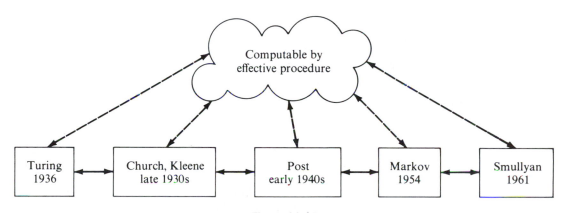

Figure 11.10

the names associated with one of the equivalent formulations of Turing computability). This invocation means that the author presumably could, if pressed, produce a Turing machine to compute the function, but, again, the Church–Turing Thesis is so universally accepted that no one bothers with these details anymore.

Although the Church–Turing Thesis is stated in terms of number-theoretic functions, it can be interpreted more broadly. Any effective procedure in which a finite set of symbols is manipulated can be translated into a number-theoretic function by a suitable encoding of the symbols as nonnegative integers. Thus, if we accept the Church–Turing Thesis, we can in general say that if there is an effective procedure to do a symbol manipulation task, there is a Turing machine to do it. We will accept and make use of the Church–Turing Thesis in Section 11.2.

By accepting the Church–Turing Thesis, we have accepted the Turing machine as the ultimate model of an effective computational device. Its capabilities exceed those of any actual computer, which, after all, does not have the unlimited tape storage of a Turing machine. It is remarkable that Turing proposed this concept in 1936, well before the advent of the modern computer.

✔ Checklist

Definitions

tape alphabet (*p. 470*)
Turing machine (*p. 470*)
final state (*p. 473*)
recognition (acceptance) of a set by a Turing machine (*p. 473*)
number-theoretic function (*p. 476*)
partial function (*p. 476*)
total function (*p. 476*)
Turing-computable function (*p. 477*)

Techniques

Describe the action of a given Turing machine on a given initial tape.

Construct a Turing machine to recognize a given set.

Construct a Turing machine to compute a given number-theoretic function.

Main Ideas

Turing machines with their deterministic mode of operation, their ability to reread and rewrite input, and their unbounded auxiliary memory.

A finite-state machine is a special case of a Turing machine.

Turing machines can be used as set recognizers and as function computers.

The existence of an effective procedure to generate the members of a set does not mean an effective procedure to decide membership in the set exists.

The Church–Turing Thesis equates a function computable by an effective procedure with a Turing-computable function. Because this thesis expresses a relationship between an intuitive idea and a formally defined one, it can never be proved but has nonetheless been widely accepted.

Exercises Section 11.1

★ 1. Given the Turing machine

$(0, 0, 0, 0, L)$

$(0, 1, 0, 1, R)$

$(0, b, b, 0, L)$

$(1, 0, 0, 1, R)$

$(1, 1, 0, 1, R)$

(a) What is its behavior when started on the tape

		b	1	0	0	1	1	b		
$\cdots$									$\cdots$	

(b) What is its behavior when started on the tape

		b	0	0	1	1	1	b		
$\cdots$									$\cdots$	

2. Given the Turing machine

$(0, 1, 1, 0, R)$

$(0, 0, 0, 1, R)$

$(1, 1, 1, 1, R)$

$(1, b, 1, 2, L)$

$(2, 1, 1, 2, L)$

$(2, 0, 0, 2, L)$

$(2, b, 1, 0, R)$

(a) What is its behavior when started on the tape

$$\cdots \;|\; b \;|\; 1 \;|\; 0 \;|\; 1 \;|\; 0 \;|\; b \;|\; \cdots$$

(b) What is its behavior when started on the tape

$$\cdots \;|\; b \;|\; 1 \;|\; 0 \;|\; 1 \;|\; b \;|\; \cdots$$

3. Given the set of quintuples describing a Turing machine T and the initial tape configuration, create a computer program that will write out the sequence of successive tape configurations. Assume that T is described by ≤ 100 quintuples, that the number of cells used on the tape is ≤ 70, and that T always halts.

4. Find a Turing machine that replaces every 0 in a string of 0s and 1s with a 1 and every 1 with a 0.

5. Find a Turing machine that recognizes the set of all strings of 0s and 1s containing at least one 1.

⋆ 6. Find a Turing machine that recognizes the set of all unary strings consisting of an even number of 1s (this includes the empty string).

7. Find a Turing machine that recognizes 0*10*1.

8. Find a Turing machine to accept the set of nonempty strings of well-balanced parentheses. (Note that (( )(( ))) is well balanced and (( )(( )) is not.)

⋆ 9. Find a Turing machine that recognizes $\{0^{2n}1^n2^{2n} \mid n \geq 0\}$.

10. Find a Turing machine that recognizes $\{w * w^R \mid w \in \{0, 1\}^*$ and w^R is the reverse of the string $w\}$.

11. Find a Turing machine that recognizes $\{w_1 * w_2 \mid w_1, w_2 \in \{0, 1\}^*$ and $w_1 \neq w_2\}$.

12. Find a Turing machine that recognizes the set of palindromes on $\{0, 1\}^*$, that is, the set of all strings in $\{0, 1\}^*$ that read the same forwards and backwards, such as 101.

13. Find a Turing machine that converts a string of 0s and 1s representing a nonzero binary number into a string of that number of 1s. As an example, the machine should, when started on a tape containing $\cdots b\,1\,0\,0\,b\cdots$, halt on a tape containing $\cdots b\,1\,1\,1\,1\,b\cdots$.

14. Find a Turing machine that, given an initial tape containing a non-empty string of 1s, marks the right end of the string with a $*$ and puts a copy of the string to the right of the $*$. As an example, the machine should, when started on a tape containing $\cdots b\,1\,1\,1\,b\cdots$, halt on a tape containing $\cdots b\,1\,1\,1 * 1\,1\,1\,b\cdots$.

⋆ 15. What number-theoretic function of three variables is computed by the Turing machine given below?

$(0, 1, b, 0, R)$

$(0, *, b, 1, R)$

$(1, 1, 1, 2, R)$

$(2, *, *, 3, R)$

$(3, 1, 1, 2, L)$

$(2, 1, 1, 4, R)$

$(4, 1, 1, 4, R)$

$(4, *, 1, 5, R)$

$(5, 1, b, 5, R)$

$(5, b, b, 6, R)$

16. Find a Turing machine to compute the function

$$f(n) = \begin{cases} n & \text{if } n \text{ is even} \\ n + 1 & \text{if } n \text{ is odd} \end{cases}$$

★ 17. Find a Turing machine to compute the function

$$f(n) = \begin{cases} 1 & \text{if } n = 0 \\ 2 & \text{if } n \neq 0 \end{cases}$$

18. Find a Turing machine to compute the function

$$f(n) = 2n$$

19. Find a Turing machine to compute the function

$$f(n) = \begin{cases} \dfrac{n}{3} & \text{if 3 divides } n \\ \text{undefined} & \text{otherwise} \end{cases}$$

★ 20. Find a Turing machine to compute the function

$$f(n_1, n_2) = n_1 + n_2$$

21. Find a Turing machine to compute the function

$$f(n_1, n_2) = \begin{cases} n_1 - n_2 & \text{if } n_1 \geq n_2 \\ 0 & \text{otherwise} \end{cases}$$

22. Find a Turing machine to compute the function

$$f(n_1, n_2) = \max(n_1, n_2)$$

23. Do Exercise 18 again, this time making use of the machines T_1 and T_2 of Exercises 14 and 20, respectively, as "procedures." (Formally, the states of these machines would have to be renumbered as the quintuples are inserted into the "main program," but you may omit this tiresome detail and merely "invoke T_1" or "invoke T_2.")

24. Describe verbally the actions of a Turing machine that computes the function $f(n_1, n_2) = n_1 \cdot n_2$, that is, design the algorithm but do not bother to create all the necessary quintuples. You may make use of Exercises 14 and 20.

★25. (a) In this section, an effective procedure was described for generating the members of a set S recognized by a Turing machine T. Why does this procedure require that we have more than one copy of T available?

   (b) Describe an effective procedure for generating the members of the range set of a Turing-computable function.

26. We can prove that there exist functions $f : \mathbb{N} \to \mathbb{N}$ that are not computable by any effective procedure. We will need three results from set theory:

   1. Any set equivalent to $\mathbb{N}$, that is, any denumerable set, is said to have **cardinality** $\aleph_0$ (cardinality is roughly a measure of the number of elements in a set).
   2. The cardinality of the set of all functions from $\mathbb{N}$ to $\mathbb{N}$ is $\aleph_0^{\aleph_0}$.
   3. $\aleph_0^{\aleph_0} > \aleph_0$.

   Show that the cardinality of the set of all algorithms is $\aleph_0$, thus proving that there are more functions than algorithms. This argument is nonconstructive; as soon as we try to describe a function that is not computable by an algorithm, say by giving an equation, we are in fact describing an algorithm for computing the function and thus making it effectively computable!

SECTION 11.2   THE UNIVERSAL TURING MACHINE—SIMULATION II—AND UNSOLVABILITY

The Universal Turing Machine

The Turing machine, according to the Church–Turing Thesis, is a mathematical structure simulating algorithms or effective procedures. Once again, we see Simulation I at work, one structure serving as a model to capture the common features of a variety of examples. Simulation II considers whether one instance of the structure can simulate another instance of that structure. Given a particular Turing machine, can another one be found to do the same job? There are several levels at which we can consider this question, but the answer is yes in any case.

Suppose we have a machine T, and we view T as a function computer for the function f (the same ideas hold if we view T as a recognizer). One trivial way to produce a new machine T' that also computes f is to add extraneous quintuples to the definition of T, quintuples that represent state–input pairs that will never be encountered while processing f. Clearly, this produces an infinite number of new machines simulating T. Another possibility is to develop some new algorithm for the computation of f. As an example, a Turing machine to compute $f(n) = n + 1$ could proceed by first copying $\bar{n}$ over and then erasing n 1s. (This procedure is not the most efficient one, but it works.) Finally, we could use a machine that computes f but operates on a different encoding of the input, say in binary rather than unary form.

As we have seen, given a Turing machine T, it is easy to find another machine T' simulating T. We will concentrate on a more sweeping idea. Can we possibly construct one machine that will simulate all others, a sort of super machine? Notice that we ourselves have been able to simulate the actions of any given machine by simply running it according to the instructions programmed into its quintuples. The process of "look at tape symbol and present state, consult the list of quintuples until you find the right one, and do what it says; if you can't find the right quintuple, halt" is an effective procedure. If we invoke the Church–Turing Thesis, then because there exists an effective procedure that runs any Turing machine on any input, there must be a single Turing machine that will simulate the actions of any Turing machine on any input. Such a machine is called a **universal Turing machine**, denoted by U. We'll give a brief discussion of how U can be designed.

For U to simulate the actions of machine T on input string α, U must know both the description of T, that is, what its quintuples are, and what the string α is. Thus, the ordered pair (T, α) serves as input to U, but we must decide how to encode (T, α) onto U's tape. One possibility is simply to write the quintuples of T followed by α on the tape. So to simulate T, given by the quintuples $(0, 0, 0, 1, R)$ and $(0, 1, 1, 1, L)$, acting on the string $\cdots b\ 0\ 1\ b \cdots$, we might write

$$\cdots b\ 0\ 0\ 0\ 1\ R * 0\ 1\ 1\ 1\ L * * 0\ 1\ b \cdots$$

on U's tape. However, U is to be a single Turing machine with a finite tape alphabet that can simulate *any* Turing machine T acting on *any* input string α. If we allow every state of every T to be its own representation on U's tape, we have no bound on the number of tape symbols that U may need. The same thing happens if we allow every symbol in every possible input string α to go directly onto U's tape.

Instead, we must encode T's quintuples and α's symbols into, say, a unary representation where n is again represented by $n + 1$ 1s. If we encode L as 1 and R as 11 and use $*$ to separate symbols, $**$ to separate quintuples, and $***$ to separate T's quintuples from α, then the string

$$\cdots b\ 1 * 1 * 1 * 11 * 11 * * 1 * 11 * 11 * 11 * 1 * * * 1 * 11\ b \cdots$$

will represent our (T, α). The machine U must also know the current position of T's read–write head. We will mark this with a P just before the tape symbol being read. Finally, U needs some working space to keep track of T's present state and symbol read. Thus, the string of Figure

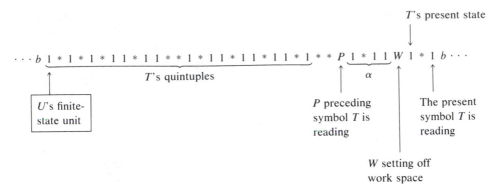

Figure 11.11

11.11 could be on U's tape at the beginning of its simulation of T acting on α.

The activities of U are carried on in cycles, each cycle representing a single step in T's computation on α. A single cycle involves the following actions. U searches through T's quintuples, comparing the first two symbols of each quintuple with the two symbols in its work space. (Here we are thinking of unencoded symbols; a single symbol in a quintuple can be more than one symbol when encoded onto U's tapes.) If it finds no match after searching all of T's quintuples, then T must halt, so U halts also. If a match is found in a given quintuple, U will change the symbol following P to the third symbol in that quintuple, this being the new symbol printed on T's tape. It will then change the symbol following W to the fourth symbol in that quintuple, this being T's new state. It will move P some number of cells left or right, depending upon the last symbol in the given quintuple. Finally, it copies the symbol that now follows P into the second block behind W to show what new tape symbol T is reading. After one cycle of our example case, U's tape would contain the string shown in Figure 11.12.

The machine U must be prepared at any time to create more room in the middle of its tape by moving some symbols farther to the right. This step will be necessary if, for example, T's new state has a longer unary representation than the old one and we need more room right after W (this happened above), or if T moves off the original nonblank portion of its tape and we need more room between ∗∗∗ and W.

We have described the general actions we want U to carry out. To write the actual quintuples for U so that it will perform these actions would be tedious but not very difficult. For example, we can see that the first phase of a cycle, searching for the correct quintuple, would involve

$\cdots b\ 1 * 1 * 1 * 1 1 * 1 1 * * 1 * 1 1 * 1 1 * 1 1 * 1 * * * 1\ P\ 1 1\ W\ 1 1 * 1 1\ b \cdots$

Figure 11.12

much sweeping back and forth across U's tape, setting markers to keep our place and then replacing the markers with the original symbols.

The universal Turing machine provides us with a single machine capable of simulating any other Turing machine when presented with that machine's program (its set of quintuples) and data. Thus, rather than building an individual Turing machine for each job we do, we can simply build the universal Turing machine and then program it for various jobs. Here is a case where a theoretical idea developed in the late 1930s by Turing served as a blueprint for reality a decade later. Early computers were individually wired to do distinct tasks. To perform a new job, a computer had to have new circuitry and essentially become a new machine. John von Neumann in 1947 turned the idea of the universal Turing machine into the real-life, stored-program computer, a single computer that could accept programs as part of its input, thereby simulating the actions of each of the individually wired machines. Today's computers are, of course, stored-program computers.

Decision Problems

We have spent quite a bit of time discussing what Turing machines can do. By the Church–Turing Thesis, they can do a great deal indeed, although not very efficiently. It is even more important, however, to consider what Turing machines *cannot* do. Because a Turing machine's abilities to perform tasks exceed those of an actual computer, if we find something no Turing machine can do, then a real computer cannot do it either. In fact, by invoking the Church–Turing Thesis, no algorithm exists to do it. The type of task we generally have in mind here is determining whether individual statements from some large class of statements are true. The question of whether an algorithm exists to perform this type of task is called a decision problem.

11.14
Definition

A **decision problem** asks if an algorithm exists to decide whether individual statements from some large class of statements are true.   □

The solution to a decision problem answers the question of whether an algorithm exists. A **positive solution** consists of proving that such an algorithm exists, and it is generally given by actually producing an algorithm that works. A **negative solution** consists of proving that no such algorithm exists. Note that this statement is much stronger than simply saying that a lot of people have tried but no one has come up with an algorithm—this might simply mean that the algorithm is hard. It must be shown that it is impossible for anyone ever to come up with an algorithm. When a negative solution to a decision problem has been found, the problem is said to be **unsolvable**, or **undecidable**. This terminology can be confusing because the decision problem itself—the question of whether an algorithm exists to do a task—has been solved; what must forever be unsolvable is the task itself.

Examples

We will look at some decision problems that have been answered.

11.15
Example

Does an algorithm exist to decide, given integers a, b, and c, whether $a^2 = b^2 + c^2$? Clearly, this question is a solvable decision problem. The algorithm consists of multiplying b by itself, multiplying c by itself, adding the two results, and comparing it with the result of multiplying a by itself.

$\square$

Obviously, Example 11.15 is a rather trivial decision problem. Historically, much of mathematics has concerned itself at least indirectly with finding positive solutions to decision problems, that is, producing algorithms. Negative solutions to decision problems arose only in the twentieth century.

11.16
Example

One of the earliest decision problems to be formulated was Hilbert's Tenth Problem, tenth in a list of problems David Hilbert posed to the International Congress of Mathematicians in 1900. The problem is: Does an algorithm exist to decide for any polynomial equation $P(x_1, x_2, \ldots, x_n) = 0$ with integral coefficients whether it has integral solutions? For polynomial equations of the form $ax + by + c = 0$, where a, b, and c are integers, it is known that integer solutions exist if and only if the greatest common divisor of a and b also divides c. Thus, for particular subclasses of polynomial equations, there might be algorithms to decide whether integer solutions exist, but the decision problem as stated applies to the whole class of polynomial equations with integer coefficients. When this problem was posed and for some time after, the general belief was that surely an algorithm existed and the fact that no one had found such an algorithm merely implied that it must be difficult. In the mid-1930s, results such as those in the next example began to cast doubt on this view. It was not until 1970, however, that the problem was finally shown to be unsolvable.

$\square$

11.17
Example

In Section 1.2, we discussed mathematical logic. As we saw there, in formal logic systems certain strings of symbols are identified as *axioms*, and *rules of inference* are given whereby a new string can be obtained from old strings. Any string that is the last one in a finite list of strings consisting of either axioms or strings obtainable by the rules of inference from earlier strings in the list is said to be a *theorem*. The decision problem for a formal theory is: Does an algorithm exist to decide whether a given string in the formal theory is a theorem of the theory?

The work of Church and Kurt Gödel showed that any formal theory that axiomatizes properties of arithmetic (making commutativity of addition an axiom, for example) and is not completely trivial (not everything is a theorem) is undecidable. Their work can be considered good news for mathematicians because it means that ingenuity in answering number theory questions will never be replaced by a mechanical procedure.   □

11.18 Example

A group can be defined by giving an alphabet of symbols (strings or words from the alphabet are the group members) together with a set of relations specifying certain words as equivalent. Thus, an abelian group could have the relation $ab \sim ba$. Using this relation, the word $\alpha = abab$ can be transformed to the word $\beta = aabb$. The **word problem** for groups asks: Does an algorithm exist to decide, given the alphabet and relations for a group and two words from the group, whether one word can be transformed to the other? The word problem was shown to be unsolvable in 1955.   □

11.19 Example

A particular Turing machine T begun on a tape containing a string α will either eventually halt or never halt. The **halting problem** for Turing machines is a decision problem: Does an algorithm exist to decide, given a Turing machine T and string α, whether T begun on a tape containing α will eventually halt? Turing proved the unsolvability of the halting problem in the late 1930s.   □

The Halting Problem

We will prove the unsolvability of the halting problem after two observations. First, it might occur to us that "run T on α," or equivalently, "run the universal machine on an encoding of (T, α)," would constitute an algorithm to see whether T halted on α. If within 25 steps of T's computation T has halted, then we know T halts on α. But if within 25,000 steps T has not halted, what can we conclude? T may still eventually halt; how long should we wait? This so-called algorithm will not always give us the answer to our question.

A second observation is that the halting problem asks for one algorithm to be applied to a large class of statements. The halting problem asks: Does an algorithm exist to decide, for any given (T, α) pair, whether T halts when begun on a tape containing α? The algorithm comes first, and that single algorithm has to give the correct answer for all (T, α) pairs. Consider the following statement, which seems very similar: Given a particular (T, α) pair, does an algorithm exist to decide whether T halts when begun on a tape containing α? Here, the (T, α) pair comes first and an algorithm is chosen based on the particular (T, α); for a different (T, α), there can be a different algorithm. This problem is solvable. Suppose someone gives us a (T, α). Two effective procedures are (1) "say yes" and (2) "say no." Since T acting on α either does or does not halt, one of these two algorithms correctly answers the question. This solution may seem trivial or even sneaky, but consider again the problem statement: Given a particular (T, α) pair, does an algorithm exist to decide, and so forth. There *does exist* an algorithm; it is either to say yes or to say no—we are not required to choose which one is correct!

This turnabout of words changes the unsolvable halting problem into a trivially solvable problem. It also points out the character of a decision problem, asking whether a single algorithm exists to solve a large class of problems. An unsolvability result has both a good side and a bad side. That no algorithm exists to solve a large class of problems guarantees jobs for creative thinkers who cannot be replaced by Turing machines. But that the class of problems considered is so large might make the result too general to be of interest; in practical terms, we don't necessarily expect or want a single algorithm to be able to solve such a wide scope of problems. (This is in spite of the fact that the universal Turing machine provides us with a single algorithm to *execute* all other problem-solving algorithms.)

We will state the halting problem again and then prove its unsolvability.

11.20
Definition

Halting Problem
Does an algorithm exist to decide, given any Turing machine T and string α, whether T begun on a tape containing α will eventually halt?   □

11.21
Theorem

The halting problem is unsolvable.

Proof: Assume that the halting problem is solvable and that a single algorithm exists that can act on any (T, α) pair as input and eventually decide whether T acting on α halts. We have already seen in building the universal Turing machine how to encode T as a string, call it s_T, of symbols. By the Church–Turing Thesis, we are assuming the existence of a single Turing machine X that acts on a tape containing (s_T, α) for any T and α and that eventually halts; at the same time the machine tells us whether T on α halts. To be definite, suppose that X begun on (s_T, α) halts with a 1 left on the tape if and only if T begun on α halts, and X begun on (s_T, α) halts with a 0 left on the tape if and only if T begun on α fails to halt; these are the only two possibilities.

Now we add to X's quintuples to create a new machine Y. Machine Y modifies X's behavior so that whenever X halts with a 1 on its tape, Y goes to a state that moves Y endlessly to the right so that it never halts. If X halts with a 0 on its tape, so does Y. Finally, we modify Y to get a new machine Z that acts on any input β by first copying β and then turning the computation over to Y so that Y acts on (β, β).

Now by the way Z is constructed, if Z acting on s_Z halts, it is because Y acting on (s_Z, s_Z) halts, and that happens because X acting on (s_Z, s_Z) halts with a 0 on the tape; but if this happens, it implies that Z begun on s_Z fails to halt! Therefore,

$$Z \text{ on } s_Z \text{ halts} \to Z \text{ on } s_Z \text{ fails to halt} \qquad (1)$$

This implication is very strange; let's see what happens if Z on s_Z does not halt. By the way Z is constructed, if Z acting on s_Z does not halt, neither does Y acting on (s_Z, s_Z), and this happens exactly when X acting on (s_Z, s_Z) halts with a 1 on the tape; but this result implies that Z begun on s_Z halts! Therefore

$$Z \text{ on } s_Z \text{ fails to halt} \to Z \text{ on } s_Z \text{ halts} \qquad (2)$$

Together, implications (1) and (2) provide an airtight contradiction, so our assumption that the halting problem is solvable is incorrect.   □

The proof of the unsolvability of the halting problem depends on two ideas. One is that of encoding a Turing machine into a string description, and the other is that of having a machine look at and act upon its own description. Notice also that neither (1) nor (2) alone in the proof is sufficient to prove the result. Both are needed to contradict the original assumption of the solvability of the halting problem.

We have previously encountered another proof of this nature, where the observation that makes the proof work is self-contradictory. You might want to review here the proof of Cantor's Theorem, Theorem 3.50.

Reducibility

As we have mentioned, proving that a decision problem is unsolvable may be somewhat difficult, since it involves showing that an algorithm to do a particular thing cannot possibly exist. Once a given problem has been found to be unsolvable (as we have now shown the halting problem to be), this result may be used to prove that related problems are also unsolvable. Suppose that A and B are two decision problems. If by assuming that B is solvable we can deduce that A is solvable also, we say that A is **reducible** to B. If we already know that A is unsolvable, then so is B.

**11.22
Example**

Consider the blank-tape halting problem: Does an algorithm exist to decide, given any Turing machine T, whether T begun on a blank tape will eventually halt?

Because the class of questions we are asking is more restricted than before, we might think that an algorithm could exist to do this job. However, we can show that the halting problem is reducible to the blank-tape halting problem. Suppose we had an algorithm P that was the solution to the blank-tape halting problem. We want to show the existence of an algorithm to solve the halting problem. For any machine T and string α, we can tack some quintuples on the front of T to create a machine T_α that first prints α on its tape (regardless of what was already there) and then turns the computation over to T acting on α. Thus T_α halts when started on a blank tape if and only if T halts when started on a tape containing α. An algorithm to solve the halting problem is: Given a (T, α) pair, create T_α (this procedure is effective) and apply algorithm P. Algorithm P will then tell us whether T acting on α halts. Therefore, the halting problem is reducible to the blank-tape halting problem, and the blank-tape halting problem is unsolvable. $\qquad\square$

Example 11.22 shows that the halting problem is reducible to the blank-tape halting problem. It is easy to see that the blank-tape halting problem is reducible to the halting problem because an algorithm to decide whether any (T, α) pair halts can clearly make this decision when α is the empty string. If two decidability problems are each reducible to the other, then they are said to be of the same **degree of unsolvability**. There are different degrees of unsolvability, a fact suggesting levels of difficulty in finding answers to the corresponding decision problems.

An unsolvability result of more practical interest concerns protection in a computer operating system. Suppose we want to prevent unauthorized access by one computer user to files created by another user. Here the decision problem asks: Does an algorithm exist to decide, given a protection system in a certain status and a particular accessing mechanism, whether the system is safe against that accessing mechanism? There is a model of a protection system for which this decision problem has been proved unsolvable by showing that the halting problem reduces to it (Michael A. Harrison, Walter L. Ruzzo, and Jeffrey D. Ullman, "Protection in Operating Systems," *Communications of the ACM* 19, no. 8 (August 1976):461–470).

**11.23
Practice**

Consider the following decision problem. Does an algorithm exist to decide, given a Turing machine T and string α, whether T begun on a tape containing α will eventually halt with the tape blank? Prove that this problem is unsolvable by showing that the halting problem is reducible to it. $\qquad\square$

✔ Checklist

Definitions

universal Turing machine (*p. 487*)
decision problem (*p. 489*)
positive solution to a decision problem (*p. 489*)
negative solution to a decision problem (*p. 489*)
unsolvable (undecidable) decision problem (*p. 489*)
word problem (*p. 491*)
halting problem (*p. 491*)

reducible decision problem (*p. 493*)
degree of unsolvability (*p. 494*)

Techniques

Prove that a decision problem is unsolvable by showing that a decision problem already known to be unsolvable is reducible to it.

Main Ideas

The universal Turing machine simulates the actions of any Turing machine on any tape.

A decision problem asks if an algorithm exists to decide whether individual statements from a large class of statements are true; if no algorithm exists, the decision problem is unsolvable.

Unsolvable decision problems have arisen in a number of contexts in this century.

The proof that the halting problem is unsolvable.

Reducibility of one decision problem to another.

Exercises Section 11.2

★ 1. The **state problem** asks: Does an algorithm exist to decide, given any Turing machine T and string α, whether T begun on a tape containing α will ever enter a particular state m? Prove that the state problem is unsolvable. (Hint: show that the halting problem is reducible to the state problem.)

2. Let s be a fixed symbol. The **printing problem** asks: Does an algorithm exist to decide, given any Turing machine T and string α, whether T begun on a tape containing α will ever print s on the tape? Prove that the printing problem is unsolvable. (Hint: show that the halting problem is reducible to the printing problem.)

3. The **uniform halting problem** asks: Does an algorithm exist to decide, given any Turing machine T, whether T halts for every input tape? Prove that the uniform halting problem is unsolvable. (Hint: show that the blank-tape halting problem is reducible to the uniform halting problem.)

4. Suppose we have a Turing machine M. The **equivalence problem** asks: Does an algorithm exist to decide, given any Turing machine T, whether T computes the same number-theoretic function of one variable as M; in other words, whether $T(\overline{n}) = M(\overline{n})$ for all $n \in \mathbb{N}$? Prove that the

equivalence problem is unsolvable. (Hint: show that the blank-tape halting problem is reducible to the equivalence problem.)

SECTION 11.3   COMPUTATIONAL COMPLEXITY

In the last section, we learned that there are unsolvable problems, problems for which a solution algorithm does not exist. In this section, we concentrate on problems that do have algorithmic solutions, but the algorithms are not very efficient.

Time Complexity

Several algorithms can exist to solve the same problem. To judge the relative merits of these algorithms, we can analyze the efficiency of each algorithm. An algorithm's efficiency is measured by getting a rough count of the number of important operations the algorithm performs. By "rough count" we mean an order-of-magnitude function (see Definition 4.35) that gives the number of operations that the algorithm might carry out as a function of the size of the input set on which the algorithm acts. By "important operations" we mean the basic steps the algorithm performs (comparisons, additions, etc.) to carry out the main task, not counting things like incrementing loop indices or initializing one or two variables. We analyzed three graph algorithms in Chapter 4—an algorithm to determine the existence of an Euler path in a graph, an algorithm to find the shortest path between two nodes in a graph, and an algorithm to find a minimal spanning tree in a graph. Under the worst of circumstances, each of these three algorithms requires $O(n^2)$ operations, where n is the number of nodes in the graph. We also analyzed the binary search algorithm in Section 1.5 and found, by solving a recurrence relation, that in the worst case $O(\log n)$ operations are required, where n is the number of elements in the list.

Suppose algorithms A and A' both solve the same problem. We will consider algorithm A' to be more efficient (faster) than algorithm A if A' requires fewer operations than A on the same input. The operations that A and A' perform must be comparable, and we must compare A and A' in the same environment. If, for example, we have expressed A as a description of a Turing machine and A' as pseudocode for instructions in a high-level programming language, then comparing the number of operations each algorithm performs is meaningless.

In this chapter, through the Church–Turing Thesis, we have equated algorithms with Turing machines. Every algorithm (effective procedure) has a representation as a set of quintuples for a Turing machine. Therefore we will use Turing machine computations as our environment, and the operations that we count will be the steps (one per clock pulse) that a Turing machine carries out.

The efficiency of an algorithm is also known as its *complexity,* although this reference is not to the amount of convolution in its logic but merely to the amount of work the algorithm must do. Thus, an algorithm whose logic is straightforward could still require a very large number of steps to carry out on a Turing machine.

Remember that we have used Turing machines for two kinds of jobs, function computation and set recognition. Suppose that the number-theoretic function f is Turing computable and is computed by a Turing machine T. We will consider only cases where T halts on all input, since we want to count the number of steps in T's computation of f. As T does a computation of f, we encode the input in some way on T's tape, start T in standard initial configuration, and count the number of steps in the computation until T halts. We expect that the number of steps required for T to process f at some value will be a function of the length of the input, that is, of the number of nonblank cells on the initial tape.

11.24
Example

Consider the Turing machine T of Example 11.9, which computes the total function $f(n) = n + 1$. For any number m, m is encoded as a string of $m + 1$ 1s. The machine T passes over each existing 1 and adds a new 1 at the end. Thus, T requires $(m + 1) + 1$ steps. For inputs other than a string of 1s, T halts immediately. Therefore the maximum number of steps in any computation by T as a function of the length n of the input to the computation is given by $t(n) = n + 1$. ☐

11.25
Definition

Let T be a Turing machine that halts on all input. If the maximum number of steps for a computation by T on any input of length n is $t(n)$, then T is of **time complexity** $t(n)$. ☐

Thus, the Turing machine of Example 11.9 is of time complexity $t(n) = n + 1$.

11.26
Practice

Find an expression for the time complexity of the Turing machine of Example 11.11. ☐

Time complexity also applies to a Turing machine used as a set recognizer; again we assume that the machine halts on all inputs.

11.27
Example

Consider the Turing machine of Example 11.5, which recognizes $S = \{0^n 1^n \mid n \geq 0\}$. The number of steps in a computation is a maximum when the input belongs to S. Suppose an input of length n is a member of S. The machine first moves the read–write head beyond the right end of the input, which requires n steps. The read–write head then sweeps back and forth across that part of the input not replaced by X's. This process requires successively $n, n - 1, n - 2, \ldots, 1$ steps. Recognition requires one final step. The total number of steps is thus

$$n + (n + (n - 1) + (n - 2) + \cdots + 1) + 1$$

which equals

$$n + \frac{n(n + 1)}{2} + 1 \qquad \text{(see Practice 1.55)}$$

$$= (n + 1)\left(1 + \frac{n}{2}\right)$$

$$= \frac{n^2 + 3n + 2}{2}$$

This Turing machine is of time complexity

$$t(n) = \frac{n^2 + 3n + 2}{2} \qquad \qquad \square$$

11.28
Practice

Verify that the Turing machine of Example 11.5 requires $t(4)$ steps to recognize the input $\cdots b\,0\,0\,1\,1\,b\cdots$, but uses less than $t(4)$ steps to reject $\cdots b\,0\,0\,0\,1\,b\cdots$ where

$$t(n) = \frac{n^2 + 3n + 2}{2} \qquad \qquad \square$$

Another measure of efficiency, which we won't consider here, is the *space complexity* of a Turing machine, a measure of the amount of tape the machine uses as a function of input length.

Intractable Problems

In the examples above, we developed precise expressions for the time complexity of Turing machines (or algorithms). We need not always be quite this accurate; in general, we will be content with an expression for the order of magnitude of the complexity of an algorithm. The algorithm given in Example 11.5 is $O(n^2)$, and that given in Example 11.9 is $O(n)$. Of course, these algorithms do two different jobs, so it is not of much interest to compare them.

However, suppose we have two algorithms to do the same job and their time complexities are of different orders of magnitude, say A is $O(n)$ and A' is $O(n^2)$ (see Exercise 1 in this section). Even if each step in a computation takes only 0.0001 second, this difference will affect total computation time as n grows larger. Figure 11.13 compares total computation time for A and A' under various values of input length. Now suppose a third algorithm A'' exists whose time complexity is not even given by a polynomial function but by an exponential function, say 2^n. Figure 11.14 adds the exponential case to Figure 11.13.

Algorithm	Order	Total computation time		
		Size of Input		
		10	50	100
A	n	0.001 s	0.005 s	0.01 s
A'	n^2	0.01 s	0.25 s	1 s

Figure 11.13

Algorithm	Order	Total computation time		
		Size of Input		
		10	50	100
A	n	0.001 s	0.005 s	0.01 s
A'	n^2	0.01 s	0.25 s	1 s
A''	2^n	0.1024 s	3570 years	4×10^{16} centuries

Figure 11.14

Note that the exponential case grows at a fantastic rate! Even if we assume that each computation step takes much less time than 0.0001 second, the relative growth rates between polynomial and exponential functions still follow this same pattern. Because of this immense growth rate, algorithms not of polynomial order are generally not useful for large values of n. In fact, problems for which no polynomial time algorithms exist are called **intractable**.

There may be extenuating circumstances for algorithms that are of exponential order. For small values of n, an exponential function can have smaller values than a polynomial function of large degree. Also, when an algorithm has time complexity $t(n) = 2^n$, for instance, at least one input of length n requires 2^n steps, but the average case may run much faster. This is what happens with the **simplex method**, a well-known method for solving linear programming problems. In linear programming problems, values are chosen for a number of variables so that they satisfy a set of linear inequalities while minimizing (or maximizing) a linear function of these variables. The applications of linear programming are varied, including routing problems in communications networks, scheduling problems in manufacturing processes, and layout problems in garment manufacturing. Many of these problems involve thousands of variables. The simplex method is of exponential order in the worst case but is of polynomial-order running time in many common applications. Now it appears that N. Karmarkar at AT&T Bell Laboratories has found an algorithm that is of polynomial order in the worst case and runs faster than the simplex method in average cases. This breakthrough will allow linear programming techniques to be applied to problems with millions of variables.

Despite extenuating circumstances, an algorithm of polynomial order is preferable to one of exponential order. In choosing between algorithms for a given problem or attempting to improve a given algorithm, we should concentrate on the order of magnitude of the time complexity functions.

The Set P

11.29
Definition

P is the collection of all sets recognizable by Turing machines of polynomial time complexity (*P* comes from "*polynomial time*").   □

Consideration of set recognition in Definition 11.29 is not as restrictive as it may seem. First, since a Turing machine for which a time complexity can be determined must halt on all input, it *decides* membership in a set. Many problems can be posed as set decision problems by suitably encoding the objects involved in the problem.

For example, we may describe a given graph by somehow encoding each of its nodes and then encoding the ordered pairs of nodes connected by arcs. This encoding would result in a string of symbols effectively describing the graph. We could then pose the Hamiltonian circuit problem (Section 4.3) as a set decision problem by considering the set of all strings of allowable symbols that are descriptions of graphs having Hamiltonian circuits. If we can build a Turing machine to decide membership in this set, we will then have an algorithm to solve the decision problem that asks whether an arbitrary graph has a Hamiltonian circuit.

The particular encoding scheme we use determines the length of the input string for a given instance of a problem and thus may affect the time complexity of an algorithm to solve the problem. However, if a given problem has two encodings such that input under each encoding can be transformed in polynomial time to corresponding input under the other encoding, then if one encoding results in a set belonging to *P*, so does the other. Thus, we can speak of a *problem* belonging to *P*, meaning that a polynomial time algorithm exists for its solution, without having to specify the details of the encoding of the problem. (Do not infer from what we've said here that the Hamiltonian circuit problem belongs to *P*; we have merely said that the problem can be formulated so that it is possible to *ask* whether it belongs to *P*; no one yet knows the answer.)

Nondeterminism

The decision problem for Hamiltonian circuits, unlike problems such as the halting problem, is not unsolvable. An algorithm exists to test whether an arbitrary graph has a Hamiltonian circuit, namely, the trial-and-error approach of testing all possible paths. As we trace out these possible paths, we may have a choice of next moves every time we come to a node. We can simulate this type of behavior by using a **nondeterministic Turing machine (NDTM)**. An NDTM is defined just like an ordinary Turing machine except that for each state–input pair there is a set of applicable quintuples and so, possibly, a choice for the Turing machine's behavior at that point. Each choice (each quintuple) specifies the symbol to be printed, the next state, and the direction of motion of the read–write head. We can think of the NDTM as pursuing all of its possible sequences of action in parallel. An NDTM T **recognizes**, or **accepts**, a string α of tape symbols if T, begun in standard configuration on α, has some sequence of moves leading to a halt in a final state. T recognizes the set of all recognized strings.

11.30
Definition

Let T be an NDTM. For every recognized input string α of length n, there is at least one sequence of moves leading to a final state; for each accepted string, consider only the shortest sequence of moves leading to acceptance. If the maximum number of steps used in any such sequence accepting a string of length n is $t(n)$, then T is of **time complexity** $t(n)$.   □

11.31
Definition

NP is the collection of all sets recognizable by NDTMs of polynomial time complexity. (*NP* comes from "*nondeterministic polynomial* time.")
   □

Any ordinary (deterministic) Turing machine is a trivial NDTM, so it is clear that $P \subseteq NP$. Whether P is a proper subset of NP is a question that occupies us for the rest of this section.

We should note that time complexity for an NDTM gives an upper bound on time required to *accept* any (acceptable) string α of length n, and time complexity for an ordinary Turing machine gives an upper bound on the time required to *process* any string α of length n. In other words, in an ordinary (deterministic) Turing machine, any input can be fed in and the answer "yes—accepted" or "no—rejected" will be obtained within $t(n)$ units of time. In an NDTM any input can be fed in, and if it is an accepted input, the answer "yes—accepted" will be obtained within $t(n)$ units of time; however, if it is a nonaccepted string, there is no bound on the amount of processing time.

Is Nondeterminism Faster?

As in the corresponding situation for finite-state machines (Theorem 9.40), any set recognizable by an NDTM T can also be recognized by a deterministic Turing machine T^*. We can think of T^* acting on a given input α as simulating one after another the possible sequences of moves T could make on α until α is accepted or all possible sequences have been tried and α is rejected. Therefore, although nondeterminism gains us no new capabilities, we would expect it to gain us some lower time complexity.

Thus, if the time complexity for T is $t(n)$, we would expect the time complexity $t^*(n)$ for T^* to be higher for two reasons. T^* cannot execute sequences of moves in parallel as T can; it must do them serially. Also, T^* gives us more information about an input α of length n; although T may give an answer within $t(n)$ units of time only if α is accepted, T^* always gives us an answer (yes or no) about any input α within $t^*(n)$ units of time. There is one detail we glossed over a moment ago in discussing T^*'s simulation of T on α. Machine T may have sequences of moves that do not halt; if T^* begins simulating one of these sequences, how does it know when to give up and try another sequence? If T has time complexity $t(n)$, then T^* need not pursue any sequence of moves for longer than $t(n)$ units of time. If α is accepted by T, there is some sequence that will do the job within this time. We can imagine T's possible actions on a given α as something like the tree shown in Figure 11.15; as T^* simulates T, it need not look below $t(n)$ levels, and it can trace out each branch of the tree that far. Because there is a bound b on the maximum number of possible moves T can make at any point, there are at most b branches of the tree at any node. Thus, the tree can have at most $b^{t(n)}$ separate paths, each of length at most $t(n)$, so we would expect some exponential expression such as $t(n)b^{t(n)}$ to be the time complexity for T^*. We should never need more time than this, but probably some input of length n for some n would require this much time.

The above argument seems to convince us that, in most cases, if a set is recognized by an NDTM of time complexity $t(n)$, then any deterministic machine for recognizing that set will probably have time complexity of the form $t(n)b^{t(n)}$, a function of a higher order of magnitude.

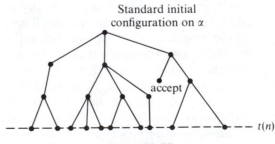

Figure 11.15

Such a result has not been proven, however. No one has found any set S recognizable by an NDTM with time complexity $t(n)$ for which no deterministic machine of complexity $t(n)$ exists to recognize S. Although there are certainly sets for which such a deterministic machine has not been found, it has not been established that one cannot exist. In particular, whether P is a proper subset of NP is an open question.

There are many quite famous problems, such as the Hamiltonian circuit problem, that have been shown to be in NP (that is, they can be represented as NP sets) but for which no polynomial-time-bounded, deterministic solution algorithm has been found. This fact lends weight to the speculation that P is indeed a proper subset of NP. This view is the prevailing one among computer scientists today, strengthened by work begun in 1971 on a class of problems known as **NP-complete problems**. Roughly, if a problem is NP-complete, it is NP and at least as hard to solve as any other NP problem in that if it could be shown to belong to P, then every NP problem would belong to P and P would equal NP.

Since 1971 many problems from many different fields (graph theory, number theory, algebra, data storage, network theory, VLSI [very large scale integration] circuit design, etc.) have been shown to be NP-complete. The Hamiltonian circuit problem is NP-complete, as is the problem of deciding whether two arbitrary regular expressions represent the same set (Section 9.3). The NP-complete problems are quite diverse, and the search for efficient (polynomial-time-bounded) solution procedures has been extensive. Remember that if an efficient procedure could be found to solve any one of them, such a procedure would exist for all the others as well. In view of the so-far unsuccessful search for an efficient solution procedure for even one such problem, it seems likely that $P \neq NP$. On the practical side, perhaps we should not look too long for a quick and easy algorithm to solve an NP problem that may be encountered.

✔ Checklist

Definitions

time complexity for a Turing machine (*p. 497*)
intractable problem (*p. 499*)
simplex method (*p. 499*)
P (*p. 500*)
nondeterministic Turing machine (NDTM) (*p. 501*)
recognition (acceptance) of a set by an NDTM (*p. 501*)
time complexity for an NDTM (*p. 501*)
NP (*p. 501*)
NP-complete problem (*p. 503*)

Main Ideas

Time complexity for a Turing machine is a measure of its efficiency in processing inputs.

Problems for which no polynomial-time solution algorithms exist are considered intractable even though special cases may exist that can be efficiently solved.

Time complexity for an NDTM is a measure of its efficiency in recognizing problem solutions.

Nondeterminism offers parallel processing, which suggests a gain in efficiency over a deterministic solution, but such a gain has not been proved.

$P \subseteq NP$, but it is unknown whether $P \subset NP$.

The *NP*-complete problems, of which there are many, are as difficult to solve efficiently as any *NP* problem.

Exercises Section 11.3

★ 1. The algorithm given in Example 11.9 that computes $f(n) = n + 1$ is of order n. Another Turing machine to compute $f(n)$ could proceed by first copying $\bar{n}$ over and then erasing n 1s. Show that the algorithm of this Turing machine is of order n^2.

In Exercises 2 through 6, explain intuitively why nondeterminism might be expected to improve the efficiency of a solution algorithm for the given problem. (Each problem is *NP*-complete.)

2. The **satisfiability problem**: Given a set of variables and a Boolean expression over these variables (see Definition 6.1), does a truth assignment to the variables exist that makes the expression true? (This problem was essentially the first ever shown to be *NP*-complete.)

3. The **clique problem**: Given a graph G and a positive integer k, does G contain a complete subgraph with k vertices?

4. The **set-packing problem**: Given a collection C of finite sets and a positive integer k, does C contain at least k mutually disjoint sets?

5. The **graph-coloring problem**: Given a graph G and a positive integer k, does there exist a coloring of G using k colors (a coloring of a graph assigns a color to each node such that adjacent nodes do not have the same color)?

6. The **bin-packing problem**: Given k bins, each of capacity 1, and n objects, each of size between 0 and 1, can the objects all fit into k bins?

Chapter 12

Formal Languages

Section 12.1 presents several classes of formal grammars, which are rules for generating "legal sentences" in a "language." Section 12.2 discusses recognizing devices for languages. If a formal grammar describes the syntax for a high-level programming language, which frequently happens, then the compiler serves as a recognizing device for that language.

SECTION 12.1 CLASSES OF LANGUAGES

Grammars and Languages

Suppose we come upon the English language sentence "The walrus talks loudly." Although we might be surprised at the meaning, or *semantics*, of the sentence, we accept its form, or *syntax*, as valid, meaning that the various parts of speech (noun, verb, etc.) are strung together in a reasonable way. In contrast, we reject "Loudly walrus the talks" as an illegal combination of parts of speech, or as syntactically incorrect. Our feeling for correct syntax in English is just that—a feeling—based on years of experience. We must also worry about correct syntax in programming languages, but in these, unlike natural languages (English, French, etc.), legal combinations of symbols are specified in detail. Let's give a formal definition of "language"; the definition will be general enough to include both natural and programming languages.

12.1 Definition

An **alphabet,** or **vocabulary,** V is a finite, nonempty set of symbols. A **word** over V is a finite-length string of symbols from V. The set V^* is the set of all words over V. A **language** over V is any subset of V^*. ☐

Because the set V^* is denumerable (see Example 7.21), a language L over V can be infinite. How can we describe a given L, that is, specify exactly those words belonging to L? If L is finite, we can just list its members, but if L is infinite, can we find a finite description of L? Not always—there are many more languages than possible finite descriptions. Although we will consider only languages that can be finitely described, we can still think of two possibilities. We may be able to describe an algorithm to *decide* membership in L; that is, given any word in V^*, we could apply our algorithm and receive a yes or no answer as to whether the word belongs to L. Or we may be able to describe a procedure allowing us only to *generate* members of L, that is, crank out one at a time a list of all the members of L.

12.2
Definition

A language L over V is **recursive** if there exists an effective procedure to decide membership in L. A language L is **recursively enumerable** (r.e.) if there exists an effective procedure to generate a list of members of L.

□

Because a language is just a subset of V^*, we will also speak of *sets* as being recursive or recursively enumerable.

Any recursive language is also recursively enumerable because a procedure for generating the language is to generate V^* (an r.e. set) and select those members of V^* belonging to L. However, an r.e. language may not be recursive, as we will see in the next section. At any rate, we will be interested in r.e. languages, and we will describe such a language by defining its generative process, or giving a *grammar* for the language.

12.3
Definition

A **phrase-structure grammar (type 0 grammar)** G is a 4-tuple, $G = (V, V_T, S, P)$, where

$V = $ a vocabulary
$V_T = $ a nonempty subset of V called the set of **terminals**
$S = $ an element of $V - V_T$ called the **start symbol**
$P = $ a finite set of **productions** of the form $\alpha \rightarrow \beta$ where α is a word over V containing at least one nonterminal symbol and β is a word over V □

12.4 Example

Here is a very simple grammar: $G = (V, V_T, S, P)$ where $V = \{0, 1, S\}$, $V_T = \{0, 1\}$, and $P = \{S \rightarrow 0S, S \rightarrow 1\}$.   □

The productions of a grammar allow us to transform some words over V into others; the productions can be called rewriting rules.

12.5
Definition

Let G be a grammar, $G = (V, V_T, S, P)$, and let w_1 and w_2 be words over V. Then w_1 **directly generates (directly derives)** w_2, written $w_1 \Rightarrow w_2$, if $\alpha \to \beta$ is a production of G, w_1 contains an instance of α, and w_2 is obtained from w_1 by replacing that instance of α with β. If $w_1, w_2, \ldots, w_n$ are words over V and $w_1 \Rightarrow w_2, w_2 \Rightarrow w_3, \ldots, w_{n-1} \Rightarrow w_n$, then w_1 **generates (derives)** w_n, written $w_1 \overset{*}{\Rightarrow} w_n$. (By convention, $w_1 \overset{*}{\Rightarrow} w_1$.)   □

12.6 Example

In the grammar of Example 12.4, $00S \Rightarrow 000S$ because the production $S \to 0S$ has been used to replace the S in $00S$ with $0S$. Also $00S \overset{*}{\Rightarrow} 00000S$.   □

12.7 Practice

Show that in the grammar of Example 12.4, $0S \overset{*}{\Rightarrow} 00001$.   □

12.8
Definition

Given a grammar G, the **language L generated by G**, sometimes denoted $L(G)$, is the set

$$L = \{w \in V_T^* \,|\, S \overset{*}{\Rightarrow} w\}$$

In other words, L is the set of all strings of terminals generated from the start symbol.   □

Notice that once a string w of terminals has been obtained, no productions can be applied to w, and w cannot generate any other words. The language L of a given grammar is an r.e. set because the following effective procedure generates a list of the members of L. Begin with the start symbol S and systematically apply some sequence of productions until a string w_1 of terminals has been obtained; $w_1 \in L$. Go back to S and repeat this procedure using a different sequence of productions to generate another word $w_2 \in L$, and so forth. Actually, this procedure doesn't quite work because we might get started on an infinite sequence of direct derivations that never leads to a string of terminals and so never contributes a word to our list. Instead, we need to run a number of derivations from S simultaneously, checking on each one after each step and, for any that terminate, adding the final word to the list of members of L. That way we cannot get stuck waiting indefinitely while unable to do anything else.

12.9 Practice

Describe the language generated by the grammar G of Example 12.4. □

Languages derived from grammars such as we have defined are called **formal languages**. If the grammar is defined first, the language will follow as an outcome of the definition. Alternatively, the language, as a well-defined set of strings, may be given first and we can then seek a grammar that generates it.

12.10
Example

Let L be the set of all nonempty strings consisting of an even number of 1s. Then L is generated by the grammar $G = (V, V_T, S, P)$ where $V = \{1, S\}$, $V_T = \{1\}$, and $P = \{S \rightarrow SS, S \rightarrow 11\}$. A language can be generated by more than one grammar. L is also generated by the grammar $G' = (V', V'_T, S', P')$ where $V' = \{1, S\}$, $V'_T = \{1\}$, and $P' = \{S \rightarrow 1S1, S \rightarrow 11\}$. □

12.11
Practice

(a) Find a grammar that generates the language $L = \{0^n 10^n \mid n \geq 0\}$.

(b) Find a grammar that generates the language $L = \{0^n 10^n \mid n \geq 1\}$. □

Trying to describe concisely the language generated by a given grammar and defining a grammar to generate a given language can both be quite difficult. We'll look at another example where the grammar is a bit more complicated than any we've seen so far.

12.12
Example

Let $L = \{a^n b^n c^n \mid n \geq 1\}$. A grammar generating L is $G = (V, V_T, S, P)$ where $V = \{a, b, c, S, B, C\}$, $V_T = \{a, b, c\}$, and P consists of the following productions:

1. $S \rightarrow aSBC$
2. $S \rightarrow aBC$
3. $CB \rightarrow BC$
4. $aB \rightarrow ab$
5. $bB \rightarrow bb$
6. $bC \rightarrow bc$
7. $cC \rightarrow cc$

It is fairly easy to see how to generate any particular member of L using these productions. Thus, a derivation of the string $a^2 b^2 c^2$ is

$$S \Rightarrow aSBC$$
$$\Rightarrow aaBCBC$$
$$\Rightarrow aaBBCC$$
$$\Rightarrow aabBCC$$
$$\Rightarrow aabbCC$$
$$\Rightarrow aabbcC$$
$$\Rightarrow aabbcc$$

In general, $L \subseteq L(G)$ where the outline of a derivation for any $a^n b^n c^n$ is given below; the numbers refer to the productions used.

$$S \overset{*}{\underset{1}{\Rightarrow}} a^{n-1} S (BC)^{n-1}$$

$$\underset{2}{\Rightarrow} a^n (BC)^n$$

$$\overset{*}{\underset{3}{\Rightarrow}} a^n B^n C^n$$

$$\underset{4}{\Rightarrow} a^n b B^{n-1} C^n$$

$$\underset{5}{\overset{*}{\Rightarrow}} a^n b^n C^n$$

$$\underset{6}{\Rightarrow} a^n b^n c C^{n-1}$$

$$\underset{7}{\overset{*}{\Rightarrow}} a^n b^n c^n$$

We must also show that $L(G) \subseteq L$, which involves arguing that some productions must be used before others and that the general derivation shown above is the only sort that will lead to a string of terminals. $\qquad\square$

In trying to invent a grammar to generate the L of Example 12.12, we first might try to use productions of the form $B \rightarrow b$ and $C \rightarrow c$ instead of productions 4 through 7. Then we would indeed have $L \subseteq L(G)$, but $L(G)$ would also include words such as $a^n(bc)^n$. In devising a grammar to generate a given language, we may have to be a bit tricky.

Formal languages are our last example of Simulation I. They were developed in the 1950s by Noam Chomsky in an attempt to model natural languages, such as English, with an eye toward automatic translation. However, since a natural language already exists and is quite complex, defining a formal grammar to generate a natural language is very difficult. Attempts to do this for English have been only partially successful.

12.13 Example

We can describe a formal grammar that will generate a very restricted class of English sentences. The terminals in the grammar are the words "the," "a," "legal," "river," "walrus," "talks," "flows," "loudly," and "swiftly," and the nonterminals are the words "sentence," "noun-phrase," "verb-phrase," "article," "noun," "verb," and "adverb." The start symbol is "sentence" and the productions are:

    sentence → noun-phrase verb-phrase

    noun-phrase → article noun

    verb-phrase → verb adverb

    article → the

    article → a

    noun → river

    noun → walrus

    verb → talks

    verb → flows

    adverb → loudly

    adverb → swiftly

Here is a derivation of "the walrus talks loudly" in this grammar:

$$
\begin{aligned}
\text{sentence} &\rightarrow \text{noun-phrase verb-phrase} \\
&\rightarrow \text{article noun verb-phrase} \\
&\rightarrow \text{the noun verb-phrase} \\
&\rightarrow \text{the walrus verb-phrase} \\
&\rightarrow \text{the walrus verb adverb} \\
&\rightarrow \text{the walrus talks adverb} \\
&\rightarrow \text{the walrus talks loudly}
\end{aligned}
$$

A few other sentences making various degrees of sense, such as "a walrus flows loudly," are also part of the language defined by this grammar. The difficulty of specifying a grammar for English as a whole is more apparent when we consider that a phrase such as "time flies" can be an instance of either a noun followed by a verb or of a verb followed by a noun! This situation is "ambiguous" (see Exercise 16 in this section). □

Programming languages are less complex than natural languages, and some programming languages, such as ALGOL, have been defined as formal languages.

12.14 Example

A section of formal grammar to generate identifiers in a programming language could be presented as follows:

$$
\begin{aligned}
\text{identifier} &\rightarrow \text{letter} \\
\text{identifier} &\rightarrow \text{identifier letter} \\
\text{identifier} &\rightarrow \text{identifier digit} \\
\text{letter} &\rightarrow a \\
\text{letter} &\rightarrow b \\
&\;\;\vdots \\
\text{letter} &\rightarrow z \\
\text{digit} &\rightarrow 0 \\
\text{digit} &\rightarrow 1 \\
&\;\;\vdots \\
\text{digit} &\rightarrow 9
\end{aligned}
$$

Here, the set of terminals is $\{a, b, \ldots , z, 0, 1, \ldots , 9\}$ and "identifier" is the start symbol. A shorthand that avoids listing all these productions is called **Backus–Naur form (BNF).** The productions listed above can be given in BNF by three lines:

$\langle\text{identifier}\rangle ::= \langle\text{letter}\rangle \,|\, \langle\text{identifier}\rangle\langle\text{letter}\rangle \,|\, \langle\text{identifier}\rangle\langle\text{digit}\rangle$

$\langle\text{letter}\rangle ::= a\,|\,b\,|\,c\,|\,\cdots\,|\,z$

$\langle\text{digit}\rangle ::= 0\,|\,1\,|\,\cdots\,|\,9$

In BNF, nonterminals are identified by ⟨ ⟩, the production arrow becomes ::=, and | stands for "or," identifying various productions having the same left-hand symbol. □

Classes of Grammars

We will identify several types of grammars. First, we'll look at one more example.

12.15 Example

Let L be the empty string λ together with the set of all strings consisting of an odd number n of 0s, $n \geq 3$. The grammar $G = (V, V_T, S, P)$ generates L where $V = \{0, A, B, E, F, W, X, Y, Z, S\}$, $V_T = \{0\}$, and the productions are

$S \rightarrow FA$	$FX \rightarrow F0W$
$S \rightarrow FBA$	$YA \rightarrow Z0A$
$FB \rightarrow F0EB0$	$W0 \rightarrow 0W$
$EB \rightarrow 0$	$0Z \rightarrow Z0$
$EB \rightarrow XBY$	$WBZ \rightarrow EB$
$0X \rightarrow X0$	$F \rightarrow \lambda$
$Y0 \rightarrow 0Y$	$A \rightarrow \lambda$

The derivation $S \Rightarrow FA \overset{*}{\Rightarrow} \lambda\lambda = \lambda$ produces λ. The derivation

$$S \Rightarrow FBA$$
$$\Rightarrow F0EB0A$$
$$\Rightarrow F0XBY0A$$
$$\overset{*}{\Rightarrow} FX0B0YA$$
$$\overset{*}{\Rightarrow} F0W0B0Z0A$$
$$\overset{*}{\Rightarrow} F00WBZ00A$$
$$\Rightarrow F00EB00A$$
$$\Rightarrow F00000A$$
$$\overset{*}{\Rightarrow} 00000$$

produces five 0s. Notice how X and Y, and also W and Z, march back and forth across the strings of 0s, adding one more 0 on each side. This activity is highly reminiscent of a Turing machine read–write head sweeping back and forth across its tape and enlarging the printed portion of the tape. □

The above grammar allows the erasing productions $F \rightarrow \lambda$ and $A \rightarrow \lambda$. To generate any language containing λ, we have to be able to erase somewhere. In the following grammar types, we will limit erasing, if it occurs at all, to a single production of the form $S \to \lambda$ where S is the start symbol, and if this production occurs we will not allow S to appear on the right-hand side of any other productions. This restriction allows us to crank out λ from S as a special case and then get on with other derivations, none of which allow any erasing. Let's call this the **erasing convention**. The following definition defines three special types of grammars by further restricting the productions allowed.

12.16 Definition

A grammar G is **context-sensitive (type 1)** if it obeys the erasing convention and for every production $\alpha \to \beta$ (except $S \to \lambda$), the word β is at least as long as the word α. A grammar G is **context-free (type 2)** if it obeys the erasing convention and for every production $\alpha \to \beta$, α is a single nonterminal. A grammar G is **regular (type 3)** if it obeys the erasing convention and for every production $\alpha \to \beta$ (except $S \to \lambda$), α is a single nonterminal and β is of the form t or tW where t is a terminal symbol and W is a nonterminal symbol. This hierarchy of grammars, from type 0 to type 3, is also called the **Chomsky hierarchy**. □

In a context-free grammar, a single, nonterminal symbol on the left of a production can be replaced wherever it appears by the right side of the production. In a context-sensitive grammar, a given nonterminal symbol might be replaceable only if it is part of a particular string, or context— hence the names context-free and context-sensitive. It is clear that any regular grammar is also context-free, and any context-free grammar is also context-sensitive. The grammar of Example 12.4 is regular, both grammars of Example 12.10 are context-free but not regular, and the grammar of Example 12.12 is context-sensitive but not context-free. The grammars of Example 12.13 and of Example 12.14 are context-free but not regular. And the grammar of Example 12.15 is a type 0 grammar, but it is not context-sensitive.

12.17 Definition

A language is **type 0 (context-sensitive, context-free,** or **regular)** if it can be generated by a type 0 (context-sensitive, context-free, or regular) grammar. □

Because of the relationships among the four grammar types, we can classify languages as shown in Figure 12.1. Thus, any regular language is also context-free because any regular grammar is also a context-free grammar, and so on. However, although it turns out to be true, we do not know from what we have done that these sets are properly contained in one another. For example, the language L described in Example 12.15 was generated there by a grammar that was type 0 but not context-sensitive, but that does not imply that L itself falls into that category. Different grammars can generate the same language.

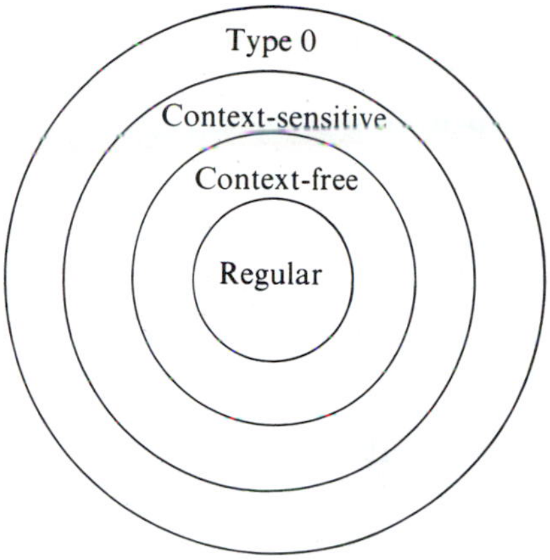

Figure 12.1

12.18
Definition

Two grammars are **equivalent** if they generate the same language.   □

12.19
Example

The language L of Example 12.15 can be described by the regular expression $\lambda \vee (000)(00)^*$, so L is a regular set. Example 12.15 gave a grammar G to generate L. We will give three more grammars equivalent to G.

$G_1 = (V, V_T, S, P)$ where $V = \{0, A, B, S\}$, $V_T = \{0\}$, and the productions are

$$S \rightarrow \lambda$$
$$S \rightarrow ABA$$
$$AB \rightarrow 00$$
$$0A \rightarrow 000A$$
$$A \rightarrow 0$$

G_1 is context-sensitive but not context-free.

$G_2 = (V, V_T, S, P)$ where $V = \{0, A, S\}$, $V_T = \{0\}$, and the productions are

$$S \rightarrow \lambda$$
$$S \rightarrow 00A$$
$$A \rightarrow 00A$$
$$A \rightarrow 0$$

G_2 is context-free but not regular.

$G_3 = (V, V_T, S, P)$ where $V = \{0, A, B, C, S\}$, $V_T = \{0\}$, and the productions are

$$S \rightarrow \lambda$$
$$S \rightarrow 0A$$

$$A \rightarrow 0B$$
$$B \rightarrow 0$$
$$B \rightarrow 0C$$
$$C \rightarrow 0B$$

G_3 is regular.

Thus, L is a regular language. That a regular set turned out to be a regular language is not coincidental, as we will see in the next section.   □

12.20
Practice

Give the derivation of 00000 in G_1, G_2, and G_3.   □

Context-Free Grammars

We will elaborate on the proper containment of the sets of Figure 12.1 in the next section. Meanwhile, we'll concentrate on context-free grammars for three reasons. Context-free grammars seem to be the easiest to work with, since they allow replacing only one symbol at a time. Furthermore, many programming languages are defined such that sections of syntax, if not the whole language, can be described by context-free grammars. Finally, a derivation in a context-free grammar has a lovely graphical representation called a **parse tree**.

12.21
Example

The grammar of Example 12.14 is context-free. The word $d2q$ can be derived as follows: identifier $\Rightarrow$ identifier letter $\Rightarrow$ identifier digit letter $\Rightarrow$ letter digit letter $\Rightarrow$ d digit letter $\Rightarrow$ $d2$ letter $\Rightarrow$ $d2q$. We can represent this derivation as a tree with the start symbol for the root. When a production is applied to a node, that node is replaced at the next lower level of the tree by the symbols in the right-hand side of the production used. A tree for the above derivation appears in Figure 12.2.   □

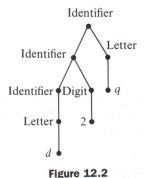

Figure 12.2

12.22
Practice

Draw a parse tree for the word $m34s$ in the grammar of Example 12.14.

□

   Suppose that a context-free grammar G describes a programming language. The programmer uses the rules of G to generate legitimate strings of symbols, that is, words in the language. Here we may think of a word as corresponding to a program instruction. Thus, a word consists of various subwords, for example, identifiers, operators, and key words for the language (such as IF and DO in FORTRAN, and "repeat" and "until" in Pascal). The program instructions are fed into the compiler for the language so that the program can be translated into machine language code for the computer. The compiler must decide whether each program instruction is legitimate in the language. This question really entails two questions: are the subwords themselves legitimate strings, and is the program instruction a legitimate way of grouping the subwords together?

   Usually the set of legitimate subwords of a language can be described by a regular expression, and then a finite-state machine can be used to detect the subwords; the lexical analyzer or scanner portion of the compiler handles this phase of compilation. If all goes well, the scanner then passes the program instruction, in the form of a string of legitimate subwords, to the syntax analyzer. The syntax analyzer determines whether the string is correct by trying to parse it (construct its parse tree).

   Various parsing techniques, which we won't go into, have been devised. One obvious approach is to construct a tree by beginning with the start symbol, applying productions, and ending with the string to be tested. This procedure is called **top-down parsing.** The alternative is to begin with the string, see what productions were used to create it, apply productions "backwards," and end with the start symbol. This process is called **bottom-up parsing.** The trick to either approach is to decide exactly which productions should be used.

12.23
Example

Consider the context-free grammar G given by $G = (V, V_T, S, P)$ where $V = \{a, b, c, A, B, C, S\}$, $V_T = \{a, b, c\}$, and the productions are

$$S \to B \qquad S \to A \qquad A \to abc$$
$$B \to C \qquad B \to ab \qquad C \to c$$

Suppose we want to test the string abc. A derivation for abc is $S \Rightarrow A \Rightarrow abc$. If we try a top-down parse, we might begin with

$$
\begin{array}{c}
S \\
| \\
B
\end{array}
$$

Then we have to detect that this will not work and try something else. If we try a bottom-up parse, we might begin with

$$BC$$
$$|$$
$$abc$$

Then we have to detect that this will not work and try something else. Parsing techniques automate this process. □

Overall, we see once again a distinction between *generating* members of a set, which the programmer does, and *deciding* membership in a set, which the compiler does. Since we ask the compiler to decide membership in a set, the set must be a recursive language. It turns out that context-free languages are indeed recursive, another point in their favor.

Checklist

Definitions

 alphabet (vocabulary) (*p. 505*)
 word (*p. 505*)
 language (*p. 505*)
 recursive language (*p. 506*)
 recursively enumerable language (*p. 506*)
 phrase-structure (type 0) grammar (*p. 506*)
 terminal (*p. 506*)
 start symbol (*p. 506*)
 production (*p. 506*)
 direct generation (derivation) of a word (*p. 507*)
 generation (derivation) of a word (*p. 507*)
 language generated by a grammar *G* (*p. 507*)
 formal language (*p. 507*)
 Backus–Naur form (BNF) (*p. 510*)
 erasing convention (*p. 512*)
 context-sensitive (type 1) grammar (*p. 512*)
 context-free (type 2) grammar (*p. 512*)
 regular (type 3) grammar (*p. 512*)
 Chomsky hierarchy (*p. 512*)
 type 0 language (*p. 512*)
 context-sensitive language (*p. 512*)
 context-free language (*p. 512*)
 regular language (*p. 512*)
 equivalent grammars (*p. 513*)
 parse tree (*p. 514*)
 top-down parsing (*p. 515*)
 bottom-up parsing (*p. 515*)

Techniques

> Describe $L(G)$ for a given grammar G.
>
> Define a grammar to generate a given language L.
>
> Construct parse trees in a context-free grammar.

Main Ideas

> A grammar G is a generating mechanism for its language $L(G)$; $L(G)$ is thus an r.e. language.
>
> Formal languages were developed in an attempt to describe correct syntax for natural languages; although this attempt has largely failed because of the complexity of natural languages, it has been quite successful for high-level programming languages.
>
> Special classes of grammars are defined by restricting the allowable productions.
>
> Derivations in context-free grammars can be illustrated by parse trees.
>
> A compiler for a context-free programming language checks correct syntax by parsing.

Exercises Section 12.1

$\star$ 1. Describe $L(G)$ for each of the following grammars G.

(a) $G = (V, V_T, S, P)$ where $V = \{a, A, B, C, S\}$, $V_T = \{a\}$, and P consists of

$$S \rightarrow A$$
$$A \rightarrow BC$$
$$B \rightarrow A$$
$$A \rightarrow a$$
$$aC \rightarrow \lambda$$

(b) $G = (V, V_T, S, P)$ where $V = \{0, 1, A, B, S\}$, $V_T = \{0, 1\}$, and P consists of

$$S \rightarrow 0A$$
$$S \rightarrow 1A$$
$$A \rightarrow 1BB$$
$$B \rightarrow 01$$
$$B \rightarrow 11$$

(c) $G = (V, V_T, S, P)$ where $V = \{0, 1, A, B, S\}$, $V_T = \{0, 1\}$, and P consists of

$$S \to 0$$
$$S \to 0A$$
$$A \to 1B$$
$$B \to 0A$$
$$B \to 0$$

(d) $G = (V, V_T, S, P)$ where $V = \{0, 1, A, S\}$, $V_T = \{0, 1\}$, and P consists of

$$S \to 0S$$
$$S \to 11A$$
$$A \to 1A$$
$$A \to 1$$

2.   (a) Which of the grammars of Exercise 1 are regular? Which are context-free?
   (b) Find regular grammars to generate each of the languages of Exercise 1.

★ 3. Describe $L(G)$ for the grammar $G = (V, V_T, S, P)$ where $V = \{a, b, A, B, S\}$, $V_T = \{a, b\}$, and P consists of

$$S \to AB$$
$$AB \to AAB$$
$$AB \to ABB$$
$$A \to a$$
$$B \to b$$

What type of grammar is G? Find a regular grammar G' that generates $L(G)$.

4. Write the productions of the following grammars in BNF:
   (a) G_3 in Example 12.19.
   (b) G in Exercise 1b.
   (c) G in Exercise 1c.
   (d) G in Exercise 1d.

5. Find a grammar that generates the set of all strings of well-balanced parentheses.

6. A word w in V^* is a palindrome if $w = w^R$, where w^R is the reverse of the string w. A language L is a palindrome language if L consists entirely of palindromes.
   (a) Find a grammar that generates the set of all palindromes over the alphabet $\{a, b\}$.
   (b) Let L be a palindrome language. Prove that $L^R = \{w^R \mid w \in L\}$ is a palindrome language.
   (c) Let w be a palindrome. Prove that the language described by the regular expression w^* is a palindrome language.

★ 7. Find a grammar that generates the language $L = (0 \lor 1)*01$.

8. Find a context-free grammar that generates the language $L = \{0^n 1^n \mid n \geq 0\}$.

9. Find a grammar that generates the language $L = \{0^{2^i} \mid i \geq 0\}$.

10. Find a context-free grammar that generates the language L where L consists of the set of all nonempty strings of 0s and 1s with an equal number of 0s and 1s.

11. Find a context-free grammar that generates the language L where L consists of the set of all nonempty strings of 0s and 1s with twice as many 0s as 1s.

★ 12. Find a context-free grammar that generates the language $L = \{ww^R \mid w \in \{0, 1\}*$ and w^R is the reverse of the string $w\}$.

13. Find a grammar that generates the language $L = \{ww \mid w \in \{0, 1\}*\}$.

14. Find a grammar that generates the language $L = \{a^{n^2} \mid n \geq 1\}$. (By Exercise 12 of Section 12.2, L is not a context-free language, so your grammar cannot be too simple.)

15. Draw parse trees for the following words:
    ★ (a) The word 111111 in the grammar G of Example 12.10
    (b) The word 111111 in the grammar G' of Example 12.10
    (c) The word 011101 in the grammar of Exercise 1b
    (d) The word 00111111 in the grammar of Exercise 1d

★ 16. Consider the context-free grammar $G = (V, V_T, S, P)$ where $V = \{0, 1, A, S\}$, $V_T = \{0, 1\}$, and P consists of

$$S \to A1A$$
$$A \to 0$$
$$A \to A1A$$

Draw two distinct parse trees for the word 01010 in G. A grammar in which a word has more than one parse tree is **ambiguous**.

17. Show that for any context-free grammar G there exists a context-free grammar G' in which for every production $\alpha \to \beta$, β is a longer string than α, $L(G') \subseteq L(G)$, and $L(G) - L(G')$ is a finite set.

SECTION 12.2   LANGUAGE RECOGNIZERS

Recognizers for Regular Languages

In Chapter 9 we defined regular sets and showed that a set is regular if and only if it can be recognized by some finite-state machine. In this chapter we have defined regular languages. Next we will see that a set is regular if and only if it is a regular language. It will follow that

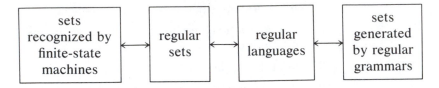

Therefore, regular grammars serve as generating devices for exactly those sets for which finite-state machines serve as recognition devices.

First, we show, given a finite-state machine M, how to construct a regular grammar G such that M recognizes the set $L(G)$. By throwing in a new start state, we could create a new machine $\overline{M}$ that recognizes the same set as M but whose start state is never revisited. Thus, we may as well assume that M has this property. If M recognizes λ, then the start state S is a final state. The grammar G uses the states and input symbols of M as its vocabulary, with the set of input symbols as the set of terminals. The start state S of M is the start symbol of G, and the productions of G are defined so that derivations in G simulate computations by M. Thus, $A \rightarrow a$ belongs to P if and only if M in state A with input symbol a goes to a final state, and $A \rightarrow aB$ is in P if and only if M in state A with input symbol a goes to state B. The production $S \rightarrow \lambda$ is in P if and only if S is a final state in M. The grammar G is then regular, and it is not hard to see that $L(G)$ is the set recognized by M.

12.24
Example

Let M be given by the state graph of Figure 12.3. Following the procedure above, let $G = (V, V_T, S, P)$ where $V = \{S, A, B, C, 0, 1\}$, $V_T = \{0, 1\}$, and P consists of

$S \rightarrow \lambda$	$A \rightarrow 1C$	$B \rightarrow 1$	$C \rightarrow 1C$
$S \rightarrow 0$	$A \rightarrow 0C$	$B \rightarrow 1B$	$C \rightarrow 0C$
$S \rightarrow 0A$		$B \rightarrow 0C$	
$S \rightarrow 1$			
$S \rightarrow 1B$			

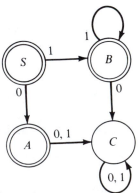

Figure 12.3

The grammar G is regular. The machine M recognizes the regular set $0 \vee 1^*$, and this is also $L(G)$. For example, a derivation of 111 in G is $S \Rightarrow 1B \Rightarrow 11B \Rightarrow 111$. □

12.25
Practice

Let M be given by the state graph of Figure 12.4. Find a regular grammar generating the set recognized by M. Describe the set by a regular expression. □

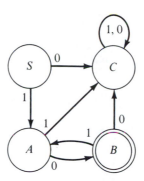

Figure 12.4

Conversely, suppose we are given a regular grammar $G = (V, V_T, S, P)$, and we want to construct a finite-state machine M to recognize $L(G)$. We first construct a nondeterministic machine $\overline{M}$, which by Theorem 9.40 is all that's necessary. We let the states of $\overline{M}$ be the nonterminals of G together with a new final state F. If the production $S \to \lambda$ is in P, then we also let state S be final. The set V_T is the input alphabet of $\overline{M}$, and S is the start state. If $\overline{M}$ is in state F with input symbol a, then the set of next states of $\overline{M}$ is empty. If $\overline{M}$ is in any state $A \neq F$ with input symbol a, then the set of next states of $\overline{M}$ includes F if $A \to a$ is a production in P, and it includes all states B such that $A \to aB$ is a production in P. Computations in $\overline{M}$ simulate derivations in G, and again it can easily be shown that $\overline{M}$ recognizes the set $L(G)$.

12.26
Example

Let $G = (V, V_T, S, P)$ where $V = \{S, A, B, 0, 1\}$, $V_T = \{0, 1\}$, and P consists of

$$S \to \lambda$$
$$S \to 0A$$
$$A \to 1$$
$$A \to 0A$$
$$A \to 1B$$
$$B \to 0B$$

The state graph of a nondeterministic machine $\overline{M}$ recognizing $L(G)$ is given in Figure 12.5. The set $L(G) = \lambda \vee 00^*1$. □

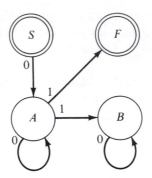

Figure 12.5

12.27
Practice

Let $G = (V, V_T, S, P)$ where $V = \{S, A, B, 0, 1\}$, $V_T = \{0, 1\}$, and P consists of

$$S \rightarrow 1A$$
$$A \rightarrow 1A$$
$$A \rightarrow 0B$$
$$B \rightarrow 0B$$
$$B \rightarrow 0$$

(a) Find the state graph of a (nondeterministic) finite-state machine recognizing $L(G)$.
(b) Describe $L(G)$ by a regular expression. ☐

We have now demonstrated the following theorem.

12.28
Theorem

Any set recognized by a finite-state machine is generated by a regular grammar, and conversely. Hence, regular sets coincide with regular languages. ☐

We know that $S = \{0^n1^n \mid n \geq 0\}$ is not a regular set, so by Theorem 12.28 it is not a regular language. By Exercise 8 of Section 12.1, S is a context-free language, which proves that the set of regular languages is a proper subset of the set of context-free languages.

Recognizers for Type 0 Languages

From Theorem 12.28 we know that the most restricted class of languages coincides with the class of sets recognized by a computational device of limited capacity. Let's consider the other end of the spectrum. The most general language is a type 0 language, and the most general computational device is the Turing machine. How do sets recognized by Turing machines compare with type 0 languages? As it happens, these also coincide. This result is easy to see in one direction if we employ the Church–Turing Thesis. A type 0 language L is one generated by a phrase-structure grammar $G = (V, V_T, S, P)$. As discussed in the previous section, L is an r.e. set, and there is an effective procedure for generating the members of L. By the Church–Turing Thesis, there must be a Turing machine T_1 generating L in the sense that T_1 begins on a blank tape and every so often signals that the word currently on its tape is a member of L. We can then use a second Turing machine T_2 that, when begun with a string $\alpha \in V^*$ on its tape, constantly compares α with the contents of T_1's tape. Whenever α matches a word on T_1's tape that T_1 says is a member of L, T_2 halts in a final state; otherwise, T_2 does not halt. Thus, T_2 recognizes the set L.

In the other direction, we want to show that given any Turing machine T, there is a phrase-structure grammar G such that $L(G)$ is exactly the set recognized by T. We will not give the proof, although it is not very difficult and involves creating productions of G that manipulate symbols in a string to simulate the action of T on a tape containing that string. The result we've been discussing is stated below.

12.29
Theorem
Any set recognized by a Turing machine is generated by a type 0 grammar, and conversely. □

Recursive and R.E. Sets

As we have just noted, an r.e. set is recognized by some Turing machine. We also know from Chapter 11 that a set recognized by a Turing machine is r.e. The r.e. sets therefore coincide with sets recognized by Turing machines. However, we have hinted that there is a distinction between a Turing machine generating a set and a Turing machine deciding membership in a set or, equivalently, between a set being r.e. and a set being recursive. Now we'll prove this distinction.

A Nonrecursive R.E. Set

First, we let S be a subset of V^* for some vocabulary V. Then we note that S is a recursive set if and only if both S and the complement of S, $S' = V^* - S$, are r.e. If S is recursive, then S is r.e. However, we also have an effective procedure for generating S'. We generate the members of V^* and, given $x \in V^*$, apply the process to decide whether $x \in S$. If the answer is no, then $x \in S'$. Therefore, S' is r.e. Now suppose both S and S' are r.e., and we are given a member $x \in V^*$. We generate members of S and of S' and compare x against these lists as they are generated. Eventually x must appear in the list for S or the list for S', and at that point we know whether $x \in S$. This process gives us an effective procedure for deciding membership in S, so S is recursive.

One final idea we need is that of an *effective enumeration* of Turing machines. In discussing the universal Turing machine, we saw how to effectively describe any Turing machine T as a finite string of symbols from the set $\{*, 1\}$. If we order these symbols so that $* < 1$, then we can enumerate the set of all finite strings over $\{*, 1\}$. We can eliminate those strings that are not the proper form for representing the quintuples of any Turing machine. The result is an enumeration $T_0, T_1, T_2, \ldots$, of all Turing machines. Furthermore, it is an effective enumeration; that is, given a nonnegative integer i, we can construct the quintuples for Turing machine T_i, and, given any Turing machine T, we can find an index i in the enumeration such that $T = T_i$. Now we have everything we need to prove our next theorem.

12.30
Theorem

There is an r.e. set that is not recursive.

Proof: Let $V = \{1\}$. For each $x \in V^*$, if x is a string of i 1s, we let T_x denote the Turing machine T_i in the above enumeration. Let the set S be defined by

$$S = \{x \mid x \in V^* \text{ and } T_x \text{ halts in a final state when begun on a tape containing } x\}$$

Then S is a subset of V^*. The set S is also r.e.; an effective procedure to generate S consists of generating V^* and, for each $x \in V^*$, constructing the Turing machine T_x. The machine T_x is then begun on a tape containing x. If T_x ever halts in a final state, then $x \in S$. Once again we resort to our usual trick to avoid waiting forever behind some one, nonhalting computation; that is, we run several computations at once, going back to check on the progress of old ones and adding new ones.

Now we consider S'.

$$S' = \{x \mid x \in V^* \text{ and } T_x(x) \text{ does not halt in a final state}\}$$

If we assume that S' is r.e., then S' is recognized by some Turing machine, say T_y for some $y \in V^*$. By the definition of recognition,

$$T_y(y) \text{ halts in a final state} \leftrightarrow y \in S'$$

But by the definition of S',

$$y \in S' \leftrightarrow T_y(y) \text{ does not halt in a final state}$$

Since together these two conclusions are contradictory, S' is not an r.e. set. (Notice that this argument is closely related to the one used to prove the unsolvability of the halting problem.) Therefore, S is r.e. but not recursive. □

Note that no algorithm exists to generate the members of the set S' in the proof of Theorem 12.30. Thus, although we defined S' as a whole in a single, finite phrase, S' has no finite description that specifies its elements. (Recall in the beginning of this chapter we said there were more languages—sets—than finite descriptions.)

Context-Sensitive Languages Are Recursive

Any type 0 language L is an r.e. set, but if L is not a recursive set, it will not be suitable as a programming language because the compiler cannot decide whether an arbitrary string of symbols belongs to the language. Any context-free language is recursive. In fact, any context-sensitive language is recursive.

12.31
Theorem

Let L be a context-sensitive language; then L is recursive.

Proof: Let L be generated by the context-sensitive grammar $G = (V, V_T, S, P)$. We want an effective procedure to decide, given any word $x \in V^*$, whether $x \in L$. If $x = \lambda$, we check whether the production $S \to \lambda$ is in P. If it is, $x \in L$; otherwise, $x \notin L$. Now suppose $x \neq \lambda$, and the length of x, $|x|$, is some positive integer k. Now we define a sequence of sets W_i as follows:

$$W_0 = \{S\}$$

and for each positive integer i

$$W_i = W_{i-1} \cup \{w \mid w \in V^*, |w| \leq k, \text{ and there is some word } v \text{ in } W_{i-1} \text{ such that } v \Rightarrow w\}$$

Therefore W_i consists of all words in L of length $\leq k$ derivable from S in no more than i steps. It is clear that $x \in L$ if and only if $x \in W_i$ for some i.

For each i, $W_{i-1} \subseteq W_i$. If this containment is always proper, the W sets continue to have more and more elements. But since V is a finite set, there is a limit to the number of words in V^* (hence in L) of length $\leq k$. Therefore, there is a j such that $W_{j-1} = W_j$. Once this happens, then clearly $W_{j-1} = W_j = W_{j+1} = \cdots$. To determine if $x \in L$, we compute $W_0 = \{S\}, W_1, W_2, \ldots$ until two sets W_{j-1} and W_j are equal (which always happens). Then $x \in L$ if and only if x has appeared in one of the finite sets in this finite collection. $\square$

A pseudocode description of the procedure given in the proof of Theorem 12.31 follows.

Algorithm CSLDecision

if $x = \lambda$
then begin
  if $S \to \lambda$ is in P
  then $x \in L$
  else $x \notin L$
  end of then
else begin
  new-set $\leftarrow \{S\}$
  repeat
    current-set $\leftarrow$ new-set
    generate all words of length $\leq k$ from current-set
    new-set $\leftarrow$ union of these words with current-set
  until new-set = current-set
  if $x \in$ new-set
  then $x \in L$
  else $x \notin L$
end else

12.32
Example

Consider the context-sensitive grammar of Example 12.12 where the productions are

$$S \to aSBC$$
$$S \to aBC$$
$$CB \to BC$$
$$aB \to ab$$
$$bB \to bb$$
$$bC \to bc$$
$$cC \to cc$$

Let's use the proof of Theorem 12.31 to test whether $x = abbc$ is a member of $L(G)$; $|x| = 4$. (Of course, since we already know that $L(G) = \{a^n b^n c^n\}$, we already know the answer.) We construct the W sets.

$$W_0 = \{S\}$$
$$W_1 = \{S, aSBC, aBC\}$$
$$W_2 = \{S, aSBC, aBC, abC\}$$
$$W_3 = \{S, aSBC, aBC, abC, abc\}$$
$$W_4 = \{W_3\}$$

Since x has not appeared, $x \notin L$. $\square$

12.33
Practice

Consider the context-sensitive grammar G_1 of Example 12.19 with productions

$$S \to \lambda$$
$$S \to ABA$$
$$AB \to 00$$
$$0A \to 000A$$
$$A \to 0$$

Use the procedure of Theorem 12.31 to determine whether $0000 \in L$.

□

Although every context-sensitive language is recursive, there are recursive languages that are not context-sensitive. The proof is rather similar to that of Theorem 12.30 and is left as an exercise (see Exercise 11 in this section).

The Grand Finale

We conclude by adding more detail to the language diagram of Figure 12.1. The only proper containment shown in Figure 12.6 that we have not mentioned is that the context-free languages are properly contained in the context-sensitive languages. However, the context-sensitive language $L = \{a^n b^n c^n \mid n \geq 1\}$ of Example 12.12 is not context-free (see Exercise 12 of this section).

There are also computational devices with capabilities midway between those of finite-state machines and those of Turing machines; these devices recognize exactly the context-free languages and the context-sensitive languages, respectively. The type of device that recognizes the context-free languages is called a **pushdown automaton**, or **pda**. A pda consists of a finite-state unit that reads input from a tape and controls activity in a stack. The stack operates like a set of plates stacked on a spring in a cafeteria. Symbols from some alphabet can be added to the top of the stack by a *push* instruction, which pushes previously stored items farther down in the stack. Only the topmost item on the stack is accessible at any moment, and it is removed from the stack by a *pop* instruction.

The finite-state unit, as a function of the input symbol read, the present state, and the top symbol on the stack, has a finite number of possible next moves. The moves are of the following types: go to a new state, pop the top symbol off the stack, and read the next input symbol; go to a new state, pop the top symbol off the stack, push a finite number of symbols onto the stack, and read the next input symbol; ignore the

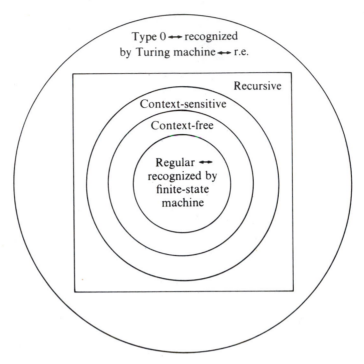

Figure 12.6

input symbol being read, manipulate the stack as above, but do not read the next input symbol. Because there is a choice of moves, the pda is a nondeterministic device, and unlike the situation for finite-state machines, nondeterminism adds to the capabilities of this type of machine. A pda recognizes the set of all inputs for which a sequence of moves exists that causes the pda to empty its stack. It can be shown that any set recognized by a pda is a context-free language, and conversely.

The type of device that recognizes the context-sensitive languages is called a **linear bounded automaton**, or **lba**. An lba is a Turing machine whose read–write head is restricted to that portion of the tape containing the original input, and that at each step has a choice of possible next moves. An lba is therefore a nondeterministic device, and it recognizes the set of all inputs for which a sequence of moves exists that causes it to halt in a final state. Any set recognized by an lba can be shown to be a context-sensitive language, and conversely.

✔ **Checklist**

Definitions

> pushdown automaton (pda) (*p. 527*)
> linear bounded automaton (lba) (*p. 528*)

Techniques

Given a finite-state machine M, find a regular grammar generating the set recognized by M.

Given a regular grammar G, construct a (nondeterministic) finite-state machine to recognize $L(G)$.

Given a context-sensitive language $L(G)$ over a vocabulary V, decide whether $x \in V^*$ is a member of $L(G)$.

Main Ideas

A language is regular if and only if it is recognized by a finite-state machine.

A language is type 0 if and only if it is recognized by a Turing machine and if and only if it is an r.e. set.

There is an r.e. set that is not recursive.

Any context-sensitive language is recursive, but not conversely.

Exercises Section 12.2

★ 1. Let M be given by the state graph of Figure 12.7. Find a regular grammar generating the set recognized by M. Describe the set by a regular expression.

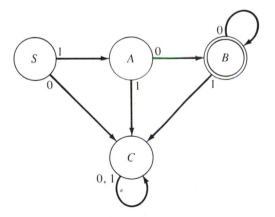

Figure 12.7

2. Let M be given by the state graph of Figure 12.8. Find a regular grammar generating the set recognized by M. Describe the set by a regular expression.

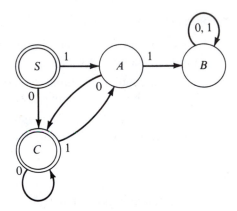

Figure 12.8

3. Let M be given by the state graph of Figure 12.9. Find a regular grammar generating the set recognized by M. Describe the set by a regular expression.

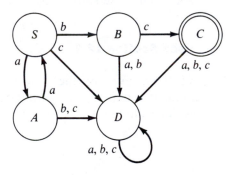

Figure 12.9

4. Let $G = (V, V_T, S, P)$ where $V = \{S, A, B, 0, 1\}$, $V_T = \{0, 1\}$, and P consists of

$S \rightarrow 0S$	$A \rightarrow 1$	$B \rightarrow 1B$
$S \rightarrow 1A$	$A \rightarrow 1B$	$B \rightarrow 0B$
$S \rightarrow 1B$	$A \rightarrow 0B$	

Find the state graph of a (nondeterministic) finite-state machine recognizing $L(G)$. Describe $L(G)$ by a regular expression.

★ 5. Let $G = (V, V_T, S, P)$ where $V = \{S, A, B, C, 0, 1\}$, $V_T = \{0, 1\}$, and P consists of

$S \rightarrow \lambda$	$A \rightarrow 0B$	$B \rightarrow 0A$	$C \rightarrow 1$
$S \rightarrow 0A$	$A \rightarrow 0$		$C \rightarrow 0C$
$S \rightarrow 1C$			

Find the state graph of a (nondeterministic) finite-state machine recognizing $L(G)$. Describe $L(G)$ by a regular expression.

6. Let $G = (V, V_T, S, P)$ where $V = \{S, A, B, C, D, E, a, b, c\}$, $V_T = \{a, b, c\}$, and P consists of

$$
\begin{array}{ll}
S \rightarrow \lambda & A \rightarrow bC \\
S \rightarrow a & C \rightarrow cD \\
S \rightarrow aA & C \rightarrow bE \\
S \rightarrow aB & D \rightarrow cD \\
B \rightarrow aB & D \rightarrow bE \\
B \rightarrow a & E \rightarrow a
\end{array}
$$

Find the state graph of a (nondeterministic) finite-state machine recognizing $L(G)$. Describe $L(G)$ by a regular expression.

★ 7. It is stated in the proof of Theorem 12.31 that because V is finite there is a limit to the number of words in V^* of length $\leq k$. Suppose V has n elements. What is the number of words in V^* of length $\leq k$?

8. Where does the proof of Theorem 12.31 break down if L is not context-sensitive?

9. Let $G = (V, V_T, S, P)$ where $V = \{S, A, B, 0, 1\}$, $V_T = \{0, 1\}$, and P consists of

$$
\begin{array}{lll}
S \rightarrow 0 & 1A \rightarrow 1B & B \rightarrow 0B1 \\
S \rightarrow 1A & & B \rightarrow 1
\end{array}
$$

Use the proof of Theorem 12.31 to decide whether 1010 belongs to $L(G)$.

10. Let $G = (V, V_T, S, P)$ where $V = \{S, A, +, a, b, c\}$, $V_T = \{+, a, b, c\}$, and P consists of

$$
\begin{array}{l}
S \rightarrow S + S \\
+S \rightarrow Sa \\
S \rightarrow a \\
S \rightarrow b \\
S \rightarrow bA \\
bA \rightarrow bc
\end{array}
$$

Use the proof of Theorem 12.31 to decide whether the following words belong to $L(G)$:

★ (a) cb
(b) bbc

11. This exercise is to prove that the set of context-sensitive languages is properly contained in the set of recursive languages.

Let x_i denote the ith string in an enumeration of all finite strings over $\{0, 1\}$. Also, we can encode all phrase-structure grammars with terminals from $\{0, 1\}$ as finite strings in $\{0, 1\}^*$. For example, we can let the terminal 0 be represented as 01, the terminal 1 as 01^2, the production arrow by 01^3, a comma by 01^4, and variables by 01^5, 01^6, and so on. A grammar is represented by the string consisting of the code for productions separated by commas. We can eliminate those strings in $\{0, 1\}^*$ that are not of the proper form to represent context-sensitive grammars. The result is an effective enumeration G_1, G_2, . . . of context-sensitive grammars. Now define a set $S = \{x_i \mid x_i \notin L(G_i)\}$. Show that S is a recursive language in $\{0, 1\}^*$ that is not context-sensitive.

12. The following is the **pumping lemma** for context-free languages. Let L be any context-free language. Then there exists some constant k such that for any word w in L with $|w| \geq k$, w can be written as the string $w_1 w_2 w_3 w_4 w_5$ with $|w_2 w_3 w_4| \leq k$ and $|w_2 w_4| \geq 1$. Furthermore, the word $w_1 w_2^i w_3 w_4^i w_5 \in L$ for each $i \geq 0$.

   (a) Use the pumping lemma to show that $L = \{a^n b^n c^n \mid n \geq 1\}$ is not context-free.

   (b) Use the pumping lemma to show that $L = \{a^{n^2} \mid n \geq 1\}$ is not context-free.

Answers to Practice Problems

Note to student: Finish all parts of a practice problem before turning to the answers.

CHAPTER 1

1.2 false, false, false

1.3

A	B	$A \vee B$
T	T	T
T	F	T
F	T	T
F	F	F

1.5 (a) antecedent: The eggs are fresh.
        consequent: They will not spoil.
    (b) antecedent: The Great American Novel is to be written.
        consequent: A word processor is to be used.
    (c) antecedent: Susan will pass her physics course.
        consequent: She is bright and studies hard.
    (d) antecedent: high gasoline mileage
        consequent: good combustion

1.6

A	B	$A \rightarrow B$
T	T	T
T	F	F
F	T	T
F	F	T

1.7

A	A'
T	F
F	T

533

1.10 (a)

A	B	$A \to B$	$B \to A$	$(A \to B) \leftrightarrow (B \to A)$
T	T	T	T	T
T	F	F	T	F
F	T	T	F	F
F	F	T	T	T

(b)

A	B	A'	B'	$A \lor A'$	$B \land B'$	$(A \lor A') \to (B \land B')$
T	T	F	F	T	F	F
T	F	F	T	T	F	F
F	T	T	F	T	F	F
F	F	T	T	T	F	F

(c)

A	B	C	B'	$A \land B'$	C'	$(A \land B') \to C'$	$((A \land B') \to C')'$
T	T	T	F	F	F	T	F
T	T	F	F	F	T	T	F
T	F	T	T	T	F	F	T
T	F	F	T	T	T	T	F
F	T	T	F	F	F	T	F
F	T	F	F	F	T	T	F
F	F	T	T	F	F	T	F
F	F	F	T	F	T	T	F

(d)

A	B	A'	B'	$A \to B$	$B' \to A'$	$(A \to B) \leftrightarrow (B' \to A')$
T	T	F	F	T	T	T
T	F	F	T	F	F	T
F	T	T	F	T	T	T
F	F	T	T	T	T	T

1.13

A	1	A'	$A \lor A'$	$A \lor A' \leftrightarrow 1$
T	T	F	T	T
F	T	T	T	T

1.15 (a) false   (b) true

1.16 For example:
   (a) The domain is the collection of licensed drivers in the United States; $P(x)$ is the property that x is older than 14.
   (b) The domain is the collection of all fish; $P(x)$ is the property that x has five legs.
   (c) No; if all objects in the domain have property P, then (since the domain must contain objects) there is an object in the domain with property P.

(d) The domain is all the people who live in Boston; $P(x)$ is the property that x is a male.

1.20 Let $x = 1$; then x is positive and any integer less than x is ≤ 0, so the truth value of the statement is true. For the second interpretation, let $A(x)$ be "x is even," $B(x, y)$ be "$x < y$," and $C(y)$ be "y is odd"; the statement is false because no even integer has the property that all larger integers are odd.

1.21 (a) $(\forall x)\, (S(x) \to I(x))$
(b) $(\exists x)\, (I(x) \wedge S(x) \wedge M(x))$
(c) $(\forall x)\, (M(x) \to S(x) \wedge (I(x))')$

1.24 Invalid. In the interpretation where the domain consists of the integers, $P(x)$ is "x is odd" and $Q(x)$ is "x is even," the antecedent is true (every integer is even or odd), but the consequent is false (it is not the case that every integer is even or that every integer is odd).

1.27 1. P' (hypothesis)
     2. $P' \to (Q' \to P')$ (Axiom 1)
     3. $Q' \to P'$ (1, 2, modus ponens)
     4. $(Q' \to P') \to (P \to Q)$ (Axiom 3)
     5. $P \to Q$ (3, 4, modus ponens)

1.29 1. $P \to Q$ (hypothesis)
     2. $Q \to R$ (hypothesis)
     3. P (hypothesis)
     4. Q (1, 3, modus ponens)
     5. R (2, 4, modus ponens)

1.32 The argument is $((P \to M) \wedge (P \vee C) \wedge M') \to C$. The proof sequence is:

     1. $P \to M$ (hypothesis)
     2. $(P \to M) \to (M' \to P')$ (tautology)
     3. $M' \to P'$ (1, 2, modus ponens)
     4. M' (hypothesis)
     5. P' (3, 4, modus ponens)
     6. $P \vee C$ (hypothesis)
     7. C (tautology $P' \wedge (P \vee C) \to C$)

1.34 1. $(\forall x)P(x)$ (hypothesis)
     2. $P(x)$ (1, Axiom 5, modus ponens)
     3. $(\exists x)P(x)$ (2, Axiom 7, modus ponens)

1.37 The argument is $(\forall x)\, (R(x) \to L(x)) \wedge (\exists x)R(x) \to (\exists x)L(x)$

     1. $(\forall x)\, (R(x) \to L(x))$ (hypothesis)
     2. $(\exists x)R(x)$ (hypothesis)
     3. $R(a)$ (2, Axiom 6, modus ponens)
     4. $R(a) \to L(a)$ (1, Axiom 5, modus ponens)
     5. $L(a)$ (3, 4, modus ponens)
     6. $(\exists x)L(x)$ (5, Axiom 7, modus ponens)

1.40 Possible answers:
(a) a whale
(b) input from a terminal

1.42 Hypothesis: x is divisible by 6
     $x = k \cdot 6$ for some integer k
     $2x = 2(k \cdot 6)$

$$2x = (2 \cdot k)6$$
$$2x = (k \cdot 2)6$$
$$2x = k(2 \cdot 6)$$
$$2x = k \cdot 12$$
$$2x = k(3 \cdot 4)$$
$$2x = (k \cdot 3)4$$

$k \cdot 3$ is an integer

Conclusion: $2x$ is divisible by 4

1.45 (a) If the eggs will spoil, then they are not fresh.
   (b) If it is false that a word processor is to be used, then it is false that the Great American Novel is to be written.
   (c) If Susan is not bright or does not study hard, then she will not pass her physics course.
   (d) If combustion is not good, then there is not high gasoline mileage.

1.48 (a) If the eggs will not spoil, then they are fresh.
   (b) If a word processor is to be used, then the Great American Novel is to be written.
   (c) If Susan is bright and studies hard, she will pass her physics course.
   (d) If there is good combustion, then there is high gasoline mileage.

1.52 Let $x = 2m + 1$ and $y = 2n + 1$, and assume that xy is even. Then

$$(2m + 1)(2n + 1) = 2k \qquad \text{for some integer } k$$
$$4mn + 2m + 2n + 1 = 2k$$
$$1 = 2(k - 2mn - m - n) \qquad \text{where } k - 2mn - m - n \text{ is an integer}$$

This is a contradiction since 1 is not even.

1.55 $P(1)$: $1 = \dfrac{1(1 + 1)}{2}$, true

Assume $P(k)$: $1 + 2 + \cdots + k = \dfrac{k(k + 1)}{2}$

Show $P(k + 1)$: $1 + 2 + \cdots + (k + 1) = \dfrac{(k + 1)((k + 1) + 1)}{2}$

$$1 + 2 + \cdots + (k + 1) = 1 + 2 + \cdots + k + (k + 1)$$
$$= \frac{k(k + 1)}{2} + (k + 1) = (k + 1)\left(\frac{k}{2} + 1\right)$$
$$= (k + 1)\left(\frac{k + 2}{2}\right) = \frac{(k + 1)((k + 1) + 1)}{2}$$

1.59 $P(2)$: $2^{2+1} < 3^2$, or $8 < 9$, true
Assume $P(k)$: $2^{k+1} < 3^k$ and $k > 1$
Show $P(k + 1)$: $2^{k+2} < 3^{k+1}$

$$3^{k+1} = 3(3^k)$$
$$> 3(2^{k+1}) \qquad \text{(by the inductive hypothesis)}$$
$$> 2(2^{k+1}) \qquad \text{(since } 3 > 2)$$
$$= 2^{k+2}$$

1.62 $P(0)$: $j_0 = x + i_0$, true since $j = x$, $i = 0$ before the loop is entered
Assume $P(k)$: $j_k = x + i_k$
Show $P(k + 1)$: $j_{k+1} = x + i_{k+1}$

$$j_{k+1} = j_k + 1 \qquad \text{(by the assignment } j \leftarrow j + 1)$$
$$= (x + i_k) + 1 \qquad \text{(by inductive hypothesis)}$$
$$= x + (i_k + 1)$$
$$= x + i_{k+1} \qquad \text{(by the assignment } i \leftarrow i + 1)$$

At loop termination, $i = y$ and $j = x + i = x + y$.

1.65 1, 4, 7, 10, 13

1.66 *Algorithm T(n)*

1. if $n = 1$
2. then $T \leftarrow 1$
3. else $T \leftarrow T(n - 1) + 3$

1.68 $T(n) = T(n - 1) + 3$
$$= (T(n - 2) + 3) + 3 = T(n - 2) + 2 * 3$$
$$= (T(n - 3) + 3) + 2 * 3 = T(n - 3) + 3 * 3$$

In general, we guess that

$$T(n) = T(n - k) + k * 3$$

When $n - k = 1$, that is, $k = n - 1$,

$$T(n) = T(1) + (n - 1) * 3 = 1 + (n - 1) * 3$$

Now prove by induction that $T(n) = 1 + (n - 1) * 3$.

$T(1)$: $T(1) = 1 + (1 - 1) * 3 = 1$, true
Assume $T(k)$: $T(k) = 1 + (k - 1) * 3$
Show $T(k + 1)$: $T(k + 1) = 1 + k * 3$

$$T(k + 1) = T(k) + 3$$
$$= 1 + (k - 1) * 3 + 3$$
$$= 1 + k * 3$$

CHAPTER 2

2.2 (a) {4, 5, 6, 7}
(b) {April, June, September, November}
(c) {0, 1, 8}

2.3 (a) $\{x \mid x = y^2 \text{ for } y \in \{1, 2, 3, 4\}\}$
(b) $\{x \mid x \text{ is one of the Three Men in a Tub}\}$
(c) $\{x \mid x \text{ is a prime number}\}$

2.4 $x \in B$

2.6 (a), (b), (d), (e), (h), (i), (l)

2.8 Let $x \in A$. Then $\cos (x/2) = 0$. But $\cos (x/2) = 0$ if and only if $x/2 = \pm \pi/2$, $\pm 3\pi/2$, $\pm 5\pi/2$, . . . or $x = \pm \pi$, $\pm 3\pi$, $\pm 5\pi$, . . . and, for any multiple of π, the sine function is 0. Thus $x \in B$.

2.9 $\mathcal{P}(A) = \{\emptyset, \{1\}, \{2\}, \{3\}, \{1, 2\}, \{1, 3\}, \{2, 3\}, \{1, 2, 3\}\}$.

2.10 2^n

2.11 By the definition of equality for ordered pairs,

$$2x - y = 7 \quad \text{and} \quad x + y = -1$$

Solving the system of equations, $x = 2$, $y = -3$

2.12 $(3, 3), (3, 4), (4, 3), (4, 4)$

2.19 (a) S is not closed under division.
(c) 0^0 is not defined.
(f) $x^\#$ is not unique for, say, $x = 4$ ($2^2 = 4$ and $(-2)^2 = 4$).

2.24 $x \in A$ and $x \in B$

2.27

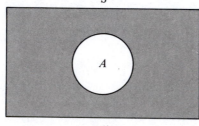

2.28

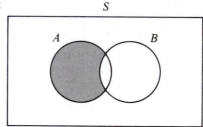

2.30 (a) $\{1, 2, 3, 4, 5, 7, 8, 9, 10\}$
(b) $\{1, 2, 3\}$
(c) $\{1, 3, 5, 10\}$

2.32 Show set inclusion in each direction. To show $A \cup \emptyset \subseteq A$, let $x \in A \cup \emptyset$. Then $x \in A$ or $x \in \emptyset$, but since $\emptyset$ has no elements, $x \in A$. To show $A \subseteq A \cup \emptyset$, let $x \in A$. Then $x \in A$ or $x \in \emptyset$, so $x \in A \cup \emptyset$.

2.34 (a) $(C \cap (A \cup B)) \cup ((A \cup B) \cap C')$
$= ((A \cup B) \cap C) \cup ((A \cup B) \cap C')$ (1b)
$= (A \cup B) \cap (C \cup C')$ (3b)
$= (A \cup B) \cap S$ (5a)
$= A \cup B$ (4b)
(b) $(C \cup (A \cap B)) \cap ((A \cap B) \cup C') = A \cap B$

2.36 (a) $A \times B = \{(1, 3), (1, 4), (2, 3), (2, 4)\}$
(b) $B \times A = \{(3, 1), (3, 2), (4, 1), (4, 2)\}$

(c) $A^2 = \{(1, 1), (1, 2), (2, 1), (2, 2)\}$

(d) $A^3 = \{(1, 1, 1), (1, 1, 2), (1, 2, 1), (1, 2, 2), (2, 1, 1), (2, 1, 2), (2, 2, 1),$
$(2, 2, 2)\}$

2.38 An enumeration of the even positive integers is 2, 4, 6, 8, 10, 12, ...

2.40 $\frac{1}{4}$

2.46 $4(8)(5) = 160$

2.51 $7(5) + 9 = 44$

2.55 $P(20, 2) = \dfrac{20!}{18!} = 380$

2.56 $6! = 720$

2.60 $C(12, 3) = \dfrac{12!}{3!9!} = 220$

2.62 $(a + b)^3 = (a + b)(a^2 + 2ab + b^2)$
$\qquad = a^3 + 2a^2b + ab^2 + ba^2 + 2ab^2 + b^3$
$\qquad = a^3 + 3a^2b + 3ab^2 + b^3$

The coefficients are 1, 3, 3, and 1 and

$$C(3, 0) = \frac{3!}{0!3!} = 1 \qquad C(3, 1) = \frac{3!}{1!2!} = 3$$

$$C(3, 2) = \frac{3!}{2!1!} = 3 \qquad C(3, 3) = \frac{3!}{3!0!} = 1$$

2.63 $(a + b)^{k+1}$
$= C(k, 0)a^{k+1}b^0 + C(k, 1)a^kb^1 + \cdots + C(k, k - 1)a^2b^{k-1} + C(k, k)a^1b^k$
$\quad + C(k, 0)a^kb^1 + C(k, 1)a^{k-1}b^2 + \cdots + C(k, k - 1)a^1b^k + C(k, k)a^0b^{k+1}$
(from previous work)
$= C(k, 0)a^{k+1}b^0 + (C(k, 0) + C(k, 1))a^kb^1 + (C(k, 1) + C(k, 2))a^{k-1}b^2$
$\quad + \cdots + (C(k, k - 1) + C(k, k))a^1b^k + C(k, k)a^0b^{k+1}$
(collecting like terms)
$= C(k, 0)a^{k+1}b^0 + C(k + 1, 1)a^kb^1 + C(k + 1, 2)a^{k-1}b^2 + \cdots$
$\quad + C(k + 1, k)a^1b^k + C(k, k)a^0b^{k+1}$ \qquad (by the given identity)
$= C(k + 1, 0)a^{k+1}b^0 + C(k + 1, 1)a^kb^1 + C(k + 1, 2)a^{k-1}b^2 + \cdots$
$\quad + C(k + 1, k)a^1b^k + C(k + 1, k + 1)a^0b^{k+1}$
(because $C(k, 0) = 1 = C(k + 1, 0)$
and $C(k, k) = 1 = C(k + 1, k + 1)$)

2.65 $(x + 1)^5 = C(5, 0)x^5 + C(5, 1)x^4 + C(5, 2)x^3 + C(5, 3)x^2$
$\qquad\qquad + C(5, 4)x + C(5, 5)$
$\qquad\quad = x^5 + 5x^4 + 10x^3 + 10x^2 + 5x + 1$

2.66 $C(7, 4)x^3y^4$

CHAPTER 3

3.6 (a) $(3, 2) \in \rho$ \qquad (b) $(2, 4), (2, 6) \in \rho$ \qquad (c) $(3, 4), (5, 6) \in \rho$
$\quad$ (d) $(2, 1), (5, 2) \in \rho$

3.7 a subset of $S_1 \times S_2 \times \cdots \times S_n$

3.8 $\rho + \sigma = \rho \cup \sigma$
$\rho \cdot \sigma = \rho \cap \sigma$
ρ' (here $'$ denotes an operation on a binary relation) $= \rho'$ (here $'$ denotes the set operation of complementation)

3.9 (a) $x(\rho + \sigma)y \leftrightarrow x \leq y$
(b) $x\rho'y \leftrightarrow x \neq y$
(c) $x\sigma'y \leftrightarrow x \geq y$
(d) $\rho \cdot \sigma = \emptyset$

3.14 (a) reflexive, symmetric, transitive
(b) reflexive, antisymmetric, transitive
(c) reflexive, symmetric, transitive
(d) antisymmetric
(e) reflexive, symmetric, antisymmetric, transitive
(f) antisymmetric (this property is tricky—recall the truth table for implication), transitive
(g) reflexive, symmetric, transitive
(h) reflexive, symmetric, transitive

3.16 (a) (1, 1), (1, 2), (2, 2), (1, 3), (3, 3), (1, 6), (6, 6), (1, 12), (12, 12), (1, 18), (18, 18), (2, 6), (2, 12), (2, 18), (3, 6), (3, 12), (3, 18), (6, 12), (6, 18),
(b) 1, 2, 3
(c) 2, 3

3.18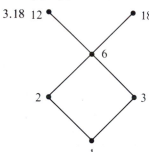

3.19 $y \in S$ is a greatest element if $x \leq y$ for all $x \in S$.
$y \in S$ is a maximal element if there is no $x \in S$ with $y < x$.

3.21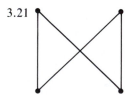

3.25 Let $q \in [x]$. Then $x \rho q$. Because $x \rho z$, by symmetry, $z \rho x$. By transitivity, $z \rho x$ together with $x \rho q$ gives $z \rho q$. Therefore, $q \in [z]$.

3.26 For any $x \in S$, x is in the same subset as itself, so $x \rho x$. If $x \rho y$, x is in the same subset as y, so y is in the same subset as x, or $y \rho x$. If $x \rho y$ and $y \rho z$, then x is in the same subset as y and y is in the same subset as z, so x is in the same subset as z, or $x \rho z$.

3.29 (a) The equivalence classes are sets consisting of lines in the plane with the same slope.

(b) $[n] = \{n\}$; the equivalence classes are all of the singleton sets of elements of $\mathbb{N}$.

(c) $[1] = [2] = \{1, 2\}, [3] = \{3\}$

3.32 $x \equiv_n y$ if $x - y$ is an integral multiple of n.

3.33 $[0] = \{\ldots, -15, -10, -5, 0, 5, 10, 15, \ldots\}$
$[1] = \{\ldots, -14, -9, -4, 1, 6, 11, 16, \ldots\}$
$[2] = \{\ldots, -13, -8, -3, 2, 7, 12, 17, \ldots\}$
$[3] = \{\ldots, -12, -7, -2, 3, 8, 13, 18, \ldots\}$
$[4] = \{\ldots, -11, -6, -1, 4, 9, 14, 19, \ldots\}$

3.35 (a) not a function; $2 \in S$ has two values associated with it   (b) function

(c) not a function; for values 0, 1, 2, 3 of the domain, the corresponding $h(x)$ values fall outside the codomain

(d) not a function; not every member of S has a Social Security number

(e) function (not every value in the codomain need be used)

(f) function   (g) function

(h) not a function; $5 \in N$ has two values associated with it

3.36 16, ± 3

3.42 (b), (f), (g)

3.45 (e), (g)

3.49 one possibility: $\{(0, 0), (1, 1), (-1, 2), (2, 3), (-2, 4), (3, 5), (-3, 6), \ldots\}$

3.53 $(g \circ f)(x) = g(f(x)) = g((2.3)^2) = g(5.29) = 5$
$(f \circ g)(x) = f(g(x)) = f(2) = 2^2 = 4$

3.54 Let $(g \circ f)(s_1) = (g \circ f)(s_2)$. Then $g(f(s_1)) = g(f(s_2))$ and because g is injective, $f(s_1) = f(s_2)$. Because f is injective, $s_1 = s_2$.

3.57 (a) $(1, 4, 5) = (4, 5, 1) = (5, 1, 4)$ (b) $\begin{pmatrix} 1 & 2 & 3 & 4 & 5 \\ 1 & 4 & 2 & 5 & 3 \end{pmatrix}$

3.58 (a) $g \circ f = (1, 3, 5, 2, 4) = (3, 5, 2, 4, 1) = \cdots$

(b) $g \circ f = \begin{pmatrix} 1 & 2 & 3 & 4 & 5 \\ 4 & 2 & 5 & 1 & 3 \end{pmatrix}$

3.59 $(1, 2, 4) \circ (3, 5)$ or $(3, 5) \circ (1, 2, 4)$.

3.60 Let $t \in T$. Then $(f \circ g)(t) = f(g(t)) = f(s) = t$.

3.63 $f^{-1} : \mathbb{R} \to \mathbb{R}, f^{-1}(x) = (x - 4)/3$

3.67 1, -7, -6

3.73 $2\mathbf{A} + \mathbf{B} = \begin{bmatrix} 6 & 14 \\ 3 & 10 \\ 9 & 16 \end{bmatrix}$

3.76 $\mathbf{A} \cdot \mathbf{B} = \begin{bmatrix} 15 & 22 \\ 12 & 28 \end{bmatrix}$ $\mathbf{B} \cdot \mathbf{A} = \begin{bmatrix} 39 & 0 \\ 27 & 4 \end{bmatrix}$

3.78 $\mathbf{I} \cdot \mathbf{A} = \begin{bmatrix} 1(a_{11}) + 0(a_{21}) & 1(a_{12}) + 0(a_{22}) \\ 0(a_{11}) + 1(a_{21}) & 0(a_{12}) + 1(a_{22}) \end{bmatrix} = \begin{bmatrix} a_{11} & a_{12} \\ a_{21} & a_{22} \end{bmatrix} = \mathbf{A}$

Similarly, $\mathbf{A} \cdot \mathbf{I} = \mathbf{A}$.

CHAPTER 4

4.4 the middle one

4.5 one possible picture:

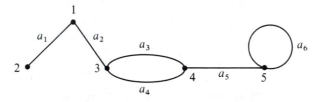

4.6 (a) yes   (b) no   (c) yes   (d) 3, a_5, 5, a_6, 6; 3, a_3, 4, a_4, 5, a_6, 6
(e) 3, a_3, 4, a_4, 5, a_5, 3   (f) a_3, a_4, or a_5 (g) a_1, a_2, a_6, or a_7

4.7 (a) 1   (b) 2   (c) 4   (d) 2

4.11(a)

(b)

(c)

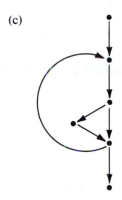

4.13

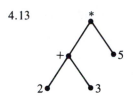

4.15 $\mathbf{A} = \begin{bmatrix} 1 & 1 & 0 & 1 \\ 1 & 0 & 1 & 0 \\ 0 & 1 & 0 & 2 \\ 1 & 0 & 2 & 0 \end{bmatrix}$

4.16 $\mathbf{A}^2 = \begin{bmatrix} 3 & 1 & 3 & 1 \\ 1 & 2 & 0 & 3 \\ 3 & 0 & 5 & 0 \\ 1 & 3 & 0 & 5 \end{bmatrix}$

4.17 (a) number of possible paths of length n from n_i to n_j (b) induction

4.19 $\mathbf{R} = \begin{bmatrix} 3 & 4 & 3 & 2 \\ 0 & 2 & 3 & 2 \\ 0 & 2 & 2 & 1 \\ 0 & 1 & 2 & 1 \end{bmatrix}$

4.21

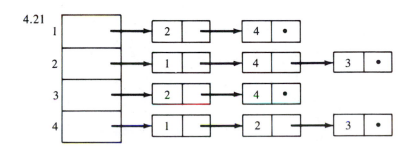

4.25

	Left child	Right child
1	0	2
2	3	4
3	0	5
4	0	0
5	0	0

4.26 There are two distinct nodes, 3 and 4, with 3 ρ 4 and 4 ρ 3

4.28 (a) no   (b) yes

4.31 (a) no, four odd nodes   (b) yes, no odd nodes

4.32 no

4.34

$$\begin{array}{c} \\ A \\ B \\ C \\ D \end{array} \begin{array}{cccc} A & B & C & D \\ \begin{bmatrix} 0 & 2 & 2 & 1 \\ 2 & 0 & 0 & 1 \\ 2 & 0 & 0 & 1 \\ 1 & 1 & 1 & 0 \end{bmatrix} \end{array}$$

After row C, $odd = 3$, the loop terminates, and there is no path.

4.37 $n^2 + n + 3 \leq 1(3n^2 + 4)$ for $n \geq 1$ and $3n^2 + 4 \leq 3(n^2 + n + 3)$ for $n \geq 1$

4.38 (a) no   (b) yes

4.40 $IN = \{x\}$

	1	2	3	y
d:	1	∞	4	∞
s:	x	x	x	x

$p = 1$
$IN = \{x, 1\}$

	1	2	3	y
d:	1	4	2	6
s:	x	1	1	1

$p = 3$
$IN = \{x, 1, 3\}$

	1	2	3	y
d:	1	4	2	5
s:	x	1	1	3

$$p = 2$$
$$IN = \{x, 1, 3, 2\}$$

	1	2	3	y
d:	1	4	2	5
s:	x	1	1	3

$$p = y$$
$$IN = \{x, 1, 3, 2, y\}$$

	1	2	3	y
d:	1	4	2	5
s:	x	1	1	3

path: $x, 1, 3, y$ length: 5

4.43

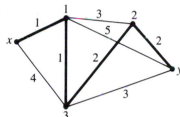

4.45 a e d b c i f g h l k m j

4.47 a e f d i b c g h j k m l

4.51 a b e f c d g i h
      e b f a c i g d h
      e f b c i g h d a

4.53

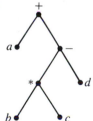

prefix notation: $+$ a $-$ $*$ b c d
postfix notation: a b c $*$ d $-$ $+$

CHAPTER 5

5.1 1a. $x + y = y + x$
     2a. $(x + y) + z = x + (y + z)$
     3a. $x + (y \cdot z) = (x + y) \cdot (x + z)$
     4a. $x + 0 = x$
     5a. $x + x' = 1$

     1b. $x \cdot y = y \cdot x$
     2b. $(x \cdot y) \cdot z = x \cdot (y \cdot z)$
     3b. $x \cdot (y + z) = x \cdot y + x \cdot z$
     4b. $x \cdot 1 = x$
     5b. $x \cdot x' = 0$

5.6 (a) $A \vee A = A$ (b) $A \cup A = A$

5.7 (a) $x + 1 = x + (x + x')$ (5a, complements)

$\qquad\qquad = (x + x) + x'$ (2a, associative property)

$\qquad\qquad = x + x'$ (idempotent property)

$\qquad\qquad = 1$ (5a, complements)

    (b) $x \cdot 0 = 0$

5.9 $0 + 1 = 1$ (Practice 5.7a)

$\quad\; 0 \cdot 1 = 1 \cdot 0$ (1b)

$\qquad\quad = 0$ (Practice 5.7b)

    Therefore $1 = 0'$ by Theorem 5.8.

5.10 Let

$$\begin{bmatrix} a_{11} & 0 \\ 0 & a_{22} \end{bmatrix} \quad \text{and} \quad \begin{bmatrix} b_{11} & 0 \\ 0 & b_{22} \end{bmatrix}$$

be elements of $M_2^D(\mathbb{Z})$. Then

$$\begin{bmatrix} a_{11} & 0 \\ 0 & a_{22} \end{bmatrix}\begin{bmatrix} b_{11} & 0 \\ 0 & b_{22} \end{bmatrix} = \begin{bmatrix} a_{11}b_{11} & 0 \\ 0 & a_{22}b_{22} \end{bmatrix} \in M_2^D(\mathbb{Z})$$

so the product of two members of $M_2^D(\mathbb{Z})$ is a unique member of $M_2^D(\mathbb{Z})$.

5.11 Let

$$\begin{bmatrix} a_{11} & 0 \\ 0 & a_{22} \end{bmatrix} \quad \text{and} \quad \begin{bmatrix} b_{11} & 0 \\ 0 & b_{22} \end{bmatrix}$$

be members of $M_2^D(\mathbb{Z})$. Then

$$\begin{bmatrix} a_{11} & 0 \\ 0 & a_{22} \end{bmatrix}\begin{bmatrix} b_{11} & 0 \\ 0 & b_{22} \end{bmatrix} = \begin{bmatrix} a_{11}b_{11} & 0 \\ 0 & a_{22}b_{22} \end{bmatrix}$$

$$= \begin{bmatrix} b_{11}a_{11} & 0 \\ 0 & b_{22}a_{22} \end{bmatrix} = \begin{bmatrix} b_{11} & 0 \\ 0 & b_{22} \end{bmatrix}\begin{bmatrix} a_{11} & 0 \\ 0 & a_{22} \end{bmatrix}$$

5.12 0

5.13 Yes; for any real numbers x and y, $x + y = y + x$

5.14 $\mathbf{X} \cdot \mathbf{X}^{-1} = \mathbf{X}^{-1} \cdot \mathbf{X} = \mathbf{I}$

5.16 (a) Yes; the product of two positive real numbers is a positive real number.

    (b) Yes; for all $x, y, z \in \mathbb{R}^+$, $(x \cdot y) \cdot z = x \cdot (y \cdot z)$.

    (c) Yes; for all $x, y \in \mathbb{R}^+$, $x \cdot y = y \cdot x$.

    (d) If i denotes an identity element, then for any $x \in \mathbb{R}^+$, $x \cdot i = i \cdot x = x$; 1 is an identity element.

    (e) For each $x \in \mathbb{R}^+$, there is an element $x^{-1} \in \mathbb{R}^+$ such that $x \cdot x^{-1} = x^{-1} \cdot x = 1$; the inverse of x is $1/x$.

5.17 Answer (e) will change. The element $0 \in \mathbb{R}$ has no inverse, since there is no real number y such that $0 \cdot y = y \cdot 0 = 1$.

5.21 (a) $f(-4 + 7) = f(3) = 2^3$; top and right

    (b) $f(-4) \cdot f(7) = 2^{-4} \cdot 2^7 = 2^3$; left and bottom

5.22 (a) $f(x \cdot y) = f(x) * f(y)$ (b) $f(x') = [f(x)]''$

5.24 (b)

$$B \times B \xrightarrow{+} B$$
$$f \times f \downarrow \qquad \downarrow f$$
$$b \times b \xrightarrow{\&} b$$

(c)

$$B \times B \xrightarrow{\cdot} B$$
$$f \times f \downarrow \qquad \downarrow f$$
$$b \times b \xrightarrow{*} b$$

(d)

$$B \xrightarrow{'} B$$
$$f \downarrow \qquad \downarrow f$$
$$b \xrightarrow{''} b$$

5.25 (a) $f(0 + a) = f(a) = \{1\} = \varnothing \cup \{1\} = f(0) \cup f(a)$
(b) $f(a + a') = f(1) = \{1, 2\} = \{1\} \cup \{2\} = f(a) \cup f(a')$
(c) $f(a \cdot a') = f(0) = \varnothing = \{1\} \cap \{2\} = f(a) \cap f(a')$
(d) $f(1') = f(0) = \varnothing = \{1, 2\}' = [f(1)]'$

5.28 $f\left(\begin{bmatrix} a_{11} & 0 \\ 0 & a_{22} \end{bmatrix} \cdot \begin{bmatrix} b_{11} & 0 \\ 0 & b_{22} \end{bmatrix} \right) = f\left(\begin{bmatrix} a_{11}b_{11} & 0 \\ 0 & a_{22}b_{22} \end{bmatrix} \right)$

$= a_{11}b_{11} = f\left(\begin{bmatrix} a_{11} & 0 \\ 0 & a_{22} \end{bmatrix} \right) \cdot f\left(\begin{bmatrix} b_{11} & 0 \\ 0 & b_{22} \end{bmatrix} \right)$

CHAPTER 6

6.5 (a) 2^n (b) $2^4 = 16$ (c) 2^{2n}

6.8 (a)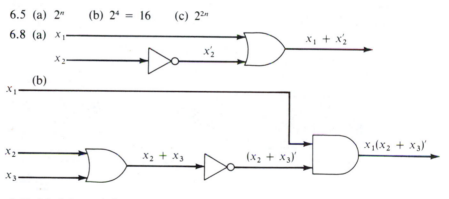

(b)

6.10 (a) $(x_1' + x_2)x_3'$
(b)

x_1	x_2	x_3	$(x_1' + x_2)x_3'$
1	1	1	0
1	1	0	1
1	0	1	0
1	0	0	0
0	1	1	0
0	1	0	1
0	0	1	0
0	0	0	1

6.13 (a) $x_1x_2x_3 + x_1x_2'x_3 + x_1x_2'x_3' + x_1'x_2'x_3 + x_1'x_2'x_3'$

(b)

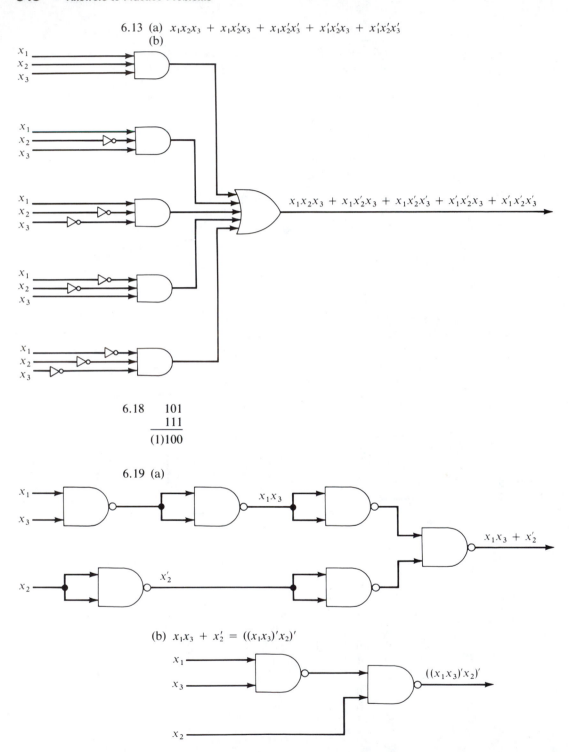

6.18   101
        111
     ‾‾‾‾‾‾‾
      (1)100

6.19 (a)

(b) $x_1x_3 + x_2' = ((x_1x_3)'x_2)'$

6.21 (a)

x_1	x_2	$f(x_1, x_2)$
1	1	0
1	0	1
0	1	1
0	0	0

(b) One possibility is the canonical sum-of-products form, $x_1x_2' + x_1'x_2$.

(c)

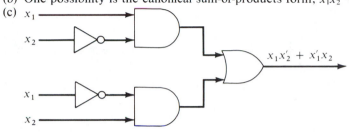

6.23 (a)

$$x_1x_2 + x_1'x_2 = x_2x_1 + x_2x_1'$$
$$= x_2(x_1 + x_1')$$
$$= x_2 \cdot 1$$
$$= x_2$$

(b)

$$x_1 + x_1'x_2 = x_1 \cdot 1 + x_1'x_2$$
$$= x_1(1 + x_2) + x_1'x_2 \quad \text{(See Practice 5.7)}$$
$$= x_1 + x_1x_2 + x_1'x_2$$
$$= x_1 + x_2(x_1 + x_1')$$
$$= x_1 + x_2 \cdot 1 = x_1 + x_2$$

6.25

	x_1	x_1'
x_2		1
x_2'		1

The reduced expression is x_1'.

6.28 x_1x_3 (4 squares) and $x_1'x_2x_3'$ (2 squares)

6.35 $x_1x_2'x_4 + x_1x_3'x_4 + x_2'x_3'$ (See following figure.)

	x_1x_2	x_1x_2'	$x_1'x_2'$	$x_1'x_2$
x_3x_4		1		
x_3x_4'				
$x_3'x_4'$		1	1	
$x_3'x_4$	1	1	1	

6.38 The reduction table is shown in the following figure.

x_1	x_2	x_3			x_1	x_2	x_3
1	1	1	1		1	1	–
1	1	0	1,2		1	–	0
1	0	0	2,3		–	0	0
0	0	1	4				
0	0	0	3,4		0	0	–

The comparison table is shown in the following figure.

	111	110	100	001	000
11–	✓	✓			
1–0		✓	✓		
–00			✓		✓
00–				✓	✓

Essential terms are 11– and 00–. Either 1–0 or –00 can be used as the third reduced term. The minimal sum-of-products form is

$$x_1 x_2 + x_1' x_2' + x_1 x_3' \qquad \text{or} \qquad x_1 x_2 + x_1' x_2' + x_2' x_3'$$

CHAPTER 7

7.7 All operations are associative binary operations on their respective sets.

7.8 (a) all   (b) all but $[\mathbb{R}^+, +]$; 0, 1, 0, 1, $\begin{bmatrix} 0 & 0 \\ 0 & 0 \end{bmatrix}$

7.11 $[\mathbb{R}, +]$, $[\mathbb{R}^+, \cdot]$, $[\mathbb{Z}, +]$, $[M_2(\mathbb{Z}), +]$

7.13 (a) $f(x) + g(x) = g(x) + f(x)$
     $(f(x) + g(x)) + h(x) = f(x) + (g(x) + h(x))$
     (b) the zero polynomial, 0
     (c) $-7x^4 + 2x^3 - 4$

7.15 (a)

$+_5$	0	1	2	3	4
0	0	1	2	3	4
1	1	2	3	4	0
2	2	3	4	0	1
3	3	4	0	1	2
4	4	0	1	2	3

$\cdot_5$	0	1	2	3	4
0	0	0	0	0	0
1	0	1	2	3	4
2	0	2	4	1	3
3	0	3	1	4	2
4	0	4	3	2	1

   (b) 0; 1   (c) 3   (d) all except 0

7.16 (a)

$\cdot_6$	0	1	2	3	4	5
0	0	0	0	0	0	0
1	0	1	2	3	4	5
2	0	2	4	0	2	4
3	0	3	0	3	0	3
4	0	4	2	0	4	2
5	0	5	4	3	2	1

(b) 1 and 5

7.18 Let $f, g, h \in S$. Then for any $x \in A$, $((f \circ g) \circ h)(x) = (f \circ g)(h(x)) = f(g(h(x)))$ and $(f \circ (g \circ h))(x) = f((g \circ h)(x)) = f(g(h(x)))$. Hence, $(f \circ g) \circ h = f \circ (g \circ h)$.

7.20 (a)

$\circ$	α_1	α_2	α_3	α_4	α_5	α_6
α_1	α_1	α_2	α_3	α_4	α_5	α_6
α_2	α_2	α_1	α_6	α_5	α_4	α_3
α_3	α_3	α_5	α_1	α_6	α_2	α_4
α_4	α_4	α_6	α_5	α_1	α_3	α_2
α_5	α_5	α_3	α_4	α_2	α_6	α_1
α_6	α_6	α_4	α_2	α_3	α_1	α_5

(b) no

7.22 (a) no   (b) no

7.24 (a) $(10, 9)$ (b) $(1, 1)$ (c) $(3, 2)$

7.25 (a) For $(g, h) \in G \times H$, $(1_G, 1_H) \cdot (g, h) = (1_G \circ g, 1_H * h) = (g, h)$, and $(g, h) \cdot (1_G, 1_H) = (g \circ 1_G, h * 1_H) = (g, h)$.
(b) $(g, h)^{-1} = (g^{-1}, h^{-1})$ because $(g^{-1}, h^{-1}) \cdot (g, h) = (g^{-1} \circ g, h^{-1} * h) = (1_G, 1_H)$, and $(g, h) \cdot (g^{-1}, h^{-1}) = (g \circ g^{-1}, h * h^{-1}) = (1_G, 1_H)$.
(c) Yes; $(g_1, h_1) \cdot (g_2, h_2) = (g_1 \circ g_2, h_1 * h_2) = (g_2 \circ g_1, h_2 * h_1) = (g_2, h_2) \cdot (g_1, h_1)$.

7.26 $i_1 = i_1 i_2$ because i_2 is an identity
$i_1 i_2 = i_2$ because i_1 is an identity

7.28 Let y and z both be inverses of x. Let i be the identity. Then $y = y \cdot i = y \cdot (x \cdot z) = (y \cdot x) \cdot z = i \cdot z = z$.

7.31 $7^{-1} = 5, 3^{-1} = 9$; so $10^{-1} = (7 +_{12} 3)^{-1} = 3^{-1} +_{12} 7^{-1} = 9 +_{12} 5 = 2$

7.33 $z \cdot x = z \cdot y$ implies
$$z^{-1} \cdot (z \cdot x) = z^{-1} \cdot (z \cdot y)$$
$$(z^{-1} \cdot z) \cdot x = (z^{-1} \cdot z) \cdot y$$
$$i \cdot x = i \cdot y$$
$$x = y$$

7.37 $x = 1 +_8 (3)^{-1} = 1 +_8 5 = 6$

7.38

∘	1	a	b	c	d
1	1	a	b	c	d
a	a	b	c	d	1
b	b	c	d	1	a
c	c	d	1	a	b
d	d	1	a	b	c

7.39 (a) $[\mathbb{Z}_{18}, +_{18}]$ (b) $[S_3, \circ]$

7.43 (a) and (c)

7.44 Yes. For $x, y \in A, x \cdot y = y \cdot x$ because x and y belong to S. Commutativity is inherited.

7.48 (a) and (b); note that (c) is not a submonoid by our definition because the set does not contain (1, 1), but (0, 1) is an identity for this set under componentwise multiplication.

7.50 (a) A is closed under $\cdot$.
(b) $i \in A$
(c) Every $x \in A$ has an inverse element in A.

7.53 (a) Closure holds; $0 \in \{0, 2, 4, 6\}$; $0^{-1} = 0, 4^{-1} = 4$, and 2 and 6 are inverses of each other.
(b) Closure holds; $1 \in \{1, 2, 4\}$; $1^{-1} = 1$, and 2 and 4 are inverses of each other.

7.54 $[\{\alpha_1, \alpha_5, \alpha_6\}, \circ], [\{\alpha_1\alpha_2\}, \circ], [\{\alpha_1, \alpha_3\}, \circ], [\{\alpha_1, \alpha_4\}, \circ]$

7.55 To show f is one-to-one, let α and β belong to A_n and suppose $f(\alpha) = f(\beta)$. Then $\alpha \circ (1, 2) = \beta \circ (1, 2)$. By the cancellation law available in the group S_n, $\alpha = \beta$. To show f is onto, let $\gamma \in O_n$. Then $\gamma \circ (1, 2) \in A_n$ and $f(\gamma \circ (1, 2)) = \gamma \circ (1, 2) \circ (1, 2) = \gamma$.

7.57 For $nz_1, nz_2 \in n\mathbb{Z}$, $nz_1 + nz_2 = n(z_1 + z_2) \in n\mathbb{Z}$, so closure holds; $0 = n \cdot 0 \in n\mathbb{Z}$; for $nz \in n\mathbb{Z}$, $-nz = n(-z) \in n\mathbb{Z}$.

7.60 $f(x) + f(y)$

7.64 (a) Yes; for $x, y \in \mathbb{N}$, $f(x + y) = 2(x + y) = 2x + 2y = f(x) + f(y)$.
(b) No; for $x, y \in \mathbb{N}$, $f(x + y) = 2(x + y) = 2x + 2y$, but $f(x) \cdot f(y) = 2x \cdot 2y = 4xy$. These are not always equal.
(c) Yes; for

$$\begin{bmatrix} a & b \\ c & d \end{bmatrix} \quad \text{and} \quad \begin{bmatrix} e & f \\ g & h \end{bmatrix}$$

in $M_2(\mathbb{Z})$,

$$f\left(\begin{bmatrix} a & b \\ c & d \end{bmatrix} + \begin{bmatrix} e & f \\ g & h \end{bmatrix}\right) = f\left(\begin{bmatrix} a+e & b+f \\ c+g & d+h \end{bmatrix}\right) = a + e$$

$$= f\left(\begin{bmatrix} a & b \\ c & d \end{bmatrix}\right) + f\left(\begin{bmatrix} e & f \\ g & h \end{bmatrix}\right)$$

(d) Yes; for $a_2x^2 + a_1x + a_0$ and $b_2x^2 + b_1x + b_0$ in $\mathbb{N}^2[x]$, $f(a_2x^2 + a_1x + a_0 + b_2x^2 + b_1x + b_0) = f((a_2 + b_2)x^2 + (a_1 + b_1)x + (a_0 + b_0)) = a_2 + b_2 = f(a_2x^2 + a_1x + a_0) + f(b_2x^2 + b_1x + b_0)$.

7.65 Let $f(x) \in f(S)$. Then $f(x) + f(i_S) = f(x \cdot i_S) = f(x)$ and $f(i_S) + f(x) = f(i_S \cdot x) = f(x)$.

7.66 $f(x^{-1}) + f(x) = f(x^{-1} \cdot x) = f(i_S)$
$f(x) + f(x^{-1}) = f(x \cdot x^{-1}) = f(i_S)$

7.68 Let $f(x)$ and $f(y)$ belong to $f(S)$. Then $f(x) + f(y) = f(x \cdot y) = f(y \cdot x) = f(y) + f(x)$.

7.72 Clearly f is onto. f is also one-to-one: let $f(x) = f(y)$, then $5x = 5y$ and $x = y$. f is a homomorphism: for $x, y \in \mathbb{Z}$, $f(x + y) = 5(x + y) = 5x + 5y = f(x) + f(y)$.

7.73 (a) Composition of bijections is a bijection, and for $x, y \in S$, $(g \circ f)(x \cdot y)$
$= g(f(x \cdot y)) = g(f(x) + f(y)) = (g \circ f)(x) * (g \circ f)(y)$.
(b) $S \simeq S$ by the identity mapping. If f is an isomorphism from S to T, then f^{-1} is an isomorphism from T to S. If $S \simeq T$ and $T \simeq V$, then by part (a), $S \simeq V$.

7.75 To show that α_g is an onto function, let $y \in G$. Then $g^{-1} \cdot y$ belongs to G and $\alpha_g(g^{-1} \cdot y) = g(g^{-1} \cdot y) = (g \cdot g^{-1})y = y$. To show that α_g is one-to-one, let $\alpha_g(x) = \alpha_g(y)$. Then $g \cdot x = g \cdot y$, and by cancellation, $x = y$.

7.76 (a) For $\alpha_g \in P$, $\alpha_g \circ \alpha_1 = \alpha_{g \cdot 1} = \alpha_g$ and $\alpha_1 \circ \alpha_g = \alpha_{1 \cdot g} = \alpha_g$.
(b) $\alpha_g \circ \alpha_{g^{-1}} = \alpha_{g \cdot g^{-1}} = \alpha_1$, and $\alpha_{g^{-1}} \circ \alpha_g = \alpha_{g \cdot g^{-1}} = \alpha_1$.

7.77 (a) Let $f(g) = f(h)$. Then $\alpha_g = \alpha_h$ and, in particular, $\alpha_g(1) = \alpha_h(1)$, or $g \cdot 1 = h \cdot 1$ and $g = h$.
(b) For $g, h, \in G$, $f(g \cdot h) = \alpha_{g \cdot h} = \alpha_g \circ \alpha_h = f(g) \circ f(h)$.

7.79 Reflexive: $f(x) = f(x)$. Symmetric: if $f(x) = f(y)$, then $f(y) = f(x)$. Transitive: if $f(x) = f(y)$ and $f(y) = f(z)$, then $f(x) = f(z)$.

7.81 $f(\mathbb{Z}) = \mathbb{Z}_3 = \{0, 1, 2\}$. The distinct members of $\mathbb{Z}/f$ are

$$[0] = \{. . . , -9, -6, -3, 0, 3, 6, 9, . . .\} = [-12]$$
$$[1] = \{. . . , -8, -5, -2, 1, 4, 7, 10, . . .\} = [16]$$
$$[2] = \{. . . , -7, -4, -1, 2, 5, 8, 11, . . .\} = [8]$$

7.82 Let $[x], [y], [z], \in S/f$. Then $([x] * [y]) * [z] = [x \cdot y] * [z] = [(x \cdot y) \cdot z]$ and $[x] * ([y] * [z]) = [x] * [y \cdot z] = [x \cdot (y \cdot z)]$, but $(x \cdot y) \cdot z = x \cdot (y \cdot z)$.

7.84 Computations (b) and (c) are correct.

7.85 $g([x]) = f(x)$

7.86 $g([x] * [y]) = g([x \cdot y]) = f(x \cdot y) = f(x) + f(y) = g[x] + g[y]$

7.89 For $[x] \in S/f$, $[x] * [i_S] = [x \cdot i_S] = [x]$ and $[i_S] * [x] = [i_S \cdot x] = [x]$.

7.91 For $[x] \in S/f$, $[x^{-1}] \in S/f$ and $[x] * [x^{-1}] = [x \cdot x^{-1}] = [i_S]$; similarly, $[x^{-1}] * [x] = [i_S]$.

7.94 (a) $g([2] * [3]) = g([2 + 3]) = g([5]) = g([1]) = 2$
(b) $g([2]) +_8 g([3]) = 4 +_8 6 = 2$

7.97 (a) $K = \{. . . , -9, -6, -3, 0, 3, 6, 9, . . .\} = 3\mathbb{Z}$
(b) $K = \{0, 2, 4, 6, 8, 10\}$

7.98 $x \in K$ implies $f(x) = i_H$; $y \in K$ implies $f(y) = i_H$. Then $f(x \cdot y^{-1}) = f(x) \cdot f(y^{-1}) = f(x) \cdot (f(y))^{-1} = i_H \cdot i_H^{-1} = i_H \cdot i_H = i_H$, and $x \cdot y^{-1} \in K$.

7.100 (a) $0 + K = 3\mathbb{Z} = \{. . . , -9, -6, -3, 0, 3, 6, 9, . . .\}$
$1 + K = 1 + 3\mathbb{Z} = \{. . . , -8, -5, -2, 1, 4, 7, 10, . . .\}$

$$2 + K = 2 + 3\mathbb{Z} = \{\ldots, -7, -4, -1, 2, 5, 8, 11, \ldots\}$$

(b) $0 + K = \{0, 2, 4, 6, 8, 10\}$, $1 + K = \{1, 3, 5, 7, 9, 11\}$

7.101 Let $y \in [x]$. Then $f(y) = f(x) = i_H \cdot f(x)$; $f(y) \cdot [f(x)]^{-1} = i_H$ or $f(y \cdot x^{-1})$ $= i_H$ and $y \cdot x^{-1} \in K$, or $y \cdot x^{-1} = k$ and $y = kx$. This proves that $[x] \subseteq Kx$. Let $kx \in Kx$. Then $f(kx) = f(k) \cdot f(x) = i_H \cdot f(x) = f(x)$, or $kx \in [x]$ and $Kx \subseteq [x]$. Therefore, $[x] = Kx$.

7.103 (a) $(7 + 3\mathbb{Z}) + (5 + 3\mathbb{Z}) = 12 + 3\mathbb{Z} = 3\mathbb{Z}$

(b) $-(4 + 3\mathbb{Z}) = 2 + 3\mathbb{Z}$ because $(4 + 3\mathbb{Z}) + (2 + 3\mathbb{Z}) = 0 + 3\mathbb{Z}$

7.106 (a) $6 + S = \{6 +_{12} 0, 6 +_{12} 4, 6 +_{12} 8\} = \{6, 10, 2\}$

(b) $(1, 2, 3)S = \{(1, 2, 3) \circ i, (1, 2, 3) \circ (2, 3)\} = \{(1, 2, 3), (1, 2)\}$
$(1, 3, 2)S = \{(1, 3, 2) \circ i, (1, 3, 2) \circ (2, 3)\} = \{(1, 3, 2), (1, 3)\}$
$S(1, 2, 3) = \{i \circ (1, 2, 3), (2, 3) \circ (1, 2, 3)\} = \{(1, 2, 3), (1, 3)\}$
$S(1, 3, 2) = \{i \circ (1, 3, 2), (2, 3) \circ (1, 3, 2)\} = \{(1, 3, 2), (1, 2)\}$

7.107 Reflexive: $x = x \cdot i_G$, $i_G \in S$. Symmetric: if $x \rho y$ then $y = xs$, so $x = ys^{-1}$, $s^{-1} \in S$, and $y \rho x$. Transitive: if $x \rho y$ and $y \rho z$, then $y = xs_1$ and $z = ys_2$, so $z = xs_1s_2$, $s_1s_2 \in S$, and $x \rho z$.

7.111 Associativity: $(xS \cdot yS) \cdot zS = (xyS) \cdot zS = (xy)zS = x(yz)S = xS \cdot (yzS)$ $= xS \cdot (yS \cdot zS)$. Identity: $i_GS = S$ because $xS \cdot i_GS = xS = i_GS \cdot xS$. Inverses: $(xS)^{-1} = x^{-1}S$ because $xS \cdot x^{-1}S = i_GS = x^{-1}S \cdot xS$.

7.115 f is clearly onto. For $x, y \in G$, $f(xy) = xyS = xSyS = f(x)f(y)$.

7.117 f is clearly onto since $s \in S$ has xs as a preimage. If $f(xs_1) = f(xs_2)$, then $s_1 = s_2$ and $xs_1 = xs_2$, so f is one-to-one.

7.120 (a) $|G| = 2 \cdot 3 \cdot 4 = 24$. Possible orders are the divisors of 24: 1, 2, 3, 4, 6, 8, 12, 24.

(b) Yes, because G is a finite commutative group.

CHAPTER 8

8.2 $H(X, Y) = 2$

8.3 The answer to this problem is so obvious it's hard to write anything, so we won't.

8.8 (a) (10110)   (b) (10110)

8.12 Each code word added to itself is 0, and any $X +_2 0 = X$.

(01111) $+_2$ (10101) $= 11010$
(01111) $+_2$ (11010) $= 10101$
(10101) $+_2$ (11010) $= 01111$

8.15 straightforward matrix multiplication

8.17 $00 \rightarrow 00000$
$01 \rightarrow 01111$
$11 \rightarrow 11010$

8.20 $0000 \rightarrow 0000000$ $1000 \rightarrow 1000110$
$0001 \rightarrow 0001101$ $1001 \rightarrow 1001011$
$0010 \rightarrow 0010111$ $1010 \rightarrow 1010001$
$0011 \rightarrow 0011010$ $1011 \rightarrow 1011100$
$0100 \rightarrow 0100011$ $1100 \rightarrow 1100101$

$$0101 \rightarrow 0101110 \quad 1101 \rightarrow 1101000$$
$$0110 \rightarrow 0110100 \quad 1110 \rightarrow 1110010$$
$$0111 \rightarrow 0111001 \quad 1111 \rightarrow 1111111$$

8.22 $00 \rightarrow 00000$
$ 01 \rightarrow 01111$
$ 11 \rightarrow 11010$

8.26 $[11011]\mathbf{H} = [001]$
$ [10100]\mathbf{H} = [001]$
$ [01110]\mathbf{H} = [001]$
$ [00001]\mathbf{H} = [001]$

8.29 The syndrome is 010, so it is decoded to $1000100 + 0000010 = 1000110$.

8.32 01100 or 10010

CHAPTER 9

9.3 000110

9.4

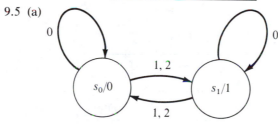

Present state	Next state		Output
	Present input		
	0	1	
s_0	s_0	s_3	0
s_1	s_0	s_2	1
s_2	s_3	s_3	1
s_3	s_1	s_3	2

9.5 (a)

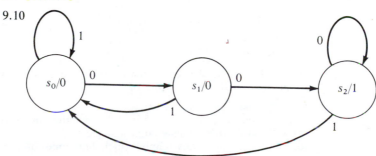

(b) 01011

9.6 (a) s_1 (b) s_1

9.7 11001011

9.10

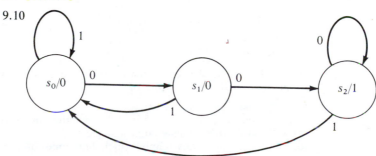

9.12 A has sent message 1, and that is the message that B is expecting, but the channel has lost the message.

9.16

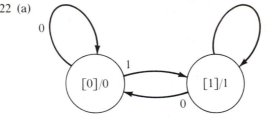

$$(2, 0) \xrightarrow{\text{Corresponding state}} (b, 0)$$

Next-state function in M | Next-state function in M'

$$0 \xrightarrow{\text{Corresponding state}} a$$

9.17 As α is processed, corresponding states will proceed to corresponding states, and corresponding states have the same output. (A formal proof would use induction on the length of the string α.)

9.19 bijection

9.20 $g^{-1}: \ S' \to S$ and g^{-1} is a bijection. For $s' \in S'$ and $i \in I$, let $s' = g(s)$. Then the equation $g(f_s(s, i)) = f'_s(g(s), i)$ can be rewritten as

$$f_s(g^{-1}(s'), i) = g^{-1}(f'_s(s', i)).$$

The equation $f_o(s) = f'_o(g(s))$ can be rewritten as

$$f_o(g^{-1}(s')) = f'_o(s').$$

Therefore, g^{-1} is a homomorphism.

9.22 (a)

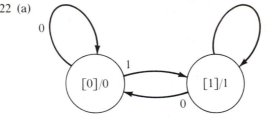

(b) $h([0]) = a$ $h([1]) = b$

9.25

Present state	Next state		Output
	Present input		
	0	1	
A	A	A	0
B	C	A	1
C	C	B	1

$g: \quad s_0 \to A$
$s_1 \to B$
$s_2 \to A$
$s_3 \to C$
$s_4 \to C$

9.27 A simulates B by the function

$$a \to 0$$
$$b \to 0$$

B cannot simulate A because there is no state in B for state 1 of A to map to; no state in B has an output of 1.

9.32 (a) set consisting of a single 0
(b) set consisting of any number of 0s (including none) followed by 10
(c) set consisting of a single 0 or a single 1
(d) set consisting of any number (including none) of pairs of 1s

9.35 (b) does not belong

9.36 (a) 0   (b) 0*10   (c) $0 \vee 1$ (d) (11)*

9.42

Present state	Next state		Output
	Present input		
	0	1	
$\{s_0\}$	$\{s_0\}$	$\{s_1\}$	0
$\{s_1\}$	$\{s_1\}$	$\{s_0, s_1\}$	1
$\{s_0, s_1\}$	$\{s_0, s_1\}$	$\{s_0, s_1\}$	1

CHAPTER 10

10.2 s_2, s_3

10.4 A state s produces the same output as itself for any input. If s_i produces the same output as s_j, then s_j produces the same output as s_i. Transitivity is equally clear.

10.5 Condition (1) is satisfied because all states in the same class have the same output strings for any input string, including the empty input string. To see that (2) is satisfied, assume s_i and s_j are equivalent states proceeding under the input symbol i to states s_i' and s_j' that are not equivalent. Then there is an input string α such that $f_o(s_i', \alpha) \neq f_o(s_j', \alpha)$. Thus, for the input string $i\alpha$, s_i and s_j produce different output strings, contradicting the equivalence of s_i and s_j.

10.12 (a) Equivalent states of M are $A = \{0, 1, 3\}$, $B = \{2\}$, and $C = \{4\}$. The reduced machine is

Present state	Next state		Output
	Present input		
	0	1	
A	B	A	1
B	C	A	0
C	A	A	0

(b) Equivalent states of M are $\{0\}$, $\{1\}$, $\{2\}$, and $\{3\}$. M is already minimal.

10.16 Clearly the network behaves properly.

10.17 First, write the state table:

Present state	Next state		Output
	Present input		
	0	1	
s_0	s_0	s_1	1
s_1	s_1	s_0	0

The states can be encoded by a single delay element as shown:

	d
s_0	0
s_1	1

The truth functions are

$x(t)$	$d(t)$	$y(t)$	$d(t+1)$
0	0	1	0
1	0	1	1
0	1	0	1
1	1	0	0

The canonical sum-of-products forms are

$$y(t) = d'$$
$$d(t+1) = xd' + x'd$$

and the sequential network is

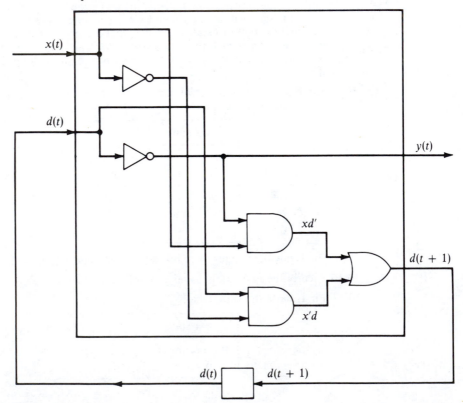

10.20

M_p	Input (0, 0)	(0, 1)	(1, 0)	(1, 1)	Output
(A, a)	(A, b)	(A, a)	(A, b)	(A, a)	$(1, 1)$
(A, b)	(A, a)	(A, b)	(A, a)	(A, b)	$(1, 1)$
(B, a)	(B, b)	(B, a)	(A, b)	(A, a)	$(0, 1)$
(B, b)	(B, a)	(B, b)	(A, a)	(A, b)	$(0, 1)$

10.22 (a) $g(f_p((A, c), 1)) = g(A, b) = 1$ (b) $f_S(g(A, c), 1) = f_S(2, 1) = 1$

10.25 (a) One example is:

M_1	Input 0	1	Output
$A = \{0, 1\}$	A	B	1
$B = \{2, 3\}$	B	A	0

M_2	Input 0	1	Output
$a = \{0, 3\}$	b	a	0
$b = \{1, 2\}$	a	b	1

(b) g: $(A, a) \to 0$ g': $(0, 0) \to 1$
        $(A, b) \to 1$ $(0, 1) \to 0$
        $(B, a) \to 3$ $(1, 0) \to 0$
        $(B, b) \to 2$ $(1, 1) \to 1$

(Other output functions g' and other output assignments can be used.)

10.27

M_s	Input 0	1	Output
(A, a)	(A, b)	(A, a)	(A, a)
(A, b)	(A, a)	(A, b)	(A, b)
(B, a)	(B, a)	(A, b)	(B, a)
(B, b)	(B, b)	(A, b)	(B, b)

10.29 Corresponding state in M_s for 4 is (C, b). Under input 1, $(C, b) \to (C, a)$ and $g(C, a) = 2$.

10.31

M_1	Input 0	1	Output
$A = \{0, 2\}$	B	A	A
$B = \{1, 3\}$	B	A	B

M_2	Input				Output
	$(A, 0)$	$(A, 1)$	$(B, 0)$	$(B, 1)$	
$a = \{0, 1\}$	b	a	a	a	a
$b = \{2, 3\}$	b	b	b	a	b

g: $(A, a) \rightarrow 0$ g': $(A, a) \rightarrow 1$
$\quad (A, b) \rightarrow 2$ $\qquad (A, b) \rightarrow 0$
$\quad (B, a) \rightarrow 1$ $\qquad (B, a) \rightarrow 1$
$\quad (B, b) \rightarrow 3$ $\qquad (B, b) \rightarrow 0$

10.40 (a) $\mathbb{Z}_{12}/\{0, 3, 6, 9\} \simeq \mathbb{Z}_3$
$\quad \{0, 3, 6, 9\}/\{0, 6\} \simeq \mathbb{Z}_2$
$\quad \{0, 6\}/\{0\} \simeq \mathbb{Z}_2$

(b) $\mathbb{Z}_{12}/\{0, 2, 4, 6, 8, 10\} \simeq \mathbb{Z}_2$
$\quad \{0, 2, 4, 6, 8, 10\}/\{0, 4, 8\} \simeq \mathbb{Z}_2$
$\quad \{0, 4, 8\}/\{0\} \simeq \mathbb{Z}_3$
and
$\quad \mathbb{Z}_{12}/\{0, 2, 4, 6, 8, 10\} \simeq \mathbb{Z}_2$
$\quad \{0, 2, 4, 6, 8, 10\}/\{0, 6\} \simeq \mathbb{Z}_3$
$\quad \{0, 6\}/\{0\} \simeq \mathbb{Z}_2$

10.42 (a)

	i	0	1	2
		10		
	11	00		
S	λ	0	1	01
0	0	0	1	1
1	1	0	0	1

(b)

$\circ$	i	0	1	2
i	i	0	1	2
0	0	0	2	2
1	1	0	i	2
2	2	0	0	2

10.44 g: $i \rightarrow 0$
$\quad 0 \rightarrow 0$
$\quad 1 \rightarrow 1$
$\quad 2 \rightarrow 1$

10.48 $[g^{*'}g] = [g^{*'}][g] = [g^*][g] = [g^*g] = [(g^*g)']$. Thus, $g^{*'}g = s_1(g^*g)'$ for some $s_1 \in S$ and $(g^{*'}g)((g^*g)')^{-1} = s_1 \in S$.

CHAPTER 11

11.3 (a) $\ldots b \; 0 \; 0 \; b \ldots$
(b) The machine cycles endlessly over the two nonblank tape squares.
(c) The machine changes the two nonblank squares to 0 1 and then moves endlessly to the right.

11.6 (a) $b \; X \; X \; 1 \; X \; X \; b$ halts without accepting
$\qquad\qquad \uparrow$
$\qquad\qquad 5$

(b) $b\ X\ X\ X\ X\ X\ b$ halts without accepting
$$\uparrow$$
$$2$$

(c) $b\ X\ X\ X\ 0\ X\ X\ b$ halts without accepting
$$\uparrow$$
$$2$$

11.7 State 3 is the only final state.

$(0, 1, 1, 0, R)$
$(0, 0, 0, 1, R)$
$(1, 0, 0, 2, R)$
$(1, 1, 1, 0, R)$
$(2, 0, 0, 2, R)$
$(2, 1, 1, 0, R)$
$(2, b, b, 3, R)$

11.8 Change $(2, 1, X, 3, L)$ to $(2, 1, X, 7, L)$ and add $(7, 1, X, 3, L)$.

11.12 One machine that works, together with a description of its actions:

$(0, 1, 1, 1, R)$ reads first 1
$(1, b, 1, 6, R)$ $n = 0$, changes to 1 and halts
$(1, 1, 1, 2, R)$ reads second 1
$(2, b, b, 6, R)$ $n = 1$, halts
$(2, 1, 1, 3, R)$ $n \geq 2$
$\left.\begin{array}{l}(3, 1, 1, 3, R)\\(3, b, b, 4, L)\end{array}\right\}$ finds right end of $\bar{n}$
$\left.\begin{array}{l}(4, 1, b, 5, L)\\(5, 1, b, 6, L)\end{array}\right\}$ erases two 1s from $\bar{n}$ and halts

11.23 For any Turing machine T we can effectively create a machine T^* that acts like T until T reaches a halting configuration, then erases the tape and halts. Then T^* halts with a blank tape when started on a tape containing α if and only if T halts when started on a tape containing α. Assume that there exists an algorithm P to solve this decision problem. We now have an algoriithm to solve the halting problem: given a (T, α) pair, create T^* and apply algorithm P.

11.26 $t(n) = 2n + 2$

11.28 straightforward computation

CHAPTER 12

12.7 $0S \Rightarrow 00S \Rightarrow 000S \Rightarrow 0000S \Rightarrow 00001$

12.9 $L = \{0^n 1 \mid n \geq 0\}$

12.11 For example:

(a) $G = (V, V_T, S, P)$ where $V = \{0, 1, S\}$, $V_T = \{0, 1\}$, and $P = \{S \rightarrow 1,$
$S \rightarrow 0S0\}$

(b) $G = (V, V_T, S, P)$ where $V = \{0, 1, S, M\}$, $V_T = \{0, 1\}$, and $P = \{S \rightarrow 0M0, M \rightarrow 0M0, M \rightarrow 1\}$

12.20 In G_1: $S \Rightarrow ABA \Rightarrow 00A \Rightarrow 0000A \Rightarrow 00000$
In G_2: $S \Rightarrow 00A \Rightarrow 0000A \Rightarrow 00000$
In G_3: $S \Rightarrow 0A \Rightarrow 00B \Rightarrow 000C \Rightarrow 0000B \Rightarrow 00000$

12.22

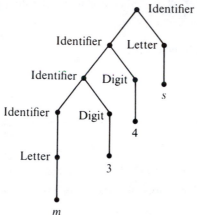

12.25 $G = (V, V_T, S, P)$ where $V = \{S, A, B, L, 0, 1\}$, $V_T = \{0, 1\}$, and P consists of

$S \rightarrow 1A$
$S \rightarrow 0C$
$A \rightarrow 0$
$A \rightarrow 0B$
$A \rightarrow 1C$
$B \rightarrow 1A$
$B \rightarrow 0C$
$C \rightarrow 0C$
$C \rightarrow 1C$
$L(G) = 10(10)^*$

12.27

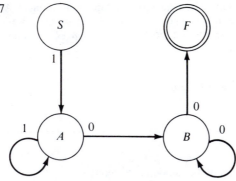

$L(G) = 11^*00^*0$

12.33 $W_0 = \{S\}$
$W_1 = \{S, \lambda, ABA\}$
$W_2 = \{S, \lambda, ABA, 00A, 0BA, AB0\}$
$W_3 = \{S, \lambda, ABA, 00A, 0BA, AB0, 000, 0B0\}$
$W_4 = W_3$
$0000 \notin L$

Answers to Selected Exercises

Section 1.1

1. (a), (c), (e), (f)

3. (a) antecedent: sufficient water
   consequent: healthy plant growth
   (b) antecedent: further technological advances
   consequent: increased availability of microcomputers
   (c) antecedent: errors will be introduced
   consequent: there is a modification of the program
   (d) antecedent: fuel savings
   consequent: good insulation or all windows are storm windows

7. (a) A: prices go up; B: housing will be plentiful; C: housing will be expensive

   $(A \rightarrow B \land C) \land (C' \rightarrow B)$

   (b) A: going to bed; B: going swimming; C: changing clothes

   $(A \lor B) \rightarrow C) \land (C \rightarrow B)'$

   (c) A: it will rain; B: it will snow

   $(A \lor B) \land (A \land B)'$

   (d) A: Janet wins; B: Janet loses; C: Janet will be tired

   $(A \lor B) \rightarrow C$

   (e) A: Janet wins; B: Janet loses; C: Janet will be tired

   $A \lor (B \rightarrow C)$

8. (a)

A	B	$A \rightarrow B$	A'	$A' \lor B$	$(A \rightarrow B) \leftrightarrow A' \lor B$
T	T	T	F	T	T
T	F	F	F	F	T
F	T	T	T	T	T
F	F	T	T	T	T

tautology

(b)

A	B	C	$A \wedge B$	$(A \wedge B) \vee C$	$B \vee C$	$A \wedge (B \vee C)$	$(A \wedge B) \vee C$ $\rightarrow A \wedge (B \vee C)$
T	T	T	T	T	T	T	T
T	T	F	T	T	T	T	T
T	F	T	F	T	T	T	T
T	F	F	F	F	F	F	T
F	T	T	F	T	T	F	F
F	T	F	F	F	T	F	T
F	F	T	F	T	T	F	F
F	F	F	F	F	F	F	T

9. $2^{2^4} = 2^{16}$

13. (a)

A	A'	$A \vee A'$
T	F	T
F	T	T

(c)

A	B	$A \wedge B$	$A \wedge B \rightarrow B$
T	T	T	T
T	F	F	T
F	T	F	T
F	F	F	T

14. (a) Assign

$B' \wedge (A \rightarrow B)$ true
A' false

From the second assignment, A is true. From the first assignment, B' is true (so B is false) and $A \rightarrow B$ is true. If $A \rightarrow B$ is true and A is true, then B is true. B is thus both true and false, and $(B' \wedge (A \rightarrow B)) \rightarrow A'$ is a tautology.

23. (a) true (pick $y = 0$)
    (b) true (pick $y = 0$)
    (c) true (pick $y = -x$)
    (d) false

25. (a) True: The domain is the integers; $A(x)$ is "x is even" and $B(x)$ is "x is odd."

        False: The domain is the positive integers; $A(x)$ is "$x > 0$" and $B(x)$ is "$x \geq 1$."

27. (a) $(\forall x)(D(x) \rightarrow S(x))$
    (b) $(\exists x)(D(x) \wedge (R(x))')$
    (c) $(\forall x)[D(x) \wedge S(x) \rightarrow (R(x))']$

28. (a) $(\forall x)(C(x) \wedge F(x))'$
    (c) $(\forall x)(\forall y)[(P(y) \wedge S(x, y)) \rightarrow C(x)]$

29. (a) John is handsome and Kathy loves John.
    (b) All men are handsome.

31. (a) The domain is the integers; $A(x)$ is "x is even" and $B(x)$ is "x is odd."

Section 1.2

1. 1. hypothesis
   2. hypothesis
   3. Axiom 2
   4. 2, 3, modus ponens
   5. 1, 4, modus ponens

3. 1. $(P')'$ (hypothesis)
   2. $(P')' \rightarrow [P' \rightarrow ((P')')')]$ (1.27)
   3. $P' \rightarrow ((P')')'$ (1, 2, modus ponens)
   4. $[P' \rightarrow ((P')')'] \rightarrow [(P')' \rightarrow P]$ (Axiom 3)
   5. $(P')' \rightarrow P$ (3, 4, modus ponens)
   6. P (1, 5, modus ponens)

9. Eliminating the connective $\vee$, we want to prove $P' \wedge (P' \rightarrow Q) \rightarrow Q$.

   1. P' (hypothesis)
   2. $P' \rightarrow Q$ (hypothesis)
   3. Q (1, 2, modus ponens)

13. The argument is

    $$(R \wedge (F' \vee N)) \wedge N' \wedge (A' \rightarrow F) \rightarrow A \wedge R$$

    A proof sequence is

    1. $R \wedge (F' \vee N)$ (hypothesis)
    2. N' (hypothesis)
    3. $A' \rightarrow F$ (hypothesis)
    4. R (1, tautology $A \wedge B \rightarrow A$, modus ponens)
    5. $F' \vee N$ (1, tautology $A \wedge B \rightarrow B$, modus ponens)
    6. $(F' \vee N) \wedge N'$ ($A \wedge B$ can be deduced from A, B)
    7. $(F' \vee N) \wedge N' \rightarrow F'$ (tautology)
    8. F' (6, 7, modus ponens)
    9. $(A' \rightarrow F) \rightarrow (F' \rightarrow A)$ (tautology)
    10. $F' \rightarrow A$ (3, 9, modus ponens)
    11. A (8, 10, modus ponens)
    12. $A \wedge R$ ($A \wedge B$ can be deduced from A, B)

15. 1. $(\forall x)P(x)$ (hypothesis)
    2. $P(x)$ (1, Axiom 5, modus ponens)
    3. $P(x) \rightarrow P(x) \vee Q(x)$ (tautology)
    4. $P(x) \vee Q(x)$ (2, 3, modus ponens)
    5. $(\forall x)(P(x) \vee Q(x))$ (4, generalization—note that $P(x) \vee Q(x)$ was deduced from $(\forall x)P(x)$, in which x is not free)

19. 1. $(\exists x)(A(x) \wedge B(x))$ (hypothesis)
    2. $A(a) \wedge B(a)$ (1, Axiom 6, modus ponens)
    3. $A(a)$ (2, tautology $A \wedge B \rightarrow A$, modus ponens)
    4. $B(a)$ (2, tautology $A \wedge B \rightarrow B$, modus ponens)
    5. $(\exists x)A(x)$ (3, Axiom 7, modus ponens)
    6. $(\exists x)B(x)$ (4, Axiom 7, modus ponens)
    7. $(\exists x)A(x) \wedge (\exists x)B(x)$ ($A \wedge B$ can be deduced from A, B)

25. The argument is

$$(\forall x)(M(x) \rightarrow I(x) \vee G(x)) \wedge (\forall x)(G(x) \wedge L(x) \rightarrow F(x))$$
$$\wedge (I(j))' \wedge L(j) \rightarrow (M(j) \rightarrow F(j))$$

A proof sequence is

1. $(\forall x)(M(x) \rightarrow I(x) \vee G(x))$	(hypothesis)
2. $(\forall x)(G(x) \wedge L(x) \rightarrow F(x))$	(hypothesis)
3. $M(j) \rightarrow I(j) \vee G(j)$	(1, Axiom 5, modus ponens)
4. $G(j) \wedge L(j) \rightarrow F(j)$	(2, Axiom 5, modus ponens)
5. $M(j)$	(hypothesis)
6. $I(j) \vee G(j)$	(3, 5, modus ponens)
7. $(I(j))'$	(hypothesis)
8. $(I(j))' \wedge (I(j) \vee G(j)) \rightarrow G(j)$	(tautology)
9. $G(j)$	(6, 7, 8, modus ponens)
10. $L(j)$	(hypothesis)
11. $G(j) \wedge L(j)$	($A \wedge B$ can be deduced from A, B)
12. $F(j)$	(4, 11, modus ponens)

Section 1.3

1. (a) converse: Healthy plant growth implies sufficient water.
   contrapositive: If there is not healthy plant growth, then there is not sufficient water.
   (b) converse: Increased availability of microcomputers implies further technological advances.
   contrapositive: If there is not increased availability of microcomputers, then there are no further technological advances.
   (c) converse: If there is a modification of the program, then errors will be introduced.
   contrapositive: No modification of the program implies that errors will not be introduced.
   (d) converse: Good insulation or all windows are storm windows implies fuel savings.
   contrapositive: Poor insulation and some windows not storm windows implies no fuel savings.

5. Let $x = 2m + 1$ and $y = 2n + 1$, where m and n are integers. Then

$$x + y = (2m + 1) + (2n + 1)$$
$$= 2m + 2n + 2$$
$$= 2(m + n + 1) \quad \text{where } m + n + 1 \text{ is an integer}$$

so $x + y$ is even.

9. If $x < y$, then multiplying both sides of the inequality by the positive numbers x and y in turn gives $x^2 < xy$ and $xy < y^2$, and therefore $x^2 < xy < y^2$ or $x^2 < y^2$.

   If $x^2 < y^2$, then

$y^2 - x^2 > 0$	(definition of $<$)
$(y + x)(y - x) > 0$	(factoring)
$y + x < 0$ and $y - x < 0$	(a positive number is the
or	product of two negatives
$y + x > 0$ and $y - x > 0$	or two positives)

But it cannot be that $y + x < 0$ because y and x are both positive, therefore $y - x > 0$ and $y > x$.

11. Let $x = 2n + 1$; then

$$x^2 = (2n + 1)^2 = 4n^2 + 4n + 1$$
$$= 4n(n + 1) + 1$$

But $n(n + 1)$ is even (Exercise 6), so $n(n + 1) = 2k$ for some integer k. Therefore $x^2 = 4(2k) + 1 = 8k + 1$.

17. Proof: If x is even, then $x = 2n$ and

$$x(x + 1)(x + 2) = (2n)(2n + 1)(2n + 2)$$
$$= 2((n)(2n + 1)(2n + 2))$$

which is even. If x is odd, then $x = 2n + 1$ and

$$x(x + 1)(x + 2) = (2n + 1)(2n + 2)(2n + 3)$$
$$= 2((2n + 1)(n + 1)(2n + 3))$$

which is even.

Section 1.4

1. $P(1)$: $4 \cdot 1 - 2 = 2(1)^2$ or $2 = 2$, true
   Assume $P(k)$: $2 + 6 + 10 + \cdots + (4k - 2) = 2k^2$
   Show $P(k + 1)$: $2 + 6 + 10 + \cdots + (4(k + 1) - 2) = 2(k + 1)^2$

$2 + 6 + 10 + \cdots + (4(k + 1) - 2)$	(left side of $P(k + 1)$)
$= 2 + 6 + 10 + \cdots + (4k - 2) + (4(k + 1) - 2)$	
$= 2k^2 + 4(k + 1) - 2$	(using $P(k)$)
$= 2k^2 + 4k + 2$	
$= 2(k^2 + 2k + 1)$	
$= 2(k + 1)^2$	(right side of $P(k + 1)$)

3. $P(1)$: $1 = 1(2 \cdot 1 - 1)$, true
   Assume $P(k)$: $1 + 5 + 9 + \cdots + (4k - 3) = k(2k - 1)$
   Show $P(k + 1)$: $1 + 5 + 9 + \cdots + (4(k + 1) - 3) =$
   $$(k + 1)(2(k + 1) - 1)$$

$1 + 5 + 9 + \cdots + (4(k + 1) - 3)$	(left side of $P(k + 1)$)
$= 1 + 5 + 9 + \cdots + (4k - 3) + (4(k + 1) - 3)$	
$= k(2k - 1) + 4(k + 1) - 3$	(using $P(k)$)
$= 2k^2 - k + 4k + 1$	
$= 2k^2 + 3k + 1$	
$= (k + 1)(2k + 1)$	
$= (k + 1)(2(k + 1) - 1)$	(right side of $P(k + 1)$)

9. $P(1)$: $1^2 = \dfrac{1(2 - 1)(2 + 1)}{3}$, true

   Assume $P(k)$: $1^2 + 3^2 + \cdots + (2k - 1)^2 = \dfrac{k(2k - 1)(2k + 1)}{3}$

   Show $P(k + 1)$: $1^2 + 3^2 + \cdots + (2(k + 1) - 1)^2 =$
   $$\frac{(k + 1)(2(k + 1) - 1)(2(k + 1) + 1)}{3}$$

$1^2 + 3^2 + \cdots + (2(k + 1) - 1)^2$	(left side of $P(k + 1)$)
$= 1^2 + 3^2 + \cdots + (2k - 1)^2 + (2(k + 1) - 1)^2$	
$= \dfrac{k(2k - 1)(2k + 1)}{3} + (2(k + 1) - 1)^2$	(using $P(k)$)
$= \dfrac{k(2k - 1)(2k + 1)}{3} + (2k + 1)^2$	

$$= (2k + 1)\left(\frac{k(2k - 1)}{3} + 2k + 1\right)$$

$$= (2k + 1)\left(\frac{2k^2 - k + 6k + 3}{3}\right)$$

$$= \frac{(2k + 1)(2k^2 + 5k + 3)}{3}$$

$$= \frac{(2k + 1)(k + 1)(2k + 3)}{3}$$

$$= \frac{(k + 1)(2(k + 1) - 1)(2(k + 1) + 1)}{3} \qquad \text{(right side of } P(k + 1))$$

13. $P(1)$: $\dfrac{1}{1 \cdot 2} = \dfrac{1}{1 + 1}$, true

Assume $P(k)$: $\dfrac{1}{1 \cdot 2} + \dfrac{1}{2 \cdot 3} + \cdots + \dfrac{1}{k(k + 1)} = \dfrac{k}{k + 1}$

Show $P(k + 1)$: $\dfrac{1}{1 \cdot 2} + \dfrac{1}{2 \cdot 3} + \cdots + \dfrac{1}{(k + 1)(k + 2)} = \dfrac{k + 1}{k + 2}$

$$\frac{1}{1 \cdot 2} + \frac{1}{2 \cdot 3} + \cdots + \frac{1}{(k + 1)(k + 2)} \qquad \text{(left side of } P(k + 1))$$

$$= \frac{1}{1 \cdot 2} + \frac{1}{2 \cdot 3} + \cdots + \frac{1}{k(k + 1)} + \frac{1}{(k + 1)(k + 2)}$$

$$= \frac{k}{k + 1} + \frac{1}{(k + 1)(k + 2)} \qquad \text{(using } P(k))$$

$$= \frac{k(k + 2) + 1}{(k + 1)(k + 2)}$$

$$= \frac{k^2 + 2k + 1}{(k + 1)(k + 2)}$$

$$= \frac{(k + 1)^2}{(k + 1)(k + 2)}$$

$$= \frac{k + 1}{k + 2} \qquad \text{(right side of } P(k + 1))$$

15. $P(2)$: $2^2 > 2 + 1$, true
Assume $P(k)$: $k^2 > k + 1$
Show $P(k + 1)$: $(k + 1)^2 > k + 2$

$$(k + 1)^2 = k^2 + 2k + 1$$
$$> (k + 1) + 2k + 1 \qquad \text{(using } P(k))$$
$$= 3k + 2 > k + 2$$

18. $P(4)$: $2^4 < 4!$, or $16 < 24$, true
Assume $P(k)$: $2^k < k!$
Show $P(k + 1)$: $2^{k+1} < (k + 1)!$

$$2^{k+1} = 2^k \cdot 2 < k!(2) < k!(k + 1)$$
$$= (k + 1)! \qquad \text{(first inequality by } P(k); \text{ second by}$$
$$k \geq 4)$$

23. $P(1)$: $2^3 - 1 = 8 - 1 = 7$ and $7 \,|\, 7$, true
Assume $P(k)$: $7 \,|\, 2^{3k} - 1$ so $2^{3k} - 1 = 7m$ or $2^{3k} = 7m + 1$
for some integer m

Show $P(k + 1)$: $2^{3(k+1)} - 1$ is divisible by 7

$$2^{3(k+1)} - 1 = 2^{3k+3} - 1 = 2^{3k} \cdot 2^3 - 1$$
$$= (7m + 1)2^3 - 1$$
$$= 7(2^3m) + 8 - 1 = 7(2^3m + 1) \qquad \text{(where } 2^3m + 1 \text{ is an integer)}$$

31. $P(1)$: $10 + 3 \cdot 4^3 + 5 = 10 + 192 + 5 = 207 = 9 \cdot 23$, true
    Assume $9 \,|\, 10^k + 3 \cdot 4^{k+2} + 5$ so $10^k + 3 \cdot 4^{k+2} + 5 = 9m$ or
    $10^k = 9m - 3 \cdot 4^{k+2} - 5$ for some integer m
    Show $P(k + 1)$: $10^{k+1} + 3 \cdot 4^{k+3} + 5$ is divisible by 9

$$10^{k+1} + 3 \cdot 4^{k+3} + 5 = 10 \cdot 10^k + 3 \cdot 4^{k+3} + 5$$
$$= 10(9m - 3 \cdot 4^{k+2} - 5) + 3 \cdot 4^{k+3} + 5$$
$$= 9(10m) - 30 \cdot 4^{k+2} - 50 + 3 \cdot 4^{k+2} \cdot 4 + 5$$
$$= 9(10m) - 45 - 3 \cdot 4^{k+2}(10 - 4)$$
$$= 9(10m - 5) - 18 \cdot 4^{k+2}$$
$$= 9(10m - 5 - 2 \cdot 4^{k+2}) \qquad \text{(where } 10m - 5 - 2 \cdot 4^{k+2}$$
$$\text{is an integer)}$$

43. $P(0)$: $j_0 = (i_0 - 1)!$, true
    (since $j = 1$ and $i = 2$ before loop is entered, and $1 = 1!$)
    Assume $P(k)$: $j_k = (i_k - 1)!$
    Show $P(k + 1)$: $j_{k+1} = (i_{k+1} - 1)!$

$$j_{k+1} = j_k \cdot i_k = (i_k - 1)! i_k = (i_k)! = (i_{k+1} - 1)!$$

At loop termination, $i = x + 1$ and $j = x!$

Section 1.5

1. 10, 20, 30, 40, 50

5. 1, 1, 2, 3, 5

9. (a), (b), (e)

11. (a) $A(1) = 50,000$
       $A(n) = 3A(n - 1)$ \qquad for $n \geq 2$
    (b) 4

15. *Algorithm* $S(n)$
    if $n = 1$
    then $S \leftarrow 1$
    else begin
       $S \leftarrow S(n - 1) + (n - 1)$
       end

19. $S(n) = S(n - 1) + 5$
    $\qquad = (S(n - 2) + 5) + 5 = S(n - 2) + 2 * 5$
    $\qquad = (S(n - 3) + 5) + 2 * 5 = S(n - 3) + 3 * 5$

    In general,

    $$S(n) = S(n - k) + k * 5$$

    When $n - k = 1$ then $k = n - 1$ and

    $$S(n) = S(1) + (n - 1) * 5$$
    $$= 5 + (n - 1) * 5 = n * 5$$

    Now prove by induction that $S(n) = n * 5$.

$S(1)$: $S(1) = 1 * 5 = 5$, true
Assume $S(k)$: $S(k) = k * 5$
Show $S(k + 1)$: $S(k + 1) = (k + 1) * 5$

$$S(k + 1) = S(k) + 5$$
$$= k * 5 + 5 = (k + 1) * 5$$

CHAPTER 2

Section 2.1

1. (a)

3. (a) $\{0, 1, 2, 3, 4\}$
   (b) $\{3, 7, 10\}$

5. If $A = \{x \mid x = 2^n$ for n a positive integer$\}$, then $16 \in A$. But if $A = \{x \mid x = 2 + n(n - 1)$ for n a positive integer$\}$, then $16 \notin A$.

7. (a) F; $\{1\} \in S$ but $\{1\} \notin R$ (b) T   (c) F; $\{1\} \in S$, not $1 \in S$
   (d) F; 1 is not a set; the correct statement is $\{1\} \subseteq U$ (e) T
   (f) F; $1 \notin S$

8. (a) T   (b) F   (c) F   (d) T   (e) T   (f) F

15. (a) binary operation   (b) no; $0 \circ 0 \notin N$ (c) binary operation

19. (a) $\{1, 2, 4, 5, 6, 8, 9\}$
    (b) $\{4, 5\}$
    (c) $\{2, 4\}$
    (h) $\{0, 1, 2, 3, 6, 7, 8, 9\}$

21. (a) True
    (c) False; let $A = \{1, 2, 3\}$, $B = \{1, 3, 5\}$, and $S = \{1, 2, 3, 4, 5\}$. Then $(A \cap B)' = \{2, 4, 5\}$, but $A' \cap B' = \{4, 5\} \cap \{2, 4\} = \{4\}$.
    (e) False; take A, B, and S as in (c). Then $A - B = \{2\}$ but $(B - A)' = \{1, 2, 3, 4\}$.

22. (a) $B \subseteq A$

29. (a) (1) $(A \cup B) \cap (A \cup B') = A \cup (B \cap B')$ (3a)
    $\qquad\qquad\qquad\qquad\qquad = A \cup \varnothing$ (5b)
    $\qquad\qquad\qquad\qquad\qquad = A$ (4a)

31. An enumeration of the set is

    $1, 3, 5, 7, 9, 11, \ldots$.

Section 2.2

1. $5 \cdot 3 \cdot 2 = 30$

2. $4 \cdot 2 \cdot 2 = 16$

7. $45 \cdot 13 = 585$

9. $26 \cdot 26 \cdot 26 \cdot 1 \cdot 1 = 17{,}576$

11. $26 + 26 \cdot 10 = 286$

13. $2^8 = 256$

15. $1 \cdot 2^7$ (begin with 0) $+ 1 \cdot 2^6 \cdot 1$ (begin with 1, end with 0) $= 2^7 + 2^6 = 192$

21. $52 \cdot 52 = 2704$

23. There are $4 \cdot 4 = 16$ ways to get two of a kind for a particular "kind"; there are 13 distinct "kinds," so by the Addition Principle, $16 + 16 + \cdots + 16 = 13 \cdot 16 = 208$.

Section 2.3

1. (a) 42

5. There are 5! total permutations, and 3! arrangements of the 3 Rs for each distinct permutation. Therefore there are $5!/3! = 5 \cdot 4 = 20$ distinct permutations.

7. $P(15, 3) = \dfrac{15!}{12!} = 15 \cdot 14 \cdot 13 = 2730$

11. (a) 120

13. $C(300, 25) = \dfrac{300!}{25!275!}$

15. $C(17, 5) \cdot C(23, 7) = (6188)(245,157)$

21. $C(13, 3) \cdot C(13, 2)$

23. $C(13, 5) + C(13, 5) + C(13, 5) + C(13, 5) = 4C(13, 5)$

27. $C(12, 4)$

35. $C(12, 3)$

37. $C(7, 3)$ (no Democrats) $+ C(9, 3)$ (no Republicans) $- C(4, 3)$ (all independents, so as not to count twice) $= 115$

Section 2.4

1. $(a + b)^4 = C(4, 0)a^4 + C(4, 1)a^3b + C(4, 2)a^2b^2 + C(4, 3)ab^3 + C(4, 4)b^4$
$= a^4 + 4a^3b + 6a^2b^2 + 4ab^3 + b^4$

3. $(a + 2)^5 = C(5, 0)a^5 + C(5, 1)a^42 + C(5, 2)a^3(2)^2 + C(5, 3)a^2(2)^3$
$+ C(5, 4)a(2)^4 + C(5, 5)(2)^5$
$= a^5 + 10a^4 + 40a^3 + 80a^2 + 80a + 32$

5. $(2x + 3y)^3 = C(3, 0)(2x)^3 + C(3, 1)(2x)^2(3y) + C(3, 2)(2x)(3y)^2 + C(3, 3)(3y)^3$
$= 8x^3 + 36x^2y + 54xy^2 + 27y^3$

11. $C(9, 5)(2x)^4(-3)^5 = -489,888x^4$

13. $C(8, 8)(-3y)^3 = 6561y^8$

15. $C(5, 2)(4x)^3(-2y)^2 = 2560x^3y^2$

19. $((2x - y) + 5)^8$: $C(8, 1)(2x - y)^75^1 = C(8, 1) \cdot 5(C(7, 4)(2x)^3(-y)^4)$
$= C(8, 1)C(7, 4)(2)^35x^3y^4 = 11,200x^3y^4$

CHAPTER 3

Section 3.1

1. (a) (1, 3), (3, 3)   (b) (4, 2), (5, 3)
   (c) (5, 0), (2, 2)   (d) (1, 1), (3, 9)

3. (a)

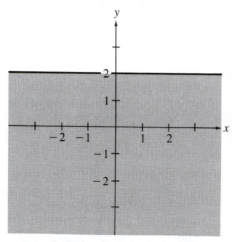

5. (a) (2, 6), (3, 17), (0, 0)   (b) (2, 12)   (c) none   (d) (1, 1), (4, 8)

7. (a) reflexive, transitive   (b) reflexive, symmetric, transitive
   (c) symmetric

11. (a) $\rho = \{(1, 1)\}$

16. (a)

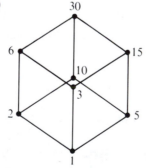

(b)

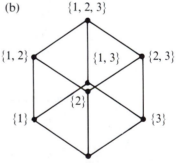

The two graphs are identical in structure.

17. (a) $\rho = \{(1, 1), (2, 2), (3, 3), (4, 4), (5, 5), (1, 3), (3, 5), (1, 5), (2, 4), (4, 5), (2, 5)\}$

21. (a) "when"; no; all but the last

(b)

Maximal elements: "a", "merry", "soul"

23. (a) $[a] = \{a, c\} = [c]$
    (b) $[3] = \{1, 2, 3\}$; $[4] = \{4, 5\}$
    (c) $[1] = \{\ldots, -5, -3, -1, 1, 3, 5, \ldots\}$
    (d) $[-3] = \{\ldots, -13, -8, -3, 2, 7, 12, \ldots\}$

Section 3.2

3. (a) function
   (b) not a function; undefined at $x = 0$
   (c) function; onto
   (d) bijection; $f^{-1}:\{p, q, r\} \to \{1, 2, 3\}$ where $f^{-1} = \{(q, 1), (r, 2), (p, 3)\}$
   (e) function; one-to-one
   (f) bijection; $h^{-1}:R^2 \to R^2$ where $h^{-1}(x, y) = (y - 1, x - 1)$

7. (a) $2^3 = 8$; 6
   (b) n^m
   (c) $n(n - 1)(n - 2) \cdots (n - (m - 1)) = \dfrac{n!}{(n - m)!}$

9. (a) $\{m, n, o, p\}$ (b) $\{n, o, p\}$; $\{o\}$

11. (a) $(g \circ f)(5) = g(f(5)) = g(6) = 18$
    (b) $(f \circ g)(5) = f(g(5)) = f(15) = 16$
    (c) $(g \circ f)(x) = g(f(x)) = g(x + 1) = 3(x + 1) = 3x + 3$
    (d) $(f \circ g)(x) = f(g(x)) = f(3x) = 3x + 1$
    (e) $(f \circ f)(x) = f(f(x)) = f(x + 1) = (x + 1) + 1 = x + 2$
    (f) $(g \circ g)(x) = g(g(x)) = g(3x) = 3(3x) = 9x$

13. (a) $(1, 3, 5, 2)$
    (b) $(1, 4, 3, 2, 5)$

15. (a) $\begin{pmatrix} 1 & 2 & 3 \\ 1 & 2 & 3 \end{pmatrix}$, $\begin{pmatrix} 1 & 2 & 3 \\ 1 & 3 & 2 \end{pmatrix}$, $\begin{pmatrix} 1 & 2 & 3 \\ 3 & 2 & 1 \end{pmatrix}$, $\begin{pmatrix} 1 & 2 & 3 \\ 3 & 1 & 2 \end{pmatrix}$, $\begin{pmatrix} 1 & 2 & 3 \\ 2 & 1 & 3 \end{pmatrix}$,
    $\begin{pmatrix} 1 & 2 & 3 \\ 2 & 3 & 1 \end{pmatrix}$

    (b) $n!$

17. (a) $(1, 2, 5, 3, 4)$
    (b) $(1, 7, 8) \circ (2, 4, 6)$
    (c) $(1, 5, 2, 4) \circ (3, 6)$
    (d) $(2, 3) \circ (4, 8) \circ (5, 7)$

19. (a) $f^{-1}(x) = x/2$
    (b) $f^{-1}(x) = \sqrt[3]{x}$
    (c) $f^{-1}(x) = 3x - 4$

Section 3.3

1. $2, -4$

3. $x = 1, y = 3, z = -2, w = 4$

5. $\begin{bmatrix} 6 & -5 \\ 0 & 3 \\ 5 & 3 \end{bmatrix}$

9. $\begin{bmatrix} 14 & -17 \\ 2 & 9 \\ 9 & 1 \end{bmatrix}$

13. $\begin{bmatrix} 21 & -23 \\ 33 & -44 \\ 11 & 1 \end{bmatrix}$

17. $\begin{bmatrix} 28 & 4 \\ 6 & 25 \end{bmatrix}$

21. $\mathbf{A(B + C)} = \mathbf{AB} + \mathbf{AC} = \begin{bmatrix} 26 & -9 \\ 40 & -23 \end{bmatrix}$

25. (a) straightforward matrix multiplication
    (b) Suppose there is a matrix $\mathbf{B}$ that is the inverse of $\mathbf{A}$. Then

$$\begin{bmatrix} 1 & 2 \\ 2 & 4 \end{bmatrix} \begin{bmatrix} b_{11} & b_{12} \\ b_{21} & b_{22} \end{bmatrix} = \begin{bmatrix} 1 & 0 \\ 0 & 1 \end{bmatrix}$$

and

$$b_{11} + 2b_{21} = 1 \qquad b_{12} + 2b_{22} = 0$$
$$2b_{11} + 4b_{21} = 0 \qquad 2b_{12} + 4b_{22} = 1$$

which is an inconsistent system of equations with no solution.

27. $\begin{bmatrix} 1 & 1 \\ -1 & -1 \end{bmatrix}$ and $\begin{bmatrix} 1 & 1 \\ -1 & -1 \end{bmatrix}$

CHAPTER 4

Section 4.1

1. (d)—the two nodes of degree 3 are not adjacent.

2. (a)

This is a tree.

   (b)

   (c) For example:

3. (a)

   (b)

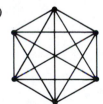

   (c)

   (d) $\dfrac{n(n-1)}{2}$

   (e) The number of arcs is

$$C(n, 2) = \frac{n(n-1)}{2}$$

   (Other proof methods include induction on the number of nodes.)

5. 4, 5, 6; length 2; for example (naming the nodes): 1, 2, 1, 2, 2, 1, 4, 5, 6

7.

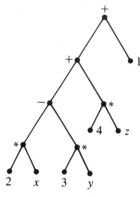

13.

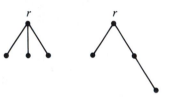

15. Consider a simple graph that is a tree. A tree is an acyclic and connected graph, so for any two nodes x and y, a path from x to y exists. If the path is not unique, then the two paths diverge at some node n_1 and converge at some node n_2, and there is a cycle from n_1 through n_2 and back to n_1, which is a contradiction.

Now consider a simple graph that has a unique path between any two nodes. The graph is clearly connected. Also, there are no cycles because the presence of a cycle produces a nonunique path between two nodes on the cycle. The graph is thus acyclic and connected and is a tree.

Section 4.2

1. $\begin{bmatrix} 1 & 1 & 0 & 0 & 2 \\ 1 & 1 & 1 & 1 & 1 \\ 0 & 1 & 0 & 1 & 0 \\ 0 & 1 & 1 & 0 & 0 \\ 2 & 1 & 0 & 0 & 0 \end{bmatrix}$

5. $\begin{bmatrix} 0 & 1 & 0 & 0 \\ 0 & 0 & 1 & 1 \\ 0 & 0 & 0 & 1 \\ 0 & 0 & 1 & 0 \end{bmatrix}$

9. $\mathbf{R} = \begin{bmatrix} 2 & 4 & 4 & 4 \\ 4 & 6 & 6 & 7 \\ \textcircled{0} & 0 & 0 & 1 \\ 0 & 0 & 0 & 0 \end{bmatrix}$

15.

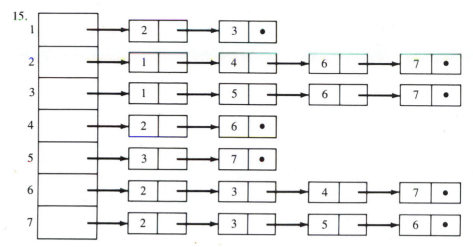

21.

Node	Pointer
1	5
2	7
3	11
4	0
5	2	6
6	3	0
7	1	8
8	2	9
9	3	10
10	4	0
11	4	0

25.

	Left child	Right child
1	2	3
2	0	4
3	5	6
4	7	0
5	0	0
6	0	0
7	0	0

27.

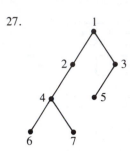

31. $\rho = \{(1, 1), (1, 2), (2, 3), (3, 4), (4, 3), (4, 2)\}$

Section 4.3

1. yes

5. $\begin{bmatrix} 0 & 1 & 1 & 1 & 1 & 0 \\ 1 & 0 & 0 & 0 & 0 & 1 \\ 1 & 0 & 0 & 0 & 0 & 1 \\ 1 & 0 & 0 & 0 & 0 & 1 \\ 1 & 0 & 0 & 0 & 0 & 1 \\ 0 & 1 & 1 & 1 & 1 & 0 \end{bmatrix}$

odd after row 2 is 0.

11. no

15. Begin at any node and take one of the arcs out from that node. Each time a new node is entered on an arc, there is exactly one unused arc on which to exit that node; because the arc is unused, it will lead to a new node or to the initial node. Upon return to the initial node, if all nodes have been used, we are done. Because the graph is connected, if there is an unused node, there is an unused path from that node to a used node, which means the used node has degree ≥ 3, a contradiction.

17. $IN = \{2\}$

	1	3	4	5	6	7	8
d:	3	2	∞	∞	∞	1	∞
s:	2	2	2	2	2	2	2

  $p = 7$
  $IN = \{2, 7\}$

	1	3	4	5	6	7	8
d:	3	2	∞	∞	6	1	2
s:	2	2	2	2	7	2	7

$p = 3$
$IN = \{2, 7, 3\}$

	1	3	4	5	6	7	8
d:	3	2	3	∞	6	1	2
s:	2	2	3	2	7	2	7

$p = 8$
$IN = \{2, 7, 3, 8\}$

	1	3	4	5	6	7	8
d:	3	2	3	3	6	1	2
s:	2	2	3	8	7	2	7

$p = 5$
$IN = \{2, 7, 3, 8, 5\}$

	1	3	4	5	6	7	8
d:	3	2	3	3	6	1	2
s:	2	2	3	8	7	2	7

path: 2, 7, 8, 5   length: 3

21. $IN = \{1\}$

	2	3	4	5	6	7
d:	2	∞	∞	3	2	∞
s:	1	1	1	1	1	1

$p = 2$
$IN = \{1, 2\}$

	2	3	4	5	6	7
d:	2	3	∞	3	2	∞
s:	1	2	1	1	1	1

$p = 6$
$IN = \{1, 2, 6\}$

	2	3	4	5	6	7
d:	2	3	∞	3	2	5
s:	1	2	1	1	1	6

$p = 3$
$IN = \{1, 2, 6, 3\}$

	2	3	4	5	6	7
d:	2	3	4	3	2	5
s:	1	2	3	1	1	6

$p = 5$
$IN = \{1, 2, 6, 3, 5\}$

	2	3	4	5	6	7
d:	2	3	4	3	2	5
s:	1	2	3	1	1	6

$p = 4$
$IN = \{1, 2, 6, 3, 5, 4\}$

	2	3	4	5	6	7
d:	2	3	4	3	2	5
s:	1	2	3	1	1	6

$p = 7$
$IN = \{1, 2, 6, 3, 5, 4, 7\}$

	2	3	4	5	6	7
d:	2	3	4	3	2	5
s:	1	2	3	1	1	6

path: 1, 6, 7   length: 5

23.

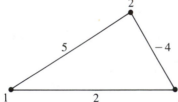

For the graph shown, the algorithm proceeds as follows to find the shortest path from 1 to 3:

$IN = \{1\}$

	2	3
d:	5	2
s:	1	1

$p = 3$
$IN = \{1, 3\}$

	2	3
d:	-2	2
s:	3	1

Thus the algorithm will select the path 1, 3 with length 2 as the shortest path from 1 to 3, although the shortest path is actually 1, 2, 3 with length 1. Allowing negative weights makes the greedy property insufficient for success, because a path with an initially high weight can later have its weight reduced by negative values, but this cannot be seen locally.

29. *IN* = {1, 8, 5, 6, 2, 7, 4, 3}

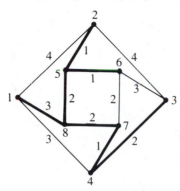

Section 4.4

1. *a* *b* *c* *e* *f* *d* *h* *g* *j* *i*

5. *e* *b* *a* *c* *f* *d* *h* *g* *j* *i*

7. *a* *b* *c* *f* *j* *g* *d* *e* *h* *k* *i*

9. *f* *c* *a* *b* *d* *e* *h* *k* *i* *g* *j*

11. *a* *b* *c* *d* *e* *g* *f* *h* *j* *i*

17. *a* *b* *c* *d* *e* *f* *g* *h* *i* *j* *k*

21. preorder: *a* *b* *d* *e* *h* *f* *c* *g*
    inorder: *d* *b* *h* *e* *f* *a* *g* *c*
    postorder: *d* *h* *e* *f* *b* *g* *c* *a*

25. preorder: *a* *b* *c* *e* *f* *d* *g* *h*
    inorder: *e* *c* *f* *b* *g* *d* *h* *a*
    postorder: *e* *f* *c* *g* *h* *d* *b* *a*

27. prefix: + / 3 4 − 2 *y*
    postfix: 3 4 / 2 *y* − +

31. prefix: + * 4 − 7 *x* *z*
    infix: 4 * (7 − *x*) + *z*

35.    Both infix and postfix traversal give
    d *c* *b* *a*

CHAPTER 5

Section 5.1

1.

+	0	1	a	a'
0	0	1	a	a'
1	1	1	1	1
a	a	1	a	1
a'	a'	1	1	a'

$\cdot$	0	1	a	a'
0	0	0	0	0
1	0	1	a	a'
a	0	a	a	0
a'	0	a'	0	a'

4. (a) $\begin{aligned} x' + x &= x + x' \quad &\text{(1a)} \\ &= 1 \quad &\text{(5a)} \end{aligned}$

and

$\begin{aligned} x' \cdot x &= x \cdot x' \quad &\text{(1b)} \\ &= 0 \quad &\text{(5b)} \end{aligned}$

Therefore $x = (x')'$ by Theorem 5.8.

(b) $\begin{aligned} x + (x \cdot y) &= x \cdot 1 + x \cdot y \quad &\text{(4b)} \\ &= x(1 + y) \quad &\text{(3b)} \\ &= x(y + 1) \quad &\text{(1a)} \\ &= x \cdot 1 \quad &\text{(Practice 5.7)} \\ &= x \quad &\text{(4b)} \end{aligned}$

$x \cdot (x + y) = x$ follows by duality.

5. (a) $\begin{aligned} x \oplus y &= x \cdot y' + y \cdot x' \quad &\text{(definition of } \oplus) \\ &= y \cdot x' + x \cdot y' \quad &\text{(1a)} \\ &= y \oplus x \quad &\text{(definition of } \oplus) \end{aligned}$

9. (a) (i) If $x \leq y$, then $x \leq y$ and $x \leq x$, so x is a lower bound of $\{x, y\}$. If $w^* \leq x$ and $w^* \leq y$, then $w^* \leq x$, so x is a greatest lower bound, and $x = x \cdot y$. If $x = x \cdot y$, then x is a greatest lower bound of $\{x, y\}$, so $x \leq y$. The proof of (ii) is very similar.

13. (a) $(x - y) - z = x - (y - z)$; no, because (for example)
    $(2 - 3) - 4 \neq 2 - (3 - 4)$
    (b) $x - y = y - x$; no, because (for example) $3 - 2 \neq 2 - 3$

14. (a) associative (b) commutative

Section 5.2

1. (a) $f(-2) = 2^{-2}$ and $f(12) = 2^{12}$. In $[\mathbb{R}^+, \cdot]$, $2^{-2} \cdot 2^{12} = 2^{10}$.
    Then $f^{-1}(2^{10}) = 10$.
    (b) $f^{-1}(4) = \log_2 4 = 2$ and $f^{-1}(8) = \log_2 8 = 3$. In $[\mathbb{R}, +]$, $2 + 3 = 5$.
    Then $f(5) = 2^5 = 32$.

3. (a)

(b)

(c)

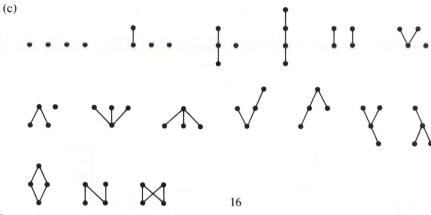

16

7. (a) For any $y \in b$, $y = f(x)$ for some $x \in B$. Then $y \& f(0) = f(x) \& f(0) = f(x + 0) = f(x) = y$, and $f(0) = \emptyset$ because the zero element in any Boolean algebra is unique (see Exercise 8, Section 5.1).

(b) $f(1) = f(0') = [f(0)]'' = \emptyset'' = 1$

CHAPTER 6

Section 6.1

1. (a)

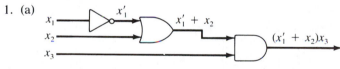

5. (a) $x_1 x_2 x_3' + x_1 x_2' x_3'$

(b)

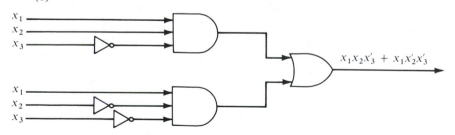

(c) $x_1 x_2 x_3' + x_1 x_2' x_3' = x_1 x_3' x_2 + x_1 x_3' x_2'$
$= x_1 x_3' (x_2 + x_2')$
$= x_1 x_3' 1$
$= x_1 x_3'$

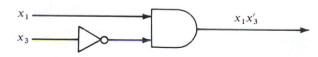

8. (a) $(x_1' + x_2')(x_1' + x_2)(x_1 + x_2')$
   (b) $(x_1' + x_2)(x_1 + x_2)$
   (c) $(x_1' + x_2' + x_3')(x_1' + x_2 + x_3)(x_1 + x_2' + x_3)(x_1 + x_2 + x_3')$
   (d) $(x_1' + x_2' + x_3')(x_1' + x_2' + x_3)(x_1 + x_2' + x_3')(x_1 + x_2 + x_3')$
   $(x_1 + x_2 + x_3)$

13. Network is represented by $(x_1'(x_2'x_3)')'$ and $(x_1'(x_2'x_3)')' = x_1 + x_2'x_3$.

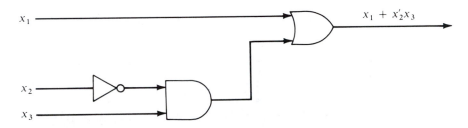

15. The truth function for $|$ is that of the NAND gate, and the truth function for $\downarrow$ is that of the NOR gate. In Section 1.1, we learned that every compound statement is equivalent to a statement using only $|$ or to a statement using only $\downarrow$; therefore any truth function can be realized by using only NAND gates or only NOR gates.

17. $x_1 =$ neutral
    $x_2 =$ park
    $x_3 =$ seat belt

x_1	x_2	x_3	$f(x_1, x_2, x_3)$
1	1	1	—
1	1	0	—
1	0	1	1
1	0	0	0
0	1	1	1
0	1	0	0
0	0	1	0
0	0	0	0

$(x_1 + x_2)x_3$

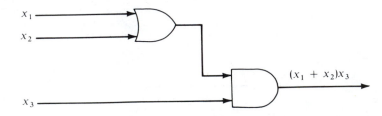

Section 6.2

1. (a)

	x_1x_2	x_1x_2'	$x_1'x_2'$	$x_1'x_2$
x_3			1	1
x_3'	1	1		1

$$x_1'x_3 + x_1x_3' + x_1'x_2 \quad \text{or} \quad x_1'x_3 + x_1x_3' + x_2x_3'$$

(c)

	x_1x_2	x_1x_2'	$x_1'x_2'$	$x_1'x_2$
x_3	1	1	1	1
x_3'	1			1

$$x_3 + x_2$$

5.

	x_1x_2	x_1x_2'	$x_1'x_2'$	$x_1'x_2$
x_3x_4			1	
x_3x_4'	1	—		1
$x_3'x_4'$	—			1
$x_3'x_4$			∧	

$$x_2x_4' + x_1'x_2'x_4$$

6.

x_1	x_2	x_3	
1	1	1	1,2,3
1	0	1	1,4
0	1	1	2,5,6
1	1	0	3,7
0	0	1	4,5
0	1	0	6,7

x_1	x_2	x_3	
1	–	1	1
–	1	1	1,2
1	1	–	2
–	0	1	1
0	–	1	1
0	1	–	2
–	1	0	2

x_1	x_2	x_3
–	–	1
–	1	–

	111	101	011	110	001	010
– –1	✓	✓	✓		✓	
–1–	✓		✓	✓		✓

– –1 and –1– are essential. The minimal form is $x_3 + x_2$.

9. (a)

x_1	x_2	x_3	x_4	
0	1	1	1	1
1	0	1	0	
0	0	1	1	1,2
0	1	0	0	3
0	0	0	1	2,4
0	0	0	0	3,4

x_1	x_2	x_3	x_4
0	–	1	1
0	0	–	1
0	–	0	0
0	0	0	–

	0111	1010	0011	0100	0001	0000
1010		✓				
0–11	✓		✓			
00–1			✓		✓	
0–00				✓		✓
000–					✓	✓

1010, 0–11, 0–00 are essential. Either 00–1 or 000– can be used as the fourth term. The minimal sum-of-products form is

$$x_1 x_2' x_3 x_4' + x_1' x_3 x_4 + x_1' x_3' x_4' + x_1' x_2' x_4$$

or

$$x_1 x_2' x_3 x_4' + x_1' x_3 x_4 + x_1' x_3' x_4' + x_1' x_2' x_3'$$

CHAPTER 7

Section 7.1

1. (a) not commutative; $a \cdot b \neq b \cdot a$
   not associative; $a \cdot (b \cdot d) \neq (a \cdot b) \cdot d$

   (b)

$\cdot$	p	q	r	s
p	p	q	r	s
q	q	r	s	p
r	r	s	p	q
s	s	p	q	r

   commutative

3. (a) f_0 = identity function
   $f_1(1) = 2 \qquad f_1(2) = 1$
   $f_2(1) = 1 \qquad f_2(2) = 1$
   $f_3(1) = 2 \qquad f_3(2) = 2$

∘	f_0	f_1	f_2	f_3
f_0	f_0	f_1	f_2	f_3
f_1	f_1	f_0	f_3	f_2
f_2	f_2	f_2	f_2	f_2
f_3	f_3	f_3	f_3	f_3

(b) f_0 and f_1 are the elements.

∘	f_0	f_1
f_0	f_0	f_1
f_1	f_1	f_0

4. (a) semigroup
   (b) not a semigroup; not associative
   (c) not a semigroup; S not closed under ·
   (d) monoid; $i = 1 + 0\sqrt{2}$
   (e) group; $i = 1 + 0\sqrt{2}$
   (f) group; $i = 1$
   (g) monoid; $i = 1$

8.

∘	R_1	R_2	R_3	F_1	F_2	F_3
R_1	R_2	R_3	R_1	F_3	F_1	F_2
R_2	R_3	R_1	R_2	F_2	F_3	F_1
R_3	R_1	R_2	R_3	F_1	F_2	F_3
F_1	F_2	F_3	F_1	R_3	R_1	R_2
F_2	F_3	F_1	F_2	R_2	R_3	R_1
F_3	F_1	F_2	F_3	R_1	R_2	R_3

Identity element is R_3; inverse for F_1 is F_1; inverse for R_2 is R_1.

11. (a) $i_L = i_L \cdot i_R = i_R$, so $i_L = i_R$ and this element is an identity in $[S, \cdot]$.

(b) For example,

·	a	b
a	a	b
b	a	b

(c) For example,

·	a	b
a	a	a
b	b	b

(d) For example, $[\mathbb{R}^+, +]$.

13. (a) $0_L = 0_L 0_R = 0_R$

    (b) For (a), zero is 0.

       For (d), zero is $0 + 0\sqrt{2}$.

       For (h), zero is 30.

17. (a) $x \rho x$ because $i \cdot x \cdot i^{-1} = x \cdot i^{-1} = x \cdot i = x$. If $x \rho y$, then for some $g \in G$, $g \cdot x \cdot g^{-1} = y$, so $g \cdot x = y \cdot g$, or $x = g^{-1} \cdot y \cdot g$, which can be written as $(g^{-1}) \cdot y \cdot (g^{-1})^{-1}$. Therefore $y \rho x$. Finally, if $x \rho y$ and $y \rho z$, then for some $g_1, g_2 \in G$, $g_1 \cdot x \cdot g_1^{-1} = y$ and $g_2 \cdot y \cdot g_2^{-1} = z$, so $g_2 \cdot g_1 \cdot x \cdot g_1^{-1} \cdot g_2^{-1} = z$, or $(g_2 \cdot g_1) \cdot x \cdot (g_2 \cdot g_1)^{-1} = z$. Therefore $x \rho z$.

    (b) Suppose G is commutative and $y \in [x]$. Then for some $g \in G$, $y = g \cdot x \cdot g^{-1} = x \cdot g \cdot g^{-1} = x \cdot i = x$. Then $[x] = \{x\}$. Conversely, suppose $[x] = \{x\}$ for each $x \in G$. Let $x, y \in G$, and let $y \cdot x \cdot y^{-1} = z$. Then $x \rho z$, so $z = x$ and $y \cdot x \cdot y^{-1} = x$ or $y \cdot x = x \cdot y$, and G is commutative.

Section 7.2

2. (a) no; not the same operation.

    (b) no; zero polynomial (identity) does not belong to P.

    (c) no; not every element of $\mathbb{Z}^*$ has an inverse in $\mathbb{Z}^*$.

3. $[\{0\}, +_{12}], [\mathbb{Z}_{12}, +_{12}], [\{0, 2, 4, 6, 8, 10\}, +_{12}], [\{0, 4, 8\}, +_{12}],$ $[\{0, 3, 6, 9\}, +_{12}], [\{0, 6\}, +_{12}]$

7. $4!/2 = 24/2 = 12$ elements

$$\alpha_1 = i \qquad\qquad \alpha_2 = (1, 2) \circ (3, 4)$$
$$\alpha_3 = (1, 3) \circ (2, 4) \qquad \alpha_4 = (1, 4) \circ (2, 3)$$
$$\alpha_5 = (1, 3) \circ (1, 2) \qquad \alpha_6 = (1, 2) \circ (1, 3)$$
$$\alpha_7 = (1, 3) \circ (1, 4) \qquad \alpha_8 = (1, 4) \circ (1, 2)$$
$$\alpha_9 = (1, 4) \circ (1, 3) \qquad \alpha_{10} = (1, 2) \circ (1, 4)$$
$$\alpha_{11} = (2, 4) \circ (2, 3) \qquad \alpha_{12} = (2, 3) \circ (2, 4)$$

9. (a) $S \cap T \subseteq G$. Closure: For $x, y \in S \cap T$, $x \cdot y \in S$ because of closure in S, $x \cdot y \in T$ because of closure in T, so $x \cdot y \in S \cap T$. Identity: $i \in S$ and $i \in T$, so $i \in S \cap T$. Inverses: For $x \in S \cap T$, $x^{-1} \in S$ and $x^{-1} \in T$, so $x^{-1} \in S \cap T$.

    (b) No. For example, $[\{0, 4, 8\}, +_{12}]$ and $[\{0, 6\}, +_{12}]$ are subgroups of $[\mathbb{Z}_{12}, +_{12}]$, but $[\{0, 4, 6, 8\}, +_{12}]$ is not a subgroup of $[\mathbb{Z}_{12}, +_{12}]$ (not closed).

10. Let $[S, \cdot]$ be a commutative monoid and let $A = \{a \,|\, a \in S, a^2 = a\}$. Then $A \subseteq S$; $i \cdot i = i$, so $i \in A$.

For $x, y \in A$, $(x \cdot y)^2 = (x \cdot y)(x \cdot y) = x \cdot x \cdot y \cdot y = x \cdot y$, so $x \cdot y \in A$.

12. Closure: Let $x, y \in B_k$. Then $(x \cdot y)^k = x^k \cdot y^k = i \cdot i = i$, so $x \cdot y \in B_k$.

Identity: $i^k = i$, so $i \in B_k$.

Inverses: For $x \in B_k$, $(x^{-1})^k = (x^k)^{-1} = i^{-1} = i$, so $x^{-1} \in B_k$.

15. Closure: Let (x, x) and (y, y) belong to A. Then $(x, x) \cdot (y, y) = (x \cdot y, x \cdot y) \in A$.

Identity: $(i, i) \in A$

Inverses: For $(x, x) \in A$, $(x, x)^{-1} = (x^{-1}, x^{-1}) \in A$.

Section 7.3

1. (a) no; $f(x + y) = 2$, $f(x) + f(y) = 2 + 2 = 4$
   (b) no; $f(x + y) = x + y + 1$, $f(x) + f(y) = x + 1 + y + 1$
   (c) yes; $f((x, y) + (p, q)) = f(x + p, y + q) = x + p + 2(y + q)$,
   $f(x, y) + f(p, q) = x + 2y + p + 2q$
   (d) no; $f(x + y) = |x + y|$, $f(x) + f(y) = |x| + |y|$
   (e) yes; $f(x \cdot y) = |x \cdot y|$, $f(x) \cdot f(y) = |x| \cdot |y|$

3. (a) $f: \mathbb{R}^+ \to \mathbb{R}$
   f is onto: for $r \in \mathbb{R}$, $b^r \in \mathbb{R}^+$ and $f(b^r) = \log_b b^r = r$
   f is one-to-one: if $f(x_1) = f(x_2)$, then $\log_b x_1 = \log_b x_2$. Let $p = \log_b x_1 = \log_b x_2$. Then $b^p = x_1$ and $b^p = x_2$, so $x_1 = x_2$.
   f is a homomorphism: for $x_1, x_2 \in \mathbb{R}^+$,
   $f(x_1 \cdot x_2) = \log_b(x_1 \cdot x_2) = \log_b x_1 + \log_b x_2 = f(x_1) + f(x_2)$
   (b) $f(64) = \log_2 64 = 6$ and $f(512) = \log_2 512 = 9$. In $[\mathbb{R}, +]$, $6 + 9 = 15$, and $f^{-1}(15) = 2^{15} = 32{,}768$.

9. $f(x \cdot y) = i = i \cdot i = f(x) \cdot f(y)$

12. Because f is an onto function, $U \neq \emptyset$. To show closure, let $x, y \in U$. Then $f(x \cdot y) = f(x) + f(y) = t + t = t$, so $x \cdot y \in U$.

13. (a) Let x be an idempotent element in $[S, \cdot]$. Then $x^2 = x$. Also $(f(x))^2 = f(x) + f(x) = f(x \cdot x) = f(x)$, and $f(x)$ is idempotent in $[T, +]$.
    (b) For any $f(x) \in T$, $f(x) + f(0) = f(x \cdot 0) = f(0)$ and $f(0) + f(x) = f(0 \cdot x) = f(0)$, so $f(0)$ is a zero in $[T, +]$.

18. (a) Let $f: G \times H \to H \times G$ be defined by $f(x, y) = (y, x)$. Then f is one-to-one and onto, and

$$f((x_1, y_1) \cdot (x_2, y_2)) = f(x_1 \cdot x_2, y_1 + y_2)$$
$$= (y_1 + y_2, x_1 \cdot x_2)$$
$$= (y_1, x_1) \cdot (y_2, x_2)$$
$$= f(x_1, y_1) \cdot f(x_2, y_2)$$

   (b) $f((x_1, y_1) \cdot (x_2, y_2)) = f(x_1 \cdot x_2, y_1 + y_2) = x_1 \cdot x_2 = f(x_1, y_1) \cdot f(x_2, y_2)$

19. Let $x, y \in G$. Then $f(x \cdot y) = (x \cdot y)^{-1} = y^{-1} \cdot x^{-1} = f(y) \cdot f(x) = f(y \cdot x)$. Therefore $x \cdot y = y \cdot x$ because f is one-to-one.

Section 7.4

1. (a) $f(a_2x^2 + a_1x + a_0 + b_2x^2 + b_1x + b_0)$
   $= f((a_2 + b_2)x^2 + (a_1 + b_1)x + (a_0 + b_0)) = a_2 + b_2$
   $= f(a_2x^2 + a_1x + a_2) + f(b_2x^2 + b_1x + b_0)$
   Clearly f is onto.
   (b) (i) and (iii)

2. (a) $f(\mathbb{Z}_{18}) = \{0, 4, 8\}$
   $\mathbb{Z}_{18}/f = \{[0], [1], [2]\}$ where

   $[0] = \{0, 3, 6, 9, 12, 15\}$
   $[1] = \{1, 4, 7, 10, 13, 16\}$
   $[2] = \{2, 5, 8, 11, 14, 17\}$

The isomorphism is

$[0] \rightarrow 0$
$[1] \rightarrow 4$
$[2] \rightarrow 8$

3. (a) $g([1] * [2]) = g([1 +_{18} 2]) = g([3]) = g([0]) = 0$
$g([1]) +_{12} g([2]) = 4 +_{12} 8 = 0$

Section 7.5

3. (a) $0 + S = \{0, 4, 8\}$
$1 + S = \{1, 5, 9\}$
$2 + S = \{2, 6, 10\}$
$3 + S = \{3, 7, 11\}$
(b) $(2 + S) + (3 + S) = 1 + S$

5. (a) $1S = \{1, 2, 4\}$
$3S = \{3, 6, 5\}$
(b) $3S \cdot 4S = 3S$

7. $K = \{x \mid f(x) = x \cdot_k 1 = 0\} = \{x \mid x \text{ is a multiple of } k\} = k\mathbb{Z}$

9. (a) $f(\mathbb{Q} \times \mathbb{Q} \times \mathbb{Q}) = \{(y, y) \mid y \in \mathbb{Q}\}$

$f((x_1, y_1, z_1) + (x_2, y_2, z_2)) = f(x_1 + x_2, y_1 + y_2, z_1 + z_2)$
$= (y_1 + y_2, y_1 + y_2) = (y_1, y_1) + (y_2, y_2)$
$= f(x_1, y_1, z_1) + f(x_2, y_2, z_2)$

(b) $K = \{(x, 0, z) \mid x, z \in \mathbb{Q}\}$
(c) $(4, 4, 4)$ and $(5, 4, 3)$
(d) yes
(e) $g : (\mathbb{Q} \times \mathbb{Q} \times \mathbb{Q})/K \rightarrow f(\mathbb{Q} \times \mathbb{Q} \times \mathbb{Q})$ defined by $g((x, y, z) + K) = (y, y)$

13. This is Exercise 17 of Section 7.3.
Let i_G and i_H denote the identity elements of G and H. Let f be an isomorphism, $f : G \rightarrow H$. Then $f(i_G) = i_H$ by Theorem 7.67, and because f is one-to-one, i_G is the only element mapping to i_H. Thus $K = \{i_G\}$. Now assume that $K = \{i_G\}$; we need to show that the homomorphism f is one-to-one. Let g_1 and g_2 be elements of G with $f(g_1) = f(g_2)$.

Then

$f(g_1) \cdot (f(g_2))^{-1} = i_H$
$f(g_1) \cdot f(g_2^{-1}) = i_H$
$f(g_1 \cdot g_2^{-1}) = i_H$

so that

$g_1 \cdot g_2^{-1} = i_G \quad \text{or} \quad g_1 = i_G \cdot g_2 = g_2$

14. Let $x \in A$ and $g \in G$; we want to show that $g^{-1}xg \in A$. But since $x \in A$, $g^{-1}xg = g^{-1}gx = x \in A$. A is normal by Theorem 7.113.

18. Let S be normal in G and let $x \in G$, $s \in S$. Because S is normal, $sx \in Sx = xS$, so that $sx = xs'$ for some $s' \in S$. Then $x^{-1}sx = x^{-1}xs' = s' \in S$.

23. For any $x \in G$, $xT = Tx$ because T is normal in G; thus $xT = Tx$ for any $x \in S$ and T is normal in S. To show that S/T is a subgroup of G/T, note that $S/T \subseteq G/T$ and let $xT, yT \in S/T$. Then $x, y \in S$ so $xy^{-1} \in S$ and $xT(yT)^{-1} = xT(y^{-1}T) = xy^{-1}T \in S/T$. To show that S/T is normal in G/T, let $sT \in S/T$ and $xT \in G/T$. Then, because S is normal, $x^{-1}sx \in S$ (Exercise 18 above) and thus $(xT)^{-1}(sT)(xT) = (x^{-1}T)(sT)(xT) = (x^{-1}sx)T \in S/T$; apply Theorem 7.113.

28. $[G:T] = \dfrac{|G|}{|T|} = \dfrac{|G|}{|S|} \cdot \dfrac{|S|}{|T|} = [G:S] \cdot [S:T]$

CHAPTER 8

Section 8.1

1. (a) $s = 12$; ALLGAULISDIVIDED
   (b) There are no occurrences of the letter E.
   (c) IBM

4. (a) yes
   (b) 152478; 152748 (error detected); 125748 (no error detected)

9. (a) H has no row of all 0s and no two rows are alike, so the code is single-error correcting.
   (b) $n = 9$, $r = 4$, $m = 9 - 4 = 5$; H can encode all of $\mathbb{Z}_2^5$:

$00000 \to 000000000$	$11100 \to 111001111$
$00001 \to 000011100$	$11010 \to 110100011$
$00010 \to 000101001$	$11001 \to 110010110$
$00100 \to 001000101$	$10110 \to 101100001$
$01000 \to 010000111$	$10101 \to 101010100$
$10000 \to 100001101$	$10011 \to 100111000$
$11000 \to 110001010$	$01011 \to 010110010$
$10100 \to 101001000$	$01101 \to 011011110$
$10010 \to 100100100$	$00111 \to 001110000$
$10001 \to 100010001$	$01110 \to 011101011$
$01100 \to 011000010$	$01111 \to 011110111$
$01010 \to 010101110$	$10111 \to 101111101$
$01001 \to 010011011$	$11011 \to 110111111$
$00110 \to 001101100$	$11101 \to 111010011$
$00101 \to 001011001$	$11110 \to 111100110$
$00011 \to 000110101$	$11111 \to 111111010$

   (c) no; $m \neq 2^r - r - 1$

11. (a) $n = 5$, $m = 3$, $r = 2$; $m \not\leq 2^r - r - 1$, so the code is neither perfect nor single-error correcting.
    (b) $n = 12$, $m = 7$, $r = 5$; $m < 2^r - r - 1$, so the code is single-error correcting but not perfect.
    (c) $n = 15$, $m = 11$, $r = 4$; $m = 2^r - r - 1$, so the code is perfect.

12. (a) 6 ($32 \leq 2^6 - 6 - 1$)
    (b) 6 ($36 \leq 2^6 - 6 - 1$)
    (c) 7 ($60 \leq 2^7 - 7 - 1$)

15.

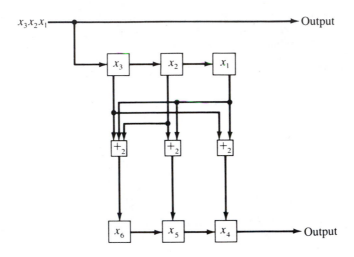

Section 8.2

1. 0011010
   1100101
   0100011   (code word)

5. If only a single error has occurred, the coset leader has weight 1. If the 1 occurs in the kth component, then the syndrome is the kth row of **H**, which is the binary representation of k. Thus computing the syndrome of X gives the binary representation of the single component in X that should be changed.

7. (a)

Coset leaders	Syndromes
000000	000
000001	001
000010	010
001000	011
000100	100
100000	101
010000	110
100010	111

   (b) 101110
       110011
       110011
       111111   (at least 2 errors have occurred, not decoded)

10. (a) One possible **H** is

$$\mathbf{H} \begin{bmatrix} 1 & 1 & 0 & 0 \\ 1 & 0 & 1 & 0 \\ 1 & 0 & 0 & 1 \\ 0 & 1 & 0 & 1 \\ 0 & 0 & 1 & 1 \\ 0 & 1 & 1 & 0 \\ 1 & 1 & 1 & 0 \\ 1 & 1 & 0 & 1 \\ 1 & 0 & 1 & 1 \\ 0 & 1 & 1 & 1 \\ 1 & 1 & 1 & 1 \\ 1 & 0 & 0 & 0 \\ 0 & 1 & 0 & 0 \\ 0 & 0 & 1 & 0 \\ 0 & 0 & 0 & 1 \end{bmatrix}$$

(b) The syndrome for 011000010111001 is 1111. The coset leader is 000000000010000, so the word is decoded as 011000010101001.

CHAPTER 9

Section 9.1

1. (a) 0001111110   (b) *aaacaaaa*   (c) 00100110

5.

| Present state | Next state | | Output |
| | Present input | | |
	0	1	
s_0	s_1	s_2	*a*
s_1	s_2	s_3	*b*
s_2	s_2	s_1	*c*
s_3	s_2	s_3	*b*

Output is *abbcbb*.

9.

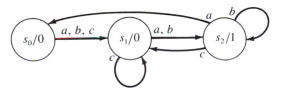

Output is 0001101.

13. (a)

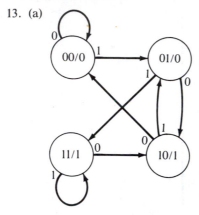

(b) The length of time required to remember a given input grows without bound and eventually would exceed the number of states.

14.

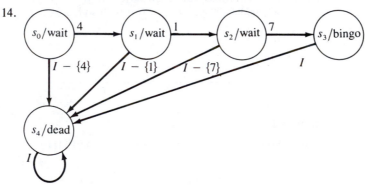

18. Once a state is revisited, behavior will be periodic, since there is no choice of paths from a state. The maximum number of inputs that can occur before this happens is $n - 1$ (visiting all n states before repeating). The maximum length of a period is n (output from all n states with the last state returning to s_0).

21. A successfully receives acknowledgment of message 0 and then sends message 1.

Section 9.2

2. (a) g: $0 \rightarrow a$ (b) g: $1 \rightarrow c$ (c) g: $0 \rightarrow a$
              $1 \rightarrow a$ $2 \rightarrow b$ $1 \rightarrow c$
              $2 \rightarrow b$ $3 \rightarrow b$ $2 \rightarrow b$
              $3 \rightarrow c$ $4 \rightarrow c$ $3 \rightarrow a$
              $4 \rightarrow c$ $4 \rightarrow d$
                                                                            $5 \rightarrow b$

3. (a) States of M/g are $[0] = \{0, 1\}$, $[2] = \{2\}$, $[3] = \{3, 4\}$.

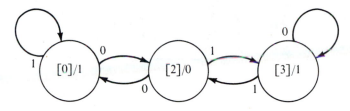

Isomorphism: [0] → a
              [2] → b
              [3] → c

Section 9.3

1. (b) 01* ∨ (110)*

3. (a) (01*)(001*)*

4. (c) 100*1

5. (a) yes   (b) no   (c) no   (d) yes

9. (a)

	Next state		
M'	**Present input**		
Present state	**0**	**1**	**Output**
{s₀}	{s₀, s₁}	{s₁}	1
{s₁}	{s₀}	{s₀, s₁}	0
{s₀, s₁}	{s₀, s₁}	{s₀, s₁}	1

(0* ∨ (0 ∨ 1)1*(0 ∨ 1))*

14. (a)

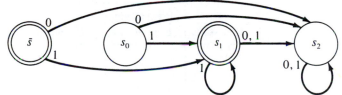

(b)

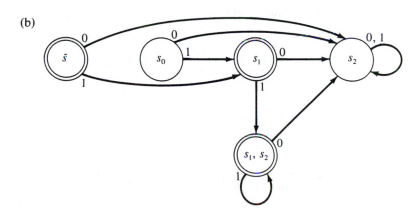

(c)

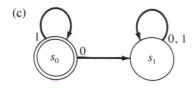

CHAPTER 10

Section 10.1

1. (a) s_1 (b) s_2

3. $A = \{0\}, B = \{1, 2, 5\}, C = \{3, 4\}, D = \{6\}$

	Next state		
	Present input		
Present state	0	1	Output
A	C	D	1
B	C	B	0
C	B	A	1
D	C	B	1

8. $A = \{0, 2\}, B = \{1, 3\}, C = \{4\}$

	Next state			
	Present input			
Present state	a	b	c	Output
A	B	C	A	1
B	C	A	B	0
C	B	A	A	0

12. (a) true   (b) false   (c) true   (d) false

Section 10.2

1. Possible answer:

	d_1	d_2
s_0	0	0
s_1	0	1
s_2	1	0
s_3	1	1

$x(t)$	$d_1(t)$	$d_2(t)$	$y(t)$	$d_1(t+1)$	$d_2(t+1)$
0	0	0	0	1	0
1	0	0	0	1	1
0	0	1	1	0	0
1	0	1	1	0	1
0	1	0	0	0	1
1	1	0	0	1	1
0	1	1	1	0	1
1	1	1	1	1	0

$$y(t) = d_1'd_2 + d_1d_2 = d_2$$
$$d_1(t+1) = x'd_1'd_2' + xd_1'd_2' + xd_1d_2' + xd_1d_2 = d_1'd_2' + xd_1$$
$$d_2(t+1) = xd_1'd_2' + xd_1'd_2 + x'd_1d_2' + xd_1d_2' + x'd_1d_2 = x(d_1' + d_2') + x'd_1$$

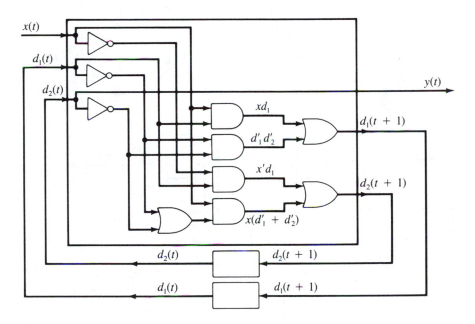

4. Possible answer:

Present state	Next state		Output
	Present input		
	0	1	
s_0	s_0	s_1	0
s_1	s_2	s_1	0
s_2	s_0	s_3	0
s_3	s_2	s_1	1

	d_1	d_2
s_0	0	0
s_1	0	1
s_2	1	0
s_3	1	1

$x(t)$	$d_1(t)$	$d_2(t)$	$y(t)$	$d_1(t+1)$	$d_2(t+1)$
0	0	0	0	0	0
1	0	0	0	0	1
0	0	1	0	1	0
1	0	1	0	0	1
0	1	0	0	0	0
1	1	0	0	1	1
0	1	1	1	1	0
1	1	1	1	0	1

$$y(t) = d_1 d_2$$
$$d_1(t+1) = x'd_1'd_2 + xd_1d_2' + x'd_1d_2 = x'd_2 + xd_1d_2'$$
$$d_2(t+1) = xd_1'd_2' + xd_1'd_2 + xd_1d_2' + xd_1d_2 = x$$

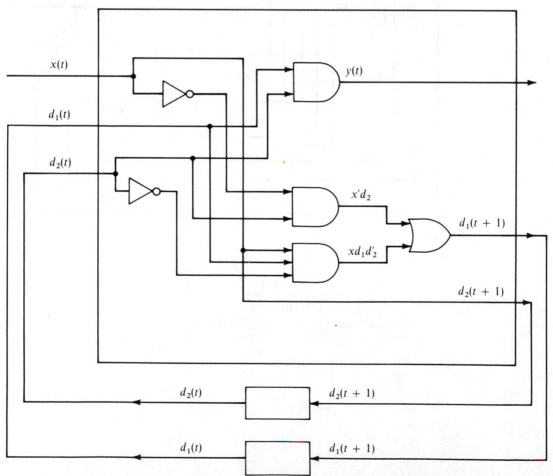

7. Possible answer:

Input	x_1	x_2
0	0	0
1	0	1
2	1	1

	d
s_0	0
s_1	1

$x_1(t)$	$x_2(t)$	$d(t)$	$y(t)$	$d(t + 1)$
0	0	0	0	0
0	1	0	0	1
1	1	0	0	1
0	0	1	1	1
0	1	1	1	0
1	1	1	1	0
1	0	0	0	0
1	0	1	1	1

$$y(t) = d$$
$$d(t + 1) = x_2 d' + x_2' d$$

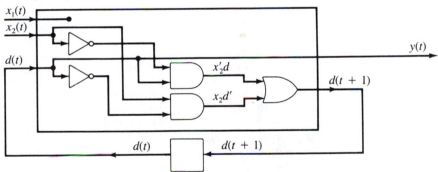

Section 10.3

1. (a) One example is

M_1	Input 0	1	Output
$A = \{0, 1, 3\}$	B	A	0
$B = \{2, 4\}$	A	A	1

M_2	Input 0	1	Output
$a = \{0, 2\}$	b	c	0
$b = \{1, 4\}$	a	b	1
$c = \{3\}$	a	a	1

$$g: \quad (A, a) \to 0 \qquad g': \quad (0, 0) \to 0$$
$$ (A, b) \to 1 \qquad (1, 0) \to 1$$
$$ (A, c) \to 3 \qquad (0, 1) \to 1$$
$$ (B, a) \to 2 \qquad (1,1) \to 0$$
$$ (B, b) \to 4$$

(b) $((B, b), 0) \to (A, a) \to 0$
$((A, a), 1) \to (A, c) \to 3$
$((A, c), 0) \to (B, a) \to 2$

5. (a) One example is

M_1	Input 0	1	Output
$A = \{0, 2, 3\}$	B	A	A
$B = \{1, 4, 5\}$	A	B	B

M_2	Input $(A, 0)$	$(A, 1)$	$(B, 0)$	$(B, 1)$	Output
$a = \{0, 1\}$	a	b	a	b	a
$b = \{2, 4\}$	c	c	c	a	b
$c = \{3, 5\}$	b	b	c	a	c

$$g: \quad (A, a) \to 0 \qquad g': \quad (A, a) \to 0$$
$$ (A, b) \to 2 \qquad (A, b) \to 1$$
$$ (A, c) \to 3 \qquad (A, c) \to 0$$
$$ (B, a) \to 1 \qquad (B, a) \to 0$$
$$ (B, b) \to 4 \qquad (B, b) \to 1$$
$$ (B, c) \to 5 \qquad (B, c) \to 0$$

(b) $((A, b), 0) \to (B, c) \to 5$
$((B, b), 1) \to (B, a) \to 1$
$((A, c), 1) \to (A, b) \to 2$

Section 10.4

3. There are many possibilities, for example,

$$\{0\} \subset \{0, 30\} \subset \{0, 6, 12, 18, 24, 30, 36, 42, 48, 54\} \subset$$
$$\{0, 2, 4, 6, 8, 10, \ldots, 58\} \subset \mathbb{Z}_{60}$$

$$\{0\} \subset \{0, 20, 40\} \subset \{0, 10, 20, 30, 40, 50\} \subset$$
$$\{0, 5, 10, 15, 20, \ldots, 55\} \subset \mathbb{Z}_{60}$$

$$\{0\} \subset \{0, 30\} \subset \{0, 15, 30, 45\} \subset \{0, 5, 10, \ldots, 55\} \subset \mathbb{Z}_{60}$$

In each case, the members of the factor set are isomorphic to $\mathbb{Z}_2$, $\mathbb{Z}_2$, $\mathbb{Z}_3$, and $\mathbb{Z}_5$.

5. (a)

i	0	1	2
	01	11	10
S	λ	0	1	00

		0	1	2
0	0	1	2	2
1	1	2	1	2
2	2	2	2	2

Table for S_M:

$\circ$	i	0	1	2
i	i	0	1	2
0	0	2	0	2
1	1	2	1	2
2	2	2	2	2

not a group

9. The present state of the cascade machine is $([3], 2)$ because $3 = 2 +_6 1$. The next state of the cascade machine is

$$([3 +_6 4], 2 +_6 \beta(4, [3])) = ([1], 2 +_6 (1 +_6 4) +_6 (-(3 +_6 4)'))$$
$$= ([1], 2 +_6 5 +_6 (-1))$$
$$= ([1], 2 +_6 5 +_6 5)$$
$$= ([1], 0)$$

and $\gamma([1], 0) = 1$ because $1 = 0 +_6 1$.

10. (a) The elements of $\mathbb{Z}_6/\{0, 3\}$ are $[0] = \{0, 3\}$, $[1] = \{1, 4\}$, and $[2] = \{2, 5\}$; let 0, 1, and 2 be the fixed coset representatives.

The present state of the cascade machine is $([2], 0)$ because $2 = 0 +_6 2$. The next state of the cascade machine is

$$([2 +_6 5], 0 +_6 \beta(5, [2])) = ([1], 0 +_6 (2 +_6 5) + (-(2 +_6 5)'))$$
$$= ([1], 0 +_6 1 +_6 (-1))$$
$$= ([1], 1 +_6 5)$$
$$= ([1], 0)$$

and $\gamma([1], 0) = 1$ because $1 = 0 +_6 1$.

CHAPTER 11

Section 11.1

1. (a) halts with final tape

	$\cdots$	b	0	0	0	0	0	b	$\cdots$

(b) does not change the tape and moves forever to the left

6. One answer: State 2 is a final state.

$(0, b, b, 2, R)$ blank tape or no more 1s
$(0, 1, 1, 1, R)$ has read odd number of 1s
$(1, 1, 1, 0, R)$ has read even number of 1s

9. One answer: State 9 is a final state.

$(0, b, b, 9, R)$ accepts blank tape

$\left.\begin{array}{l}(0, 0, 0, 0, R)\\(0, 1, X, 1, R)\end{array}\right\}$ finds first 1, marks with X

$\left.\begin{array}{l}(1, 1, 1, 1, R)\\(1, Y, Y, 1, R)\end{array}\right\}$ searches right for 2s

$\left.\begin{array}{l}(1, 2, Y, 3, R)\\(3, 2, Y, 4, L)\end{array}\right\}$ pair of 2s, marks with Ys

$\left.\begin{array}{l}(4, Y, Y, 4, L)\\(4, X, X, 4, L)\\(4, 1, 1, 4, L)\\(4, Z, Z, 4, L)\end{array}\right\}$ searches left for 0s

$\left.\begin{array}{l}(4, 0, Z, 5, L)\\(5, 0, Z, 6, R)\end{array}\right\}$ pair of 0s, marks with Zs

$\left.\begin{array}{l}(6, Z, Z, 6, R)\\(6, X, X, 6, R)\\(6, 1, X, 1, R)\end{array}\right\}$ passes right to next 1

$(6, Y, Y, 7, R)$ no more 1s

$\left.\begin{array}{l}(7, Y, Y, 7, R)\\(7, b, b, 8, L)\end{array}\right\}$ no more 2s

$\left.\begin{array}{l}(8, Y, Y, 8, L)\\(8, X, X, 8, L)\\(8, Z, Z, 8, L)\\(8, b, b, 9, L)\end{array}\right\}$ no more 0s, halts and accepts

15. $f(n_1, n_2, n_3) = \begin{cases} n_2 + 1 & \text{if } n_2 > 0 \\ \text{undefined} & \text{if } n_2 = 0 \end{cases}$

17. One answer:

$(0, 1, 1, 1, R)$

$(1, b, 1, 4, R)$ $n = 0$, add 1 and halt

$\left.\begin{array}{l}(1, 1, 1, 2, R)\\(2, b, 1, 4, R)\end{array}\right\}$ $n = 1$, add additional 1 and halt

$\left.\begin{array}{l}(2, 1, 1, 3, R)\\(3, 1, b, 3, R)\\(3, b, b, 4, R)\end{array}\right\}$ $n \geq 2$, erase extra 1s and halt

20. One answer:

$(0, 1, b, 1, R)$ erases one extra 1

$(1, *, b, 3, R)$ $n_1 = 0$

$\left.\begin{array}{l}(1, 1, b, 2, R)\\(2, 1, 1, 2, R)\\(2, *, 1, 3, R)\end{array}\right\}$ $n_1 > 0$, replaces $*$ with leftmost 1 of $\bar{n}_1$, halts

25. (a) T may run forever, processing a given input string α, and we would be unable to test other strings.

(b) Let T be a Turing machine that computes the function. Using copies T_1, T_2, . . . of T, feed input into them, and check to see which have halted with a representation $\overline{m}$ on the tape; any m so represented is in the range set.

Section 11.2

1. For any Turing machine T, we can effectively create a machine T^* that acts like T but replaces any occurrences of m as a state symbol with a new state symbol; in addition, whenever T reaches a halting configuration, T^* enters state m. Then T^* enters state m when started on a tape containing α if and only if T halts on α. Assume that an algorithm P exists to solve the state problem. An algorithm to solve the halting problem is: given a (T, α) pair, create T^* and then apply P.

Section 11.3

1. It requires $\bar{n} + 1$ moves to locate the blank beyond the $\bar{n}$ symbols and to move left to the end of $\bar{n}$. The machine can then X out successive symbols of n and add 1s on the right end. Working from the "middle" outward, this requires, successively, $1 + 2 + 3 + \cdots + 2\bar{n}$ steps. To blank out n X's and change the remaining X to 1 requires $\bar{n} + 3$ more steps. The total number of steps is thus

$$\bar{n} + 1 + (1 + 2 + \cdots + 2\bar{n}) + \bar{n} + 3 = 2\bar{n} + 4 + \frac{2\bar{n}(2\bar{n} + 1)}{2}$$
$$= 2\bar{n}^2 + 3\bar{n} + 4$$
$$= 2(n + 1)^2 + 3(n + 1) + 4$$

which is of order n^2.

CHAPTER 12

Section 12.1

1. (a) $L(G) = \{a\}$
   (b) $L(G) = \{010101, 010111, 011101, 011111, 110101, 110111, 111101, 111111\}$
   (c) $L(G) = 0(10)^*$
   (d) $L(G) = 0^*1111^*$

3. $L(G) = aa^*bb^*$. G is context-sensitive. An example of a regular grammar G' that generates $L(G)$ is $G' = (V, V_T, P, S)$ where $V = \{a, b, A, B, S\}$, $V_T = \{a, b\}$, and P consists of the productions

$S \to aA$	$A \to aA$	$B \to bB$
$S \to aB$	$A \to aB$	$B \to b$

7. For example, $G = (V, V_T, S, P)$ where $V = \{0, 1, A, S\}$, $V_T = \{0, 1\}$, and P consists of the productions

$S \to 01$	$A \to A0$
$S \to A01$	$A \to A1$
	$A \to 0$
	$A \to 1$

12. For example, $G = (V, V_T, S, P)$ where $V = \{0, 1, S, S_1\}$, $V_T = \{0, 1\}$, and P consists of the productions

$S \to \lambda$ $S_1 \to 0S_10$
$S \to S_1$ $S_1 \to 1S_11$
          $S_1 \to 00$
          $S_1 \to 11$

15. (a)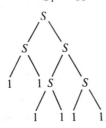

16.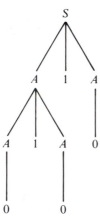

Section 12.2

1. $G = (V, V_T, S, P)$ where $V = \{S, A, B, C, 0, 1\}$, $V_T = \{0, 1\}$, and P consists of the productions

$S \to 1A$
$S \to 0C$
$A \to 0$
$A \to 0B$
$A \to 1C$
$B \to 0B$
$B \to 1C$
$B \to 0$
$C \to 0C$
$C \to 1C$

$L(G) = 100^*$

5.

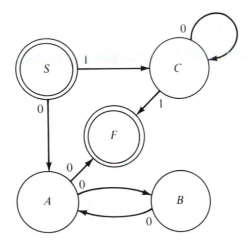

$L(G) = \lambda \lor 0(00)*0 \lor 10*1$

7. There are

   n words of length 1
   n^2 words of length 2
   .
   .
   .
   n^k words of length k

So the total number of words is $n + n^2 + \cdots + n^k$.

10. (a) $W_0 = \{S\}$
      $W_1 = \{S, a, b, bA\}$
      $W_2 = \{S, a, b, bA, bc\}$
      $W_3 = W_2$ so $cb \notin L(G)$

Index